Venous Ulcers

Venous Ulcers

SECOND EDITION

Edited by

CYNTHIA K. SHORTELL
Duke University Medical Center, Durham, NC, United States

JOVAN N. MARKOVIC
Duke University Medical Center, Durham, NC, United States

Academic Press is an imprint of Elsevier
125 London Wall, London EC2Y 5AS, United Kingdom
525 B Street, Suite 1650, San Diego, CA 92101, United States
50 Hampshire Street, 5th Floor, Cambridge, MA 02139, United States
The Boulevard, Langford Lane, Kidlington, Oxford OX5 1GB, United Kingdom

Copyright © 2023 Elsevier Inc. All rights reserved.

No part of this publication may be reproduced or transmitted in any form or by any means, electronic or mechanical, including photocopying, recording, or any information storage and retrieval system, without permission in writing from the publisher. Details on how to seek permission, further information about the Publisher's permissions policies and our arrangements with organizations such as the Copyright Clearance Center and the Copyright Licensing Agency, can be found at our website: www.elsevier.com/permissions.

This book and the individual contributions contained in it are protected under copyright by the Publisher (other than as may be noted herein).

Notices

Knowledge and best practice in this field are constantly changing. As new research and experience broaden our understanding, changes in research methods, professional practices, or medical treatment may become necessary.

Practitioners and researchers must always rely on their own experience and knowledge in evaluating and using any information, methods, compounds, or experiments described herein. In using such information or methods they should be mindful of their own safety and the safety of others, including parties for whom they have a professional responsibility.

To the fullest extent of the law, neither the Publisher nor the authors, contributors, or editors, assume any liability for any injury and/or damage to persons or property as a matter of products liability, negligence or otherwise, or from any use or operation of any methods, products, instructions, or ideas contained in the material herein.

ISBN: 978-0-323-90610-4

For information on all Academic Press publications visit our website at https://www.elsevier.com/books-and-journals

Publisher: Stacy Masucci
Acquisitions Editor: Ana Claudia A. Garcia
Editorial Project Manager: Susan E. Ikeda
Production Project Manager: Stalin Viswanathan
Cover Designer: Miles Hitchen

Typeset by TNQ Technologies

Contents

List of contributors — xvii
Preface to Venous Ulcers — xxi

SECTION 1 Chronic venous insufficiency—basic considerations

1. Pathophysiology of chronic venous disease: genetic, molecular, and biochemical mechanisms — 3
Joseph D. Raffetto and Raouf A. Khalil

- Key points — 3
- Introduction — 4
- Content — 4
- Conclusions — 13
- Acknowledgments — 13
- References — 14

2. Venous hemodynamics and microcirculation in chronic venous insufficiency — 19
John Blebea

- Introduction — 19
- Hemodynamics and anatomy — 20
- Pathophysiology—superficial venous incompetence — 22
- Deep venous dysfunction — 23
- Perforating veins of the calf — 26
- Foot and calf pump function in venous hemodynamics — 27
- Microcirculation and interstitial edema — 34
- Regulation of capillary hydrostatic pressure — 35
- Blood flow and inflammation — 36
- Conclusions — 37
- References — 37

3. Venous ulcers of the lower extremity: etiology, risks, and predictive factors — 41
Thomas F. O'Donnell

- Introduction — 41
- CEAP etiologies of CVD — 41
- Epidemiological data on CVI and its progression — 49

Progression of venous disease 53
Clinical risk factors for first VLU 54
Genetic risk factors for VLU 56
Lymphatic involvement in VLU 57
Summary 58
References 58

4. **Venous ulcers of the lower extremity: epidemiology and socioeconomic burden** — 63
   Olle Nelzén

   Role of epidemiology in the study and treatment of venous leg ulcers 63
   Incidence and prevalence of venous leg ulcers 64
   Distribution of chronic venous insufficiency and its subgroups within the population 65
   Defining venous leg ulcers 66
   Leg ulcer prevalence 66
   Etiologic spectrum of leg ulcers 67
   Venous ulcer prevalence 68
   Incidence of venous ulcers 71
   Age and sex distribution 72
   Socioeconomic aspects 73
   Venous ulcer natural history 75
   Obesity 76
   History of DVT 77
   Healing and survival 77
   Future trends 78
   References 79

SECTION 2 Clinical evaluation and diagnostic modalities

5. **Initial clinical evaluation in patients with chronic venous insufficiency** — 85
   Yana Etkin and Ruth L. Bush

   Introduction 85
   Medical history 86
   Physical exam 89
   Quality of life assessment 96
   Classification and venous severity scoring 97
   Summary 99
   References 99

6. **Ultrasound evaluation of lower extremity chronic venous disease** — 101
   Raudel Garcia and Nicos Labropoulos

   Key points — 101
   Introduction — 102
   Diagnosis of obstruction — 113
   Disclosure statement — 120
   References — 120

7. **The diagnosis of major venous outflow obstruction in chronic venous insufficiency** — 127
   Jovan N. Markovic, Martin V. Taormina and Ellen D. Dillavou

   Introduction — 127
   Diagnosis of chronic venous insufficiency and venous hypertension — 128
   Diagnostic modalities — 129
   References — 136

8. **Hypercoagulable states associated with chronic venous insufficiency** — 139
   Samuel Anthony Galea and Emma Wilton

   Protein C and protein S — 141
   Antithrombin — 142
   Factor V Leiden — 144
   Activated protein C resistance — 145
   Prothrombin G20210A gene mutations — 145
   Dysfibrinogenemia — 146
   Antiphospholipid syndrome — 146
   Paroxysmal nocturnal hemoglobinuria — 147
   Myeloproliferative neoplasms — 148
   Other hypercoagulable states that increase the risk of developing VTE and potential subsequent venous ulceration — 149
   References — 154

9. **The chronically swollen leg with ulcers—finding the cause: theory and practice** — 159
   Pier Luigi Antignani, Luca Costanzo, Giacomo Failla and Francesco Paolo Palumbo

   Introduction and epidemiology — 159
   Assessment of edema — 161
   Physical examination — 163
   Diagnostic studies — 163
   References — 165

10. Lower extremity wounds associated with mixed venous and arterial insufficiency and relevant differential diagnosis — 167
Enjae Jung and Robert B. McLafferty

Introduction	167
Evaluation	167
Treatment	170
Differential diagnosis	171
References	176

11. Assessment tools and wound documentation for patients with chronic venous insufficiency — 179
Michael Palmer, Rick Mathews, Gregory L. Moneta and Khanh P. Nguyen

Wound assessment tools	179
References	196

SECTION 3 Nonoperative management of chronic venous insufficiency and wound care

12. Compression therapy in venous leg ulcers — 201
Hugo Partsch

Introduction	201
Compression devices	202
Medical compression stockings	202
Compression bandages	204
Elastic and inelastic bandages	205
"Single-layer" and multilayer bandages	207
Intermittent pneumatic compression	209
Pelottes and pads	210
Prevention of ulcer recurrence	210
Compression techniques—practical guidelines	211
Summary	213
References	214

13. Wound healing: adjuvant therapy and treatment adherance — 217
Juliet Blakeslee-Carter and Marc A. Passman

Introduction	217
Patient and wound assessment	218

Standard therapies for wound bed management	223
Wound dressings	228
Adjuvant therapy for venous ulcer wound healing	234
Optimizing care delivery	238
References	242

14. Negative pressure wound therapy for venous leg ulcers — 249
Fedor Lurie and Richard Simman

NPWT description and mechanism of action	250
Use of NPWT in patients with VLUs	251
Case study	253
Summary	257
References	257

15. Medical therapies for chronic venous insufficiency — 261
Mark D. Iafrati

Pathophysiology	261
Lifestyle considerations	262
Compression therapy	263
Compression socks	263
Inelastic compression garments	263
Paste boots	264
Multilayer compression bandages	264
The bottom line on compression	265
Pneumatic compression	265
Dietary supplements	266
Physical activity	266
Medications	267
Animal models	267
Micronized purified flavonoid fraction	269
Calcium dobesilate	272
Pentoxifylline and sulodexide	273
Horse chestnut	274
Acetylsalicylic acid	274
Antibiotics	274
Biologic wound care products	275
Guidelines	276
Conclusions	276
References	277

16. Treatment modalities for the management of nonhealing wounds in patients with chronic venous insufficiency 283

Gregory G. Westin, John G. Maijub and Michael C. Dalsing

Introduction	283
Lifestyle modification	283
Medical therapy	284
Untreated arterial or deep venous disease	284
Compression	285
Identification and treatment of infection	285
Debridement	286
Dressings and local agents	288
External stimulants	289
Skin grafting, skin substitutes, and soft tissue substitutes	291
Conclusion	291
References	292

17. Deep vein thrombosis and prevention of postthrombotic syndrome 297

Matthew Sussman and Jose Almeida

Anatomic considerations in PTS	298
Hemodynamics of PTS	299
Collateralization and flow patterns in acute DVT	300
Microcirculation and capillary exchange	301
Anticoagulation to prevent PTS	301
Venoactive drugs	302
Elastic compression therapy	302
Thrombolysis/endovascular therapies to prevent PTS	303
Surgical thrombectomy	304
Our experience	304
A case study	305
Open surgical reconstructions	306
Ultrasound-accelerated thrombolysis	307
Venous ulcer care	307
Conclusion	309
References	311

18. Improving treatment outcomes—management of coexisting comorbidities in patients with venous ulcers 315

Giovanni Mosti and Alberto Caggiati

Introduction	315
Diabetes mellitus	316

Arterial hypertension	316
Coronary artery disease and congestive heart failure	317
Reduced muscle pumping function	317
Other clinical conditions concurring with venous leg ulcers, venous ulcer occurrence, healing failure, and recurrence	320
Obesity	321
References	322

19. Emerging modalities in local treatment of venous ulcers: advanced dressings, bioengineering, and biologics   327

Mabel Chan, Jani Lee and John C. Lantis, II

Introduction	327
Mechanism of action	329
Wound bed preparation	330
Negative pressure wound therapy	330
Matrix metalloprotease modulation	331
Debridement in venous leg ulcers	332
Topical antimicrobial therapy	333
Combination therapy	334
Cellular and tissue-based therapies	335
Living skin cellular products	335
Extracellular matrix therapy (ECM)	337
Acellular human tissue	338
Acellular xenograft tissue	339
Biosynthetic	341
Placenta-derived tissue	342
Dehydrated PDT	342
Cryopreserved and other "viable PDT"	343
Our treatment algorithm	343
Conclusion	345
References	346
Further reading	350

SECTION 4   Operative and endovascular procedures for chronic venous insufficiency

20. Benefits of superficial venous intervention (surgery or endovenous ablation) in the treatment of venous leg ulceration   353

Manjit Gohel

Introduction	353
The EVRA study	355

Implementation of ESCHAR and EVRA study results	359
Conclusions	361
References	362

21. Superficial surgery and perforator interruption in the treatment of venous leg ulcers — 363
Peter F. Lawrence

Diagnosis	363
Treatment options for incompetent superficial truncal veins	366
Treatment options for incompetent perforator veins	366
References	378
Further Reading	379

22. Endovenous techniques for superficial vein ablation for treatment of venous ulcers — 381
Monika L. Gloviczki and Peter Gloviczki

Introduction	381
Preoperative planning	381
Endovenous thermal ablation techniques	383
Nonthermal endovenous ablation techniques	388
Outcomes	388
Conclusion	397
References	397

23. Treatment of chronic venous insufficiency with foam sclerotherapy — 405
Julianne Stoughton and Sujin Lee

Introduction	405
Complications and outcomes	408
Technique	411
Summary	417
References	417

24. Ultrasound guidance for endovenous treatment — 421
Lisa Amatangelo, Kimberly Scherer, Vibhor Wadhwa, Jimmy Xia ScB and Neil Khilnani

Introduction	421
Equipment	421
Preprocedure planning	422
Venous access	423
Periprocedural monitoring	425

Tumescent anesthesia administration	428
Postprocedure monitoring	428
References	430

25. Iliac vein stenting in chronic venous leg ulcers — 433
Taimur Saleem and Seshadri Raju

Introduction	433
Pathophysiology of venous ulcers	434
Clinical assessment	435
Etiology of venous ulcers	437
Investigations	438
Diagnosis of iliac vein stenosis	441
Treatment of chronic venous ulceration	444
Poor stocking compliance	445
Judicious wound care	446
Saphenous vein ablation versus iliac vein stenting in patients with chronic venous leg ulcers: determining the ideal procedure sequence	446
Other procedures	448
Iliac vein stenting and chronic venous ulcers	448
Sizing of stents	449
Stenting technique for routine cases	450
Technical success of procedure	451
Recanalization of chronic total venous occlusion	453
Management of iliac-caval confluence	454
Reinterventions in stented limbs	455
Factors affecting ISR and stent compression	458
Anticoagulation protocol	459
Thrombophilia and venous ulcers	459
Pentoxyfylline and venous ulcers	459
Stent surveillance in standard and recanalization cases	460
Stent occlusions after recanalization	460
Stent occlusion from ISR	460
Threshold stenosis for intervention	461
Dedicated self-expanding nonbraided nitinol venous stents	461
Special considerations and techniques	462
Hybrid approaches	465
Conclusions	466
Financial disclosures	466
References	466

26. Venous valve reconstructions in patients with severe chronic venous insufficiency — 473
Oscar Maleti, Marzia Lugli and Michel Perrin

Introduction	473
Deep venous reflux etiology	474
Hemodynamic and diagnostic evaluation	474
Indication for deep venous reflux correction	476
Surgical techniques	476
Transposition	478
Valve transplantation	479
Neovalve	479
Banding	481
Results	481
Conclusion	484
References	485

SECTION 5 Special considerations

27. Treatment of recalcitrant venous ulcers with free tissue transfer for limb salvage — 491
Grant R. Darner and David A. Brown

Introduction	491
Indications and preoperative planning	494
Flap selection	495
Surgical technique	496
Postoperative care	497
Discussion	497
Conclusion	499
References	500

28. Management of venous ulcers in patients with congenital vascular malformations — 503
Jovan N. Markovic and Byung-Boong Lee

Multidisciplinary approach	505
Classification	506
Diagnostic modalities	507
Treatment	508
Special consideration	512
References	514

29. Lymphatic disorders in the pathogenesis of chronic venous insufficiency   519
Stanley G. Rockson

Clinical staging of lymphedema   520
Clinical presentation   521
Diagnosis   523
Management of lymphedema   525
References   527

Index   *531*

List of contributors

Jose Almeida
Miami Vein, Division of Vascular and Endovascular Surgery, University of Miami Miller School of Medicine, Miami, FL, United States

Lisa Amatangelo
Division of Interventional Radiology, Weill Cornell Medicine, New York Presbyterian Hospital, New York, NY, United States

Pier Luigi Antignani
Vascular Center, Nuova Villa Claudia, Rome, Italy

Juliet Blakeslee-Carter
University of Alabama at Birmingham, Division of Vascular Surgery and Endovascular Therapy, Birmingham, AL, United States

John Blebea
Department of Surgery, College of Medicine, Central Michigan University, Saginaw, MI, United States

David A. Brown
Division of Plastic, Maxillofacial, and Oral Surgery, Duke University School of Medicine, Durham, NC, United States

Ruth L. Bush
University of Houston College of Medicine, Houston, TX, United States

Alberto Caggiati
Department of Anatomy, Sapienza University of Rome, Rome, Italy

Mabel Chan
Department of Surgery, Mount Sinai West Hospital, Icahn School of Medicine, New York, NY, United States

Luca Costanzo
Angiology Unit, San Marco Hospital, Department of Cardiovascular Disease, A.O.U. "G. Rodolico-San Marco", University of Catania, Catania, Italy

Michael C. Dalsing
Division of Vascular Surgery, Indiana University, Indianapolis, IN, United States

Grant R. Darner
Duke University School of Medicine, Durham, NC, United States

Ellen D. Dillavou
Vascular Surgery, WakeMed Hospital System, Raleigh, NC, United States

Yana Etkin
Division of Vascular and Endovascular Surgery, Zucker School of Medicine at Hofstra/Northwell, Hempstead, NY, United States

Giacomo Failla
Angiology Unit, San Marco Hospital, Department of Cardiovascular Disease, A.O.U. "G. Rodolico-San Marco", University of Catania, Catania, Italy

Samuel Anthony Galea
Oxford University Hospitals, NHS Foundation Trust and Buckinghamshire Healthcare NHS Trust, Oxford, United Kingdom

Raudel Garcia
ChenMed, Miami, FL, United States

Monika L. Gloviczki
Department of Internal Medicine and Gonda Vascular Center, Mayo Clinic, Rochester, MN, United States

Peter Gloviczki
Mayo Clinic College of Medicine, Division of Vascular and Endovascular Surgery, Gonda Vascular Center, Mayo Clinic, Rochester, MN, United States

Manjit Gohel
Cambridge Vascular Unit, Cambridge University Hospitals, Cambridge, United Kingdom

Mark D. Iafrati
Department of Vascular Surgery, Vanderbilt University Medical Center, Nashville, TN, United States

Enjae Jung
Division of Vascular Surgery, Oregon Health and Science University, Portland, OR, United States

Raouf A. Khalil
Harvard Medical School, Brigham and Women's Hospital, Division of Vascular and Endovascular Surgery, Boston, MA, United States

Neil Khilnani
Division of Interventional Radiology, Weill Cornell Medicine, New York Presbyterian Hospital, New York, NY, United States

Nicos Labropoulos
Department of Surgery and Radiology, Vascular Laboratory, Division of Vascular and Endovascular Surgery, Stony Brook University Medical Center, Stony Brook, NY, United States

John C. Lantis, II
Department of Surgery, Mount Sinai West Hospital, Icahn School of Medicine, New York, NY, United States

Peter F. Lawrence
Gonda Vascular Center and Division of Vascular and Endovascular Surgery, David Geffen School of Medicine, UCLA, Los Angeles, CA, United States

Byung-Boong Lee
Department of Surgery, Division of Vascular Surgery, George Washington University, Washington, DC, United States

Jani Lee
Department of Surgery, Mount Sinai West Hospital, Icahn School of Medicine, New York, NY, United States

Sujin Lee
Harvard Medical School, Massachusetts General Hospital, Boston, MA, United States

Marzia Lugli
Department of Cardiovascular Surgery, International Centre of Deep Venous Surgery, Hesperia Hospital, Modena, Italy

Fedor Lurie
Jobst Vascular Institute of Promedica, Toledo, OH, United States; Division of Vascular Surgery, University of Michigan, Ann Arbor, MI, United States

John G. Maijub
Division of Vascular Surgery, Indiana University, Indianapolis, IN, United States

Oscar Maleti
Department of Cardiovascular Surgery, International Centre of Deep Venous Surgery, Hesperia Hospital, Modena, Italy

Jovan N. Markovic
Department of Surgery, Division of Vascular Surgery, Duke University School of Medicine, Durham, NC, United States

Rick Mathews
Oregon Health & Science University, Department of Surgery, Division of Vascular & Endovascular Surgery, Portland, OR, United States

Robert B. McLafferty
Division of Vascular Surgery, Oregon Health and Science University, Portland, OR, United States

Gregory L. Moneta
Oregon Health & Science University, Department of Surgery, Division of Vascular & Endovascular Surgery, Portland, OR, United States

Giovanni Mosti
Head Angiology Department, MD Barbantini Hospital, Lucca, Italy

Olle Nelzén
Vascular Surgery Unit & Department of Research and Development Skaraborg Hospital Skövde & Uppsala University, Sweden

Khanh P. Nguyen
Oregon Health & Science University, Department of Surgery, Division of Vascular & Endovascular Surgery, Portland, OR, United States; Oregon Health & Science University, Department of Biomedical Engineering, Portland, OR, United States; Portland VA Health Care System, Research & Development, Portland, OR, United States

Thomas F. O'Donnell
Benjamin Andrews Emeritus Professor of Surgery, Cardiovascular Center, Tufts Medical Center, Boston, MA, United States

Michael Palmer
Oregon Health & Science University, Department of Surgery, Division of Vascular & Endovascular Surgery, Portland, OR, United States

Francesco Paolo Palumbo
Surgery Unit, Villa Fiorita Clinic, Prato, Italy

Hugo Partsch
Department of Dermatology, University of Vienna, Vienna, Austria

Marc A. Passman
Department of Surgery, University of Alabama at Birmingham, Division of Vascular Surgery and Endovascular Therapy, Birmingham, AL, United States

Michel Perrin
Unité de Pathologie Vasculaire Jean Kunlin, Chassieu, France

Joseph D. Raffetto
Harvard Medical School, VA Boston Healthcare System, Brigham and Women's Hospital Boston, Boston, MA, United States

Seshadri Raju
The Rane Center for Venous and Lymphatic Diseases, Jackson, MS, United States

Stanley G. Rockson
Allan and Tina Neill Professor of Lymphatic Research and Medicine, Stanford University School of Medicine, Stanford, CA, United States

Taimur Saleem
The Rane Center for Venous and Lymphatic Diseases, Jackson, MS, United States

Kimberly Scherer
Division of Interventional Radiology, Weill Cornell Medicine, New York Presbyterian Hospital, New York, NY, United States

Richard Simman
Jobst Vascular Institute of Promedica, Toledo, OH, United States; Division of Plastic Surgery, Promedica Physician Group, Toledo, OH, United States; University of Toledo, Department of Surgery, Toledo, OH, United States

Julianne Stoughton
Harvard Medical School, Venous Program at Massachusetts General Hospital, Boston, MA, United States

Matthew Sussman
Division of Vascular and Endovascular Surgery, University of Miami Miller School of Medicine, Miami, FL, United States

Martin V. Taormina
Department of Surgery, Division of Vascular Surgery, Duke University School of Medicine, Durham, NC, United States

Vibhor Wadhwa
Division of Interventional Radiology, Weill Cornell Medicine, New York Presbyterian Hospital, New York, NY, United States

Gregory G. Westin
Division of Vascular Surgery, Indiana University, Indianapolis, IN, United States

Emma Wilton
Oxford University Hospitals, NHS Foundation Trust and Buckinghamshire Healthcare NHS Trust, Oxford, United Kingdom

Jimmy Xia ScB
Division of Interventional Radiology, Weill Cornell Medicine, New York Presbyterian Hospital, New York, NY, United States

Preface to Venous Ulcers

It is a privilege to write the preface for the new second edition of *Venous Ulcers*, the first edition of which was originally edited by the late John Bergan and its current senior editor, Cynthia K. Shortell. Like Captain Ahab in Moby Dick or Jason, the mythical Greek hero in his quest for the golden fleece, many have pursued solutions for both healing and prevention of recurrence of venous leg ulcers (VLUs) over the last 150 years. Why a new edition of the classic textbook, *Venous Ulcers*? Due to its sole focus on venous ulcers, rather than being diluted by covering the full range of vascular diseases or even various forms of venous disease, *Venous Ulcers* is a unique compendium of everything one needs to know about VLUs. The editors of the new second edition, Cynthia K. Shortell and Jovan N. Markovic from Duke Univeristy (Durham, NC, USA), recognized the need for an update of this textbook, based on the accumulation of new substantive studies on the diagnosis and treatment of venous ulcers since the publication of the original textbook a decade and a half ago in 2007. More high-quality evidence has accrued along with the refinement of minimally invasive techniques, both of which will reshape our approaches for treating this the most common chronic lower extremity wound.

What is the source of the new evidence? Prominent in providing early material and momentum for new investigations as well as reasons for an update of *Venous Ulcers* was the Pacific Vascular Symposium 6 (PVS6), convened by the American Venous Forum (AVF) in November 2009. The papers from this meeting were published a year later as a supplement to the *Journal of Vascular Surgery* and summarized this meeting—a call to action from a group of interested international experts in venous disease. Their task was to formulate a doable and achievable plan to reduce the incidence of venous leg ulcers by 50% in 10 years, coincidently around the time of this second edition. Many of the participants in that meeting are authors of chapters in *Venous Ulcer*. Another critical element for the second edition is the Society for Vascular Surgery/American Venous Forum (SVS/AVF) Evidence-Based Clinical Practice Guidelines for Management of Venous Leg Ulcers. This set of guidelines was the result of this programmatic initiative by the AVF, which emphasized the need for clinical practice guidelines for VLUs, as recognized by the SVS Document Oversight Committee. Two formal systematic reviews and meta-analyses: *A systematic Review and Meta-analysis of Surgical Interventions Versus Conservative Therapy for Venous Ulcers* and *A Comparative Systematic Review and Meta-analysis of Compression Modalities for the Promotion of Venous Ulcer Healing and Reducing Ulcer Recurrence,* were authorized by these two societies to provide some of the evidence base for these guidelines. Rather than a clinical practice guideline siloed into one society, the authors of this guideline enlisted review by the then American College of Phlebology (now The American Vein and Lymphatic Society), and endorsement by the American College of Wound Healing and Tissue Repair, as well as the International Union of Phlebology (UIP), a catholic approach adopted for subsequent vascular clinical practice guidelines.

Since the publication of the first edition of *Venous Ulcers*, several additional key systematic reviews, as well as other technology assessments, examined wound dressings for venous ulcers. Although venous stenting is the recommended treatment of choice for venous outflow obstruction, systematic reviews and meta-analysis of venous stenting are limited due to the paucity of larger randomized controlled trials (RCTs) as well as the lack of focus solely on this intervention for VLU. Focusing on the key outcome measures for VLUs: ulcer healing, ulcer recurrence and improved quality of life, the ESCHAR study clearly showed that ligation and stripping of the axial veins reduced VLU recurrence. It was not until the recent EVRA study, however, that early intervention by ablation of the saphenous vein(s) was associated with improved healing of VLU over delayed intervention (this RCT did not compare to a control group of "conservative care"—wound dressings and compression). Despite improved time to healing, no improvement in the quality of life by the Aberdeen Varicose Vein Questionnaire, EQ-5D-5L, and SF-36 was observed. Cost, however, was reduced for the early intervention group due to an increased ulcer-free time, a combination of a shorter time to VLU healing and a lower rate of recurrence (0.11 per person-year in the early-intervention group compared with 0.16 per person-year in the deferred-intervention).

One of the guiding principles in writing a Preface to a textbook is linking your own experience back to the subject matter of the book. Venous ulcers became personal to me at a young age and would be a lifelong interest. My introduction to the problem of VLUs was as a 12-year-old watching my father daily dress a wound on his darkly bronzed leg and then wrap a compression bandage over the wound. My father was an All-American football player in the late 1920s and played semiprofessional football, during which he sustained "dings to his legs." He religiously dressed his VLU before he set off to his job as a high school principal. As I learned about VLUs from Norman Browse at Saint Thomas's Hospital, I thought that the pathogenesis of my father's VLUs was consistent with the famous Irish phlebologist, George Fegan's, theory on the pathogenesis of VLUs—repeated blows to the shin, which caused episodes of thrombophlebitis in the veins. Fegan theorized that the thrombosis spilled over into the deep veins initiating the post-thrombotic syndrome. My father also may have been a victim of the Celtic gene, which predisposes to VLUs through defects in iron metabolism, as his parents immigrated to the United States from Spiddal (ironically Irish for hospital), Ireland in the early 1900s.

Besides a personal acquaintance with venous ulcers, my surgical training on the Harvard Surgical Service at Boston City Hospital exposed me to treatment methods, which appeared back in the early1970s frozen in time and unchanged over a century. Patients with venous ulcers would come to a special wound clinic in this "safety net hospital" for their weekly leg wash and wrap. I remember one hospital elevator operator that I treated over the course of my 7-year surgical residency who never healed. During my Post-Surgical residency year in London, UK, at Saint Thomas's Hospital with the late Professor Norman Browse and my research colleague Kevin Burnand, I was exposed to potential surgical treatments for VLUs, but most importantly was imbued with a critical evidentiary approach in evaluating these interventions. In our clinical studies, we identified the influential role of deep venous disease that was associated with lack of healing and recurrence of VLUs, which today remains unsolved. At that time the incompetent perforating vein was a putative cause of VLUs in the genesis of venous hypertension, so that ablation of this connection was thought to be fundamental. This was not a novel concept, as Robert Linton had posed it 30 years earlier and today its value has not yet been proven with a rigorous RCT.

I had the opportunity to preview the chapters of the new second edition of *Venous Ulcers*, which has been expanded from the original first edition by four sections to five as well as by 130 pages. Twenty-two of the 29 chapters are by new authors all of whom have been responsible for the seminal studies in their particular area. While all chapters are superior, I will highlight a selected number of these chapters.

The pathophysiology of VLUs has been consolidated into one chapter by Joseph Raffetto with an emphasis on subcellular mechanisms and genetics, while a chapter on hemodynamics authored by Nicos Labropoulos, whose laboratory conducted many of the key studies, has been added. Unique to this chapter is practical "how to do it" technical advice provided to the reader. This practical approach is a theme throughout the book. Two chapters explore the etiology, risk, and predictive factors for VLUs and are replete with excellent summary tables. These chapters replace the previous single chapter on the epidemiology of venous ulcers. However, Olle Nelzen of the Skaraborg leg ulcer study returns to update the findings of his landmark epidemiologic and treatment study.

I enjoyed the chapter on compression by Hugo Partsch, who challenged the concept of standard graduated compression stockings with higher pressure distally. His studies showed compression stockings with higher pressure *proximally* are more effective hemodynamically and also led to improved patient compliance. Like other chapters, Partsch includes some practical guidelines for the physician. Mark Passman's chapter on adjuvant therapy for wound healing has an important section on treatment adherence, which is not usually addressed in most texts. With the recent FDA approval of certain venoactive drugs, Mark Iafrati's timely review of the proposed mechanisms of action and data from both human and animal trials helps to define their benefits.

John Lantis, who has conducted numerous RCTs for advanced dressings, outlines the appropriate indications for these dressings and does not pull punches. In his candid opinion, there is presently neither a clear need nor clinical indication for the use of amniotic tissue-derived dressings to treat VLUs in contrast to diabetic foot wounds. This chapter provides a helpful treatment algorithm for using these advanced dressings, which is based on the size of the ulcer.

Nowhere since the publication of the first edition has the evidence for the treatment of VLUs leaped forward as has a surgical intervention. Manjit Gohel, the lead participant in the two major large RCTs, the ESCHAR and the EVRA trials, provides support for a surgical approach. In comparison to standard compression and wound care, reduction of superficial hypertension by treating the saphenous vein lowers recurrence rates in patients with superficial venous involvement alone or with superficial and limited deep venous reflux. Moreover, the recent EVRA trial showed that earlier endovenous ablation of axial veins leads to reduced time to healing and a month longer of ulcer-free time. It should be noted, however, that this trial did not compare endovenous ablation directly to conservative therapy, but rather to the timing of the intervention.

In keeping with the practical approach of this textbook Khilnani and his team provide a step-by-step method for the utilization of ultrasound with the endovenous treatment of superficial venous hypertension. In the absence of a properly conducted RCT comparing the treatment of incompetent perforators alone in patients with VLU to conservative therapy, Lawrence and colleagues draw upon their extensive experience and on observational data for the treatment of incompetent perforating veins. Besides a review of the rationale for this approach, they provide a sequence of treatments for the superficial anatomy associated with venous hypertension. In keeping with the practical hints in this text, Lawrence outlines with great detail down to the type of room where the procedure should be carried out.

Describing the new frontiers of venous surgery, correction of infrainguinal deep valvular incompetence, Maleti and Perrin update the evidence for valve repair, as deep venous pathology represents an appreciable segment of the VLU population. In the final chapter, Rockson introduces the role of the lymphatics, which are embryologically and pathologically related to the venous system. He focuses on the failure of lymphatic transport to keep up with the increased filtration induced by venous hypertension. This results in a rise of glycoproteins within the extracellular space as well as the cellular elements of chronic inflammation. The characteristic clinical sequelae of advanced chronic venous insufficiency ensue.

Venous Ulcers will be a "go to" book on this common disease for all physicians.

Thomas F. O'Donnell Jr., MD, FACS

SECTION 1

Chronic venous insufficiency—basic considerations

CHAPTER 1

Pathophysiology of chronic venous disease: genetic, molecular, and biochemical mechanisms

Joseph D. Raffetto[1] and Raouf A. Khalil[2]

[1]Harvard Medical School, VA Boston Healthcare System, Brigham and Women's Hospital Boston, Boston, MA, United States; [2]Harvard Medical School, Brigham and Women's Hospital, Division of Vascular and Endovascular Surgery, Boston, MA, United States

Key points

1. Venous reflux and obstruction lead to venous hypertension
2. Genetic predisposition and various candidate genes and gene polymorphisms, as well as environmental factors, are important in the development of chronic venous disease
3. Inflammatory cells have a central role in the pathophysiology of chronic venous disease and venous leg ulcers
4. Changes in the glycocalyx and shear stress lead to dysfunctional endothelium and expression of adhesions molecules that attract leukocytes and initiate an inflammatory response
5. Cytokines and matrix metalloproteinases are expressed in chronic venous disease particularly in venous leg ulcers and are necessary for wound closure, but are also responsible for persistently impaired wound healing. Identifying the sequence of molecules necessary for healing will provide biomarkers for disease progression and healing as well as potential targets for therapy
6. Changes in structural proteins such as collagen and elastin are key features in varicose veins likely due to posttranslational modifications by MMPs, resulting in vein wall weakness and decreased elasticity
7. Novel areas of research that could provide important insights into the pathophysiology of chronic venous disease and venous leg ulcer include the identification of metabolic signatures and the role of connexins, both having the potential for directed therapy
8. Impaired iron metabolism and the generation of reactive oxygen and nitrogen species lead to significant cellular damage, and the physiological changes in critical enzyme activities to restore cellular function are compromised. Genome-wide association studies have further confirmed a genetic predisposition with characteristic phenotypes and chronic venous disease. Additionally, there are new findings of cytokines, growth factors, proteolytic (ADAMTS, MT-MMP), and degradomic molecules in venous leg

ulcers that could affect the progression and healing process. Understanding these genetic, cellular, and molecular pathways can lead to new therapeutic targets to prevent venous leg ulcer and enhance its healing.

Introduction

Chronic venous disease (CVD) is a debilitating condition of the lower extremity that affects millions of individuals worldwide. CVD can result in varicose veins (VV), or advance to severe skin changes and venous leg ulceration. Both venous reflux and obstruction account for the pathophysiology of CVD. Venous reflux has a much higher prevalence in patients presenting with the different stages of CVD including venous leg ulcers (VLU), but obstruction has a higher rate in patients developing VLU and has a much more rapid progression of the disease.[1-5] Whether reflux or obstruction, or both are the cause for the patient's clinical presentation and symptomatology, both processes lead to increased ambulatory venous pressure. Genetic and environmental factors also influence predisposition to CVD and VLU. Inflammatory cells are activated upon changes in shear stress resulting in endothelial cell disruption, glycocalyx damage, and expression of adhesion molecules. Matrix metalloproteinases (MMP) are activated and cause changes in the structural components of the vein wall including collagen and elastin, and promote degradation of extracellular matrix proteins, resulting in CVD as seen in VV, skin changes, and VLU. The fundamental basis for CVD and venous ulceration is inflammation within the venous circulation that is subjected to increased hydrostatic pressure, resulting in increased ambulatory venous pressure, increased inflammation within the vein wall and valve leaflet, and extravasation of inflammatory cells and molecules into the interstitium.[6-8] The inflammatory response involves leukocytes in particular macrophages and monocytes, as well as T-lymphocytes and mast cells, inflammatory modulators and chemokines, cytokine expression, growth factors, metalloproteinase activity, and many regulatory pathways that perpetuate inflammation and the resultant changes seen with CVD (Fig. 1.1).[9-14]

Content

Genetic predisposition for CVD and VLU

The pathophysiology of primary venous disease is a complex entity with genetic and environmental factors, changes in venous endothelium, inflammatory biomolecules, and structural wall changes, that lead to dilated tortuous veins, dysfunctional valves with insufficiency, venous hypertension, and the associated clinical manifestations seen with CVD.[13] Several epidemiologic studies have assessed the associated risk factors. Genetic and environmental factors influence the predisposition and perpetuation of developing primary venous disease. Other contributing factors include family history, female

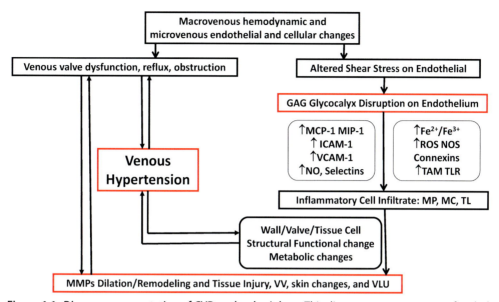

Figure 1.1 *Diagram representation of CVD pathophysiology.* This diagram represents areas of pathology that lead to varicose veins and venous leg ulcers. Both the macrovenous and microvenous components of the venous system are affected. In the macrovenous, there are several abnormalities including venous valve dysfunction and obstruction that have a common pathway leading to venous hypertension and skin changes including venous leg ulcers. Leukocytes and matrix metalloproteinases (MMPs) have direct involvement with the pathology seen in venous structures (indicated by bidirectional arrows). In the microvenous circulation, endothelial dysfunction, glycocalyx injury, and activation chemokines (e.g., MCP-1 and MIP-1) and adhesion molecules (e.g., ICAM-1, VCAM-1, selectins) and endothelial regulators (NO) are potent molecules to allow for leukocytes migration within the venous wall and valve and eventually in the interstitium. In addition, through oxidative stress (oxygen and nitrogen reactive species), iron activation, and innate immunity receptors and their ligands, leads to further expression and activation of leukocyte activity (MP macrophage, MC mast cells, TL T-lymphocytes). A variety of cytokines are expressed by leukocytes, with both direct and indirect effects, leading to continuous state of proinflammatory and inflammatory environment, in addition to the proteolytic activation of matrix metalloproteinases (MMPs), which have been demonstrated to both cause endothelial-smooth muscle relaxation, venous wall dilation, proteolytic degradation, and wound formation in venous leg ulcers. Cellular (endothelial, smooth muscle cell, fibroblast) metabolic changes occur, leading to loss of integrity of the venous wall and valves, which is directly linked with the microcirculation resulting in venous hypertension (indicated by bidirectional arrows). *CVD*, chronic venous disease; Fe^{2+}/Fe^{3+}, ferrous/ferric ions; *GAG*, glycosaminoglycans; *ICAM-1*, intercellular adhesion molecule; *MCP-1*, monocyte chemoattractant protein; *MIP-1*, macrophage inflammatory protein; *NO*, nitric oxide; *NOS*, nitrogen oxidative species; *ROS*, reactive oxygen species; *TAM*, Tyro Axl MerTK receptor family tyrosine kinase; *TLR*, toll-like receptors; *VCAM-1*, vascular cell adhesion molecule; *VLU*, venous leg ulcers; *VV*, varicose veins.

gender, pregnancy, and estrogen, the latter three are associated with VV clinically, as well as prolonged standing and sitting postures, and obesity.[9] Genetic disorders such as Klippel–Trenaunay syndrome, Ehlers–Danlos syndrome, cerebral autosomal dominant arteriopathy with subcortical infarcts and leukoencephalopathy (CADASIL), Forkhead

Box C2 (FOXC2) gene mutation, and dysregulated desmulin are associated with early onset of VV.[9,15] Of note, most individuals with the primary venous disease do not have these rare genetic syndromes as the trait is autosomal dominant with variable penetrance, and the specific genetic underpinning of primary venous disease and VV has not been clearly defined.[15,16] On the other hand, studies of individuals with CVD and VLU have suggested an association between polymorphic or mutated genes and the clinical phenotype. In a cohort of 2701 CVD patients in Germany, inherited genetic disorders were identified in approximately 17% of cases,[17] representing a substantial proportion of individuals with a genetic link to venous disease. Also, a genome-wide association study (GWAS) of 2269 CVD patients and 7765 control subjects showed an association between EFEMP1, KCNH8, and SKAP2 gene variants and susceptibility to CVD.[18] Given that these genes participate in the regulation of extracellular matrix proteins, potassium channels, and intracellular signaling, make it more likely that polymorphisms or mutations in these genes are involved in the pathophysiology of CVD.[18] Other genes of significance in the development of CVD include FOXC2, HFE, and MMPs, and have shown strong associations with VV and advanced stages of chronic venous insufficiency.[19] Additional studies in CVD patients have shown genetic polymorphism in the development and healing potential of VLU.[20] Hemochromatosis C282Y (HFE) gene mutation and certain Factor XIII V34L gene variants have been identified in patients with VV, and have been associated with increased risk of advanced stages of CVD, and the size of VLU.[21,22] Factor XIII (FXIII) is a cross-linking protein essential for ulcer healing.[23] Among CVD patients undergoing venous surgery, specific FXIII genotypes had favorable ulcer healing rates, but HFE gene mutation did not influence healing time despite its role in increasing the risk of developing VLU.[24] Further research is needed to identify the genetic basis of CVD to better understand the mechanisms of the disease and develop new and effective therapies.

Venous structure, endothelium and glycocalyx, inflammatory cells and molecules

Biochemical, immunohistochemical, and functional studies suggest that both vein wall dysfunction and valve disruption are important primary events leading to CVD. Whether vein wall dysfunction precedes valve insufficiency or whether valve dysfunction causes vein wall distension is unclear.[9] Perturbed microcirculation is also a critical component in the pathophysiology of CVD. The endothelium is a key regulator of vascular tone, hemostasis, and coagulation. The endothelium can be adversely affected by genetic and environmental factors, smoking, flow-induced injury, infection, immune disease, and diabetes. Failure of the endothelium to compensate leads to endothelial cell damage and disruption of the integrity of the vein wall. In CVD, persistently elevated ambulatory venous pressure leads to venous hypertension and deleterious effects on venous microcirculation. Altered shear stress in the venous microcirculation promotes endothelial cells to release vasoactive factors, selectins, inflammatory cytokines and chemokines, and prothrombotic precursors.[25,26] Endothelial cells sense changes in blood flow, shear stress, and vein wall stretch via intercellular adhesion molecule-1 (ICAM-1, CD54), vascular cell adhesion molecule-1 (VCAM-1, CD-106), endothelial leukocyte adhesion molecule-1 (ELAM-1, CD-62, E-selectin), and the mechanosensitive transient receptor potential vanilloid channels (TRPVs).[26,27] CVD patients show increased ICAM-1, VCAM-1, and ELAM-1 in endothelial cells.[12,28–30] Increases in shear stress cause perturbation in endothelial nitric oxide production, the release of vasoactive factors, expression of ICAM-1, VCAM-1, macrophage chemoattractant protein-1 (MCP-1), ELAM-1, L-selectin, and E-selectin, leading to recruitment of leukocytes and transmigration into the vein wall and valve, and initiation of inflammatory cascade with increased release of chemokines (IL-8), cytokines (TGF-β1, TNF-α, IL-1), and MMPs.[11,26,30] Also, the glycocalyx is an important macromolecule composed of glycoproteins, proteoglycans, and glycosaminoglycan, functioning as a mechanical sensor on the surface of endothelial cells. The endothelial glycocalyx prevents leukocyte adhesion, inflammation, and thrombosis, but altered shear stress and mechanical forces on the vein wall cause endothelial cell injury, and loss of the glycocalyx, leukocyte adhesion, and inflammation.[31,32] Heparanase and MMPs could degrade the glycocalyx and cause alterations in the levels of glycosaminoglycans. The wall of VV shows endothelial glycocalyx disruption and increased levels of degraded sulfated glycosaminoglycans.[33]

MMPs regulation, cellular changes, and CVD/VLU

An important component of inflammation in CVD and VLU is the overexpression of MMPs, which could have marked effects on the venous valve, vein wall, the endothelium and glycocalyx, as well as surrounding tissues including the dermal and subcutaneous structures leading to skin changes and VLU (Fig. 1.2).[11,34,35]

Several MMPs have been detected in vein specimens from CVD patients. MMPs cause degradation of extracellular matrix proteins and collagen bundles and elastin in the medial layer and likely the adventitia, and could have additional venous dilation effects.[9,30,35] MMPs are released in the vein wall in response to mechanical stretch and venous hypertension and in turn affect the different components of the vein wall including the endothelium, vascular smooth muscle (VSM), and the adventitia.[36] In rat veins, MMPs cause venous dilation through vein wall hyperpolarization and inhibition of Ca^{2+} entry through surface membrane channels, and these processes may be regulated by hypoxia-inducible factor (HIF).[37–39] HIF-1α and HIF-2α transcriptional factors are overexpressed in VV compared with control veins, suggesting that hypoxia and induction of HIF are contributing factors to the pathogenesis of CVD.[40] MMPs are present in large amounts in VLU and in the wound fluid, and increased proteinase activity is correlated with poor VLU healing. MMP release and activity are regulated by multiple factors including cytokines, urokinase-type plasminogen activator (uPA), extracellular MMP inducer (EMMPRIN,

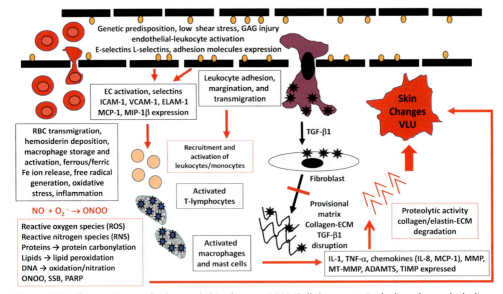

Figure 1.2 Inflammation and advanced skin changes/VLU. Cellular events including the endothelium, glycocalyx, adhesion molecules, leukocytes and red blood cells, cellular activation and inflammation pathways, and events in the interstitial space affecting the subcutaneous and dermal structures. There are predisposing factors and changes in the glycocalyx, leading to adhesion molecule activation (selectins, ICAM, VCAM) and leukocyte activation with chemokine expression (e.g., MCP-1, MIP-1, IL-8). Inflammatory cascade initiates with several cytokine mediators expressed. Matrix metalloproteinases (MMPs, especially MMP-2, MMP-7, MMP-9, MMP-8, MMP12), membrane-type MMPs, and a disintegrin and metalloproteinase with thrombospondin motif (ADAMTS) are involved in the process of tissue degradation including collagen, elastin, and the extracellular matrix (ECM). Tissue inhibitors of MMP (TIMP) can be overexpressed or underexpressed depending on the proteolytic and degradomics conditions within the interstitium and subcutaneous-dermal structures. Transforming growth factor (TGF) is an important cytokine and growth factor with inflammatory and structural signaling. TGF is blocked and/or degraded delaying provisional and structural fibronectin, tenascin, collagen, elastin, and ECM components from synthesizing. Unmitigated inflammation and proteolytic activity results in tissue destruction, skin changes, and venous leg ulcer (VLU) formation. Red blood cells (RBC) transmigrate and breakdown products include hemosiderin and free iron (ferrous and ferric). Skin changes including hyperpigmentation and lipodermatosclerosis (LDS) are explained by extravasation and disruption of erythrocytes, followed by decomposition of hemoglobin. The resulting excessive tissue iron (Fe^{2+} ferrous ion and Fe^{3+} ferric ion) is stored as hemosiderin. Macrophages with ingested hemosiderin are toxic and cyclic oxidation and reduction reactions result in the generation of toxic free radicals (reactive oxygen species ROS, reactive nitrogen species, RNS) via the fenton reaction and nitric oxide (NO) and superoxide ($O2-$). Usually, iron is protein bound in heme and iron-binding proteins (ferritin and transferrin). ONOO, peroxynitrite; SSB, single strand breaks of DNA; PARP, poly-ADP ribose polymerase. PARP is rapidly activated by DNA SSBs and cleaves NAD+ to nicotinamide and ADP-ribose. PARP1 then attaches ADP-ribose to suitable protein acceptors near the DNA nicks and builds a branched poly-ADP-ribose (PAR) polymer to initiate the repair process. The physiological signaling roles of ROS and RNS require redox sensor proteins such as activator protein 1 (AP-1, a heterodimer of c-Jun and c-Fos), heat shock factor 1 (HSF-1), or Keap-1. An extensive antioxidant defense system consisting of small molecular and enzymatic antioxidants (e.g., glutathione, superoxide dismutase, catalase, and glutathione-peroxidase, respectively) operates to prevent tissue damage caused by ROS/RNS. Peroxynitrite is a potent oxidizing and nitrating agent that damages mitochondria, tissue, DNA, lipid peroxidation, posttranslational modifications of many proteins, protein oxidizer and nitration, and enzyme inactivation, causing significant cellular dysfunction. ONOO also decreases the function of superoxide dismutase (SOD) increasing ROS production.

CD147) platelet-derived growth factor isoform AA, and mitogen-activated protein kinase.[41–52] Cytokines play important roles in different stages of CVD by contributing to the inflammatory process and its propagation in the interstitial space and the VLU bed.[11,33] In a study on CVD patients, untreated VLU displayed high levels of proinflammatory cytokines such as interleukins, TNF-α, and interferon-γ (IFN-γ), but after 4 weeks of compression therapy, the cytokines levels markedly decreased, the levels of transforming growth factor (TGF-β1) increased and the VLU began to heal. Importantly, VLUs with higher levels of IL-1 and IFN-γ healed much better (defined as 40% or greater reduction in wound surface area) than VLUs with lower levels of these cytokines before compression therapy.[53] In a related study, the authors showed marked increases in the levels of MMP-1, 2, 3, 8, 9, 12, and 13 in VLUs compared to healthy tissue, and 4 weeks compression therapy was associated with reductions in MMP-3, 8, and 9 levels. Study also showed a correlation between reductions in MMP-1, 2, and 3 levels and higher VLU healing rates.[54] Also, studies on inflammatory and granulating VLUs showed marked differences in the levels of cytokines, chemokines, granulocyte-monocyte colony-stimulating factor, and growth factors in the wound fluid depending on the wound environment. Notably, marked differences in the levels of MMPs and tissue inhibitors of metalloproteases (TIMPs) were also observed depending on the stage of the VLU wound (inflammatory vs. granulating).[55,56] These studies show marked differences in the wound environment in the healing versus nonhealing VLU and provide important information regarding the mechanisms underlying wound progression, and potential biomarkers for VLU targeted therapy and prognosis. In addition to venous hypertension, inflammation, vein wall remodeling, and increased expression of cytokines and MMPs, a fibrin cuff comprising complex fibrin and collagen deposition is often identified in the postcapillary venule.[11] The postcapillary cuff has many components including collagen I and III, fibronectin, vitronectin, laminin, tenascin, fibrin, TGF-β1, and α2-macroglobulin and represent a major abnormality in the dermal microcirculation.[11,57] Interestingly, macrophages and mast cells have also been identified in the fibrin cuff and may represent a major source of the increased cytokines and MMPs levels and the consequent pathological changes associated with CVD, skin changes, and VLU.

Structural proteins and alterations in CVD and VLU

Analysis of structural proteins in VV from patients with CVD showed an overall increase in collagen and a decrease in both elastin and laminin.[58] Also, VSM cells and dermal fibroblasts from varicose vein patients show increased collagen type I and decreased collagen type III when compared to cells from control nonvaricose vein subjects.[58,59] These changes in collagen profile may represent a systemic condition influenced by potential genetic factors. Collagen I mainly confers rigidity while collagen III takes part in tissue extensibility, and changes in collagen I/III ratio could contribute to the vein wall weakness and decreased elasticity in VV. Interestingly, the collagen III gene transcription is normal in VSM cells from VV, but MMP-3 activity is increased suggesting posttranslational modification and degradation of collagen III, which can be reversed by MMP-3 inhibitors.[60] Thus, MMPs participate in the different processes during the development of primary venous disease; with implications both in the early events affecting the endothelium–VSM interactions and venodilation, as well as the late events involving ECM degradation, changes in the vein wall structural proteins, venous tissue remodeling and fibrosis, and interstitial tissue proteolysis and damage leading to VLU formation.[9]

Microvenous valves and CVD

Microvenous valves have been studied using retrograde resin injection and vein casting in amputated lower limbs of patients with CVD as well as control subjects without CVD.[61] The network of tributaries was divided into six sequential generations before reaching the small venous networks. The valves in the greater saphenous vein and major tributaries were assigned generation 0, and the valves in each subsequent tributary were assigned a consecutively numbered generation (1–5). In regions with incompetent microvalves out to the third-generation tributary ("the boundary"), the injected resin penetrated deeper into the microvenous networks of the dermis. In limbs with VV and VLUs, reflux into the small venous networks and capillary loops was extensive with more dense networks and greater tortuosity. Thus, in addition to superficial axial saphenous vein insufficiency, microvenous valve insufficiency also exists, and once the third-generation microvalves are compromised, there is a greater risk for the development of dermal venous ulceration. This may explain why some patients with longstanding VV do not develop VLUs, since the microvalves at the third-generation network are intact and prevent clinical deterioration.[61] This may also explain why skin changes consistent with CVD such as hyperpigmentation and even small skin ulceration may be seen clinically in patients with normal duplex ultrasound of the superficial, deep, and perforator venous systems (likely because of compromised third-generation microvalves). Further research on the factors regulating shear stress, vein wall stretch and microvalve function, and the presence of cytokines and proteinases will help define specific targets to restore the integrity of the venous microcirculation and treat the spectrum of CVD.

Hypoxia, apoptosis, metabolic abnormalities, and connexins in CVD/VLU

Other potential mechanisms for the development of CVD and venous insufficiency are hypoxia and apoptosis of the vein wall. Studies have suggested an association between hypoxia, apoptotic changes in the vein wall, and CVD, but the results showed significant variability and were not sufficiently conclusive.[62,63] Metabolomics is the comprehensive study of metabolism in biological systems under normal conditions and in response to pathophysiologic stimuli and genetic modifications. Metabolic abnormalities could also play a role in venous dysfunction and lead to CVD. Metabolic products such as creatine, lactate, and myoinositol metabolites are increased in VV compared to control non-VV patients.[64] Also, increased levels of valine and choline metabolites, and triglyceride moieties were identified in isolated rat inferior vena cava subjected to prolonged stretch for 18 h compared to nonstretched vein. When interpreting these findings in the context of CVD, the increased levels of the branched-chain amino acid valine and cell membrane constituent choline could indicate increased muscle breakdown. The increased levels of triglyceride moieties in stretched vein segments suggest that high venous pressure may induce an inflammatory response. These observations are consistent with the pathology observed in VV and CVD and provide insight into the underlying mechanisms, metabolic pathways, and potential therapeutic targets. Other studies have shown higher concentrations of glutamate, taurine, myo-inositol, creatine, and inosine in aqueous extracts and phosphatidylcholine, phosphatidylethanolamine and sphingomyelin in lipid extracts in vein specimens from VV patients compared with control subjects. Pathway analysis indicated an association of phosphatidylcholine and sphingomyelin with inflammation and of myo-inositol with cell proliferation, thus implicating major metabolic pathways in the pathogenesis of CVD.[65] Further analysis of cellular metabolism and signature end products in six studies on CVD and two studies in VLU has provided key information to the metabolic basis of the disease processes. Upregulated metabolites in veins from patients with CVD include lipids, branched-chain amino acids (BCAA), glutamate, taurine, lactate, and myo-inositol. Upregulated metabolites in VLU wound fluid and ulcer biopsies include lactate, BCAA, lysine, 3-hydroxybutyrate, and glutamate.[66] Further research into the metabolic profiles would identify molecular targets for the prevention and improved targeted therapy of CVD and VLU.

Gap junctions have emerged as a new area of research with implications in the pathophysiology and potential therapy of CVD. Gap junctions are involved in the different processes associated with the pathogenesis of chronic wounds including inflammation, edema formation, and fibrosis. Connexins are the channel-forming components of gap junctions, facilitating electrical propagation between excitable cells and may allow small molecules to pass between cells' cytoplasm. Connexins may play a role in the inflammatory response associated with CVD and in VLU healing. Connexin43 is abnormally elevated in the wound margin of VLU.[66] ACT1, a peptide inhibitor of connexin43, accelerates fibroblast proliferation and epithelialization in animal models.[67] Also, in a study of VLU patients randomized to compression plus ACT1 gel application versus compression alone, VLU treated with ACT1 gel showed greater mean percent reepithelialization at 12 weeks, and reduced median time to 100% ulcer healing.[67] Further studies of the different connexins and their contribution to the entire spectrum of CVD would highlight the connexin cellular pathway as a novel target for CVD treatment and altering disease progression.

Machine learning GWAS and genetics, iron, oxidative stress, reactive oxygen/nitrogen species, biomarkers cytokines/proteolytic, and MT-MMP/ADAMTS in CVD/VLU

Recent genetic analysis has identified important gene loci and phenotypic changes associated with VV and VLU. An extensive study of nearly half a million subjects (VV and control), utilizing machine learning for risk factors as well as GWAS, determined that advanced age, female sex, obesity, pregnancy, deep venous thrombosis, increased height, and leg bioimpedance as risk factors for VV.[68] GWAS identified 30 gene loci strongly associated with VV, including genes encoding for blood pressure, vascular mechanosensing and channels (e.g., glycocalyx, calcium channels, TPRV), vascular maturation, development and integrity, and genes near the hemochromatosis gene that are associated with VLU.[68] Another study evaluating the relationship between gene expression and prediction of VLU healing identified 14 candidate genes (WounD 14—WD14 signature), and when examined in a prospective blinded study, the WD14 signature could predict ulcers likely to heal.[69] These observations have both clinical and socio-economic implications as they would highlight potential target genes for future therapeutic interventions and gene therapy, and would identify patients with potentially difficult-to-heal VLU and requiring additional support. Iron deposits are increased in CVD and VLU and could be involved in the development of skin changes and ulceration.[70] Erythrodiapedesis is a process in which red blood cells exit the capillaries and pericapillary network and enter into the interstitial tissue space, leading to erythrocyte disruption, hemoglobin degradation, and storing of ferric iron as hemosiderin. Erythrodiapedesis has been detected in CVD patients with skin changes (lipodermatosclerosis—LDS) and VLU.[71] In patients with VLU oxidative stress is elevated and can lead to slow-healing or nonhealing VLU.[72] The properties of ferrous iron (Fe^{2+}) have significant effects on stimulating macrophages, and activation of cytokine and chemokine release. These effects lead to an inflammatory state, with oxidative stress, hemolysis of red cells, and perpetuates skin changes and VLU development.[73] The net effect of a highly oxidative state within tissue and VLU is the activation of several oxidative and nitrating processes. Specifically, there is generation of reactive oxygen species, reactive nitrogen species, protein carbonylation, lipid peroxidation, and DNA oxidation and nitration resulting in damage by single-strand breaks.[74] In addition to free iron in its Fe^{2+} form, ferric ion (Fe^{3+}) is extremely toxic and has been detected in tissue biopsies of patients with advanced CVD with LDS and VLU, but not in VV or only edema and hyperpigmentation CVD patients.[75] Peroxynitrite ($ONOO^-$) is a potent oxidizing and nitrating agent that causes irreparable damage to the mitochondria, DNA, lipid peroxidation, and protein oxidation and nitration, leading to modification of many proteins and inactivation of multiple enzymes, and ultimately disruption of cellular functions.[76] $ONOO^-$ effect on DNA is to cause single-strand breaks, which activates poly-ADP ribose polymerase (PARP) and builds at the DNA nicks a branched poly-ADP-ribose (PAR) polymer to initiate the repair process.[74] In the first study to evaluate $ONOO^-$ and PARP activity, the researchers evaluated tissue biopsies of VLU versus normal tissue as control. The study found both elevated PAR (PARP, DNA damage/repair) and elevated nitrotyrosine (an indirect measure of ONOO$^-$ via metabolite).[74] This indicated that the pathophysiology of VLU in part involves the generation of ONOO$^-$ with its many destructive properties including DNA damage. Future work should examine if inhibition of ONOO$^-$ formation can alter the progression of VLU and improve healing. Of note, ONOO$^-$ and its damage of DNA could potentially lead to mutations and carcinogenic transformation of VLU as is seen with squamous cell carcinoma.[76,77]

Accumulating research evidence has demonstrated several biomarkers including cytokines, chemokines, growth factors, proteolytic enzymes (MMP, EMMPRIN, TIMP) in tissue specimens, serum, and wound fluid from patients with VLU.[78] In addition, changes in the levels of membrane type-MMPs (MT-MMP), and a disintegrin and metalloproteinase with thrombospondin motif (ADAMTS) have been associated with various stages of CVD and VLU.[76,79] While the presence of these inflammatory biomarkers could be an epiphenomenon, they could have an integral function in the pathophysiology of CVD and VLU, providing important areas for further research studies and future clinical treatments.

Conclusions

The pathophysiology of CVD and VLU is complex and multifactorial involving predisposing genetic and environmental factors, alterations of key functional proteins and enzymes, changes in shear stress, injury to the glycocalyx, and activation of several adhesion molecules. These events lead to leukocyte activation and transmigration into the vein wall, valves, and interstitium, with the release of many cytokines, chemokines, growth factors, proteinases, and regulatory molecules. Various metabolites have been identified and are important regulators and signatures of metabolic cellular dysfunction. Inflammation of the tissue and cellular dysfunction within the lower limb ensues, leading to changes observed clinically as CVD and VLU. Further escalating the inflammatory process is oxidative stress driven by iron deposition from erythrodiapedesis and red blood cell degradation, resulting in toxic levels of free iron release and generation of reactive oxygen and nitrogen species. These oxidative and nitrating compounds disrupt many cellular functions and cause tissue damage. Ultimately, progressive inflammation leads to advanced forms of CVD, skin changes, and VLU. Further research of the pathophysiology and mechanistic pathways will provide a better understanding of the complex disease process and help develop innovative and targeted therapies of CVD and VLU.

Acknowledgments

Dr. R.A. Khalil was supported by BRI Fund to Sustain Research Excellence from Brigham Research Institute, and grants from National Heart, Lung, and Blood Institute (HL111775, R56HL147889, and R01HL147889-A1).

References

1. Labropoulos N, Leon M, Nicolaides AN, et al. Superficial venous insufficiency: correlation of anatomic extent of reflux with clinical symptoms and signs. *J Vasc Surg*. 1994;20:953−958.
2. Meissner MH, Moneta G, Burnand K, et al. The hemodynamics and diagnosis of venous disease. *J Vasc Surg*. 2007;46(Suppl S):4S−24S.
3. Labropoulos N, Leon M, Nicolaides AN, et al. Venous reflux in patients with previous deep venous thrombosis: correlation with ulceration and other symptoms. *J Vasc Surg*. 1994;20:20−26.
4. Labropoulos N, Gasparis AP, Tassiopoulos AK. Prospective evaluation of the clinical deterioration in post-thrombotic limbs. *J Vasc Surg*. 2009;50:826−830.
5. Labropoulos N, Gasparis AP, Pefanis D, et al. Secondary chronic venous disease progresses faster than primary. *J Vasc Surg*. 2009;49:704−710.
6. Eberhardt RT, Raffetto JD. Chronic venous insufficiency. *Circulation*. 2014;130:333−346.
7. Raffetto JD. Pathophysiology of wound healing and alterations in venous leg ulcers-review. *Phlebology*. 2016;31(1 Suppl):56−62.
8. Chi YW, Raffetto JD. Venous leg ulceration pathophysiology and evidence based treatment. *Vasc Med*. 2015;20:168−181.
9. Raffetto JD, Khalil RA. Mechanisms of varicose vein formation: valve dysfunction and wall dilation. *Phlebology*. 2008;23:85−98.
10. Deroo S, Deatrick KB, Henke PK. The vessel wall: a forgotten player in post thrombotic syndrome. *Thromb Haemost*. 2010;104:681−692.
11. Raffetto JD. Inflammation in chronic venous ulcers. *Phlebology*. 2013;28(Suppl 1):61−67.
12. Ono T, Bergan JJ, Schmid-Schönbein GW, et al. Monocyte infiltration into venous valves. *J Vasc Surg*. 1998;27:158−166.
13. Raffetto JD, Mannello F. Pathophysiology of chronic venous disease. *Int Angiol*. 2014;33:212−221.
14. Mannello F, Ligi D, Raffetto JD. Glycosaminoglycan sulodexide modulates inflammatory pathways in chronic venous disease. *Int Angiol*. 2014;33:236−242.
15. Anwar MA, Georgiadis KA, Shalhoub J, et al. A review of familial, genetic, and congenital aspects of primary varicose vein disease. *Circ Cardiovasc Genet*. 2012;5:460−466.
16. Cornu-Thenard A, Boivin P, Baud JM, et al. Importance of the familial factor in varicose disease. Clinical study of 134 families. *J Dermatol Surg Oncol*. 1994;20:318−326.
17. Fiebig A, Krusche P, Wolf A, et al. Heritability of chronic venous disease. *Hum Genet*. 2010;127:669−674.
18. Ellinghaus E, Ellinghaus D, Krusche P, et al. Genome-wide association analysis for chronic venous disease identifies EFEMP1 and KCNH8 as susceptibility loci. *Sci Rep*. 2017;7:45652.
19. Bharath V, Kahn SR, Lazo-Langner A. Genetic polymorphisms of vein wall remodeling in chronic venous disease: a narrative and systematic review. *Blood*. 2014;124:1242−1250.
20. Zamboni P, Gemmati D. Clinical implications of gene polymorphisms in venous leg ulcer: a model in tissue injury and reparative process. *Thromb Haemost*. 2007;98:131−137.
21. Zamboni P, Tognazzo S, Izzo M, et al. Hemochromatosis C282Y gene mutation increases the risk of venous leg ulceration. *J Vasc Surg*. 2005;42:309−314.
22. Tognazzo S, Gemmati D, Pallazzo A, et al. Prognostic role of factor XIII gene variants in nonhealing venous leg ulcers. *J Vasc Surg*. 2006;44:815−819.
23. Zamboni P, De Mattei M, Ongaro A, et al. Factor XIII contrasts the effects of metalloproteinases in human dermal fibroblast cultured cells. *Vasc Endovasc Surg*. 2004;38:431−438.
24. Gemmati D, Tognazzo S, Catozzi L, et al. Influence of gene polymorphisms in ulcer healing process after superficial venous surgery. *J Vasc Surg*. 2006;44:554−562.
25. Schmid-Shonbein GW, Takase S, Bergan JJ. New advances in the understanding of the pathophysiology of chronic venous insufficiency. *Angiology*. 2001;52(Suppl 1):S27−S34.
26. Bergan JJ, Schmid-Shonbein GW, Coleridge Smith PD, et al. Chronic venous disease. *N Engl J Med*. 2006;355:488−498.
27. Chen YS, Lu MJ, Huang HS, et al. Mechanosensitive transient receptor potential vanilloid type 1 channels contribute to vascular remodeling of rat fistula veins. *J Vasc Surg*. 2010;52:1310−1320.

28. Takase S, Pascarella L, Lerond L, et al. Venous hypertension, inflammation and valve remodeling. *Eur J Vasc Endovasc Surg*. 2004;28:484–493.
29. Takase S, Bergan JJ, Schmid-Schönbein G. Expression of adhesion molecules and cytokines on saphenous veins in chronic venous insufficiency. *Ann Vasc Surg*. 2000;14:427–435.
30. Castro-Ferreira R, Cardoso R, Leite-Moreira A, et al. The role of endothelial dysfunction and inflammation in chronic venous disease. *Ann Vasc Surg*. 2017;pii: S0890–5096(17):30843–30849.
31. Mannello F, Raffetto JD. Matrix metalloproteinase activity and glycosaminoglycans in chronic venous disease: the linkage among cell biology, pathology and translational research. *Am J Transl Res*. 2011;3: 149–158.
32. Mannello F, Medda V, Ligi D, et al. Glycosaminoglycan sulodexide inhibition of mmp-9 gelatinase secretion and activity: possible pharmacological role against collagen degradation in vascular chronic diseases. *Curr Vasc Pharmacol*. 2013;11:354–365.
33. Mannello F, Ligi D, Canale M, et al. Omics profiles in chronic venous ulcer wound fluid: innovative applications for translational medicine. *Expert Rev Mol Diagn*. 2014;14:737–762.
34. Serra R, Grande R, Butrico L, et al. Effects of a new nutraceutical substance on clinical and molecular parameters in patients with chronic venous ulceration. *Int Wound J*. 2016;13:88–96.
35. Chen Y, Peng W, Raffetto JD, et al. Matrix metalloproteinases in remodeling of lower extremity veins and chronic venous disease. *Prog Mol Biol Transl Sci*. 2017;147:267–299.
36. Raffetto JD, Qiao X, Koledova VV, et al. Prolonged increases in vein wall tension increase matrix metalloproteinases and decrease constriction in rat vena cava: potential implications in varicose veins. *J Vasc Surg*. 2008;48:447–456.
37. Raffetto JD, Ross RL, Khalil RA. Matrix metalloproteinase 2-induced venous dilation via hyperpolarization and activation of K+ channels: relevance to varicose vein formation. *J Vasc Surg*. 2007;45: 373–380.
38. Raffetto JD, Barros YV, Wells AK, et al. MMP-2 induced vein relaxation via inhibition of [Ca2+]e-dependent mechanisms of venous smooth muscle contraction. Role of RGD peptides. *J Surg Res*. 2010; 159:755–764.
39. Lim CS, Qiao X, Reslan OM, et al. Prolonged mechanical stretch is associated with upregulation of hypoxia-inducible factors and reduced contraction in rat inferior vena cava. *J Vasc Surg*. 2011;53: 764–773.
40. Lim CS, Kiriakidis S, Paleolog EM, et al. Increased activation of the hypoxia inducible factor pathway in varicose veins. *J Vasc Surg*. 2012;55:1427–1439.
41. Trengove NJ, Bielefeldt-Ohmann H, Stacey MC. Mitogenic activity and cytokine levels in non-healing and healing chronic leg ulcers. *Wound Repair Regen*. 2000;8:13–25.
42. Tian YW, Stacey MC. Cytokines and growth factors in keratinocytes and sweat glands in chronic venous leg ulcers. An immunohistochemical study. *Wound Repair Regen*. 2003;11:316–325.
43. Gohel MS, Windhaber RA, Tarlton JF, et al. The relationship between cytokine concentrations and wound healing in chronic venous ulceration. *J Vasc Surg*. 2008;48:1272–1277.
44. Wysocki AB, Staiano-Coico L, Grinnell F. Wound fluid from chronic leg ulcers contains elevated levels of metalloproteinases MMP-2 and MMP-9. *J Invest Dermatol*. 1993;101:64–68.
45. Weckroth M, Vaheri A, Lauharanta J, et al. Matrix metalloproteinases, gelatinase and collagenase in chronic leg ulcers. *J Invest Dermatol*. 1996;106:1119–1124.
46. Herouy Y, May AE, Pornschlegel G, et al. Lipodermatosclerosis is characterized by elevated expression and activation of matrix metalloproteinases. Implications for venous ulcer formation. *J Invest Dermatol*. 1998;111:822–827.
47. Herouy Y, Trefzer D, Hellstern MO, et al. Plasminogen activation in venous leg ulcers. *Br J Dermatol*. 2000;143:930–936.
48. Norgauer J, Hildenbrand T, Idzko M, et al. Elevated expression of extracellular matrix metalloproteinase inducer (CD147) and membrane-type matrix metalloproteinases in venous leg ulcers. *Br J Dermatol*. 2002;147:1180–1186.
49. Mwaura B, Mahendran B, Hynes N, et al. The impact of differential expression of extracellular matrix metalloproteinase inducer, matrix metalloproteinase-2, tissue inhibitor of matrix metalloproteinase-2 and PDGF-AA on the chronicity of venous leg ulcers. *Eur J Vasc Endovasc Surg*. 2006;31:306–310.

50. Meyer FJ, Burnand KG, Abisi S, et al. Effect of collagen turnover and matrix metalloproteinase activity on healing of venous leg ulcers. *Br J Surg.* 2008;95:319–325.
51. Raffetto JD, Vasquez R, Goodwin DG, et al. Mitogen-activated protein kinase pathway regulates cell proliferation in venous ulcer fibroblasts. *Vasc Endovasc Surg.* 2006;40:59–66.
52. Raffetto JD, Gram CH, Overman KC, et al. Mitogen-activated protein kinase p38 pathway in venous ulcer fibroblasts. *Vasc Endovasc Surg.* 2008;42:367–374.
53. Beidler SK, Douillet CD, Berndt DF, et al. Inflammatory cytokine levels in chronic venous insufficiency ulcer tissue before and after compression therapy. *J Vasc Surg.* 2009;49:1013–1020.
54. Beidler SK, Douillet CD, Berndt DF, et al. Multiplexed analysis of matrix metalloproteinases in leg ulcer tissue of patients with chronic venous insufficiency before and after compression therapy. *Wound Repair Regen.* 2008;16:642–648.
55. Ligi D, Mosti G, Croce L, et al. Chronic venous disease - Part I: inflammatory biomarkers in wound healing. *Biochim Biophys Acta.* 2016;1862:1964–1974.
56. Ligi D, Mosti G, Croce L, et al. Chronic venous disease - Part II: proteolytic biomarkers in wound healing. *Biochim Biophys Acta.* 2016;1862:1900–1908.
57. Pappas PJ, DeFouw DO, Venezio LM, et al. Morphometric assessment of the dermal microcirculation in patients with chronic venous insufficiency. *J Vasc Surg.* 1997;26:784–795.
58. Sansilvestri-Morel P, Rupin A, Badier-Commander C, et al. Imbalance in the synthesis of collagen type I and collagen type III in smooth muscle cells derived from human varicose veins. *J Vasc Res.* 2001;38:560–568.
59. Sansilvestri-Morel P, Rupin A, Jaisson S, et al. Synthesis of collagen is dysregulated in cultured fibroblasts derived from skin of subjects with varicose veins as it is in venous smooth muscle cells. *Circulation.* 2002;106:479–483.
60. Sansilvestri-Morel P, Rupin A, Jullien ND, et al. Decreased production of collagen Type III in cultured smooth muscle cells from varicose vein patients is due to a degradation by MMPs: possible implication of MMP-3. *J Vasc Res.* 2005;42:388–398.
61. Vincent JR, Jones GT, Hill GB, et al. Failure of microvenous valves in small superficial veins is a key to the skin changes of venous insufficiency. *J Vasc Surg.* 2011;54(6 Suppl):62S–69S.
62. Lim CS, Davies AH. Pathogenesis of primary varicose veins. *Br J Surg.* 2009;96:1231–1242.
63. Lim CS, Gohel MS, Shepherd AC, et al. Venous hypoxia: a poorly studied etiological factor of varicose veins. *J Vasc Res.* 2011;48:185–194.
64. Anwar MA, Shalhoub J, Vorkas PA, et al. In-vitro identification of distinctive metabolic signatures of intact varicose vein tissue via magic angle spinning nuclear magnetic resonance spectroscopy. *Eur J Vasc Endovasc Surg.* 2012;44:442–450.
65. Anwar MA, Adesina-Georgiadis KN, Spagou K, et al. A comprehensive characterisation of the metabolic profile of varicose veins; implications in elaborating plausible cellular pathways for disease pathogenesis. *Sci Rep.* 2017;7:2989.
66. Onida S, Tan MKH, Kafeza M, et al. Metabolic phenotyping in venous disease: the need for standardization. *J Proteome Res.* 2019;18:3809–3820.
67. Ghatnekar GS, Grek CL, Armstrong DG, et al. The effect of a connexin43-based Peptide on the healing of chronic venous leg ulcers: a multicenter, randomized trial. *J Invest Dermatol.* 2015;135:289–298.
68. Fukaya E, Flores AM, Lindholm D, et al. Clinical and genetic determinants of varicose veins. *Circulation.* 2018;138:2869–2880.
69. Bosanquet DC, Sanders AJ, Ruge F, et al. Development and validation of a gene expression test to identify hard-to-heal chronic venous leg ulcers. *Br J Surg.* 2019;106:1035–1042.
70. Zamboni P. Is leg ulceration a defending mechanism against toxic iron accumulation. *Acta Haematol.* 2016;135:122–123.
71. Caggiati A, Franceschini M, Heyn R, Rosi C. Skin erythrodiapedesis during chronic venous disorders. *J Vasc Surg.* 2011;53:1649–1653.
72. Yeoh-Ellerton S, Stacey MC. Iron and 8-isoprostane levels in acute and chronic wounds. *J Invest Dermatol.* 2003;121:918–925.

73. Wlaschek M, Singh K, Sindrilaru A, Crisan D, Scharffetter-Kochanek K. Iron and iron-dependent reactive oxygen species in the regulation of macrophages and fibroblasts in non-healing chronic wounds. *Free Radic Biol Med*. 2019;133:262–275.
74. Bodnár E, Bakondi E, Kovács K, et al. Redox profiling reveals clear differences between molecular patterns of wound fluids from acute and chronic wounds. *Oxid Med Cell Longev*. 2018;2018:5286785.
75. Caggiati A, Rosi C, Casini A, et al. Skin iron deposition characterises lipodermatosclerosis and leg ulcer. *Eur J Vasc Endovasc Surg*. 2010;40:777–782.
76. Raffetto JD, Khalil RA. Mechanisms of lower extremity vein dysfunction in chronic venous disease and implications in management of varicose veins. *Vessel Plus*. 2021;5:36.
77. Shalhout SZ, Kaufman HL, Sullivan RJ, Lawrence D, Miller DM. Immune checkpoint inhibition in marjolin ulcer: a case series. *J Immunother*. 2021;44:234–238.
78. Raffetto JD, Ligi D, Maniscalco R, Khalil RA, Mannello F. Why venous leg ulcers have difficulty healing: overview on pathophysiology, clinical consequences, and treatment. *J Clin Med*. 2020;10:29.
79. Serra R, Gallelli L, Butrico L, et al. From varices to venous ulceration: the story of chronic venous disease described by metalloproteinases. *Int Wound J*. 2017;14:233–240.

CHAPTER 2

Venous hemodynamics and microcirculation in chronic venous insufficiency

John Blebea
Department of Surgery, College of Medicine, Central Michigan University, Saginaw, MI, United States

Introduction

Our understanding of venous hemodynamics has a long history. Andre Vesalius described the anatomy of veins in 1543, William Harvey in 1628 defined the direction of blood flow through the veins back to the heart utilizing their unidirectional valves, and Antonio Valsalva in 1710 elucidated the pumping effect of muscles on venous flow. Fabricius d'Aquapendente first attributed varicose veins to valvular dysfunction but it took until 1842 for Paul Briquet to propose that abnormal retrograde flow from deep veins via the perforators was involved in the development of varicose veins. In the modern era, we have benefited greatly from noninvasive imaging and functional information gained from Duplex ultrasound, computed tomography, magnetic resonance imaging, and plethysmographic modalities. Through the use of intravenous pressure measurements, ultrasound, and venography, we have attained a more quantitative appreciation of venous hemodynamics at the macrolevel that expands our prior understanding of its dysfunction and role in the development of chronic venous insufficiency (CVI).

The microcirculation is composed of the microvasculature of arterioles, capillaries, venules, and lymphatic capillaries along with the associated interstitial tissue space. The vessels are generally less than 25 μm in diameter and are usually considered as belonging to the tissue to which they are connected. In the case of CVI, microcirculation plays a crucial role in controlling skin blood flow, vascular permeability, and immunologic responses associated with ulceration. The venous vessels at this level are very different than those in the macrocirculation, and their hemodynamics are much more complex. Their small size limits their visualization and measurements to microscopic analysis, laser Doppler fluximetry and imaging, as well ex vivo tissue culture and molecular analysis.[1] Although many hypotheses have been proposed describing microcirculatory dysfunction in the development of CVI, there is still much to be learned. In this chapter, we will review the role of both venous hemodynamics at the macrocirculatory level as well as at the microlevel in the context of CVI.

By accepted definition, chronic venous insufficiency refers to functional abnormalities of the venous system that result in producing moderate to severe edema (C3) but more commonly is applied to conditions where skin changes (C4) or ulceration (C5-6) have taken place, as defined in the Clinical, Etiological, Anatomical, and Pathophysiological (CEAP) classification system.[2,3] Edema, the accumulation of fluid in the interstitial space leading to leg swelling, is frequently the first clinical symptom or sign associated with CVI and reflects underlying hemodynamic dysfunction.[4] Although a multitude of disorders and medications can induce leg swelling, it is estimated that CVI-associated edema represents 90% of all lower limb edema because of the prevalence of venous insufficiency.[5]

Hemodynamics and anatomy

To understand venous hemodynamics, an appreciation of the lower extremity venous system is necessary.[6,7] The interplay between the three venous systems of the leg, the deep and superficial saphenous systems and the connecting perforators, are important in the hemodynamic changes that occur in disease states. While the deep system accompanying the major arteries of the leg return approximately 80%—90% of the lower limb venous blood, and the superficial system is most closely associated with visible varicosities and is the one most frequently subject to therapeutic interventions, it is the more than 150 perforating veins of the legs that provide the collateral pathways for pathologic retrograde flow that are hemodynamically very important in the development of the venous ulcers CVI.[8] Of these perforators, the most well described and of greatest clinical relevance have been the medial calf perforators involving the posterior tibial veins, but any group of perforators can be causally important in the treatment of specific patients.

Under normal physiologic conditions, unidirectional and cephalad venous blood flow is maintained by a healthy system of venous valves, lower limb muscular pump compressive action, and negative intraabdominal and intrathoracic pressure. During walking, contraction of calf muscles within their fascially enclosed compartments generates an ambulatory pressure gradient between the lower leg and thigh, which leads to displacement of blood in an antigravity direction toward the heart with the assistance of competent valves. This calf pump is most important because it contains the largest venous capacitance within the soleal and gastrocnemius sinusoids and generates the highest pressures. Intramuscular pressures generated in the gastrocnemius and soleus muscles can increase up to 250 mmHg from 9 to 15 mmHg in their relaxed state.[9] Valvular function, in all three venous systems, is important in the maintenance of antegrade venous flow. It is most helpful that modern ultrasound technology allows us both to directly visualize the valves and quantify their function as exemplified by valve closure (Fig. 2.1). The initial studies by van Bemmelen and associates established a duplex-derived valve closure time for the diagnosis of superficial reflux of 0.5 s.[10] For the deep system, 1 s or more

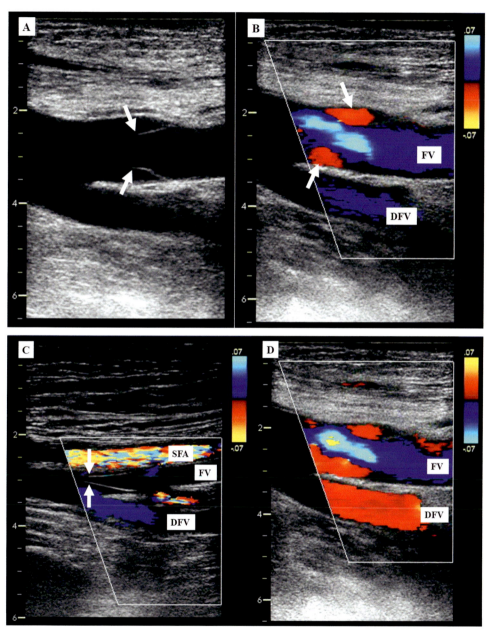

Figure 2.1 (A) A longitudinal B mode ultrasound illustrates a normal venous valve in the femoral vein with leaflets (arrows) open to allow normal antegrade cephalad flow of blood. (B) Color duplex ultrasound demonstrates the antegrade blue flow of blood toward the transducer in both the femoral (FV) and deep femoral (DFV) veins. Vortex retrograde red flow in the cusp of the leaflets (arrows) is a normal physiologic finding. (C) Cyclic cessation of antegrade flow induces closure of the valve leaflets (arrows) in the femoral vein (FV) while cephalad flow continues in the deep femoral vein (DFV). The overlying superficial femoral artery (SFA) is also seen in this image with the pedal flow away from the probe. (D) In a different patient (note that the color assignment of flow has been reversed), there is an incompetent valve with the blue retrograde pedal flow in the femoral vein (FV) while normal red cephalad flow continues in the deep femoral vein (DFV).

is suggestive of valvular dysfunction while 0.35 s of reversed flow in combination with a diameter of greater than 3.5 mm is generally accepted as abnormal for leg perforators.[11] Duplex ultrasound has high sensitivity and specificity with an overall 94% accuracy for detecting reflux when performed appropriately.

Pathophysiology—superficial venous incompetence

In pathologic states, be it due to primary valvular dysfunction or secondarily due to prior episodes of superficial thrombophlebitis or deep venous thrombosis, venous reflux with retrograde blood flow takes place.[12] Valvular incompetence of the superficial venous system is thought to most commonly be due to a primary weakness in the vein wall that leads to dilation of the vein and separation of the valves cusps resulting in valvular dysfunction. Numerous biochemical abnormalities have been reported within the venous wall which have an impact on its distensibility. Varicose veins have abnormal elastic properties with increased collagen content, elastin fiber fragmentation, and degradation and accumulation of extracellular matrix.[13,14] These abnormalities have supported either an initial deficiency in wall integrity or an induction of structural degradation. Akroyd et al.[15] showed that the valve ring and its leaflets had far greater tensile strength than the vein wall itself, further supporting the theory that valvular incompetence is secondary to a defect in the vein wall. There is also a genetic basis for primary valvular dysfunction with mutations of the FOXC2 gene being associated with venous valve failure in both the superficial and deep veins of the lower limb.[16]

Secondary valvular dysfunction is seen following episodes of superficial thrombophlebitis. As in the deep system, thrombolysis and recanalization can take place to reopen the clotted vein. However, the associated inflammatory process, easily visible by the focal area of redness, pain, and tenderness because of the superficial location of the vein just below the skin surface, leads to fibrosis and restricted movement of the valve leaflets. This results in valvular incompetence and reflux. The vein wall is also affected by the development of thickening and calcification.

Gravitational effects upon standing induce an increase in the intraluminal hydrostatic pressure and thereby increase the outward wall tension and distension of the vein, something that patients with varicose veins commonly relate to physicians during their first office visit. Superimposed on a structurally weakened wall, this additional force and increased vessel diameter separate the valve leaflets even further leading to an exacerbation of reflux and increased symptoms at the end of the day. The finding that reflux progresses from a more distal high-pressure location to a more proximal lower-pressure venous segment supports the hypothesis that gravitational pressure has a contributing effect to superficial venous reflux.[17] As the severity of reflux increases, the increased venous pressure on standing cannot be relieved by walking or exercise. Whereas in normal limbs the measured superficial venous pressure in the dorsum of the foot decreased from an average of 87–22 mmHg with standing and exercise, in those with varicose veins, it decreased to only 44 mmHg. In addition, the plethysmographic recovery refilling time of 31 s in normal patients decreased markedly to only 3 s in patients with great saphenous reflux.[18] This reflects the reflux and increased pressure transmitted when an in-line column of fluid is present without the protective pressure separation of closed valves. When expressed in terms of leg blood volume rather than pressure, the ejection fraction is less than 65% and the residual volume fraction is greater than 30% as measured by air plethysmography.[19]

Our treatment recommendations for venous insufficiency have been based on our understanding of the physiologic hemodynamic and pressure changes of superficial venous incompetence. The original open surgical proximal saphenofemoral ligation and great saphenous vein stripping sought to eliminate the entire axial pathway for reflux and venous hypertension. Present-day minimally invasive segmental endovenous ablation has demonstrated that equivalent clinical improvement can be achieved by the limited closure of only proximal incompetent segments. In addition, the CHIVA procedure, in which the great saphenous vein is spared and only refluxing collateral branches are disrupted, has been successful in relieving symptoms of superficial incompetence.[20] This suggests that decreasing the volume and pressure overload from contributing branches can be sufficient to improve saphenous vein distension and valvular dysfunction. Finally, relief of venous edema (C3) can be achieved through the use of class II (20–30 mm Hg) and class III (30–40 mmHg) compression stockings. These not only decrease the potential volume of the leg in which interstitial fluid can accumulate, but are able to compress the veins and thereby help control reflux and venous hypertension in the leg. Interestingly, at least in some positions with less diseased legs, the deep veins are compressed more than the superficial by low-compression stockings.[21]

Deep venous dysfunction

Occlusion of the deep veins due to acute deep venous thrombosis (DVT) has a much more significant hemodynamic impact because of the associated venous outflow obstruction. Acute DVT of the femoral, common femoral or iliac veins can limit blood outflow to such an extent that arterial inflow to the leg is diminished leading to resultant phlegmasia cerulea dolens and possible ischemic limb loss. Fortunately, a variety of thrombolytic, mechanical, and interventional procedures are available to treat such extensive acute venous occlusions. Although rapid thrombus resolution has been found to be associated with a higher incidence of valve competency[22] a large prospective randomized clinical trial did not demonstrate the expected clinical benefit for most patients.[23] More significant leg edema will occur if the thrombus is above the confluence of the deep femoral or great saphenous veins which act as collateral channels for occlusions involving the femoral and more distal veins.

Fortunately, with extended anticoagulation therapy, approximately half of venous thrombi resolve completely within 6 months of presentation. The anatomical location of the thrombus occlusion is somewhat predictive with the femoral vein more likely to remain occluded, whereas partial to full recanalization is more commonly found in the external iliac, common femoral, and popliteal veins. This may be a result of higher flow rates at these sites as well as the presence of collateral channels. After the initial thrombotic event, intrinsic thrombolysis and recanalization allow for blood flow to resume within the previously occluded vein. Recanalization alone, however, is always hemodynamically incomplete and results in a relative obstruction to flow and resultant secondary valvular incompetence (Fig. 2.2).[24] The inflammatory and fibrotic processes that take place on the valve cusp restrict the movement of the valve leaflets resulting in only a partially mobile leaflet or a completely frozen valve. In addition, inflammation of the nonvalvular segments of the vein can lead to thickening and calcification in the vein wall. It is unclear to what degree this loss of elasticity and distensibility affects venous flow hemodynamics but, at a minimum, it must decrease volume flow in those segments because of the diminished luminal diameter. The result from these processes is valvular incompetence and reflux.[25]

If early thrombus resolution does not occur, the thrombus organizes and is replaced by fibrous tissue filling the lumen and causing a complete and permanent venous obstruction. This has important hemodynamic effects by inducing a progressive increase in

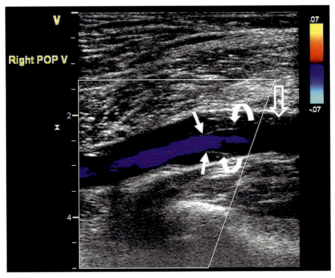

Figure 2.2 Postthrombotic recanalization of the popliteal vein showing the decreased diameter of luminal blood flow (blue color) with organized thrombus on the wall (open arrow) extending as a fibrotic process incorporating the cusps (curved arrows) of the valve leaflets (filled arrows), which are now immobile and do not close.

venous outflow through collateral vessels. These collaterals provide the required, but depending on the obstructive extent, potentially insufficient pathways for venous outflow out of the leg. The extent of the obstruction and the number of collateral pathways developed determine the severity of the hemodynamic changes and therefore the severity of the postthrombotic symptoms. With potentially fewer or less robust valves, collateral vessels themselves may become channels for retrograde reflux into the extremity. When the popliteal vein has been occluded, the calf perforating veins become important collaterals retrograde into the superficial venous system. Popliteal obstruction, either in isolation or in combination with calf vein and iliofemoral damage, is usually associated with more severe symptoms and subsequent leg ulcer development.

Incomplete thrombus in the deep venous system can also be associated with significant hemodynamic dysfunction beyond its partial obstructive effect (Fig. 2.3). Clot that is located in a valve pocket, or in direct contact with valve cusps, can irreparably damage

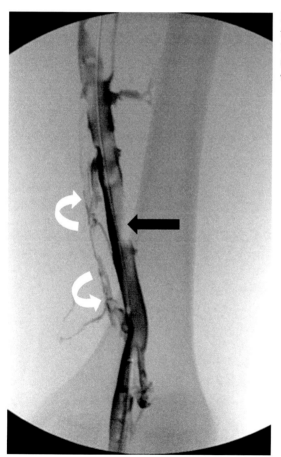

Figure 2.3 Left femoral venogram with partial thrombotic occlusion as evidenced by incomplete contrast opacification (black arrow) along the entire vein with some flow via collateral branches (curved arrows).

their function. Acutely, the valves cannot move when encased by thrombus. Moreover, lysis is more limited in the valve cusps due to the low vortex flow in this region (Fig. 2.1B).[26] The organizational fibrotic process causes retraction and shortening of the valve leaflets that limits their mobility and ability to close (Fig. 2.2). Furthermore, this is not just a simple mechanical effect. There is evidence to suggest a local neuro-hormonal sympathetic activity that controls venous wall tone and the base of the annulus.[27] Occlusion of draining vasa venarum at the base of the valve would change the local norepinephrine concentrations and further limit both valve closure and vein dilation.

Through these processes, permanent valvular incompetence develops and reflux occurs. In segments of veins where there are no valves, synechiae can develop. Synechiae are permanent endothelialized strands of residual organized thrombus, often crisscrossing the lumen of the vein and producing a cribriform meshwork that limits blood outflow. If extending to areas with valves, they can entrap the valve leaflets and bind them to the vein wall. Furthermore, in many patients, the perivenous inflammatory fibrosis that follows intraluminal thrombosis prevents venous distension and may also act as a functional obstruction limiting total blood flow even though no thrombus remains in the lumen.

Perforating veins of the calf

Although it has been estimated that there are more than 150 perforating veins in the lower extremity, the most important ones in the context of leg ulcerations are those in the calf. The perforating venous system normally allows blood to flow from the superficial to the much larger capacitance deep venous system. In circumstances of superficial venous hypertension and reflux, these effects can be transmitted to the communicating perforator veins and induce their dilatation and valvular incompetence with secondary abnormal retrograde flow from the deep to the superficial system (Fig. 2.4). Lower in the leg, perforating veins can also act as reentry points allowing blood refluxing down the saphenous system to flow back into the deep system (Fig. 2.5). After ablation of the saphenous veins and alleviation of the venous hypertension perforator valve competence is restored.[28]

Incompetent calf perforating veins are also often associated with deep vein obstruction or incompetence, where the primary abnormality is in the deep veins. The perforating veins act as safety valves or collateral pathways allowing blood under high pressure in the deep veins to escape to the superficial veins. This becomes especially problematic with calf muscle contraction which forces more flow into the superficial venous system of the calf (Fig. 2.6). This venous hypertension extends into the microcirculation with increased hydrostatic pressure in capillaries. There is a secondary enlargement of the dermal capillary bed, excessive transcapillary filtration causing interstitial edema formation with the exudation of fibrinogen and proteins into the interstitial space producing the characteristic changes of lipdermatosclerosis.[29]

Incompetence of one venous system in isolation is usually associated with minimal signs of chronic venous insufficiency. Incompetence in all three, however, is much more likely to be associated with active ulceration and higher residual calf pump volumes following muscle contraction.

Foot and calf pump function in venous hemodynamics

Beyond the structural anatomy of the venous system, the hemodynamics are more complex than on the arterial side because veins are collapsible, blood flow is intermittent and also dependent on the effects of gravity/hydrostatic pressure and extrinsic muscle compression.

Calf muscles, and to a lesser extent the foot and thigh musculature, act as physiologic pumps in returning venous blood against gravity from the lower limbs in the erect individual. The calf pump is the most important because it contains the largest venous capacitance within the soleal and gastrocnemius sinusoids and generates the highest pressures. Muscle contraction within the enclosed fascial compartments drives blood up the deep axial veins of the leg.[30]

Intramuscular pressures generated in the gastrocnemius and soleus muscles can increase up to 250 mmHg from 9 to 15 mmHg in their relaxed supine state.[9] With muscle

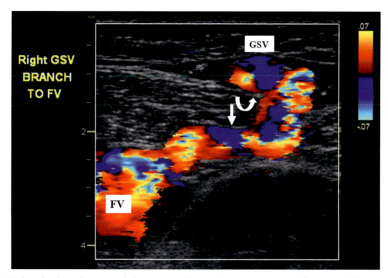

Figure 2.4 Color duplex ultrasound demonstrating an incompetent perforator with retrograde flow from the deep femoral vein (FV) passing through the compartmental fascia (straight arrow) and saphenous fascia (curved arrow) to join the great saphenous vein (GSV).

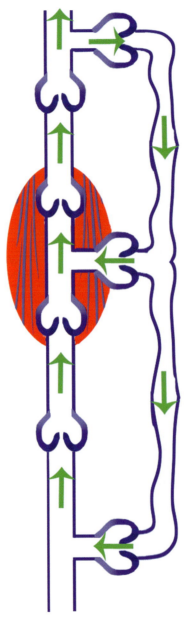

Figure 2.5 Superficial vein incompetence induces blood to reflux down the superficial veins, most commonly at the superior saphenofemoral junction. As long as the communicating perforating veins are competent, reentry can occur anywhere along the leg. The calf pump can usually cope with the additional load and reduce the foot vein pressure during exercise. This is why simple superficial varicose veins alone are an uncommon cause of venous ulceration. *(Blebea J, The pathophysiology and hemodynamics hemodynamics of chronic venous insufficiency. In: Glovizki P, ed.* Handbook of Venous Disorders. *4th ed. Boca Raton, FL: CRC Press; 2017, 51−61, with permission.)*

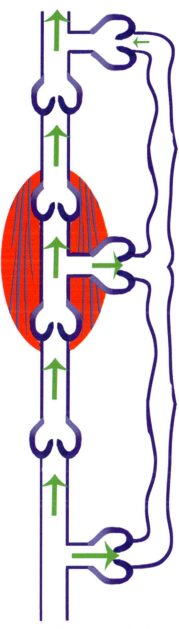

Figure 2.6 Perforating vein incompetence alone, as may develop after deep vein thrombosis, leads to dilatation and reflux of blood into the superficial compartment, exacerbated during calf muscle contraction. Perforator vein dilatation and valvular incompetence may also occur primarily or secondary to superficial hypertension. Without superficial incompetence, blood returns cephalad via the GSV and re-enters the deep venous system at the level of the common femoral vein. *(Blebea J, The pathophysiology and hemodynamics hemodynamics of chronic venous insufficiency. In: Glovizki P, ed. Handbook of Venous Disorders. 4th ed. Boca Raton, FL: CRC Press; 2017, 51–61, with permission.)*

contraction, the large pressure gradient induced on the deep calf veins and the popliteal vein induces rapid efflux of blood from the calf to the thigh with an emptying efficiency in normal subjects of about 70%. During ensuing muscle relaxation, venous pressure decreases within the compartments that allow the perforating veins to direct blood flow from superficial to deep veins. This in turn reduces the pressure in the superficial veins. Capillary venous inflow slowly refills the superficial veins over the next 20—35 s back to their original resting volume and pressure.

The foot pump is obviously of smaller size and venous capacitance compared to the calf. It is deep and intermuscular, principally comprised the lateral plantar veins that drain directly into the posterior tibial veins.[31] It also communicates via the inframalleolar perforators into the medial marginal vein which is the origin of the great saphenous vein below the ankle. This interestingly provides a reversed direction of blood flow from the deep to the superficial system rather than the opposite direction normally seen everywhere else in the leg. Finally, there are the anterior communicating veins from the plantar reservoir, which lead directly into the anterior tibial veins. There is therefore normal efflux from the foot both through the deep and venous systems. Approximately 20—30 mL of blood volume is ejected from the sole of the foot during contraction.

Venous pressure at the ankle is only around 10 mmHg in the supine position but can increase to 90 mmHg in the upright position. This is effectively decreased with walking or heel-raising exercises to less than 30 mmHg. Intermediate venous hypertension is defined as a pressure of 31—45 mmHg and severe venous hypertension when greater than 45 mmHg.[32]

The foot pump has been effectively used for the prevention of DVT in immobile postoperative patients who, because of trauma or orthopedic procedures, cannot undergo calf intermittent compression. Extrinsic mechanical compression of the plantar venous plexus produced a peak velocity of 123 ± 71 cm/s in the posterior tibial veins.[33]

Incompetence of the deep venous valves allows retrograde flow within the deep system leading to both an increase in calf volume and inefficient return of blood to the right heart. Fortunately, deep venous valvular incompetence without coexisting cephalad obstruction can be compensated by the calf and foot pumps and competent perforating veins (Fig. 2.7). On the other hand, proximal obstruction by either organized thrombus or decreased volume flow due to fibrotic recanalized deep veins, in association with perforating veins that are primarily or secondarily incompetent, negates the ability of the muscle pump in pushing blood out of the leg. Even worse, the calf pump exacerbates blood efflux out through retrograde flow via the connecting perforating veins and induces superficial venous hypertension (Fig. 2.8). In cases of deep venous outflow obstruction or severe valvular insufficiency, the pumps' inability to induce sufficient venous outflow results in persistent superficial ambulatory venous hypertension, not relieved by walking or exercise (Fig. 2.9). These abnormalities are further exacerbated when there is concomitant preexisting reflux in the superficial venous system. Similar but less severe

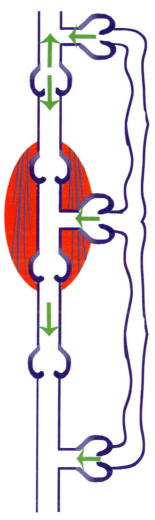

Figure 2.7 With deep venous reflux and perforator competence, the calf pump can compensate by increasing its output and still effectively pushing the blood volume cephalad during contraction. (Blebea J, The pathophysiology and hemodynamics hemodynamics of chronic venous insufficiency. In: Glovizki P, ed. Handbook of Venous Disorders. 4th ed. Boca Raton, FL: CRC Press; 2017, 51–61, with permission.)

effects are seen in the absence of deep venous pathology but when there is perforating and superficial incompetence. Persistently elevated ambulatory pressure in the leg leads to raised pressure at the venous end of the capillaries. This increased capillary hydrostatic pressure induces both transudation and exudation with a high protein content of interstitial fluid and the secondary skin changes seen with chronic venous insufficiency.

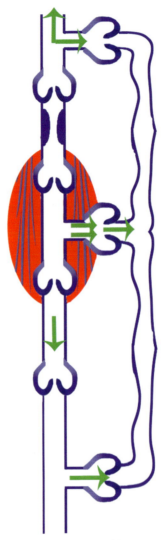

Figure 2.8 Deep venous obstruction causes upstream dilatation of the veins and secondary incompetence of the communicating veins because these veins become part of the collateral outflow tract. Blood efflux is increased through the superficial system. *(Blebea J, The pathophysiology and hemodynamics hemodynamics of chronic venous insufficiency. In: Glovizki P, ed.* Handbook of Venous Disorders. *4th ed. Boca Raton, FL: CRC Press; 2017, 51–61, with permission.)*

Dysfunction in venous valves most frequently affects all three venous systems in patients with more advanced CVI, particularly those with CEAP C6 disease with active ulcers.[34] This leads to venous hypertension, which can be quantified by elevated ambulatory pressures in the superficial venous system when measured in the pedal veins.

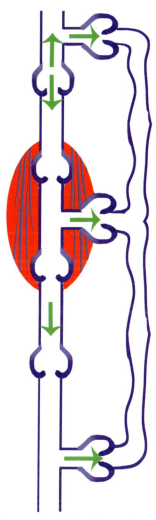

Figure 2.9 In the setting of both deep reflux and perforator incompetence, pump efficiency fails during exercise, and ambulatory hypertension is not relieved with walking. *(Blebea J, The pathophysiology and hemodynamics hemodynamics of chronic venous insufficiency. In: Glovizki P, ed.* Handbook of Venous Disorders. *4th ed. Boca Raton, FL: CRC Press; 2017, 51–61, with permission.)*

In these circumstances, venous hypertension and associated increase in hydrostatic capillary pressure leads to edema formation that can be visualized and even quantified with ultrasound (Fig. 2.10). Persistent outflow obstruction, caused by lack or incomplete recanalization of DVT in the deep system, induces venous return through collateral channels, requiring retrograde flow through perforators and overloading of the superficial veins. This, in turn, causes their secondary dilatation and associated reflux with venous hypertension. Superimposed in all of these circumstances are gravitational effects in the

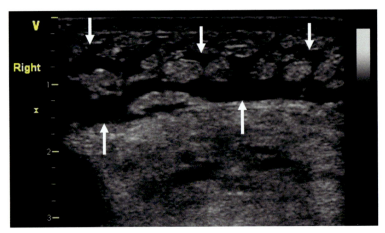

Figure 2.10 Longitudinal B-mode ultrasound image of the calf illustrating multiple subcutaneous horizontal anechoic areas of edema (arrows).

standing position which would worsen valvular incompetence and retrograde flow.[35] Macrocirculatory venous hypertension extends into the microcirculation with increased hydrostatic pressure in the capillaries that leads to wall remodeling and microvenous valvular insufficiency.[36] The sum of these hemodynamic changes causes the typical complaints of leg heaviness, achiness, swelling, throbbing, and itching. These clinical signs and symptoms can be quantified and reflected in the Venous Clinical Severity Score.[37]

Microcirculation and interstitial edema

As mentioned earlier, the first manifestation of CVI is the development of lower extremity edema. At the microcirculatory level, this reflects a dysfunction of transcapillary fluid exchange. Under physiologic conditions, total interstitial fluid volume is kept low and fairly constant because of the maintained balance between transcapillary filtration and lymphatic outflow. In 1896, the British physiologist Ernest Starling described the opposing forces between hydrostatic pressure and oncotic attraction across the semipermeable capillary wall to control fluid filtration from the intravascular to the extravascular space.[38] With venous hypertension and associated increased hydrostatic pressure, there is a secondary enlargement of the venous capillary beds and excessive transcapillary filtration, which overwhelms the lymphatic outflow capacity. The resultant interstitial fluid accumulation also includes the exudation of fibrinogen and proteins into the interstitial space, which later produces the characteristic changes of lipodermatosclerosis.[29] Plasma proteins, mainly albumin, are responsible for intravascular oncotic pressure. Increased transcapillary pressures and oncotic attraction with progressive protein accumulation in the interstitial space is thought to be the primary pathophysiologic mechanism in the development of CVI-associated edema.[39,40]

This original mechanistic understanding has been further expanded and elucidated with details of the molecular basis of this process. Almost a century after Starling, Curry, and Michel explained that transcapillary fluid movement reflects an ultrafiltration process through the interpolymer spaces of the glycocalyx layer of the endothelial cells of capillary membranes.[41] In addition, the underlying capillary basement membrane, composed of type IV collagen and laminin, and the adjacent extracellular matrix, act as second- and third-order resistance layers opposing fluid movement from the intravascular space to the interstitial space.[42] Additionally, although initially conceived of as a nonfunctional space, the interstitium contains an extensive extracellular matrix composed of collagen fibrils upon which glycoprotein molecules are attached in a web-like manner. Beyond the forces of transcapillary pressure and oncotic gradients, in inflammatory conditions such as those seen in CEAP C4-6, this extracellular matrix can physiologically affect transendothelial pressure differences and thus result in increased transcapillary fluid flow.[43,44] Any additional derangements in lymphatic drainage would only exacerbate this problem if the lymphatic capillaries had a decreased ability to clear proteins and macromolecules from this relatively hyper-oncotic environment.

Regulation of capillary hydrostatic pressure

Hydrostatic capillary pressure is regulated by the balanced changes in precapillary arteriolar and to a lesser extent postcapillary venular resistance (vasoconstriction/vasodilation). This regulatory balance helps to maintain a relatively constant interstitial fluid volume by dynamically adjusting the transcapillary fluid filtration rate. Because arteriolar resistance is four times greater than venular resistance, even small changes in venular pressure, as seen with venous outflow obstruction or with venous valvular insufficiency and secondary hypertension, can have a very profound effect on transcapillary filtration.

Any decrease in venous return from the lower limb, such as caused by deep venous outflow obstruction, multisystem valvular incompetency, or inadequate lower limb muscle contraction, can lead to an increase in the ambulatory venous pressure up to 60–90 mmHg.[45] With the excess blood volume in both the deep and superficial venous systems associated with venous hypertension, exacerbated by standing and worsening toward the end of the day, venous distension leads to anatomic distortion and progressive valvular incompetence in a vicious circle. As the veins become maximally distended, any further increase in venous blood volume can produce a large increase in intraluminal venous pressure. At the level of the microvasculature, this is associated with a resultant loss of precapillary arteriolar reflex constriction that is intended to decrease the transmission of the increased venous pressure to the capillary system.[46] This venoarteriolar response (VAR) involves the arterioles constricting to reduce blood flow and normalize the arteriovenous pressure difference in the face of venous hypertension. Reduction of this VAR contributes significantly in the formation of edema and correlates with CVI clinical severity.[39,47] Venoactive agents have had some positive effects by increasing capillary resistance, improving lymphatic drainage, and reducing capillary filtration, thus decreasing interstitial edema.[48]

Blood flow and inflammation

Another factor that plays a role in the microvascular pathophysiology of CVI is the reduction in shear stress, which is the tangential force of the blood flow on the endothelial lining of the blood vessel. Shear stress was found to be associated with regenerative functional and morphological changes in endothelial cells as well as the release of antiinflammatory and vasodilatory molecules.[49] However, the reduction in shear stress associated with venous hypertension is associated with proinflammatory changes in the vein wall and valves. This can lead to luminal gap formation between endothelial cells secondary to endothelial cells' actin/myosin filaments contraction. These gaps lead to excess permeability of plasma proteins into the interstitial space, with a reduction in oncotic pressure differences between the intra- and extravascular spaces and secondary edema formation.[50]

The combined volume effect of venous hypertension and changes in endothelial shear stress causes endothelial glycocalyx damage (degradation by proteolytic enzymes such as matrix metalloproteinase, elastase, and heparanase), which allows the interaction of the activated blood cells with receptors expressed on the surface of endothelial cells. This both diminishes the barrier function and also facilitates the extravasation of reactive leukocytes and the associated release of vasoactive substances such as chemokines, inflammatory mediators, and adhesion molecules (ICAM-1 and E-Selectin).[51] The increased release of the ICAM-1 adhesion molecules leads to increased leukocyte adherence. Due to altered shear stress, leukocytes begin sticking to the vein wall, migrate out of the capillary, and release inflammatory mediators. These mediators trigger local inflammation, which induces further remodeling of the adjacent venous wall and valves and aggravating venous hypertension.[52] This response also includes local monocyte and macrophage recruitment with their infiltration into the vein wall and valves.[53] Chronic venous hypertension increases hypoxia-inducible factors leading to increased matrix metalloproteinases expression/activity with resultant degradation of extracellular matrix proteins. This inflammatory cascade, expressed through proinflammatory mediators monocyte chemoattractant protein-1 (MCP-1), TNF-α, and IL-1β, involves both the venous system and surrounding soft tissues and ultimately leads to the classical presentation of lower limb edema, lipodermatosclerosis and leg ulcer associated with chronic venous insufficiency.[54] Whether such inflammatory responses precede or follow valvular dysfunction and venous incompetency, is not yet conclusively elucidated.[39,55] The combination of these changes contributes to endothelial cell dysfunction and venous wall and valvular damage that result in venous insufficiency.[50]

Conclusions

The underlying venous hemodynamics and microcirculatory pathophysiology of chronic venous insufficiency are complex and most clinicians would acknowledge that more progress needs to be made, especially at the cellular and biochemical levels. Most recent efforts have focused on technological advances for the treatment of superficial venous disease and, more recently, interventions within the deep venous system. These mechanistic interventions, clearly needed, are however only a first and macroscopic step in the treatment of lower extremity CVI and ulcerations. We will need in the future to use modern technology toward the more precise investigation of the microcirculatory and inflammatory abnormalities of chronic venous insufficiency to understand in greater detail the mechanisms that cause leg ulceration. This should lead us to better methods of preventing ulcers from developing and of treating them once they have developed.

The basic hemodynamic principles of transcapillary filtration and oncotic pressure leading to edema have been known for decades. Our understanding of the details of microvascular function, molecular mediators, and inflammatory processes, however, is still at a very basic level. We have a long way to go in discovering the particulars of the intermediary pathways of ulcer formation and discover more specific targeted drug therapy for treatment. This should provide fertile grounds for investigation in the years to come.

References

1. Klonizakis M. *Cutaneous Microcirculation and Lower Limb Venous Disease*. Saarbrucken, Germany: Lambert Academic Publishing; 2010.
2. Eklof B, Perrin M, Delis KT, et al. Updated terminology of chronic venous disorders: the VEIN-TERM transatlantic interdisciplinary consensus document. *J Vasc Surg*. 2009;49(2):498–501.
3. Lurie F, Passman M, Meisner M, et al. The 2020 update of the CEAP classification system and reporting standards. *J Vasc Surg: Venous Lymphat Disord*. 2020;8(3):342–352.
4. Kamel M, Blebea J. Pathophysiology of edema in patients with chronic venous insufficiency. *Phlebolymphology*. 2020;27(1):3–10.
5. Nicolaides AN. Chronic venous insufficiency and the leukocyte-endothelium interaction: from symptoms to ulceration. *Angiology*. 2005;56(Suppl 1):S11–S19.
6. Caggiati A, Bergan JJ, Gloviczki P, Jantet G, Wendell —Smith CP, Partsch H. Nomenclature of the veins of the lower limbs: an international interdisciplinary consensus statement. *J Vasc Surg*. 2002;36: 416–422.
7. Kachlik D, Pechacek V, Hnatkova G, Hnatek L, Musil V, Baca V. The venous perforators of the lower limb - a new terminology. *Phlebology*. 2019;34(10):650–668.
8. Van Limborgh J. L'anatomie du systemeveineux de l'extremiteinferieure en relation avec la pathologievariqueuse. *Folia Angiol*. 1961;8:240–257.
9. Ludbrook J. The musculovenous pumps of the human lower limb. *Am Heart J*. 1966;71(5):635–641.
10. van Bemmelen PS, Bedford G, Beach K, Strandness DE. Quantitative segmental evaluation of venous valvular reflux with duplex ultrasound scanning. *J Vasc Surg*. 1989;10(4):425–431.
11. Gloviczki P, Comerota AJ, Dalsing MC, et al. The care of patients with varicose veins and associated chronic venous diseases: clinical practice guidelines of the Society for Vascular Surgery and the American Venous Forum. *J Vasc Surg*. 2011;53:2S–48S.

12. Shchatsko A, Blebea J. Superficial venous insufficiency and varicose veins. In: Nazzal M, Blebea J, Osman M, eds. *Lange Vascular and Endovascular Surgery: Clinical Diagnosis and Management*. New York, NY: McGraw-Hill; 2022 (in press).
13. Wali MA, Eid RA. Changes of elastic and collagen fibers in varicose veins. *Int Angiol*. 2002;21(4):337−343.
14. Pocock ES, Alsaigh T, Mazor R, et al. Cellular and molecular basis of venous insufficiency. *Vasc Cell*. 2014;6(1):24.
15. Ackroyd JS, Pattison M, Browse NL. A study on the mechanical properties of fresh and preserved human femoral vein wall and valve cusps. *Br J Surg*. 1985;72:117−119.
16. Mellor RH, Brice G, Stanton AW, et al. Mutations in FOXC2 are strongly associated with primary valve failure in veins of the lower limb. *Circulation*. 2007;115(14):1912−1920.
17. Bernardini E, DeRango P, Piccioli R, et al. Development of primary superficial venous insufficiency: the ascending theory. Observational and hemodynamic data from a 9-year experience. *Ann Vasc Surg*. 2010;24(6):709−720.
18. Pollack AA, Taylor BE, Myers TT, Wood EH. The effect of exercise and body position on the venous pressure at the ankle in patients having venous valvular defects. *J Clin Invest*. 1949;28(3):559−563.
19. Nicolaides A, Christopoulos D, Vasdekis S. Progress in the investigation of chronic venous insufficiency. *Ann Vasc Surg*. 1989;3(3):278−292.
20. Gianesini S, Occhionorelli S, Menegatti E, et al. CHIVA strategy in chronic venous disease treatment: instructions for users. *Phlebology*. 2014;30(3):157−171.
21. Partsch H, Mosti G, Mosti F. Narrowing of leg veins under compression demonstrated by magnetic resonance imaging (MRI). *Int Angiol*. 2010;29(5):408−410.
22. Elsharawy M, Elzayat E. Early results of thrombolysis vs anticoagulation in iliofemoral venous thrombosis. A randomized clinical trial. *Eur J Vasc Endovasc Surg*. 2002;24(3):209−214.
23. Vedantham S, Goldhaber SZ, Julian JA, et al. Pharmacomechanical catheter-directed thrombolysis for deep-vein thrombosis. *N Engl J Med*. 2017;377:2240−2252.
24. Eberhardt RT, Raffetto JD. Chronic venous insufficiency. *Circulation*. 2014;130:333−346.
25. Blebea J. The pathophysiology and hemodynamics hemodynamics of chronic venous insufficiency. In: Glovizki P, ed. *Handbook of Venous Disorders*. 4th ed. Boca Raton, FL: CRC Press; 2017:51−61.
26. Lurie F, Kistner RL, Eklof B, Kessler D. Mechanism of venous valve closure and role of the valve in circulation: a new concept. *J Vasc Surg*. 2003;38(5):955−961.
27. Crotty TP. The venous valve agger and plasma noradrenaline-mediated venodilator feedback. *Phlebology*. 2007;22(3):116−130.
28. Labropoulos N, Mansour MA, Kang SS, Gloviczki P, Baker WH. New insights into perforator vein incompetence. *Eur J Vasc Endovasc Surg*. 1999;18(3):228−234.
29. Comerota A, Lurie F. Pathogenesis of venous ulcer. *Semin Vasc Surg*. 2015;28:6−14.
30. Raju S, Knepper J, May C, Knight A, Pace N, Jayaraj A. Ambulatory venous pressure, air plethysmography, and the role of calf venous pump in chronic venous disease. *J Vasc Surg: Venous and Lym Dis*. 2019;7:428−440.
31. Uhl JF, Gillot C. Anatomy of the foot venous pump: physiology and influence on chronic venous disease. *Phlebology*. 2012;27(5):219−230.
32. Reeder SW, Wolff O, Partsch H, et al. Expert consensus document on direct ambulatory venous pressure measurement. *Int Angiol*. 2013;32(5):453−458.
33. White JV, Katz ML, Cisek P, Kreithen J. Venous outflow of the leg: anatomy and physiologic mechanism of the plantar venous plexus. *J Vasc Surg*. 1996;24(5):819−824.
34. Thulesius O. Vein wall characteristics and valvular function in chronic venous insufficiency. *Phlebology*. 1993;8:94−98.
35. Raju S, Knight A, Lamanilao L, Pace N, Jones T. Peripheral venous hypertension in chronic venous disease. *J Vasc Surg: Venous and Lym Dis*. 2019;7:706−714.
36. Vincent JR, Jones GT, Hill GB, van Rij AM. Failure of microvenous valves in small superficial veins is a key to the skin changes of venous insufficiency. *J Vasc Surg*. 2011;54(Suppl):62S−69S.

37. Vasquez MA, Rabe E, McLafferty RB, et al. Revision of the venous clinical severity score: venous outcomes consensus statement: special communication of the American venous forum ad hoc outcomes working group. *J Vasc Surg.* 2010;52:1387–1396.
38. Starling EH. On the absorption of fluids from the connective tissue spaces. *J Physiol.* 1896;19:312–326.
39. Balance TF. Edema in venous insufficiency. *Phlebolymphology.* 2011;18(1):3–14.
40. Woodcock TE, Woodcock TM. Revised Starling equation and the glycocalyx model of transvascular fluid exchange: an improved paradigm for prescribing intravenous fluid therapy. *Br J Anaesth.* 2012; 108(3):384–394.
41. Curry FE, Michel CC. A fibre-matrix model of capillary permeability. *Microvasc Res.* 1980;20:96–99.
42. Levick JR. Fluid exchange across endothelium. *Int J Microcirc Clin Exp.* 1997;17:241–247.
43. Sarin H. Physiologic upper limits of pore size of different blood capillary types and another perspective on the dual pore theory of microvascular permeability. *J Angiogenesis Res.* 2010;2:14.
44. Heino J, Kapyla J. Cellular receptors of extracellular matrix molecules. *Curr Pharmaceut Des.* 2009;15: 1309–1317.
45. Takase S, Pascarella L, Lerond L, Bergan JJ, Schmid-schönbein GW. Venous hypertension, inflammation and valve remodeling. *Eur J Vasc Endovasc Surg.* 2004;28(5):484–493.
46. Stücker M, Schöbe MC, Hoffmann K, Schultz-ehrenburg U. Cutaneous microcirculation in skin lesions associated with chronic venous insufficiency. *Dermatol Surg.* 1995;21(10):877–882.
47. Labropoulos N, Wierks C, Golts E, et al. Microcirculatory changes parallel the clinical deterioration of chronic venous insufficiency. *Phlebology.* 2004;19(2):81–86.
48. Nicolaides A, Kakkos S, Baekgaard N, et al. Management of chronic venous disorders of the lower limbs. Guidelines According to Scientific Evidence. Part I. *Int Angiol.* 2018;37:181–254.
49. Li YS, Haga JH, Chien S. Molecular basis of the effects of shear stress on vascular endothelial cells. *J Biomech.* 2005;38(10):1949–1971.
50. Mansilha A, Sousa J. Pathophysiological mechanisms of chronic venous insufficiency and implications for venoactive drug therapy. *Int J Mol Sci.* 2018;19:1669.
51. Ligi D, Croce L, Mannello F. Chronic venous disorders: the dangerous, the good, and the diverse. *Int J Mol Sci.* 2018;19(9):2544–2563.
52. Barros BS, Kakkos SK, De Maeseneer M, Nicolaides AN. Chronic venous disease: from symptoms to microcirculation. *Int Angiol.* 2019;38:211–218.
53. Ono T, Bergan JJ, Schmid-Schonbein GW, Takase S. Monocyte infiltration into venous valves. *J Vasc Surg.* 1998;27:158–166.
54. Castro-Ferreira R, Cardoso R, Leite-Moreira A, Mansilha A. The role of endothelial dysfunction and inflammation in chronic venous insufficiency. *Ann Vasc Surg.* 2018;46:380–393.
55. Pascarella L, Penn A, Schmid-Schönbein GW. Venous hypertension and the inflammatory cascade: major manifestations and trigger mechanisms. *Angiology.* 2005;56:S3–S10.

CHAPTER 3

Venous ulcers of the lower extremity: etiology, risks, and predictive factors

Thomas F. O'Donnell
Benjamin Andrews Emeritus Professor of Surgery, Cardiovascular Center, Tufts Medical Center, Boston, MA, United States

Introduction

Venous leg ulcers (VLUs) represent 50% of all leg ulcers[1,2] and up to 80% of ulcers seen in healthcare systems.[3] VLUs are the most severe sequela of chronic venous disease (CVD), and VLU risk increases with the severity of venous insufficiency.[4–6] However, the question remains why a minority of patients with advanced CVD acquire ulcers. In light of the growing prevalence and cost burden of VLU,[7] its continued high rate of recurrence,[8] and especially the suffering and debilitation that VLU causes in the individuals who experience it,[9] there is a pressing need to provide better answers to this question than we have today. Historically, investigation of clinical risk factors—that is, those available from patient or family history, physical examination, or population studies—has focused on factors associated with VLU healing or recurrence, while a relatively small number of studies have focused specifically on the first VLU (fVLU).[5] Genetic biomarkers of fVLU have also been tentatively suggested.[10] This chapter will describe the current understanding of the etiology, as well as risk and predictive factors, for fVLU. As these and other terms used in connection with venous disease have often been used imprecisely, we begin with a set of definitions (Table 3.1).[11–13]

CEAP etiologies of CVD

VLU is the most severe expression of CVD, and the etiologies of VLU are best defined within the CEAP CVD classification system. The 2020 CEAP revision committee emphasized the importance of "E," the etiologic classification, because determining etiology guides prognosis and treatment choices as well as influencing outcomes.[14] While the earlier format of the E classification has not changed, the subgroups have been expanded. Primary etiology (E_P) was refined to indicate a degenerative process of the venous valve or venous wall leading to valve and/or vein wall weakness and dilation. This results in pathological reflux. Observations on duplex imaging do not exhibit the thickening of the vein wall or occluded segments that is characteristic of postthrombotic syndrome (PTS). The secondary etiology (E_S) has been subdivided into intravenous (E_{SI})

Table 3.1 Definitions of terms.

Term	Definition
Chronic venous disease (CVD)[11]	Chronic conditions related to or caused by veins that become diseased or abnormal
Chronic venous insufficiency (CVI)[12]	Should be limited to CEAP C3−C6, defined as morphologic abnormalities of the venous system that lead to moderate-severe symptoms and physical findings, such as edema, skin changes, and venous ulcers
Venous leg ulcer (VLU)[13]	An open skin lesion of the leg or foot that occurs in an area affected by venous hypertension
Etiology	The cause of a disease; specifically, an event, condition, or characteristic preceding a disease without which the disease would not have developed
Disease risk factor	An exposure, attribute, or element that is associated with a change, usually an increase in the risk of disease occurrence
Predictive factor	Predicts the probability of occurrence or existence of a disease, particularly in association with an intervention
Prevalence	The proportion of persons who *have* a condition at or during a particular time period
Incidence	The proportion or rate of persons who *develop* a condition during a particular time period

and extravenous (E_{SE}) causes. Finally, congenital etiology (E_C) refers to embryological abnormalities in the development of the venous system, apparent at birth or recognized later.

A detailed examination of the pathophysiology of VLU is beyond the scope of this chapter and is described elsewhere (see Chapters 1 and 2); nonetheless, a review of VLU risk factors should be grounded in a basic understanding of the mechanisms of chronic venous insufficiency (CVI) progression. Among VLU patients, approximately 70%−80% have primary CVD etiologies (E_P), while 20%−30% have CVD of secondary origin, mainly postthrombotic (E_{SI}).[4,15] Both E_P and E_{SI} contribute to an alteration of the venous wall and valve integrity resulting in reflux, pooling of blood in the extremity, and venous hypertension. Venous hypertension is the underlying pathophysiological feature common to all etiologies of VLUs. In addition to being secondary to reflux, hypertension may result from obstruction due to postthrombotic changes in the iliocaval outflow segments or microcirculation (F_{SE}) or be due to congenital venous malformations (E_C). The fundamental basis for CVI and venous ulceration is inflammation within the venous circulation when subjected to increased hydrostatic pressure.[16,17] While reflux or obstruction, individually or in combination, can lead to venous pathology, disease progression is marked by a vicious cycle of inflammation and that further damages vein walls and valves, thereby increasing hypertension.[18] Among patients with varicose veins, the risk of VLU increases with CVD progression to skin changes.[4–6].

Secondary etiologies
Intravenous etiology
It is well recognized that an episode of deep venous thrombosis (DVT) can lead to pathologic changes within the vein that produce venous hypertension. Thus, the postthrombotic state is an important *secondary* etiology of VLU (E_{si}). While postthrombotic CVI is associated with greater VLU risk compared with primary etiology,[19] reports on the incidence of VLU development following an episode of DVT are variable in evidentiary strength due to heterogeneous methodologies.[20] Unfortunately, analyses of major systematic reviews (e.g., Cochrane Reviews), while providing basic demographic features, do not characterize CEAP E, A, and P.[21] Tables 3.2 and 3.3 present results from a range of studies of secondary CVI and VLU.[19,22–31].

The anatomic site of the DVT episode relates to the probability and incidence of severe sequelae. For example, about 45 years ago this author and his colleagues from the Surgical Professorial Unit at St. Thomas' Hospital in London were among the first to document the relationship between the proximal location of the DVT and the subsequent severity of PTS. Twenty-one patients with acute thrombosis involving the common femoral or iliac vein segments as defined by phlebography were followed for 10 years. VLUs developed in 67% of cases required recurrent hospitalizations. Recurrent DVT developed for a mean of 4.5 instances per patient over the 10-year follow-up period, resulting in 112 separate hospitalizations. Fourteen patients were hospitalized >5 times. Three patients underwent below-knee amputations due to unrelenting pain and sepsis. Over the decades long follow-up period, 23% of patients in this highly selected group developed a VLU at 1 year, 50% by 3 years, 80% by 5 years, and nearly 90% at 10 years.[20]

In a serial duplex ultrasound (DUS) follow-up study conducted over 5 years, Asbeutah et al.[32] showed that thrombi resolved faster and more completely in distal DVT compared with proximal DVT. Among limbs with proximal DVT, lack of resolution noted at 1 week, 6 months, and 5 years was 66%, 15%, and 4%, respectively. For limbs with distal DVT, 33% experienced a lack of resolution at 1 week and all thrombi were resolved at 6 months. At 5 years, 54% of patients with proximal DVT were in the C4–C6 category versus 11% of those with distal DVT. Akesson et al.[33] demonstrated that 95% of patients treated with anticoagulation alone for an acute iliofemoral DVT had ambulatory venous hypertension at 5 years, with VLU development in 15% of cases.

Table 3.2 Characteristics of secondary causes of VLU.

Study	Years	N	Study type	Age (mean)	Male (%)	VLU (%)	PTS (%)	DVI/SVI	Reflux/obs (%)
Labropoulos[19]	2003–08	P-50 S-46	Retrospective			0 6.5	5 (C4) 24 (C4/6)		48/6.5
Johnson[22]	1986–94	78	Prospective, 3-yr follow up	50	53	2.5	41	88/ND	35/12
TenBrook[23]	1996–2000	1140	Systematic review	57	49		36	56	
Glovicski[24]	1993–96	146	NA SEPS registry	56	54	84	38	72/67	64/3
Marston[25]	1995–98	229	Case series	61.4	53	All C6		71/29	74/3
Ma[26]	2005–11	84	Registry	61	57	All C6	42	55/38	(23- combined)

Table 3.3 Randomized controlled trials—VLU characteristics.

Study	Years	N	Study type	Age (mean)	Male (%)	VLU (%)	PTS Hx DVT (%)	DVI/SVI	Reflux/obs (%)
ESCHAR[27]	1999–2002	500	RCT	73	41	68 C6 22 C5	8	40/60	
DUTCH SEPS[28]	1999–2001	200	RCT	65	39	ALL C6	31	53	
EVRA[29]	2013–16	450	RCT	68	45	ALL C6	7	32	
CAVENT[30]	2006–09	176	RCT	50	62	2.8[a]	ALL	All DVI	All obs
ATTRACT[31]	2009–14	691	RCT	53	62	9	ALL	ALL DEEP	All obs

[a] trophic changes

Prospective natural history studies

Using ultrasound to determine the timing of recanalization and the degree of thrombus lysis following DVT, Killewich et al.[34] showed that nearly 95% of the thrombi were lysed within a year; however, residual occlusion was found in 10%−15%. In 1995, Johnson et al.[22] studied 78 patients (83 legs) who underwent both yearly clinical and DUS follow-up examinations from 1 to 6 years (median 3 years) after an episode of DVT (Table 3.2). The majority (59%) of limbs were asymptomatic, but 41% of limbs had PTS. Among PTS limbs, 11 had trophic changes with hyperpigmentation. In this short follow-up period, two patients (2.4% based on limbs at risk) developed a VLU.

Yamaki et al.[35] performed serial venous DUS 1 year after an episode of DVT in 70 limbs during which 147 segments were examined. The initial DUS showed that DVT involved isolated segments in 35 limbs, while 35 had multi-segmental involvement. DUS examination at 1 year demonstrated that 75% of limbs had fully resolved their DVT while 5% were totally occluded. An additional 20% were partially recanalized. Again the investigators observed a difference in the condition of the vein by anatomic site: 20% of the femoral veins were occluded, while 100% of the calf veins were recanalized.

Labropoulos et al.[19] followed patients with secondary CVD (PTS group) after the first episode of proximal DVT for a minimum of 5 years and compared with an age- and gender-matched cohort of patients with primary CVD (Table 3.2). Importantly, DUS was performed to identify anatomic prognostic factors. The prevalence of trophic changes was significantly higher in the PTS group (24% vs. 6%; $P = .02$). In addition, progression to trophic changes was more rapid in the PTS group. DUS findings revealed reflux in the PTS group in 22 limbs (50%), with 19 (41%) showing combined reflux and obstruction. The combination of proximal and distal obstruction was a significant factor leading to trophic changes. Of the 46 limbs in the PTS group, 3 developed a VLU (6.5%), but none of the 50 limbs in the primary group developed ulceration. The authors called attention to several influential factors for the progression of venous disease, including the combination of reflux and obstruction, recurrent DVT, and multisegmental involvement.

Epidemiologic studies

The Mayo Clinic group carried out a population-based study that determined an annual VLU incidence rate of 18 per 100,000 person-years.[36] Incidence was higher in women (20.4/100,000) and increased with age. The investigators also suggested that approximately 10% will develop VLU within 1−2 years following an episode of DVT, with increased rates subsequently. In a 2009 review, Heit et al.[37] showed a highly variable PTS incidence (23%−60%) within 2 years of DVT. Unfortunately, the review classified later stages as severe PTS but did not specifically describe the incidence of VLU. Incidence ranged from 1.4% (3 years) to 23% (5 years) in the four prospective studies reviewed.[38−41]

The Mayo group performed another long-term cross-sectional population study to record the cumulative incidence rates of PTS and specifically VLU.[42] Medical records of 1527 patients with DVT or pulmonary embolism between 1966 and 1990 were examined. Baseline characteristics, event type (DVT with or without pulmonary embolism or pulmonary embolism alone), leg side and site of DVT (proximal with or without distal DVT vs. distal DVT alone), venous stasis syndrome, and VLU were detailed at 1, 5, 10, and 20 years. Among 245 patients with venous stasis syndrome, the cumulative incidence of venous ulcers at 20 years was 3.7%. There was a 30% increased risk for venous ulcer per decade of age at the incident of venous thromboembolism.

In the latest follow-up study on the prevalence and incidence of VLUs in Olmsted County, Minnesota, Gloviczki et al. surveyed county residents during the years 2010—11.[24] Patients were identified under the auspices of the Rochester Epidemiology Project using ICD-9-CM codes. Among 1551 individuals, a random sample of 15% (227 subjects) underwent a review of medical records. The mean age in this sample was older than in other studies (73 years), but the same slight male predominance (53%) was found. The estimated prevalence of VLUs was 210 per 100,000 person-years, while new VLUs incidence was 85 per 100,000 person-years, an incidence that appeared higher than in previous studies in this population.

Randomized trials of DVT treatment

Although there are numerous studies showing the prevalence of venous ulcers following DVT, recent randomized control trials (RCT) examining therapy for acute DVT provide contemporary and natural history data.

In the 5-year follow-up study of the Catheter-directed Venous Thrombolysis in Acute Iliofemoral Vein Thrombosis trial (CaVenT trial), Haig et al.[43] examined 176 patients (84% of the originally randomized groups). Overall, 5 (2.8%) patients developed severe advanced CVI (Villalta score >15%). Unfortunately, no specific data on VLU occurrence was provided.

The most recent The Acute Venous Thrombosis: Thrombus Removal with Adjunctive Catheter-Directed Thrombolysis study (ATTRACT trial) randomly assigned patients by an intention-to-treat protocol to pharmacomechanical treatment ($n = 336$) or a control group ($n = 355$) initially treated with heparin (followed by anticoagulation therapy and antiplatelet therapy with respective rates of 98% and 13% at 1 month, and 86% and 13% at 6 months).[31] All patients were followed for 24 months. In contrast to the Catheter-Directed Venous Thrombolysis trial (CAVENT) and other trials employing the Villalta scale, a specific description of the development of VLU was provided. The trial showed VLU development in 12 patients (4%) of the treated group and in 17 patients (5%) of the control group. As with any RCT, inclusion/exclusion criteria may have influenced the study population by selection bias. Nevertheless, the ATTRACT trial provides data on the appreciable incidence of VLU development within a 2-year period following DVT.

Extravenous etiology

The autopsy studies of May and Thurner[44] provided an anatomic mechanism to explain the propensity for a greater proportion of left lower extremities to develop DVT. These physicians identified intraluminal thickening or "venous spurs" and related this to extraluminal compression of the left common iliac vein by the right common iliac artery against the bony structure of the vertebral spine (usually the fifth lumbar vertebrae). Ten years later, Cockett et al.[45] described the relationship between iliac vein compression and the development of PTS with iliofemoral caval (IFC) obstruction and its important role in the genesis of VLUs. Ulceration can occur secondary to DVT of this segment or from compression alone by the right iliac artery (May–Thurner syndrome). In a small case series ($n = 17$) of patients undergoing mechanical thrombectomy for iliofemoral DVT, Kasirajan et al.[46] found that 59% of patients had an underlying lesion causing the occlusion. With May-Thurner syndrome, women are more frequently affected than men and typically present in their twenties or thirties with recent onset of left lower extremity swelling. At this initial presentation, DVT may not be present, but symptomatic involvement of the entire limb provides a further diagnostic clue. A high proportion of patients will have concomitant venous claudication.

Raju and Neglen[47] reviewed their extensive experience with >4000 patients with symptomatic venous disease. Among 879 severely symptomatic patients (938 limbs), intravascular ultrasound detected outflow obstruction in the vast majority of limbs (93.7%). Strikingly, 53% of these limbs had nonthrombotic iliac vein lesions, 40% were postthrombotic, and 7% were a combination. The average age was 54 with a 4:1 female-to-male ratio. In addition, they described a unique distal iliac venous lesion caused by the crossing of the hypogastric artery. Analysis of CEAP clinical class showed that the highest proportion of patients (55%) were C3, while only 14% were C5 or C6, which contrasts with the incidence following thrombotic involvement. Superficial reflux was found in 32%, deep vein reflux in 20%, and combined superficial and deep reflux in nearly 40%.

Over an 18-month period, Marston et al.[25] studied 78 C5–C6 limbs in 64 patients by computerized tomography ($n = 62$) and magnetic resonance ($n = 2$). Their purpose was to objectively document the proportion of obstruction at the IFC level. Notably, 50% of patients had a medical history of DVT. About one in three (37%) imaging studies demonstrated IFC obstruction of at least 50%; 23% had obstruction of >80%. Complete obstruction was found in seven limbs (9%), and 11 additional limbs (14%) had 80%–99% stenosis. Interestingly, the etiology of venous obstruction in limbs with >50% stenosis was external compression for 18 limbs, DVT alone in two limbs, and combined external compression and DVT in nine limbs.

In addition to having a greater risk of pulmonary embolus, IFC DVT is associated with a greater risk of PTS.[48] Neglen has stated that the prevalence of outflow obstruction in patients with recalcitrant VLUs remains unknown.[49] He suggests that this knowledge gap is related to the widespread prevalence of nonthrombotic iliac vein lesions and the absence of accepted criteria for determining the degree of venous stenosis that is considered hemodynamically significant.

Epidemiological data on CVI and its progression

As VLU is a sequela of CVI, epidemiological data on CVI provides an optimal, minimally biased starting point for the examination of risk factors for ulceration, as shown in Table 3.4.

Early point-prevalence population studies established that venous disease was the predominant cause of leg ulceration and that VLU prevalence increased with age.[58,59] Nelzén et al. compared the prevalence of venous with nonvenous leg ulcers. In a defined Swedish population, the prevalence of active leg ulcers was 0.31%, with 54% classified as venous and 46% nonvenous.[1] In a German population-based study of 31,619 patients, venous reflux was the causative factor in 47.6%, arterial insufficiency in 14.5%, and combined arterial and venous reflux in 17.6%.[2]

The overall prevalence of VLU (healed and active) in Western populations has consistently been found to be about 1%, while point prevalence (active VLU) has been estimated at 0.1%–0.7%.[36,60,61] In these studies, the presence or absence of ulcers represented a binary outcome of CVI suitable for comparisons between populations. However, venous disease is a dynamic process. The advent of ultrasonographic mapping of reflux in the superficial and deep systems facilitated the study of the evolution of venous disease through stages culminating in ulceration.

The Edinburgh Vein Study[50]

Between May 1994 and April 1996, the University of Edinburgh group conducted the first study of correlations between duplex findings and CVD in a representative cross-section of the general population. The landmark Edinburgh Vein Study screened a total of 1566 women and men randomly selected from general practice registers. Subjects were aged 18–64 years, with a mean age of 44.8 years for women and 45.8 for men. The choice of a younger population provided a cohort for follow-up studies. Nearly all subjects (98.9%) were white, reflecting the local demographics. As CEAP classification criteria were not available at the initiation of recruitment, the study relied on clinical criteria and categories of both varicose veins and CVI adopted from the Basle Study.[62] A total of 124 (7.9%) subjects were diagnosed with CVI. The prevalence of CVI correlated strongly with age and gender. Age-adjusted prevalence of CVI was 9.4% (95% CI, 7.2–11.6) in men and 6.6% (95% CI, 4.9–8.3) in women; however, among subjects aged >50 years, CVI was present in 21.2% of men and 12.0% of women. The frequency of reflux in both superficial and deep segments increased with the clinical severity of the disease.

Table 3.4 Comparison of epidemiological studies of VLU.

Study	Years	Male (%)	Age (mean)	Sample size	C5–C6 (%)	SVI (%)	DVI (%)	Limitations
Edinburgh[50]	1994–96	45	44.8	1566	0.6			No subjects >64 years
San Diego[51]	1994–98	35	59.9	2211	4.6[a]	52	22	Combined C4-6
Bonn[52]	2000–06	44		3072	0.7	73	55	
France[53]	1988–92	68	ND	409	Age 60–70 = 1 Age 70–80 = 3			No DUS
Italy[54]	2003	14	54	5241	8.6[a]	None targeted	None targeted	Combined C4-6
USA (AVF)[55]	2007	23	60	2234	2			
Sweden[1,56,57]		38	77	282	0.3 (C6) 1 (C5–C6)	34[c]	60	[b]

AVF, American Venous Forum; *DUS*, duplex ultrasound; *DVI*, deep venous insufficiency; *SVI*, superficial venous insufficiency.
[a]Includes C4 patients.
[b]Self care.
[c]SVI alone.

The CEAP classification for the chronic venous disease was published in 1995.[63] Since then, the use of CEAP criteria in varicose vein studies has facilitated epidemiological comparisons. Table 3.4 shows findings from a range of epidemiological studies, with all but the Edinburgh Vein Study employing CEAP classification. These studies provide detail beyond simple prevalence.

The San Diego population study[51]

In the US study, Criqui et al. reported on manifestations of CVD in 2211 men and women, aged 40–79 years, evaluated between 1994 and 1996. Investigators enrolled more women than men to improve power for several female-related hypotheses. Whereas 15% of men and 7% of women were visible without CVD signs in the Edinburgh study, respective rates in the San Diego study were 34% and 11%. Nonetheless, patterns of CVD progression are broadly comparable between the two studies. Criqui et al. found a marked increase in varicose veins (C2) and trophic changes (C4–6) with age, from 16.9% to 2.3%, respectively, in the <50 years group in contrast to 29.9% and 10.2%, respectively, in subjects ≥70 years. Compared with subjects <50 years, those aged 50–59 years were nearly twice as likely to exhibit skin changes (OR 1.75); these odds doubled again among subjects aged 60–69 (OR 4.16). Non-Hispanic Whites had a higher prevalence of both visible and functional disease compared with African-American, Asian, and Hispanic subjects.

Gender differences were pronounced. Spider veins, varicose veins, superficial functional disease, and superficial thrombotic events were more common in women than men (OR 5.4, 2.2, 1.9, and 1.9, respectively), but trophic changes and deep functional disease and thrombotic events were less common in women (OR 0.69 for both). For legs evidencing both trophic changes and deep functional disease, the age-adjusted prevalence of edema, superficial events, and the proportions of deep events were 48.2%, 11.3%, and 24.6%, respectively, compared with 1.7%, 0.6%, and 1.3% for legs visibly and functionally without CVD.

In the San Diego study, visible signs of CVD (varicose veins or skin changes) and ultrasonographic findings of venous dysfunction (superficial or deep reflux) were closely associated, while both were strongly associated with edema and thrombotic events. Nonetheless, Criqui et al. found discord between visible and functional findings in 8% of subjects, reinforcing the value of ultrasonographic examination.

The Bonn Vein study[52]

The Bonn Vein study, conducted between November 2000 and March 2002, evaluated >3000 subjects aged 18–80 in the city of Bonn and two rural townships in Germany. Sampling was randomized and derived from a population-registered format that included a standardized questionnaire (Short Form-36). Interestingly, the prevalence of varicose veins in women was nearly half that seen in the San Diego study (15.8% vs. 27.7%, respectively), but edema (C3) in women was threefold more prevalent (14.9% vs. 4.9%). In both studies, C4—C6 disease (Table 3.4) was more prevalent in men compared with women. The proportion of both varicose veins and ulceration increased with age.

The Polish study[64]

This 2003 cross-sectional survey included 40,095 consecutive patients (84% women) from outpatient clinics treated by primary care physicians. CEAP classes were heavily weighted toward mild disease, with 89.4% of patients documented as C0—C2. Only 0.5% had an open ulcer (C6) and 1% had a healed ulcer (C5). The prevalence of varicose veins and severe CVI (skin changes, leg ulcer) reported was similar to that observed in the other developed countries.

The French study[53]

In 2004, subjects from a French study of Raynaud's phenomenon underwent a cross-sectional study for varicose veins. By contrast with the Polish study, the majority (68%) of subjects in the French study were male. In four regions in France, random samples of 2000 subjects per site underwent telephone interviews. A subsample of 409 subjects received physical examinations. Approximately 70% of subjects were C2. Trophic skin changes were observed in 2.8% of women compared with 5.4% of men. Notably, advanced CVI was found only in subjects with varicose veins. Age and pitting ankle edema correlated best with the presence of trophic changes, while a past history of DVT was "borderline significant" for trophic changes. Interestingly, major trophic changes usually occurred in limbs with pigmentation. No active ulceration was found among 558 women and 277 men, but evidence of a healed ulcer was found in 0.7% of women and 1.4% of men.

The Italian study[54]

In 2003, investigators in Italy used television and newspaper advertising to recruit 5247 people aged 18—90 in cities throughout Italy to complete a standardized CVD questionnaire and undergo subsequent examinations. Most participants were women (85.9%). As seen previously, investigators found a linear relationship between age and the signs of CVI, and a higher representation of men in advanced stages of venous disease. Risk factors for varicose veins included living in southern Italy, pregnancy, and family history.

The American Venous Forum study[55]

In 2006, the American Venous Forum sponsored a national venous screening program that involved 83 physicians across 40 states. The screened population ($n = 2234$) had a mean age of 60, was predominantly female (77%), White (80%), and had a mean BMI of 29. The broad screening instrument included demographics, risk assessment, quality of life (QoL) assessment by the Chronic Venous Insufficiency Quality of Life Questionnaire (CIVIQ) instrument, DUS, and clinical examination. CIVIQ QoL scores correlated with increasing severity of CEAP. On DUS examination, obstruction was observed in 5% of subjects, while nearly 40% had reflux. As with other surveys, the prevalence of ulcers was low: only 0.5% had an open ulcer (C6) while 1.5% had a healed ulcer (C5). However, skin changes indicative of advanced CVI were found in 7%. Interestingly, higher CEAP classifications were noted in patients with reflux or hypertension as well as in patients with an elevated venous thromboembolism risk score.

The Swedish Skaraborg County studies[1,56,57]

Nelzén et al. conducted a series of epidemiologic studies on leg ulcers in Skaraborg County, Sweden. They included all ulcers of the lower extremity, differentiated via Doppler ultrasound and Duplex scans. A point prevalence (C6) of 0.3% and an overall (C5–6) prevalence of 1% was documented. Among a total of 463 active ulcers, the majority, 332 (72%), had a venous etiology, with deep venous involvement in 38% and pure superficial reflux in 34%. Of note, 40% of VLUs had an ankle-brachial index of <0.9, signifying peripheral artery disease. This finding called attention to the phenomenon of a "mixed" venous and arterial ulceration. The authors concluded that 40% of ulcers evaluated were amenable to some form of intervention. In a follow-up study on the impact of treatment (particularly for superficial venous and perforator incompetence), investigators reported a 23% overall reduction in all ulcers accompanied by a 46% reduction in VLU (for additional details see chapter 4).[65]

Progression of venous disease

While several epidemiologic studies address the prevalence of VLU, fewer have looked at the progression of venous disease, that is, the incidence of CVI over time with the development of VLU. In a follow-up study of the Bonn cohort at 6.6 years, Rabe et al.[66] identified the incidence of newly developed varicose veins as well as the progression of CVI among 1978 subjects from the original study. The prevalence of varicose veins rose from 23% to 25% and CVI progressed from 14.5% to 16%. Among C2 patients in the original study, 1 in 5 (19.8%) with nonsaphenous vein reflux at baseline progressed to higher C classes, while 31.8% of saphenous vein reflux progressed. Risk factors for progression in multivariate analysis were age, obesity, and arterial hypertension. The authors noted the rapid progression of venous disease.

In 2015, Lee et al. reported on CVI progression among the original Edinburgh Vein Study population at 13 years.[67] Of the original 1500 patients nearly 900 (~60%) had a follow-up examination with DUS. Strikingly, over half (57.8%) of subjects showed progression of venous disease, with an annual progression rate of 4.3%. Among 270 subjects who demonstrated solely varicose veins in the original study, over 30% had developed CVI, which paralleled increasing age. Of 64 patients with baseline CVI, 5 (7.8%) progressed to C5−6 disease. By contrast, none of the 270 patients with varicose veins and *absence of CVI* progressed to C5−6 disease. Risk factors for deterioration from varicose veins to CVI were family history of varicose veins (OR 1.85), being overweight (OR 1.85), and history of DVT (OR: 4.10).

Given its rapid progression, early intervention is clearly needed in CVI. A retrospective cohort analysis of administrative insurance claims data over 5 years compared patients who received treatment for varicose veins with those who did not. The groups were temporally divided into distinct time periods in relation to the index diagnosis. Early intervention (<2 months) was associated with a decreased overall progression of varicose vein disease (29.2%) compared with intervention at 2−6 months (42.5%, $P < .0001$) and delayed intervention at 6−24 months (52.2%, $P < .0001$). Among all preulcer patients at baseline (C < 4), delayed interventional treatment had 1.4 times the adjusted odds ($P < .0001$) of varicose veins progression to ulcers or more severe Thomas Reuters stages (>5).[68]

In summary, epidemiological studies have identified age as the most important risk factor for CVI, with greater risk beyond the 50-year mark. Women experience more C1−C2 disease, superficial venous reflux, and thrombotic events compared with men. Conversely, the prevalence of CVI and deep venous reflux is greater in men. Positive family history is a risk factor for CVI in both sexes, while BMI is also likely to play a role in disease progression.

Clinical risk factors for first VLU (Table 3.5)

In 2009, Robertson et al.[4] published the first case-control study that compared patients with VLU (C5−C6) with venous patients without ulceration (C2−C4). The groups were age- and gender-matched, with 120 subjects per group. Clinical risk factors for VLU included severity of clinical venous disease, particularly with skin changes ($P < .0001$); history of DVT ($P = .001$); and deep venous reflux ($P = .0001$). Multivariate analysis ($P > .05$) showed that patients who had progressed to lipodermatosclerosis (C4b) and corona phlebectatica (C4c) were at pronounced risk of VLU (odds ratio [OR] 8.90, OR 4.52, respectively). Other independent risk factors for ulceration included popliteal vein reflux (OR 2.82) and BMI (OR 1.08), while protective factors included good dorsiflexion of the ankle (OR 0.88) and an effective calf muscle pump (OR 0.96).

In their 2019 systematic review, Meulendijks et al.[5] assessed five case-control or cross-sectional studies comparing venous patients with and without ulceration. Methodological differences prevented statistical comparisons of results, but there was agreement among most studies (4/5) that higher age, higher BMI, low physical activity, and arterial

Table 3.5 Risk factors for fVLU.

Outcome measure	Darwin AHR (CI)[a]	Robertson OR (CI)[a]
Age	45–54: 1.316 (1.276–1.358) 55–64: 1.596 (1.546–1.648)	Matched
Gender	Men: 1.838 (1.798–1.880)	Matched
Arterial hypertension[b]	1.067 (1.040–1.093)	
Patient history of nonvenous ulceration	3.923 (3.699–4.161)	
Higher BMI		1.08 (1.01–1.15)
C2: Varicose veins	0.438 (0.418–0.458)	
CVI	1.244 (1.193–1.298)	
Chronic venous hypertension[b]	1.671 (1.440–1.939)	
C3: Edema	1.224 (1.193–1.256)	
C4a, stasis dermatitis	1.078 (0.924–1.258)	6.15 (3.45–10.97)
C4a Lipodermatosclerosis		8.90 (1.44–54.8)
C4a Corona phlebectatica		4.52 (1.81–11.3)
Deep vein reflux		2.82 (1.03–7.75); popliteal only
Dorsiflexion—index leg		0.88 (0.81–0.97)[c]
Venous pump power		0.96 (0.92–0.99)[c]

[a]Significant in multivariate analysis at $P = .05$.
[b]Identified through antihypertensive use.
[c]Protective factors.

hypertension were significantly associated with ulceration. Three of five studies found associations between VLU and deep venous reflux and DVT, respectively.

In 2021, Darwin et al.[6] published an administrative claims data analysis of longitudinal data (2005–14) specifically focused on identifying risk factors for fVLU. Data from >650,000 patients in the IBM commercial claims database was used to compare comorbidities in patients with CVI who developed their fVLU with those who did not. The database limited the analysis to patients aged 64 or younger. Venous insufficiency was identified via ICD-9 coding for various manifestations of venous disease. Adjusted hazard ratio (AHR) for comorbidities demonstrated increased risk in men (AHR 1.8) while increasing age heightened risk (45–54 years: AHR 1.32; 55–64 years: AHR 1.59). Patients with baseline CVI or edema were at equivalent increased risk of fVLU (AHR 1.24), while baseline chronic venous hypertension was a pronounced risk factor (AHR 1.67). Table 3.4 shows independent fVLU risk factors identified in the Robertson and Darwin studies.

Patients with CVI are often referred late, after tissue changes that increase risk of ulceration have already occurred.[4] Although the natural progression of venous disease to ulceration is slow, Darwin et al.[6] found that the highest incidence of fVLU occurred within 500 days of CVI diagnosis, suggesting a delay in diagnosis. Diagnosis and treatment of varicose veins were strong protective factors against VLU (AHR 0.43 and 0.45, respectively), whereas the diagnosis of CVI (AHR 1.22) and edema (1.22) was associated with higher risk. While the average age at fVLU was similar between men and women (51.9 vs. 51.5, respectively), CVI was diagnosed at an earlier age in women (50.2 vs. 49.6), suggesting that early CVI diagnosis, and presumably treatment, may have reduced fVLU onset in women.

Other novel findings in the Darwin analysis would seem to support the protective influence of early intervention. The association of pain medication with benefit (AHR 0.779) may be related to the surprising finding that arthritis and leg fractures were protective. The authors speculate that pain reduction may improve calf muscle pump action and dorsiflexion, both considered to be protective factors in VLU[44]; Other protective factors included using compression stockings (AHR 0.73) as well as prescribed statin medications (AHR 0.72).

Genetic risk factors for VLU

While venous hypertension, varicose veins, and reflux are the sine qua non of VLU formation, inflammatory mechanisms triggered by abnormal venous flow and iron deposition are increasingly seen as amplifiers of CVD.[17,69] Theoretically, DNA testing could provide prognostic markers by identifying genetic mutations that underline or exacerbate the pathogenesis of VLU during prolonged inflammation.[15]

Research on genetic risk factors in CVD has concentrated on the influence of *FOXC2* gene mutations, matrix metalloproteinases (MMPs), and hemochromatosis genes on the development of varicose veins and progression to ulceration.[10] *FOXC2*, which encodes a regulatory forkhead transcription factor, plays an important role in the development of venous and lymphatic valve function.[10,70] As reflux is secondary to valve function, *FOXC2* mutations may contribute to varicose veins and the progression of venous disease.[70] During inflammation, MMPs and tissue inhibitors of MMPs facilitate tissue remodeling through degradation and reorganization of the extracellular matrix; dysregulated MMP activity is thought to undermine vein wall integrity and contribute to ulcer formation.[10,71] As described below, extensive study of *HFE* mutations of hemochromatosis genes has clarified their important role in susceptibility to varicose veins and VLU.[10,15]

Sustained venous hypertension in the microcirculation leads to the extravasation of proteins and erythrocytes into the dermal interstitium, accompanied by significant iron deposition.[69,70] Iron deposits in CVD are readily visible in the coppery-brownish hyperpigmentation that may occur in later stages of the disease and is always present around ulcers.[69] As erythrocytes are degraded by interstitial macrophages, released iron is incorporated into ferritin, which with time changes structurally to hemosiderin.[69] These increased iron stores and interstitial proteins create a strong chronic inflammatory signal responsible for leukocyte recruitment and migration in the matrix.[69,70]

However, iron deposits in the legs of CVD patients do not always lead to ulceration. Zamboni et al.[72] hypothesized that individual differences in progression to VLU could be genetically determined and investigated the role of the *C282Y* and *H63D* mutations of the *HFE* gene, associated with hemochromatosis in Northern European populations. They found that the *C282Y* mutation increases the risk of ulcer in primary CVD by a factor of nearly 7, while patients with the *H63D* variant have an earlier age of ulcer onset by almost 10 years.[73] In 2009, the same group investigated two additional mutations in CVD patients: *FPN1 −8CG*, which promotes the ferroportin gene, and *MMP12 −82AG*, a promoter of the *MMP12* gene. The *FPN1 −8CG* had a VLU risk of 5.2, while *MMP12 −82AA* had a VLU risk of 1.96.[15]

As these investigators themselves suggest, the results of these and similar studies may be viewed as tentative prognostic indicators for VLU among CVD patients. Although these results are significant, identification of risk factors for VLU through candidate-gene studies presupposes knowledge of the pathogenesis of ulceration, which remains poorly understood.[3] Genome-wide studies (GWAS) may contribute to this understanding. In a recent GWAS examining varicose veins using the UK Biobank ($n = 9577$), and comparing them to 327,959 control samples Fukaya et al.[74] identified 30 new loci, demonstrating the highly polygenic nature of the venous disease and the multiplicity of its genetic factors.[75] Notably, investigators used machine learning to assess >2700 clinical variables, demonstrating the potential for artificial intelligence-based tools to drive large-scale research on genetic associations in CVD as the basis for future studies of causal relationships.

Lymphatic involvement in VLU

The close relationship between the venous and lymphatic systems as complementary drainage pathways has prompted the investigation of the potential role of lymphatic dysfunction in venous ulceration. Lymphatic anatomy and function have been shown to degrade from early to more advanced venous insufficiency[76]; as noted by Partsch, CVI is always a chronic venous-lymphatic insufficiency.[77] However, whether lymphatic degradation is a cause or an effect of CVI remains undetermined.

In unaffected limbs, the lymphatics are well formed, linear in nature, and present with a pulsatile unidirectional flow toward regional nodal basins in the lower extremity. In patients with varicose veins, unusual lymphatic drainage patterns emerge with early signs of lymphatic vessel dilation and segmentation. The presence of edema marks C3 venous disease, characterized by distinct segmentation of lymphatic vessels. VLUs present with a complete absence of lymphatics in the ulcer bed and a marked decrease in the number of lymphatics surrounding an ulcer, with significant damage to those remaining.[78] Employing near-infrared imaging techniques with indocyanine green (NIRFLI), Rasmussen and Sevick-Muraca[79] documented bilateral dermal backflow and abnormal lymphatic pooling in 12 VLU patients. In contrast to normal patients, contractile events within the lymphangion of VLU patients were diminished and too infrequent to be quantified. Fewer functioning lymphatic vessels were observed in subjects with the longest duration of ulceration. These and more recent findings[76] from the same study group suggest a potential role for lymphatics in disease progression.

Unfortunately, combined venous and lymphatic insufficiency (phlebolymphedema) often goes undiagnosed until VLU occurs. In a claims-based analysis of 27,000 patients diagnosed with lymphedema, the majority of patients with phlebolymphedema (10% of the study population) were coded with venous ulcer (9.6%), while only 0.8% were coded with CVI.[80] A second claims-based study yielded approximately 86,000 patients with the diagnosis of lymphedema who were followed for 1 year.[68] A similar proportion of patients (9.8%) had a diagnosis of VLU.

Summary

Whereas primary CVD is the most common VLU etiology, secondary CVD carries a higher VLU risk. Advanced CVI is the sine qua non of VLU, and fVLU risk increases with the severity of skin changes. Older age, male gender, deep venous reflux, and obesity also predispose patients to fVLU. Mutations of the *HFE* gene, associated with hemochromatosis, may substantially increase the risk of fVLU in primary CVD. In pragmatic terms, CVI is often referred late, when patients already experience advanced CVI. Although the progression of primary CVD to ulceration is slow, most fVLU occurs within 500 days of CVI diagnosis. Diagnosis and treatment of varicose veins are protective factors against VLU, whereas the diagnosis of edema and CVI is associated with higher risk.

References

1. Nelzén O, Bergqvist D, Lindhagen A. Leg ulcer etiology—a cross sectional population study. *J Vasc Surg*. 1991;14(4):557–564.
2. Körber A, Klode J, Al-Benna S, et al. Etiology of chronic leg ulcers in 31,619 patients in Germany analyzed by an expert survey. *JDDG J der Deutschen Dermatol Gesellschaft*. 2011;9(2):116–121.
3. Raffetto JD, Ligi D, Maniscalco R, Khalil RA, Mannello F. Why venous leg ulcers have difficulty healing: overview on pathophysiology, clinical consequences, and treatment. *J Clin Med*. 2021;10(1):29.
4. Robertson L, Lee AJ, Gallagher K, et al. Risk factors for chronic ulceration in patients with varicose veins: a case control study. *J Vasc Surg*. 2009;49(6):1490–1498.
5. Meulendijks A, de Vries F, van Dooren A, Schuurmans M, Neumann H. A systematic review on risk factors in developing a first-time Venous Leg Ulcer. *J Eur Acad Dermatol Venereol*. 2019;33(7): 1241–1248.

6. Darwin E, Liu G, Kirsner RS, Lev-Tov H. Examining risk factors and preventive treatments for first venous leg ulceration: a cohort study. *J Am Acad Dermatol*. 2021;84(1):76–85.
7. Rice JB, Desai U, Cummings AKG, Birnbaum HG, Skornicki M, Parsons N. Burden of venous leg ulcers in the United States. *J Med Econ*. 2014;17(5):347–356.
8. Gohel M, Taylor M, Earnshaw J, Heather B, Poskitt K, Whyman M. Risk factors for delayed healing and recurrence of chronic venous leg ulcers—an analysis of 1324 legs. *Eur J Vasc Endovasc Surg*. 2005;29(1):74–77.
9. Phillips T, Stanton B, Provan A, Lew R. A study of the impact of leg ulcers on quality of life: financial, social, and psychologic implications. *J Am Acad Dermatol*. 1994;31(1):49–53.
10. Serra R, Ssempijja L, Provenzano M, Andreucci M. Genetic biomarkers in chronic venous disease. *Biomarkers Med*. 2020;14(2):75–80.
11. American Venous Forum. What is Chronic Venous Disease? (https://www.veinforum.org/patients/what-is-vein-disease/what-is-chronic-venous-disease/).
12. Gloviczki P, Dalsing MC, Henke P, et al. Report of the society for vascular surgery and the American venous Forum on the july 20, 2016 meeting of the medicare evidence development and coverage advisory committee panel on lower extremity chronic venous disease. *J Vasc Surg Venous Lymphat Disord*. 2017;5(3):378–398.
13. O'Donnell Jr TF, Passman MA, Marston WA, et al. Management of venous leg ulcers: clinical practice guidelines of the Society for Vascular Surgery (R) and the American Venous Forum. *J Vasc Surg*. 2014;60(2 Suppl):3S–59S. https://doi.org/10.1016/j.jvs.2014.04.049.
14. Lurie F, Passman M, Meisner M, et al. CEAP classification system and reporting standard, revision. *J Vasc Surg Venous Lymphat Disord*. 2020;8:342–352.
15. Gemmati D, Federici F, Catozzi L, et al. DNA-array of gene variants in venous leg ulcers: detection of prognostic indicators. *J Vasc Surg*. 2009;50(6):1444–1451.
16. Chi Y-W, Raffetto JD. Venous leg ulceration pathophysiology and evidence based treatment. *Vasc Med*. 2015;20(2):168–181.
17. Bergan JJ, Schmid-Schonbein GW, Smith PD, Nicolaides AN, Boisseau MR, Eklof B. Chronic venous disease. *N Engl J Med*. 2006;355(5):488–498. https://doi.org/10.1056/NEJMra055289.
18. Labropoulos N. How does chronic venous disease progress from the first symptoms to the advanced stages? A review. *Adv Ther*. 2019;36(1):13–19.
19. Labropoulos N, Gasparis AP, Pefanis D, Leon Jr LR, Tassiopoulos AK. Secondary chronic venous disease progresses faster than primary. *J Vasc Surg*. 2009;49(3):704–710.
20. O'Donnell Jr TF, Lau J. A systematic review of randomized controlled trials of wound dressings for chronic venous ulcer. *J Vasc Surg*. 2006;44(5):1118–1125.
21. Norman G, Westby MJ, Rithalia AD, Stubbs N, Soares MO, Dumville JC. Dressings and topical agents for treating venous leg ulcers. *Cochrane Database Syst Rev*. 2018;6.
22. Johnson BF, Manzo RA, Bergelin RO, Strandness Jr DE. Relationship between changes in the deep venous system and the development of the postthrombotic syndrome after an acute episode of lower limb deep vein thrombosis: a one-to six-year follow-up. *J Vasc Surg*. 1995;21(2):307–313.
23. TenBrook Jr JA, Iafrati MD, O'Donnell Jr TF, et al. Systematic review of outcomes after surgical management of venous disease incorporating subfascial endoscopic perforator surgery. *J Vasc Surg*. 2004;39(3):583–589.
24. Gloviczki ML, Kalsi H, Gloviczki P, Gibson M, Cha S, Heit JA. Validity of international classification of diseases, ninth revision, clinical modification codes for estimating the prevalence of venous ulcer. *J Vasc Surg Venous Lymphat Disord*. 2014;2(4):362–367.
25. Marston W, Fish D, Unger J, Keagy B. Incidence of and risk factors for iliocaval venous obstruction in patients with active or healed venous leg ulcers. *J Vasc Surg*. 2011;53(5):1303–1308.
26. Ma H, O'Donnell Jr TF, Rosen NA, Iafrati MD. The real cost of treating venous ulcers in a contemporary vascular practice. *J Vasc Surg Venous Lymphat Disord*. 2014;2(4):355–361.
27. Barwell JR, Davies CE, Deacon J, et al. Comparison of surgery and compression with compression alone in chronic venous ulceration (ESCHAR study): randomised controlled trial. *Lancet*. 2004;363(9424):1854–1859.

28. van Gent WB, Hop WC, van Praag MC, Mackaay AJ, de Boer EM, Wittens CH. Conservative versus surgical treatment of venous leg ulcers: a prospective, randomized, multicenter trial. *J Vasc Surg*. 2006; 44(3):563–571.
29. Gohel MS, Heatley F, Liu X, et al. A randomized trial of early endovenous ablation in venous ulceration. *N Engl J Med*. 2018;378(22):2105–2114.
30. Enden T, Haig Y, Kløw N-E, et al. Long-term outcome after additional catheter-directed thrombolysis versus standard treatment for acute iliofemoral deep vein thrombosis (the CaVenT study): a randomised controlled trial. *Lancet*. 2012;379(9810):31–38.
31. Vedantham S, Goldhaber SZ, Julian JA, et al. Pharmacomechanical catheter-directed thrombolysis for deep-vein thrombosis. *N Engl J Med*. 2017;377(23):2240–2252.
32. Asbeutah AM, Riha AZ, Cameron JD, McGrath BP. Five-year outcome study of deep vein thrombosis in the lower limbs. *J Vasc Surg*. 2004;40(6):1184–1189.
33. Akesson H, Brudin L, Dahlström J, Eklöf B, Ohlin P, Plate G. Venous function assessed during a 5 year period after acute ilio-femoral venous thrombosis treated with anticoagulation. *Eur J Vasc Surg*. 1990; 4(1):43–48.
34. Killewich LA, Bedford GR, Beach KW, Strandness Jr D. Spontaneous lysis of deep venous thrombi: rate and outcome. *J Vasc Surg*. 1989;9(1):89–97.
35. Yamaki T, Nozaki M. Patterns of venous insufficiency after an acute deep vein thrombosis. *J Am Coll Surg*. 2005;201(2):231–238.
36. Heit JA, Rooke TW, Silverstein MD, et al. Trends in the incidence of venous stasis syndrome and venous ulcer: a 25-year population-based study. *J Vasc Surg*. 2001;33(5):1022–1027.
37. Ashrani AA, Heit JA. Incidence and cost burden of post-thrombotic syndrome. *J Thromb Thrombol*. 2009;28(4):465–476.
38. Franzeck UK, Schalch I, Jager KA, Schneider E, Grimm Jr, Bollinger A. Prospective 12-year follow-up study of clinical and hemodynamic sequelae after deep vein thrombosis in low-risk patients (Zurich study). *Circulation*. 1996;93(1):74–79.
39. Prandoni P, Lensing AW, Cogo A, et al. The long-term clinical course of acute deep venous thrombosis. *Ann Intern Med*. 1996;125(1):1–7.
40. Brandjes DP, Büller HR, Heijboer H, et al. Randomised trial of effect of compression stockings in patients with symptomatic proximal-vein thrombosis. *Lancet*. 1997;349(9054):759–762.
41. Stain M, Schönauer V, Minar E, et al. The post-thrombotic syndrome: risk factors and impact on the course of thrombotic disease. *J Thromb Haemost*. 2005;3(12):2671–2676.
42. Mohr DN, Silverstein MD, Heit JA, Petterson TM, O'Fallon WM, Melton III LJ. The venous stasis syndrome after deep venous thrombosis or pulmonary embolism: a population-based study. *Mayo Clin Proc*. 2000;75(12):1249–1256.
43. Haig Y, Enden T, Grøtta O, et al. Post-thrombotic syndrome after catheter-directed thrombolysis for deep vein thrombosis (CaVenT): 5-year follow-up results of an open-label, randomised controlled trial. *Lancet Haematol*. 2016;3(2):e64–e71.
44. May R, Thurner J. The cause of the predominantly sinistral occurrence of thrombosis of the pelvic veins. *Angiology*. 1957;8(5):419–427.
45. Cockett F, Thomas ML, Negus D. Iliac vein compression.–Its relation to iliofemoral thrombosis and the post-thrombotic syndrome. *Br Med J*. 1967;2(5543):14.
46. Kasirajan K, Gray B, Ouriel K. Percutaneous AngioJet thrombectomy in the management of extensive deep venous thrombosis. *J Vasc Interv Radiol*. 2001;12(2):179–185.
47. Raju S, Neglen P. High prevalence of nonthrombotic iliac vein lesions in chronic venous disease: a permissive role in pathogenicity. *J Vasc Surg*. 2006;44(1):136–144.
48. Nyamekye I, Merker L. Management of proximal deep vein thrombosis. *Phlebology*. 2012;27(2_suppl): 61–72.
49. Neglén P. Chronic deep venous obstruction: definition, prevalence, diagnosis, management. *Phlebology*. 2008;23(4):149–157.
50. Evans CJ, Allan PL, Lee AJ, Bradbury AW, Ruckley CV, Fowkes FGR. Prevalence of venous reflux in the general population on duplex scanning: the Edinburgh vein study. *J Vasc Surg*. 1998;28(5): 767–776.

51. Criqui MH, Jamosmos M, Fronek A, et al. Chronic venous disease in an ethnically diverse population: the San Diego Population Study. *Am J Epidemiol*. 2003;158(5):448−456.
52. Rabe E, Pannier-Fischer F, Bromen K, et al. Bonner venenstudie der deutschen gesellschaft für phlebologie. *Phlebologie*. 2003;32(01):1−14.
53. Carpentier PH, Maricq HR, Biro C, Ponçot-Makinen CO, Franco A. Prevalence, risk factors, and clinical patterns of chronic venous disorders of lower limbs: a population-based study in France. *J Vasc Surg*. 2004;40(4):650−659.
54. Chiesa R, Marone EM, Limoni C, Volonté M, Schaefer E, Petrini O. Chronic venous insufficiency in Italy: the 24-cities cohort study. *Eur J Vasc Endovasc Surg*. 2005;30(4):422−429.
55. McLafferty RB, Passman MA, Caprini JA, et al. Increasing awareness about venous disease: the American Venous Forum expands the national venous screening program. *J Vasc Surg*. 2008;48(2):394−399.
56. Nelzen O, Bergqvist D, Lindhagen A, Hallböök T. Chronic leg ulcers: an underestimated problem in primary health care among elderly patients. *J Epidemiol Commun Health*. 1991;45(3):184−187.
57. Nelzen O, Bergqvist D, Lindhagen A. The prevalence of chronic lower-limb ulceration has been underestimated: results of a validated population questionnaire. *Br J Surg*. 1996;83(2):255−258.
58. Baker S, Stacey M, Jopp-McKay A, Hoskin S, Thompson P. Epidemiology of chronic venous ulcers. *Br J Surg*. 1991;78(7):864−867.
59. Nelzen O, Bergqvist D, Lindhagen A. Venous and non-venous leg ulcers: clinical history and appearance in a population study. *J Br Surg*. 1994;81(2):182−187.
60. Nelzen O. Prevalence of venous leg ulcer: the importance of the data collection method. *Phlebolymphology*. 2008;15(4):143−150.
61. Rabe E, Pannier F. Epidemiology of chronic venous disorders. In: Gloviczki P, ed. *Handbook of Venous and Lymphatic Disorders, Guidelines of the American Venous Forum*. 4th ed. Boca Raton, FL: CRC Press; 2017:121−128.
62. Widmer L. Peripheral venous disorders. Prevalence and sociomedical importance. *Basel III study Bern*. 1978:43−50.
63. Porter JM, Moneta GL, on Chronic AICC, Disease V. Reporting standards in venous disease: an update. *J Vasc Surg*. 1995;21(4):635−645.
64. Jawien A, Grzela T, Ochwat A. Prevalence of chronic venous insufficiency in men and women in Poland: multicentre cross-sectional study in 40,095 patients. *Phlebology*. 2003;18(3):110−122.
65. Forssgren A, Fransson I, Nelzen O. Leg ulcer point prevalence can be decreased by broad-scale intervention: a follow-up cross-sectional study of a defined geographical population. *Acta Derm Venereol*. 2008;88(3):252−256.
66. Pannier F, Rabe E. Progression of chronic venous disorders: results from the Bonn Vein Study. *J Vasc Surg*. 2011;53(1):254−255.
67. Lee AJ, Robertson LA, Boghossian SM, et al. Progression of varicose veins and chronic venous insufficiency in the general population in the Edinburgh Vein Study. *J Vasc Surg Venous Lymphat Disord*. 2015;3(1):18−26.
68. Raju A, Mallick R, Campbell C, Carlton R, O'Donnell T, Eaddy M. Real-world assessment of interventional treatment timing and outcomes for varicose veins: a retrospective claims analysis. *J Vasc Interv Radiol*. 2016;27(1):58−67.
69. Zamboni P. The big idea: iron-dependent inflammation in venous disease and proposed parallels in multiple sclerosis. *J R Soc Med*. 2006;99(11):589−593.
70. Crawford JM, Lal BK, Durán WN, Pappas PJ. Pathophysiology of venous ulceration. *J Vasc Surg Venous Lymphat Disord*. 2017;5(4):596−605.
71. Woodside KJ, Hu M, Burke A, et al. Morphologic characteristics of varicose veins: possible role of metalloproteinases. *J Vasc Surg*. 2003;38(1):162−169.
72. Zamboni P, Tognazzo S, Izzo M, et al. Hemochromatosis C282Y gene mutation increases the risk of venous leg ulceration. *J Vasc Surg*. 2005;42(2):309−314.
73. Zamboni P, Izzo M, Tognazzo S, et al. The overlapping of local iron overload and HFE mutation in venous leg ulcer pathogenesis. *Free Radic Biol Med*. 2006;40(10):1869−1873.
74. Fukaya E, Flores AM, Lindholm D, et al. Clinical and genetic determinants of varicose veins: prospective, community-based study of ≈ 500 000 individuals. *Circulation*. 2018;138(25):2869−2880.

75. Wells QS. Varicose veins reach new heights. *Circulation*. 2018;138:2881–2883.
76. Rasmussen JC, Zhu B, Morrow JR, et al. Degradation of lymphatic anatomy and function in early venous insufficiency. *J Vasc Surg Venous Lymphat Disord*. 2021;9(3):720–730. e2.
77. Partsch H, Lee B. Phlebology and lymphology–a family affair. *Phlebology*. 2014;29(10):645–647. https://doi.org/10.1177/0268355514551514.
78. Eliska O, Eliskova M. Morphology of lymphatics in human venous crural ulcers with lipodermatosclerosis. *Lymphology*. 2001;34(3):111–123.
79. Rasmussen JC, Aldrich MB, Tan I-C, et al. Lymphatic transport in patients with chronic venous insufficiency and venous leg ulcers following sequential pneumatic compression. *J Vasc Surg Venous Lymphat Disord*. 2016;4(1):9–17.
80. Son A, O'Donnell Jr TF, Izhakoff J, Gaebler JA, Niecko T, Iafrati MA. Lymphedema-associated comorbidities and treatment gap. *J Vasc Surg Venous Lymphat Disord*. 2019;7(5):724–730. https://doi.org/10.1016/j.jvsv.2019.02.015.

CHAPTER 4

Venous ulcers of the lower extremity: epidemiology and socioeconomic burden

Olle Nelzén

Vascular Surgery Unit & Department of Research and Development Skaraborg Hospital Skövde & Uppsala University, Sweden

A leg ulcer is not a disease, but rather a clinical finding that can be caused by a variety of different disorders and is associated with several risk factors. Venous ulcers can be caused by a number of pathophysiological mechanisms related to hemodynamic aberrations in different venous anatomic segments. A population-based study conducted in Sweden, published by our group, showed that superficial and perforating vein refluxes are becoming the leading causes of venous ulcers, whereas deep vein insufficiency and/or obstruction following deep vein thrombosis (DVT) and primary deep vein disease are becoming less frequent.[1] In addition, combinations of various underlying pathophysiological mechanisms, different venous anatomic segment involvements, and/or venous reflux distribution in patients with venous diseases are far from being uncommon; this underlines the necessity for a detailed and comprehensive diagnostic approach.

Venous ulcers are not a new healthcare problem; rather, they are known to have plagued mankind since ancient times. A written description of the treatment of venous ulcers was found in the Papyrus of Eber originating from Egypt around 1550 BCE.[2] In the Milan Cathedral in Italy, there are four paintings depicting "miracles" performed by Saint Carlo for patients with leg ulcers at the beginning of the 17th century, clearly indicating that leg ulcers were already a major problem within the population (Fig. 4.1). The greatest challenge with the modern management of venous ulcers is that their clinical and economic significance is not widely appreciated, highlighting the need for more epidemiological studies.

Role of epidemiology in the study and treatment of venous leg ulcers

Epidemiology is the healthcare science of the distribution (frequency and pattern) and determinants (causes and risk factors) of diseases or events in specified populations. Understanding the epidemiology of a disease is of fundamental importance for planning appropriate care pathways and optimizing health systems and treatments. A good epidemiological survey can form a very valuable baseline to calculate the magnitude of needed

Figure 4.1 The miracle of Aurelia Degli Angeli. She had suffered from a smelly painful ulcer for 3 years when she in 1601 called on the Holy Archbishop Saint Carlo who prayed and her sores instantly closed, and she returned to full health. Painting by Giovan Battista Crespi, known as Cerano—1610. Milan cathedral, Italy. *(Copyright © 1995, Veneranda Fabbrica del Duomo di Milano.)*

healthcare resources—both financial resources and workload-related factors pertaining to healthcare providers—and serve as a useful comparison benchmark for measuring outcome changes resulting from implementing treatment changes. To accomplish the aforementioned goals, a repeat epidemiological survey must be undertaken using the same (or a similar) methodology. Further epidemiological research can be used to assess or detect possible risk factors for a disease; however, to identify risk factors with high accuracy rates, longitudinal studies are required. It is of vital importance to control for confounding factors, such as statistically significant differences in age distribution, which may be responsible for the observed changes in the prevalence of a disease rather than treatment changes. However, whether the result of such treatment changes impacts the occurrence of the targeted disease within the total population needs to be determined. An epidemiological study of this specific population is required to provide an evidence-based probability. Furthermore, epidemiological studies are used to evaluate the natural history of diseases and to ascertain the characteristics of a healthy population compared with patients affected by the disorder.

Incidence and prevalence of venous leg ulcers

To analyze and understand epidemiological papers, an understanding of the epidemiological terminology is essential (Table 4.1). The incidence of a disease or disorder is the number of patients that newly develop the disease or experience a health-related event specific to the disease during a particular time frame (usually 1 year). It is generally presented as a proportion of the total population. Prevalence refers to the total number of individuals with a certain disease or disorder within the population studied at a specific time period, usually expressed as a proportion or percentage of the population. The prevalence measurement includes both newly diseased and existing cases, in contrast to the

Table 4.1 Epidemiological terminology.

Incidence	Number of new cases per time unit and population—usually 1 year
Point prevalence	Proportion of individuals with a certain disease at any point in time—time period usually shorter than 3 months
Period prevalence	Proportion of individuals with a certain disease within a longer period of time—usually 1 year or more
Overall prevalence	Proportion of individuals that have had a certain disease—lifetime period = lifetime prevalence

incidence measurement, which only includes new cases. For chronic disorders, such as venous leg ulcers, the prevalence is generally much higher than the incidence, whereas acute illnesses (for example influenza) yield more equal proportions. There are different methods to determine the prevalence; the two most common methods are point prevalence and period prevalence. A *point prevalence* measurement involves a sample taken within a narrow time frame that is generally less than 3 months for chronic diseases. A *period prevalence* is used for longer time periods, usually 1 year or more. The *overall prevalence*, often referred to as lifetime prevalence, is used to determine the proportion of patients in a population who at some point in life had any disorder. In the case of venous leg ulcers, it includes individuals with a history of healed ulceration in addition to individuals with active ulceration.

Distribution of chronic venous insufficiency and its subgroups within the population

Today, many venous ulcers are caused by superficial and/or perforator vein insufficiency. Superficial venous insufficiency is very common in the general population and data from random population samples in Scotland (The Edinburg Vein Study) and Germany (The Bonn Study) showed that approximately one-third of participants had detectable superficial venous insufficiency, with no identified gender predilection.[2,3] According to the Clinical-Etiology-Anatomy-Pathophysiology (CEAP) classification system,[4,5] varicose veins are classified as at least class C2 and are commonly found in classes C3–C6; the latter classes are defined as chronic venous insufficiency (CVI). In The Bonn Study, participants older than 79 years were excluded, whereas in The Edinburgh Study, retired participants older than 65 years were excluded from the analysis. As a result, slightly higher CVI rates were reported from Germany, including 13.5%, 2.9%, and 0.7% for C3, C4, and C5–C6 CVI, respectively, than from Scotland, where the respective rates were 6.1%, 1.2%, and 0.6%. If the studies included older individuals, it would be reasonable to expect even higher rates, given that CVI prevalence increases with patient age. It is worth emphasizing that the abovementioned C4–C6 rates were based on a very small population sample size, which represents a major limitation for estimating the true point prevalence. In both studies, varicose vein disease progression was monitored prospectively, and approximately 4% progressed each year to a more severe disease stage.[6,7] To prevent one venous ulcer from developing, the estimated number of patients needed to treat in the C2 class is 100, based on data from Skaraborg studies.[8–13]

Defining venous leg ulcers

Patients or individuals in the general public may have a greater difficulty identifying a leg ulcer. Thus, providing the exact definition of a leg venous leg ulcer is extremely important. Many previous epidemiological studies did not provide exact definitions of the abovementioned disorders, which negatively affected the validity of such studies.[14] *Chronic leg ulcer* was defined in our Swedish studies[1,9–13,15–17] as follows: any ulcer below the knee level that did not heal within a 6-week period; other studies also used this definition.[18] To diagnose a venous ulcer, it is mandatory to document the presence of venous incompetence or obstruction using noninvasive or invasive diagnostic modalities, in addition to clinical signs. Using clinical signs alone, we showed a diagnostic accuracy of 75% for venous ulcers; the diagnostic accuracy was higher when additional investigations were used to confirm the diagnosis.[8] Hence, one out of four patients with venous ulcers is at risk of being misdiagnosed by relying on physical examination alone.

Leg ulcer prevalence

Epidemiological studies have shown a global occurrence of leg ulcers.[15,16] The most common way of estimating the impact of venous ulcers on a healthcare system is to assess the point prevalence of leg ulcers in a defined population in the healthcare system. This impact assessment has been undertaken through cross-sectional studies to identify all patients receiving professional care for leg ulcers at a specific point in time. The estimated point prevalence of chronic leg ulcers ranges from 0.1% to 0.3% of the total population with active ulcers at any time point.[15,16] In addition, some individuals with leg ulcers do not seek medical help; in Sweden, we found that the number of these individuals was as high as the number of people receiving professional care.[11,15] A later repeated population-based questionnaire study showed that the number of individuals who did not seek medical treatment had been reduced to approximately one-third.[17] The strategic steps to improve leg ulcer management, especially for venous ulcers, are shown in Table 4.2. In the initial study in Skaraborg County in Sweden, we found a point prevalence of approximately 0.6%, representing approximately 50,000 individuals with active leg ulcers in Sweden,[11] which at the time of study had a total population of approximately 9 million. As a leg ulcer is a chronic condition, healing is often followed by recurrences, which can occur even if the underlying causes of the ulcers are permanently treated.[13,15] For ulcers that are refractory to a given treatment, the diagnosis of underlying etiology and coexisting comorbidities should be reassessed. Thus, the total Swedish population with a history of leg ulcers is approximately three times higher (150,000 individuals) than the number of individuals with active leg ulcers; at the time of the study, 100,000 individuals were healed.[15]

Different studies can have different findings regarding the point prevalence of leg ulcer. The results are dependent on the research methodology used, identification of all patients with leg ulcers, correct ulcer diagnosis, quality of leg ulcer treatment in the area, and proportion of individuals who take care of their ulcers by themselves.[15,16] The true prevalence of venous ulcers and their impact on the health care system of any country cannot be accurately determined until an assessment is made of whether "self-care" is common. Epidemiological research is a valuable tool for improving leg ulcer management and assessing the effects of novel management strategies.

Etiologic spectrum of leg ulcers

Diagnosing ulcers is of vital importance, not only for epidemiological research reasons, but, more importantly, for appropriate ulcer treatment. Individualized treatment strategies have become increasingly important as there are now a variety of conservative treatment options, and several surgical or minimally invasive alternatives, which alone or in combination can provide the patient with the greatest benefit in terms of rapid ulcer healing and recurrence prevention. In some epidemiological studies, the etiological spectrum for leg ulcers has been specified.[1,8–12,17,18,20,21] Venous leg ulcers are the most common and constitute approximately 50% of all leg ulcers.[10,15] As approximately 20%–25% of all ulcers are located in the foot and are unlikely to be of venous etiology, a more accurate venous ulcers occurrence rate may be approximately 70%, if foot ulcers are excluded.[10,18] Approximately half of all venous ulcers are caused by superficial venous insufficiency

Table 4.2 Strategic steps for leg ulcer treatment improvement in Skaraborg County.[19]

(1) To facilitate early Doppler and CDU diagnostic assessments.
(2) Early use of superficial and perforator surgery.
(3) Creation of guidelines.
(4) To provide continuous repeated staff workshops regarding diagnosis, the use of Doppler, and compression treatments.
(5) The creation of organized care pathways.
(6) Multidisciplinary cooperation.
(7) Establishing a hospital-based leg ulcer center.

and/or perforator vein insufficiency[8,10,15,21]; moreover, there are reasons to believe that venous ulcers are becoming even more prevalent because of an aging population.

The most common etiologic diagnosis for foot ulcers is peripheral arterial disease, which is responsible for approximately half of all foot ulcers.[10] The detailed etiologic spectrum is complex, and many ulcers have mixed etiologies.[10,15] In more than one-third of patients, combinations of causative factors are likely to be responsible for ulcers. Mixed venous and arterial etiologies are commonly encountered combinations seen in approximately one-fifth of ulcer cases, with a venous etiology predominance.[8,10] Ulcers of multifactorial cause where venous incompetence may be involved, are common with no obviously dominant causative factor; therefore, they are more difficult to treat.[10,15,22] These ulcers are most often observed in the elderly. From our latest studies and other studies, the prevalence of multifactorial ulcers is apparently increasing, which poses new challenges to the healthcare system.[1,22]

Venous ulcer prevalence

Very few studies provide accurate data regarding venous ulcer prevalence, as only a few studies are based on imaging-confirmed venous insufficiency in contrast to the many studies assessing leg ulcers regardless of etiology.[15,16] One of the most common prevalence estimates is the overall prevalence of venous ulcers. Most studies have shown that approximately 1% of the adult population has a history of healed or active venous leg ulcers, and this estimate is consistent over the years in many countries (Table 4.3). Based on data from The Skaraborg Study in Sweden, we also found that an approximately equal proportion of participants had chronic lower limb ulceration of venous and nonvenous etiology. Thus, approximately 2% of the population can be expected to have a history of chronic lower limb ulceration.[11,15]

Only a relatively few point prevalence estimates regarding venous ulcers have been published (Table 4.3). In the latest studies, the diagnosis of venous ulcers was validated using noninvasive diagnostic modalities.[1,12,17,20,22] The lowest point prevalence of 0.024% was reported in a study from the United Kingdom.[22] However, the aforementioned point prevalence is based on a study with limitations in methodology related to the use of an extensive, time-consuming questionnaire, which likely biased the recruitment of patients negatively.[14] The highest prevalence was found in the two studies[11,25] that were based on large random population samples, and thus also included participants in a "self-care" group (patients who were not treated by healthcare professionals). In both studies, from Germany[25] and Sweden,[11] the point prevalence was 0.29%. The point prevalence based on patients who received professional care ranged from 0.05% to 0.20%. It seems reasonable to expect the point prevalence in the total population to be somewhere between 0.07% and 0.3% in most Western countries, although local variations are likely to exist (Table 4.3). Table 4.4 shows the initial weighted results from Skaraborg

Table 4.3 Prevalence estimates of venous leg ulcers.

Authors (publication year)	Country	Method	All known to healthcare	Noninvasive diagnosis	Prevalence % Overall	Prevalence % Point
Bobek et al. (1966)[23]	Czechoslovakia	Pop. study n = 15,060 Adults >15 yrs	No	No	1.0 adult pop	—
Widmer (1978)[24]	Switzerland	Selected sample n = 4529 industrial Workers 25–74 yrs	No	No	1.0 adult pop	—
Fischer (1981)[25]	West-Germany	Random pop. sample n = 4260 Adults 20–74 yrs	No	No	2.7 in sample 2.3 based on examined	0.44 in sample 0.29 based on examined
Nelzén et al. (1994)[8]	Sweden	Cross-sectional study Pop. 270.800 n = 387/827 ulcer pathology validated (randomly selected)	Yes	Yes Bidirectional Doppler arterial and venous	n.a.	0.16 total pop 0.22 adult pop
Baker et al. (1991)[18]	Australia	Cross-sectional study Pop. 238.000 n = 246/259 ulcer pathology validated	Yes	Yes Doppler + photoplethysmography	n.a.	0.06 total pop
Nelzén et al. (1996)[11]	Sweden	Random pop. sample n = 12,000 people 50–89 yrs	No	Yes Bidirectional Doppler arterial and venous	0.8 total pop 1.0 adult pop	0.29 total pop
Nelzén et al. (1996)[12]	Sweden	Selected sample n = 2785 industrial Workers 30–65 yrs	No	Yes Bidirectional Doppler arterial and venous	0.8 in sample	0.2 in sample

Continued

Table 4.3 Prevalence estimates of venous leg ulcers.—cont'd

Authors (publication year)	Country	Method	All known to healthcare	Noninvasive diagnosis	Prevalence % Overall	Prevalence % Point
Moffatt et al. (2004)[22]	United Kingdom	Cross-sectional study in health care Pop. 252,000 $n = 113$ ulcer pathology validated	Yes	Yes Doppler + photoplethysmography	n.a.	0.024 total pop
Forssgren et al. (2012)[1]	Sweden	Cross-sectional study Pop. 254,111 $n = 198/621$ ulcer pathology validated (randomly selected)		Yes Doppler + color Doppler ultrasound	n.a.	0.09 total pop
Forssgren et al. (2015)[17]	Sweden	Random pop sample $n = 10,000$ 30–89 yrs Pop. 255,042		Yes Doppler + color Doppler ultrasound	0.5 among 30–89 yrs	0.18 total pop
Nelzén et al. (2018)[20]	Sweden	Cross-sectional study Pop. 259,914 $n = 231/512$ ulcer pat. Validated			n.a.	0.05 total pop

Adult pop. = population above the age of 15; n.a. = not assessed; Pop: population.

County (with a population of 270,800), based on our results from three separate studies.[8,11,12] Given that our data from the initial studies were analyzed before the initiation of a program to reduce the occurrence rate of ulcerations, the estimated prevalence rates correlated more accurately with the prevalence rates in regions without established, structured care for patients with CVI.[11,12] The point prevalence reduction that was demonstrated in our follow-up epidemiologic study was achieved due to a significant change in management strategy, including early duplex ultrasonography diagnostic evaluation to select patients that were likely to benefit from surgical or endovenous CVI treatment.[17]

Incidence of venous ulcers

There is little information on the incidence of venous ulcers. In one study from the United States,[26] the yearly incidence of venous ulcers remained unchanged during a 25-year period; the overall age- and sex-adjusted incidence was 18 per 100,000 person-years. In a study conducted in New Zealand,[27] the annual cumulative incidence rate of lower limb ulceration was 32 per 100,000 individuals. The study did not show the proportion of patients with venous ulcers; however, a reasonable estimate would be 50%, which would be consistent with the incidence rate estimated in the study from the United States. Based on our studies from Skaraborg, we retrospectively estimated the yearly incidence to be one-tenth of the point prevalence, giving an incidence of 16 per 100,000 individuals based on patients receiving professional care, and a total incidence of approximately 30 per 100,000 if individuals in a "self-care" group are

Table 4.4 Combined baseline prevalence estimates of venous ulcers from Skaraborg County.[15]

A. Point prevalence of open venous ulcers		
	Managed by healthcare professionals	"Self-care" included
Total population	0.16	0.3
Adult population (>15 years)	0.22	0.4
Retired population (>65 years)	0.76	1.0
B. Overall prevalence of healed and open venous ulcers		
	Managed by healthcare professionals	"Self-care" included
Total population	0.5	1.0
Adult population (>15 years)	0.6	1.3
Retired population (>65 years)	2.3	3.0

Rates given as percentages

included.[15] In summary, the yearly incidence of venous ulcers is likely to range between 15 and 30 per 100,000 individuals.

Age and sex distribution

Mostly elderly patients have venous ulcers that are treated by healthcare professionals. Studies have shown that the median ages of these patients were approximately 75 and 77 years in Australia[18] and Sweden,[8] respectively. Apparently, ulcer prevalence is higher in females than in males, albeit only after 65 years.[8,18] The female predilection for venous ulcers was, however, overestimated in the past because the observed frequencies were not adjusted for age.[8] Female longevity has been the cause of such overestimation.[15] The age-adjusted male-to-female ratio in the baseline Skaraborg Study was 1:1.6.[11] Fig. 4.2 shows the age- and sex-specific point prevalence of venous ulcers.[8] Only a slight female predominance was reported in a study conducted in Australia.[18] Although there is a paucity of data concerning individuals with "self-care" for sex distribution assessment, it is clear that the age distribution is shifted toward individuals of working age.[11,12]

Two population-based studies that have thoroughly assessed venous disease within random population samples have both shown surprisingly similar occurrences for varicose veins in males and females. In The Edinburgh Vein Study from Scotland,[2] there were four times more males with a history of venous ulcers, although the numbers were small and somewhat uncertain. Nevertheless, these numbers question female predilection. In The Bonn Study from Germany,[3] the researchers found equal rates of ulceration occurrence between genders. Moreover, in a previous study conducted in San Diego, the United States,[28] varicose veins were found to be more common among females, whereas trophic skin changes (CEAP class C4) were more common among males. Thus, although it is obvious that increasing age is a risk factor for venous ulcers, it remains uncertain whether the female sex is a risk factor. Based on our experience, males tend to be

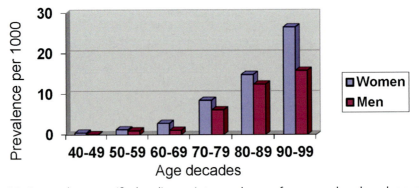

Figure 4.2 Age and sex-specific baseline point prevalence of venous ulcer based on patients receiving professional care in Skaraborg.[8]

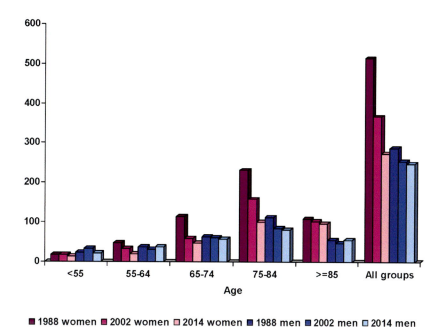

■ 1988 women ■ 2002 women □ 2014 women ■ 1988 men ■ 2002 men □ 2014 men

Figure 4.3 Age and sex distributions representing the number of patients with lower limb ulceration in the 3 cross-sectional surveys performed in Skaraborg County Sweden.

more ashamed of ulcers, and thus may avoid seeking treatment from the health care system. This leads to an overrepresentation of females in studies that included patients treated by health care professionals, not only because of female longevity.[15] Thus, the female sex is probably not a risk factor for venous leg ulcers, as supported by the above-cited studies. The abovementioned statement is also supported by results from The Skaraborg Leg Ulcer Study in 2002,[1] where the numbers of men and women were almost equal after correcting for age, and also according to the preliminary findings of our latest epidemiological study that started in 2014. One other reason for the increased prevalence of venous ulcers among male patients we believe is that males now have a greater tendency to seek professional help for leg ulcers allowing more accurate data capture.[17] Fig. 4.3 shows the reduction in the number of patients with lower limb ulceration in most age intervals and the diminishing female predominance, which was high in the initial study most likely due to the unequal patient sampling.

Socioeconomic aspects

A review published in 2000 reported that there were high costs involved in managing venous ulcers.[29] The major fraction of the total cost was nursing costs, including wages and staff travel costs, representing 50%–75% of the total cost; the costs of dressings and

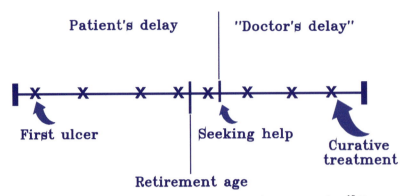

Figure 4.4 The typical scenario and history for a patient with a venous ulcer.[15] X represents ulcer episodes.

compression materials accounted for approximately 10%—30%. In addition, indirect costs related to the costs for the patient, such as income loss, accounted for approximately 15%—20% of the total cost. The relative cost proportions are likely to be similar today.[30] The most recent estimates regarding total yearly costs for managing a patient with a venous ulcer were made in Australia and the United Kingdom with very similar results. In the United Kingdom, the National Health Service mean cost of wound care over a 12-month period was estimated to be £7600 for a venous ulcer (approximately $10,740).[29] However, the cost was higher (£13,500 [$14,070]) for an ulcer that remained unhealed; the cost for a healed ulcer was £3000. In Australia, the mean cost for guideline-based care for unhealed ulcers was AU$15,355 ($11,589) per year[30]; however, in the 1990s considerably higher costs were estimated in the United States.[28] Moreover, according to a retrospective US study published in 2014, the annual cost estimate for an unhealed ulcer has increased to $33,907.[31] The mean cost of treating active venous ulcers in 84 patients from 2005 to 2011 was $15,732. This cost was estimated using data from 50 patients (60%) whose ulcers were healed without recurrence (mean healing time 122 days; range 6—379 days). Surgical interventions did not significantly increase the health care costs compared to conservative treatment ($19,503 vs. $12,304), although it resulted in a significant reduction in the venous ulcer recurrence rate when compared to conservative treatment (5% vs. 34%).

Low socioeconomic status affects the outcome of patients with venous ulcers, possibly due to reduced access to optimal management strategies, including surgical interventions, as has been suggested in studies from the United Kingdom and Brazil.[32,33] The impact of race on advanced CVI has been noted in the United States.[34,35] African-Americans present with more advanced CVI at an earlier age than Caucasians,[34] and ethnically nonwhite participants were more prevalent among patients with venous ulcers in northern Brazil.[33] In Sweden, we found that socioeconomic factors did not play a major role in

CVI occurrence.[15] Most likely, higher prevalence rates can be expected in areas with lower socioeconomic status in the world.

Venous ulcer natural history

Most venous ulcers develop for the first time in middle age[8]; however, as shown, patients tend not to seek professional help until after retirement (Fig. 4.4). Consequently, there are many young and middle-aged individuals in the community with venous ulcers that are not receiving medical care.[15] Many of these individuals (50%—75%) only have superficial and/or perforator venous incompetence, and thus can benefit from the elimination of venous reflux either using surgery or minimally invasive endovenous procedures (including endovenous laser ablation [EVLA] and radiofrequency ablation [RFA]).[11,12] In The Skövde Study,[12] where only people below retirement age (30—65 years) were included, only 25% of patients with venous ulcers had detectable deep venous incompetence (DVI), in contrast to The Skaraborg Study,[8] where the corresponding rate was 60%. This may indicate that venous ulcer caused by superficial and/or perforating vein reflux is more common among younger individuals, and that superficial and perforator venous insufficiency may progress with time, if left untreated, to also involve the deep veins.[36,37] Unfortunately, these younger patients with first-time venous ulcers rarely seek medical help, at an early stage, when their venous incompetence could be treated by surgery, EVLA, or RFA (Fig. 4.4). Fortunately, our repeat random population-based study in 2005 showed a lower number of individuals who were

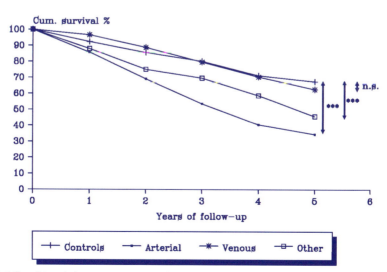

Figure 4.5 Life table of the 5-year survival for patients with leg ulcers of various origins. Wilcoxon (Gehan) test ***$P < .001$, n.s. $P = .53$.[8,14]

managing their ulcers using "self-care," with general population rates of approximately 30% compared with approximately 50% in 1991.[17] Sensitizing people about the high efficacy rates of surgery or endovenous ablation techniques and the low associated complication rates can potentially attract patients with ulcers and varicose veins to seek professional help earlier.

The chronicity of venous leg ulcers is empirically well known. Venous ulcers are more difficult to treat and have a greater recurrence tendency than other types of chronic ulcers.[8] The median duration of ulceration was 26 weeks in a study from Australia[18]; moreover, in Sweden, 54% of patients had ulcers evolving for more than 1 year.[8] Ulcers caused by DVI are the most difficult to treat,[13,35,38,39] with 64% of these ulcers having a duration of evolution of more than 1 year.[8] More than 70% of active venous ulcers are recurrent.[8,18] Ulcers caused by varicose veins tend to recur, as do ulcers caused by deep vein involvement.[8,13] Based on our previously mentioned baseline study in Sweden, 61% of patients with venous leg ulcers had their first ulcer episode before retirement, and 37% before 50 years old.[8] In Skaraborg, the median duration of the ulcer history (time since the first ulcer developed) was long: 13.5 years compared with 2.5 years for patients with ulcers from other causes.[8]

Obesity

It is well known that obesity is quite common among patients with venous ulcers, and it is currently debated whether obesity is a risk factor for venous disease and ulcers. Obesity was found to be a risk factor of CVI in The Edinburgh Study, which was a prospective follow-up study.[7] In the Skaraborg Study, we found that patients with venous ulcers were more obese than patients with nonvenous ulcers; moreover, the former group of patients were more obese than those in a matched Swedish population.[8] However, when we looked at the same problem associated with obesity in a large population sample, also including individuals in "self-care" group, we found that patients with ulcers were not more obese than age- and sex-matched controls.[15] This finding indicates that the venous ulcer potentially causes obesity in some patients, rather than vice versa. Other studies have also found that obesity is common among patients with venous ulcers and is likely to be a risk factor for venous ulceration.[34,40,41] Based on the preliminary unpublished data of our ongoing cross-sectional survey that was initiated in 2014, the median BMI among patients with venous ulcers was approximately 28 kg/m^2. A high BMI is likely to cause an increased intraabdominal pressure that negatively affects venous return, similar to the mechanism of deep vein obstruction.[41]

History of DVT

Following a clinically confirmed DVT, about 5% of patients will develop leg ulcers,[42,43] suggesting that a history of DVT is not common among patients with venous leg ulcers. In a study from Sweden, only 37% of patients with venous leg ulcers had a DVT history,[6] and a study from Australia showed a lower prevalence (17%).[16] A history of DVT was more frequently found among patients with deep venous insufficiency than among those with superficial or perforating vein incompetence alone (54 vs. 14%, $P < .0001$).[8] When all potential factors predisposing to DVT (such as pregnancy, limb fracture, major trauma, and general anesthesia) were analyzed, only 4% of Australian patients with leg ulcers had no evidence of a predisposing factor.[16] However, it seems likely that postthrombotic ulcers comprise less than 50% of all venous ulcers, with a true rate of approximately 25%–50%.[15] The latter is supported by a US study[28] that reported that only 22% of legs with trophic changes showed signs of deep vein reflux. In the second cross-sectional study in Skaraborg County, in 2002, 27% of patients with venous ulcers showed detectable postthrombotic changes on color Doppler ultrasonography. Nevertheless, four patients had received no DVT treatment, and only 19% of all patients with ulceration had a history of DVT in either leg.[1] The decreasing frequency of postthrombotic ulcer occurrence seems to be a trend, probably because of better DVT treatment and frequent DVT prophylaxis use. Very few patients who were diagnosed with DVT before the introduction of heparin treatment are still alive today; this also partially explains the abovementioned trend in the number of patients with superficial vein reflux. Our preliminary data from an ongoing cross-sectional study in Skaraborg County showed that only 13% of responders reported a DVT history.

Healing and survival

Ulcers due to a previous DVT are more difficult to heal.[35] The healing prognosis for patients in the community is rather poor, with an expected 12-week healing duration achieved in less than 30% of cases.[44] In Scotland, a nationwide randomized study[45] failed to detect any benefit from a nurse-training program focused on compression treatment of venous ulcers. The 12-week healing rate remained virtually unchanged. In only approximately 28% of patients, ulcers healed during a follow-up period of 21 months.

As part of the Skaraborg Study, we performed a 5-year follow-up of all patients to assess healing and survival rates.[13] Venous ulcers had a poor outcome, as only 54% had healed; 44% had no recurrence, and 10% recurred during the study period. At the time of the study, most venous ulcers were treated using conservative compression bandaging applied by district nurses in the community. Patients with deep vein incompetence had a significantly worse outcome than those with superficial venous incompetence alone. Regarding patient survival, we found that patients with venous ulcers had a

5-year survival rate that was equal to that of an age- and sex-matched control population (Fig. 4.5). Today, healing rates have improved considerably due to the use of more active intervention strategies for superficial and perforating vein reflux and more structured management options.[19] Based on preliminary data from a 6-year follow-up of patients diagnosed with venous ulcer during the last cross-sectional study in Skaraborg 2014/2015, 97% of the ulcers healed, 41% had no recurrence, and 69% healed at the end of follow-up. No patient underwent amputation during this follow-up for a venous ulcer, unlike the previous long-term follow-up in the 1990s where 4 patients (5%) underwent amputation.[13]

Future trends

A more comprehensive multidisciplinary approach in Skaraborg, Sweden, including interventions such as the wide use of varicose vein surgery,[19] has reduced the point prevalence of venous ulcers, and considerably shifted the etiologic spectrum, according to Skaraborg epidemiological data (Fig. 4.6). These data resulted from two repeat cross-sectional studies in 2002[1] and 2014/2015 comparing the data with those of a baseline study performed in Skaraborg area in 1988[4–6]; in 2002 and 2014 the proportions of patients with venous ulcers reduced by 46% and 71%, respectively.[20] The point prevalence for venous ulcers was reduced from 0.16% to 0.09% in 2002 and further to 0.05% in 2014.[20] Data from a repeat large-scale study (10,000 participants) validating the population-based questionnaire, in the same area in 2005, also showed similar findings.[17]

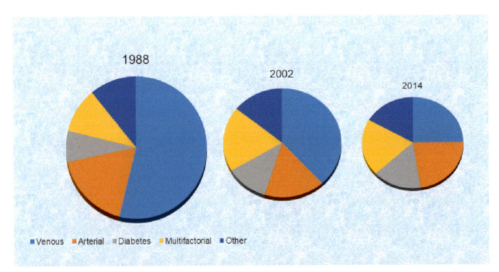

Figure 4.6 Reduction in total numbers and the relative proportions of the major etiologies for lower limb ulceration based on the 3 cross-sectional studies in Skaraborg County Sweden.

Apparently, males now, to a greater extent, seek professional help; moreover, in 2002, patients with varicose ulcers outnumbered those having ulcers with deep vein incompetence, which made it technically feasible to increasingly use surgical treatment for the correction of underlying venous hemodynamic disorders. As a result, a further reduction of the venous ulcers prevalence was achieved until 2014, although patients with ulcers caused by deep vein incompetence again slightly outnumbered those having the remaining ulcers of superficial venous etiology. It may be difficult to further reduce the point prevalence in Skaraborg because the remaining ulcers probably consist mainly of post-thrombotic, deep venous incompetence-induced, or newly developed varicose ulcers. It seems clear that a paradigm shift in the ulcer management strategy, including superficial and perforator vein interventions, results in improved healing and a lower ulcer recurrence rate.

Leg ulcers are a common healthcare problem, and their prevalence increases with age. Due to the expected increasing proportion of elderly people in most countries, an increased prevalence of chronic venous leg ulcers is expected, unless significant changes in management are undertaken. The continuous use of conservative compressive regimes to treat the majority of patients most likely results in the cure of a few patients; further, the majority of these patients experience recurrent ulceration, and some ulcers never heal. We have strong evidence that superficial venous surgery significantly reduces the risk of venous ulcer recurrence within 1 year postoperatively.[46] Furthermore, an early endovenous intervention could result in faster healing.[47] Untreated superficial venous reflux is a known risk factor for venous ulcer recurrence,[48] which is consistent with our experience in Skaraborg, where we reduced the prevalence of venous ulcers by 71% during a 25-year period, thereby proving that venous ulcer epidemiology can be changed (Fig. 4.6). These results were achieved using open surgery; endovenous thermal treatment modalities can yield similar results, although they have not yet been tested using long-term randomized controlled trials in patients with venous ulcers. Therefore, with the increased leg ulcer diagnosis rate and personalized treatment use based on individual underlying etiologies and hemodynamic characteristics (including superficial, perforator, and possibly deep vein interventions), it is likely that the scenario can be substantially improved, and the prevalence of leg ulcers can be reduced worldwide, not only in certain geographical parts of Sweden.

References

1. Forssgren A, Nelzén O. Changes in the aetiological spectrum of leg ulcers after a broad-scale intervention in a defined geographical population in Sweden. *Eur J Vasc Endovasc Surg.* 2012;44:498–503.
2. Evans CJ, Fowkes FGR, Ruckley CV, Lee AJ. Prevalence of varicose veins and chronic venous insufficiency in men and women in the general population: Edinburgh Vein Study. *J Epidemiol Community Health.* 1999;53:149–153.

3. Maurins U, Hoffman BH, Lösch C, Jöckel KH, Rabe E, Pannier F. Distribution and prevalence of reflux in the superficial and deep venous system in the general population - results from the Bonn Vein Study Germany. *J Vasc Surg*. 2008;48:680–687.
4. Lurie F, Passman M, Meisner M, et al. The 2020 update of the CEAP classification system and reporting standards. *J Vasc Surg Venous Lymphat Disord*. May 2020;8(3):342–352. https://doi.org/10.1016/j.jvsv.2019.12.075. Epub 2020 Feb 27. Erratum in: J Vasc Surg Venous Lymphat Disord. 2021 Jan; 9(1):288. PMID: 32113854.
5. Eklöf B, Rutherford RB, Bergan JJ, et al. American venous forum international ad hoc committee for revision of the CEAP classification. Revision of the CEAP classification for chronic venous disorders: consensus statement. *J Vasc Surg*. December 2004;40(6):1248–1252. https://doi.org/10.1016/j.jvs.2004.09.027. PMID: 15622385.
6. Pannier F, Rabe E. The relevance of the natural history of varicose veins and refunded care. *Phlebology*. 2012;27(Suppl1):23–26.
7. Robertsson L, Lee AJ, Evans CJ, et al. Incidence of chronic venous disease in the Edinburgh vein study. *J Vasc Surg:Venous and Lym Dis*. 2013;1:59–67.
8. Nelzén O, Bergqvist D, Lindhagen A. Venous and non-venous leg ulcers: clinical history and appearance in a population study. *Br J Surg*. 1994;81:182–187.
9. Nelzén O, Bergqvist D, Hallböök T, Lindhagen A. Chronic leg ulcers: an underestimated problem in primary health care among elderly patients. *J Epidemiol Community Health*. 1991;45:184–187.
10. Nelzén O, Bergqvist D, Lindhagen A. Leg ulcer etiology—a cross-sectional population study. *J Vasc Surg*. 1991;14:557–564.
11. Nelzén O, Bergqvist D, Lindhagen A. The prevalence of chronic lower-limb ulceration has been underestimated: results of a validated population questionnaire. *Br J Surg*. 1996;83:255–258.
12. Nelzén O, Bergqvist D, Fransson I, Lindhagen A. Prevalence and aetiology of leg ulcers in a defined population of industrial workers. *Phlebology*. 1996;11:50–54.
13. Nelzén O, Bergqvist D, Lindhagen A. Long term prognosis for patients with chronic leg ulcers: a prospective cohort study. *Eur J Vasc Endovasc Surg*. 1997;13:500–508.
14. Nelzén O. Prevalence of venous leg ulcer: the importance of the data collection method. *Phlebolymphology*. 2008;15:143–150.
15. Nelzén O. Patients with chronic leg ulcers: aspects of epidemiology, aetiology, clinical history, prognosis and choice of treatment. Comprehensive Summaries of Uppsala dissertations from the Faculty of Medicine 664. *Uppsala: Acta Univ Upsal*. 1997;1:1–88.
16. Baker SR, Stacey MC, Jopp-McKay AG, Hoskin SE, Thompson PJ. Epidemiology of chronic venous ulcers. *Br J Surg*. 1991;78:864–867.
17. Forssgren A, Nelzén O. A repeat validated population questionnaire of a defined Swedish population verifies reduction in leg ulcer prevalence over time. *Acta Derm Venereol*. 2015;95:725–729.
18. Graham ID, Harrison MB, Nelson EA, Lorimer K, Fisher A. Prevalence of lower-limb ulceration:a systematic review of prevalence studies. *Adv Skin Wound Care*. 2003;16:305–316.
19. Nelzén O. Fifty percent reduction in venous ulcer prevalence is achievable - Swedish experience. *J Vasc Surg*. November 2010;52(5 Suppl):39S–44S. https://doi.org/10.1016/j.jvs.2010.05.122. PMID: 21069938.
20. Guarnera G, Zamboni P, Nelzén O, Manello F, Andriessen A. Pending questions in venous ulcers management. Report from a symposium of the world union of wound healing societies international congress. *Veins and Lymphatics*. 2020;9:62–71.
21. Cornwall JV, Dore CJ, Lewis JD. Leg ulcers:epidemiology and aetiology. *Br J Surg*. 1986;73:693–696.
22. Moffatt CJ, Franks PJ, Doherty DC, Martin R, Blewett R, Ross F. Prevalence of leg ulceration in a London population. *Q J Med*. 2004;97:431–437.
23. Bobek K, Cajzl L, Cepelak V, Slaisova V, Opatzny K, Barcal R. Étude de la fréquence des maladies phlebologiques et de l'influence de quelques facteurs étiologiques. *Phlebologie*. 1966;19:227–230.
24. Widmer LK. *Peripheral Venous Disorders. Basle Study III*. Bern: Hans Huber. 1978.
25. Fisher H. *Venenleiden: Eine Repräsentative Untersuchung in der Bevölkerung der Bundesrepublk Deutschland (Tübinger-studie)*. München: Urban Schwartsenberg. 1981.

26. Heit JA, Rooke TW, Silverstein MD, et al. Trends in the incidence of venous stasis syndrome and venous ulcer: a 25-year population-based study. *J Vasc Surg*. 2001;33:1022−1027.
27. Walker N, Rodgers A, Birchall N, Norton R, MacMahon S. The occurrence of leg ulcers in Auckland: results of a population-based study. *New Zeeland Med J*. 2002;1151:159−162.
28. Criqui MH, Jamosmos M, Fronek A, et al. Chronic venous disease in an ethnically diverse population: the San Diego Population Study. *Am J Epidemiol*. 2003;158:448−456.
29. Nelzén O. Leg ulcers: economic aspects. *Phlebology*. 2000;15:110−114.
30. Guest JF, Fuller GW, Vowden P. Venous leg ulcer management in clinical practice in the UK: costs and outcomes. *Int Wound J*. 2018;15:29−37.
31. Ma H, O'Donnell Jr TF, Rosen NA, Iafrati MD. The real cost of treating venous ulcers in a contemporary vascular practice. *J Vasc Surg: Venous Lym Dis*. October 2014;2(4):355−361. https://doi.org/10.1016/j.jvsv.2014.04.006. Epub 2014 Jun 24. PMID: 26993537.
32. Petherick ES, Cullum NA, Pickett KE. Investigation of the effect of deprivation on the burden and manageneent of venous leg ulcers: a cohort study using the thin database. PLoS One 8(3): e58948. DOI: 10.1371/journal.pone.0058948.
33. de Sousa EM, Yoshida WB, de Melo VA, Aragao JA, Oliveira de. Ulcer due to chronic venous disease: a sociodemographic study in Northeastern Brazil. *Ann Vasc Surg*. 2013;27:571−576.
34. Dua A, Desai SS, Heller JA. The impact of race on advanced chronic venous insufficiency. *Ann Vasc Surg*. 2016;34:152−156.
35. Melikian R, ODonnel TF, Suarez L, Ifrati MD. Risk factors associated with the venous leg ulcer that fails to heal after one year of treatment. *J Vasc Surg: Venous and Lym Dis*. 2019;7:98−105.
36. Walsh JC, Bergan JJ, Beeman S, Comer TP. Femoral venous reflux abolished by greater saphenous vein stripping. *Ann Vasc Surg*. 1994;8:566−570.
37. Magnusson M, Nelzén O, Volkmann R. Leg ulcer recurrence and its risk factors: a duple ultrasound study before and after vein surgery. *Eur J Vasc Endovasc Surg*. 2006;32:453−461.
38. Skene AI, Smith JM, Dore CJ, Charlett A, Lewis JD. Venous leg ulcers: a prognostic index to predict time to healing. *Br Med J*. 1992;305:1119−1121.
39. Vlajinac H, Marinkovic J, Maksimovic M, Radak D. Factors related to venous ulceration: a cross-sectional study. *Angiology*. 2014;65:824−830.
40. Barnsbee L, Cheng Q, Tulleners R, Lee X, Brain D, Pacella R. Measuring costs and quality of life for venous leg ulcers. *Int Wound J*. 2019;16:112−121.
41. Van Rij AM, De Alwis CS, Jiang P, et al. Obesity and impaired venous function. *Eur J Vasc Endovasc Surg*. 2008;35:739−744.
42. Milne AA, Ruckley CV. The clinical course of patients following extensive deep venoue thrombosis. *Eur J Vasc Surg*. 1994;8:56−59.
43. Saarinen J, Sisto T, Laurikka J, Salenius J-P, Tarkka M. Late sequelae of acute deep venous thrombosis: evaluation five and ten years after. *Phlebology*. 1995;10:106−109.
44. Simon DA, Freak L, Kinsella A, et al. Community leg ulcer clinics: a comparative study in two health authorities. *BMJ*. 1996;312:1648. https://doi.org/10.1136/bmj.312.7047.1648.
45. Scottish Leg Ulcer Trial Participants. Effect of a national community intervention program on healing rates of chronic leg ulcer: randomized controlled trial. *Phlebology*. 2002;17:47−53.
46. Barwell JR, Davies CE, Deacon J, et al. Comparison of surgery and compression with compression alone in chronic venous ulceration (ESCHAR study): randomized controlled trial. *Lancet*. 2004;363:1854−1859.
47. Gohel MS, Heatly F, Liu XB, et al, Davies AH for the EVRA Trial Investigators. A randomized trial of early endovenous ablation in venous ulceration. *N Engl J Med*. 2018;378:2105−2114.
48. Gohel MS, Taylor M, Earnshaw JJ, Heather BP, Poskitt KR, Whyman MR. Risk factors for delayed healing and recurrence of chronic venous leg ulcers. An analysis of 1324 legs. *Eur J Vasc Endovasc Surg*. 2005;29:74−77.

SECTION 2

Clinical evaluation and diagnostic modalities

CHAPTER 5

Initial clinical evaluation in patients with chronic venous insufficiency

Yana Etkin[1] and Ruth L. Bush[2]

[1]Division of Vascular and Endovascular Surgery, Zucker School of Medicine at Hofstra/Northwell, Hempstead, NY, United States; [2]University of Houston College of Medicine, Houston, TX, United States

Introduction

Venous leg ulcers (VLUs) represent the most advanced stage of chronic venous insufficiency (CVI) and occur in 0.3% of the adult population in Western countries and the prevalence of active and healed ulcers combined is about 1%.[1,2] They account for up to 80% of lower extremity ulcerations, and are often recurrent or nonhealing for weeks to many years.[3,4] Venous ulcers are associated with an annual loss of about two million working days, the cost of treatment is approximately $3 billion per year in the United States, and CVI is among the 10 most common reasons for a person to seek medical care.[5,6]

It is critical for healthcare professionals to perform a thorough clinical evaluation as the initial step in the management of patients with this very common pathology. This will allow for an accurate diagnosis and avoid delays in treatment. Additionally, timely and accurate diagnosis can reduce the risk of severe complications such as cellulitis, osteomyelitis, or malignant deterioration.[7] The main goal of the clinical evaluation of patients with lower extremity CVI that has progressed to ulceration is to correlate symptoms with a physical exam and radiologic findings to differentiate venous disease as the main underlying pathology from the other etiologies such as peripheral arterial disease (PAD), diabetes mellitus (DM), hematologic, autoimmune and neuropathic disorders. The diagnosis of venous ulcers is mainly clinical but inconsistencies and overlap between symptoms and clinical signs may require additional diagnostic studies such as ultrasonography, plethysmography, or venography. CVI and associated venous hypertension resulting in the pooling of blood in the lower extremity venous circulation triggers an inflammatory cascade, capillary and endothelial damage, platelet aggregation, and intracellular edema,[1,3,8] thus, contributing to the cycle of VLU development and impaired wound healing. Correctly examining a VLU and performing a detailed history will advance understanding of the ulcer pathogenesis and personalize the management plan.[8]

The purpose of this chapter is to provide step-by-step clinical guidelines to healthcare professionals on how to perform the proper initial evaluation of patients with a lower extremity ulcer including a thorough medical history with an assessment of risk factors for

VLU and contributing conditions, physical examination, deep and superficial venous Duplex ultrasound, determination of additional appropriate imaging modalities, classification into the Clinical-Etiology-Anatomy-Pathophysiology (CEAP) system,[9] and assignment of a Venous Clinical Severity Score (VCSS).[10]

Medical history

Clinical evaluation of patients with venous ulcerations starts with a thorough medical history to establish the etiology of CVI, to gauge the severity of the disease, and to determine its impact on the patient's quality of life. However, physicians and other healthcare professionals must remember that VLUs do not occur simply by themselves. The primary etiology must be elucidated but leg ulcerations are usually multifactorial in nature. Other coexisting factors, such as DM and PAD (see Chapter 10) must be taken into consideration in contributing to nonhealing along with the principal cause of the ulcer.

Risk factors

It is important to assess common risk factors associated with CVI and VLUs including obesity, family history of CVI, advanced age, female sex, number of pregnancies, history of leg trauma with long-term immobilization, impaired daily immobility and ineffective pumping of the calf muscle, history of deep venous thrombosis (DVT) or phlebitis, and occupation.[8,11] A history of DVT or pulmonary embolism is critical to elicit. If the patient is still on anticoagulation it will be necessary to investigate indications for continuing anticoagulant medication. Particular attention should be paid to modifiable risk factors such as obesity, sedentariness, and/or an occupation that requires prolonged standing or sitting. It is important to pay attention to risk factors that might make another diagnosis more likely such as PAD, lymphatic obstruction, trauma, vasculitis, or infection.

Differential diagnosis of lower extremity ulcerations

Obtaining a standardized and detailed history as detailed in Society for Vascular Surgery/American Venous Forum (SVS/AVF) practice guidelines allows the clinician to differentiate venous ulcerations from other underlying pathologies that may be present.[12] A population-based study that included over 30,000 patients from Germany with chronic leg ulcers demonstrated the following etiologies: 47.6% venous, 14.5% arterial, and 17.6% mixed arterial and venous. Other less frequent causes included vasculitis (5.1%), exogenous factors (3.8%), pyoderma gangrenosum (3.0%), infection (1.4%), neoplasm (1.1%), calciphylaxis (1.1%), and drug induced (1.1%).[12] As mentioned earlier, there are several medical conditions that may cause leg ulcers and/or interfere with healing such as PAD, lymphedema, peripheral neuropathy, metabolic, hematologic such as sickle cell disease, autoimmune diseases, and skin disorders Table 5.1. Chronic infectious diseases such as hepatitis B or C can also be contributory. Evaluating for systemic symptoms,

Table 5.1 Differential diagnosis of lower extremity ulcerations.

Ulcer etiology	Pathophysiology
Venous disease	CVI, postthrombotic syndrome
Arterial disease	PAD, thromboangiitis obliterans, aneurysms, hypertensive arteriopathy
Lymphatic	Lymphedema
Microangiopathy	Diabetes mellitus, livedoid vasculopathy
Peripheral neuropathy	Diabetes mellitus, alcohol, medication, hereditary
Central neuropathy	Myelodysplasia, spina bifida, poliomyelitis, multiple sclerosis
Metabolic	Diabetes mellitus, gout, gaucher disease, amyloidosis, calciphylaxis, porphyria, hyperhomocysteinemia
Hematologic	Sickle cell anemia, thalassemia, polycythemia vera, leukemia, thrombocythemia, lymphoma, myelodysplastic disorder, disorders of coagulation factors, or fibrinolysis factors.
Autoimmune	Vasculitis, rheumatoid arthritis, polyarteritis nodosa, wegener granulomatosis, churgstraus syndrome, systemic lupus erythematosus, sjogren syndrome, scleroderma, behcet disease, cryoglobulinemia
Exogenous	Heat, cold, pressure, radiation, chemical, allergens, trauma
Neoplasia	Basal cell carcinoma, squamous cell carcinoma, melanoma, angiosarcoma, cutaneous lymphoma, papillomatosis cutis carcinoides, keratoacanthoma
Infection	Bacterial: furuncles, ecthyma, mycobacterioses, syphilis, erysipelas, anthrax, diphtheria, chronic vegetative pyoderma, tropical ulcer Viral: herpes, variola, cytomegaly Fungal: sporotrichosis, histoplasmosis, blastomycosis, coccidioidomycosis Protozoal: leishmaniasis
Medication	Hydroxyurea, leflunomide, methotrexate, halogens, vaccinations, ergotamine, infiltration of cytostatic agents
Skin disorders	Pyoderma gangrenosum necrobiosis lipoidica, sarcoidosis, perforating dermatosis, langerhans cell histiocytosis, populous maligna atrophicans, bullous skin disorders.

Modified from O'Donnell TF, Jr., Passman MA, Marston WA et al. Management of venous leg ulcers: clinical practice guidelines of the Society for Vascular Surgery and the American Venous Forum. *J Vasc Surg.* 2014;60(2 Suppl):3s–59s.

such as generalized muscle or joint pain, purpura, or petechiae, is important to be able to diagnose VLU correctly, identify contributory conditions, and treat the patient holistically.

Ulcers that have underlying venous pathophysiology are typically very painful. The location and appearance of the ulceration can also be diagnostic. VLUs are more likely in the circumferential region (so-called gaiter distribution) at the level of the medial and/or lateral malleolus. The location of VLUs correlates with reflux distribution, where Great Saphenous Vein (GSV) reflux is likely to cause medial ulcerations while Small

Saphenous Vein (SSV) reflux leads to ulcerations over the lateral malleolus. It is also important to note whether the ulcer is recurrent, how long it has been present, size and morphologic characteristics of the wound, and what were other locations of ulcerations if the patient has been previously treated.

Symptoms of CVI

For a leg ulcer to be classified as a VLU, clinical manifestations of CVI should be present. Most patients have at least one subjective symptom along with the ulcer at the time of evaluation. CVI may present with a wide range of symptoms including lower extremity pain, swelling, tiredness, heaviness, varicose veins, and skin changes such as hyperpigmentation, lipodermatosclerosis, and ulcerations. In one study of patients from 24 Italian cities, the complaint of tired and heavy legs was reported by 67% of men and 79% of women, pain was reported by 50% of men and 56% of women, and swollen legs by 39% of men and 57% of women.[13]

Pain due to CVI is usually worse with prolonged standing or sitting with feet in a dependent position and improves with leg elevation and walking. This is distinctively different from the pain associated with PAD, which tends to be worse with walking in patients with mild/moderate PAD or with limb elevation in patients with moderate/severe PAD. Pain due to CVI typically becomes worse throughout the day. If the pain is worse in the morning before the patient has stood for periods of time, the clinician should consider other etiologies, such as arthritis.

Other symptoms include limb aching, fatigue, skin, discoloration, tingling, itching, and varicosities Table 5.2. The clinician should determine the degree to which these

Table 5.2 Signs and symptoms of lower extremity chronic venous insufficiency.

Symptoms	Signs
Pain/aching	Telangiectasia
Burning	Corona phlebectatica
Throbbing	Reticular veins
Cramps	Varicose veins
Swelling	Edema
Heaviness	Skin induration/inflammation
Itching	Skin discoloration
Tiredness/fatigue	Eczema/venous stasis dermatitis
Numbness/tingling	Hyperpigmentation
Restless legs	Malleolar flair
	Atrophie blanche
	Lipodermatosclerosis
	Healed or active ulcerations

symptoms restrict a patient's activities.[10] If the symptoms are severely restrictive and the patient is unable to ambulate due to pain other etiology such as arterial claudication should be considered. The onset and how fast the symptoms progress should be determined as well. The venous disease tends to have a more indolent course as compared to arterial disease and usually gets worse over months or years.

History of prior VLU

It is important to determine the onset and duration of symptoms related to VLU, whether the ulcer is recent, how long the ulcer has been present, and if there is a history of trauma or infection. Many patients will describe an inciting event such as a minor injury, which preceded the ulceration and then, nonhealing of the wound for a prolonged period of time. Another important factor to access is the rate of ulcer growth. Venous ulcers tend to enlarge at a slower rate than arterial ulcers. A history of prior ulceration including number, size, and location as well as treatments used should be documented. Previous ulcers may be indicative of more severe and/or progressive venous disease and have been shown to be the strongest predictor of poor ulcer healing (unhealed for >10 years) as well as ulcer recurrence.[7]

Past treatment history

In addition to the general medical history, a history specific to venous disease should be obtained. It should be determined if the patient was previously evaluated for any venous pathology (i.e., varicose veins, spider veins, etc.) and if an ultrasound assessment was done. A history of any prior treatments for varicose veins, such as vein stripping, endovenous ablation, or sclerotherapy should be obtained, and vein closure and reflux elimination then documented with imaging. Compression therapy use (inelastic, elastic, intermittent pneumatic), including type/grade of stockings, compliance, tolerance, and efficacy for symptom relief should also be documented.

A thorough history should include nonsurgical treatment options for VLUs including topical antiseptics, oral medications such as pentoxifylline, aspirin, iloprost, zinc, antibiotics, and hyperbaric oxygen therapy. Any surgical management such as debridement, skin grafting, skin substitutes, and other biological dressings should be noted.

Physical exam

A complete physical examination should be performed including a detailed assessment of clinical signs of venous disease and peripheral pulse examination. A thorough inspection of the leg from groin to toes should be performed with additional assessment of the patient in a standing position if possible and in a warm and well-lit room. The location and distribution of all major findings should be noted. Many practitioners use standardized forms with drawings of the right and left lower extremities for chart documentation (Fig. 5.1).

MEDICAL CENTER
Noninvasive Vascular Laboratory

Accredited Vascular Laboratory — ICAVL
Extracranial Cerebrovascular
Peripheral Arterial
Peripheral Venous

Patient Name: ______________________

Date: ______________________

MR #: ______________________

Performed by:

Right Lower Extremity:

DVT/SVT	Negative	Positive
Perforator	______________________	
CFV	Competent	Incompetent
FV	Competent	Incompetent
Popliteal	Competent	Incompetent
SFJ	Competent	Incompetent
GSV	Competent	Incompetent
ASV	Competent	Incompetent
SPJ	Competent	Incompetent
SSV	Competent	Incompetent
REFLUX: Deep	YES	NO
Superficial	YES	NO
Ablation candidate:	**YES**	**NO**
GSV	ASV	SSV

Notes: ______________________

____

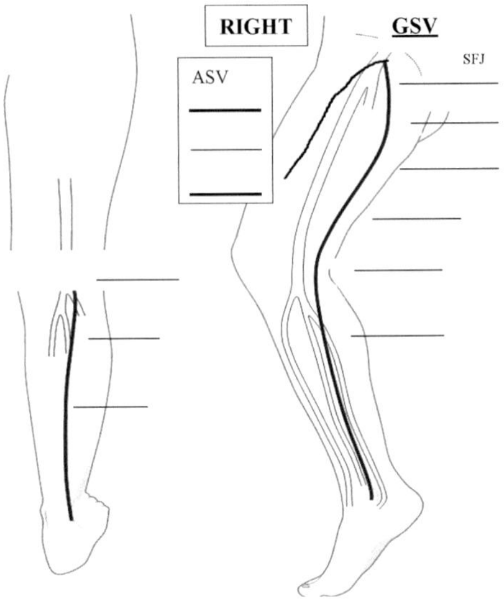

Left Lower Extremity:

DVT/SVT	Negative	Positive
Perforator	______________________	
CFV	Competent	Incompetent
FV	Competent	Incompetent
Popliteal	Competent	Incompetent
SFJ	Competent	Incompetent
GSV	Competent	Incompetent
ASV	Competent	Incompetent
SPJ	Competent	Incompetent
SSV	Competent	Incompetent
REFLUX: Deep	YES	NO
Superficial	YES	NO
Ablation candidate:	**YES**	**NO**
GSV	ASV	SSV

Notes: ______________________

____

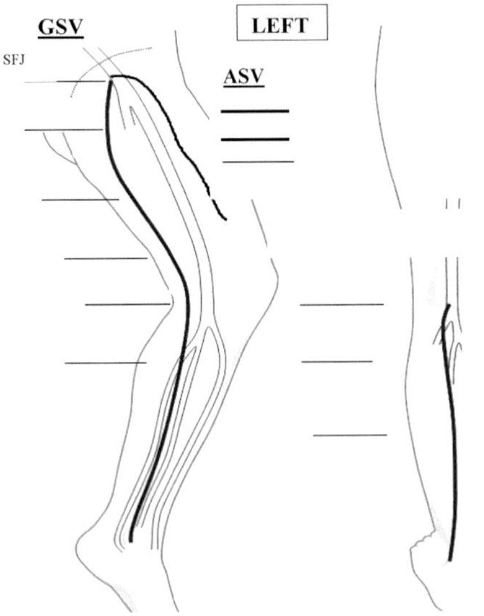

Figure. 5.1 Example of a standarized form to document exam and ultrasound findings.

Signs of CVI

Common signs of CVI and venous hypertension in addition to the VLU include varicose veins, edema, skin hyperpigmentation in the gaiter area between the ankle and the calf, venous eczema, atrophie blanche (white scar tissue), and lipodermatosclerosis (induration from fibrin deposition in the subcutaneous fat) Table 5.2.[14]

The presence of edema is not a specific finding and could be due to other underlying conditions such as renal insufficiency, heart failure, or lymphedema (see Chapter 9). Edema associated with CVI typically worsens at the end of the day after the legs have been in a dependent position for a prolonged period of time and tend to improve in the morning once the legs have been elevated during the night. CVI-related edema is limited to the lower extremity distally to the level of the malleolus and is often unilateral. This is in comparison to edema secondary to chronic medical conditions such as renal disease or heart failure which tend to present with generalized edema and bilateral lower extremity swelling.

One of the most common and earliest skin changes associated with CVI is stasis dermatitis which is characterized by a rash associated with itching, erythema, scaling, weeping, and crusting (Fig. 5.2). Skin changes are characterized by hyperpigmentation on the lower leg which most commonly affects areas over the medial malleolus (Fig. 5.3). The pigmentation is due to hemosiderin deposition due to the breakdown of red blood cells that extravasate through damaged capillaries.[15,16] Long-standing, severe CVI can lead to lipodermatosclerosis which is associated with skin induration and fibrosis around the medial ankle (Fig. 5.4). Due to tissue thickening, the leg shape can change being referred to as an "inverted champagne bottle" shape. Other skin changes associated with lipodermatosclerosis include corona phlebectatica or "ankle flare," which is a dilation of small blood vessels around the inner ankle (Fig. 5.5) and atrophie blanche that are white smooth areas of thin skin with visible capillaries (Fig. 5.6).

The distribution of signs and symptoms may vary based on sex and age. For example, men tend to have a higher incidence of varicose veins.[17] The frequency of edema is similar

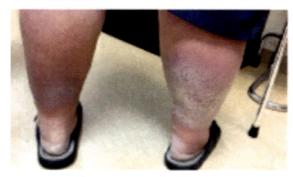

Figure 5.2 *Stasis dermatitis.* Stasis dermatitis characterized by inflammation, scaling and hyperpigmentation.

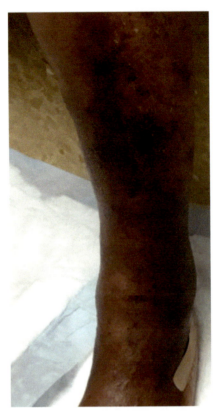

Figure 5.3 *Hyperpigmentation over the gaiter area.* Hyperpigmentation due to VI typically occurs over the gaiter area and can range from small regions of mild skin changes to extensive areas of severe skin darkening.

between the sexes but the prevalence of ulcerations is three times as common in men than in nullipara women. Overall, women have a higher prevalence of ulcers.[3,14] Symptoms are also age dependent with increased incidence of edema, varicose veins, and atrophic changes in patients over 50 years old. Chiesa et al. reported a significant increase in the frequency of venous ulcers from 0.1% before 50 years old to 0.6% over 50 years.[17]

Body mass index (BMI) is associated with the most visible signs of CVI, except for telangiectasias. A study by Musil et al. demonstrated BMI as an independent risk factor in women, but not men.[18] However, BMI together with age were significant predictors of advanced clinical grade based on the CEAP classification. Severity of obesity (BMI >35 kg/m^2) has also been shown to correlate with worse outcomes following venous treatments.[19]

Initial clinical evaluation in patients with chronic venous insufficiency   93

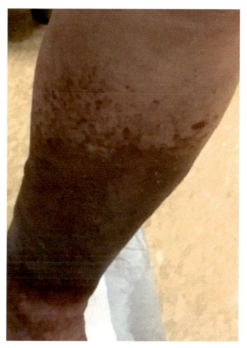

Figure 5.4 *Lipodermatosclerosis.* Lipodermatosclerosis is associated with skin thickening, discoloration, swelling, and tapering of the leg above the ankle (inverted champagne bottle shape).

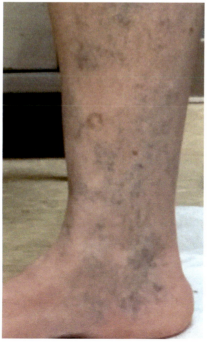

Figure 5.5 *Corona phlebectatica or "ankle flare".* A fan-shaped pattern of a small vein over the medial or lateral malleolus.

Figure 5.6 *Atrophie blanche.* Atrophic blanche is scarring that occurs as a result of healed ulcerations.

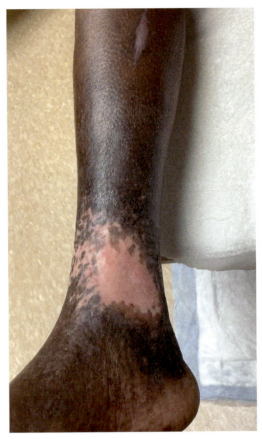

Assessment of the VLU

Venous ulcers are usually located over the gaiter area (medial or lateral malleolus), along the course of the great or small saphenous vein, and can extend circumferentially. VLUs are usually not present on the foot or above the knee. The ulcers are painful and shallow with irregular, flat borders. A mixture of granulation tissue, fibrin, and slough are commonly present at the base (Fig. 5.7). If there is a very thick slough or black eschar present, these may be indicative of concomitant arterial insufficiency. The size of the ulcer should be accessed as well and monitored throughout the treatment.

If the wound has no signs of healing after 12 weeks of compression therapy and venous closure (if indicated), a wound biopsy should be considered to rule out an underlying malignancy. The wound should be also accessed for signs of infection, which include surrounding erythema, worsening swelling, increase level of exudate, odor, and/or increased pain. Some patients may also develop systemic signs of infection including fever and tachycardia.

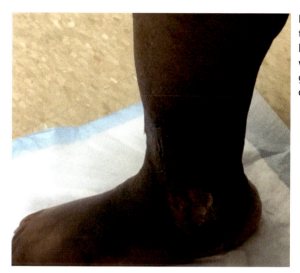

Figure 5.7 *Venous ulcer.* Venous ulcerations classically occur over the medial or lateral malleolus. These ulcers are shallow with irregular, flat borders. A mixture of granulation tissue, fibrin, and slough are commonly present at the base.

In all patients that present with lower leg ulceration assessment of perfusion is a key part of the initial assessment. Patients with the venous disease typically have warm and well-perfused extremities with palpable pedal pulses. Sometimes pulses might be hard to palpate due to severe edema and other modalities can be used such as an ankle-brachial index (ABI) or an arterial Duplex ultrasound to further assess circulation status. We recommend obtaining an ABI on all patients at first evaluation to rule out an arterial disease. It is worth emphasizing that falsely elevated ABI can be recorded in diabetic patients, elderly patients, and patients with renal disease due to calcified, noncompressible arteries in the lower extremity. In these patients, toe pressures or pulse volume recordings (PVR) may be necessary. Additionally, in patients with VLU, the application of blood pressure cuffs may not be possible, and other modalities such as toe pressures or computed tomographic angiography may be required. A baseline ABI will also assist in determining the level of compression that will be tolerated by the patient without worsening the peripheral perfusion.

It is important to differentiate venous ulcers from other underlying pathologies based on physical exam Table 5.3. Arterial ulcers are painful, appear punched out or stellate, and may have an overlying eschar. They are often present over the dorsum of the foot and the pressure points. Other signs of PAD are usually present such as diminished or absent pedal pulses, atrophic and hairless skin, which may be cool to touch, dependent rubor, and delayed capillary refill.

Diabetic or neuropathic ulcers are generally located over the pressure points such as the metatarsal heads and surrounding skin may be thickened. These ulcers tend to be painless and numb. Often, patients with a diabetic foot ulcer will have intact pedal pulses or normal toe pressures or PVRs.

Table 5.3 Clinical manifestations of lower extremity ulcers.

Characteristics	Venous ulcers	Arterial ulcer	Neuropathic ulcers
Location	Circumferential gaiter distribution at the level of the medial and/or lateral malleolus. Not present on the foot or above the knee	Dorsum of the foot, pressure points, over the toe joints, anterior shin	On plantar aspect of feet over the pressure points such as the metatarsal heads and heel
Pain around the ulcer	Yes, severe	Yes, usually mild	No
Ulcer appearance	Shallow with irregular, flat borders, mixture of granulation tissue, fibrin, and slough are commonly present at the base	Deep, punched out or stellate with sharply demarcated boarders, and may have an overlying eschar	Punched out, superficial with red based with surrounding callus
Perfusion	Warm and well-perfused palpable pulses or normal toe pressures/PVRs	Diminished or absent pedal pulses, may be cool to touch, and delayed capillary refill	Intact pedal pulses, or normal toe pressures/PVRs
Skin changes	Induration, eczema, hyperpigmentation, malleolar flair, atrophie blanche, lipodermatosclerosis	Atrophic and hairless skin, dependent rubor	Surrounding skin may be thickened
Leg pain	Worse with prolonged standing or sitting with feet in a dependent position and improves with leg elevation and walking	Pain with walking in patients with mild/moderate PAD or with limb elevation in patients with moderate/severe PAD	No

Quality of life assessment

Severe CVI that has progressed to a VLU is associated with a reduced quality of life, due to chronic pain, decreased physical function, and poor mobility. It is also associated with depression and social isolation.[14] Significant associations between quality of life and severity of the venous disease have been described based on using the 36-item Shot-Form General Health Survey (SF-36) questionary as well as a disease-specific questionnaire.[14]

Venous disease evaluation should include a health-related quality-of-life assessment. Both generic SF-36 and venous disease-specific surveys such as Chronic Venous Insufficiency Questionnaire, the Charring Cross Venous Ulceration Questionnaire, and several others have been used and validated for patients with venous disease.[12]

Classification and venous severity scoring

The clinical manifestations of CVI have wide variability between patients but some manifestation must be present for an ulcer to be classified as a VLU. To address the complexity of the clinical manifestations of CVI, as well as standardize clinical communication and implementation of practice guidelines, the CEAP classification system was initially developed in 1996 and revised several times with the latest updates in 2020.[9] The basic CEAP classification Table 5.4 is a simplified version of the more comprehensive

Table 5.4 Basic CEAP classification system.

Clinical classification	
C0	No visible/palpable sing of venous disease
C1	Telangiectasis or reticular veins
C2	Varicose vines
C3	Edema
C4a	Pigmentation and/or eczema
C4b	Lipodermatosclerosis and/or atrophie blanche
C5	Healed ulcer
C6	Active ulcer
Etiologic classification	
Ec	Congenital
Ep	Primary
Es	Secondary (postthrombotic)
En	No venous etiology identified
Anatomic classification	
As	Superficial veins
Ap	Perforator veins
Ad	Deep veins
An	No venous location identified
Pathophysiologic classification	
Pr	Reflux
Po	Obstruction
Pr,o	Reflux and obstruction
Pn	No venous pathophysiology identified

CEAP and is recommended for clinical practice, with the more comprehensive CEAP reserved for research purposes.[12] A clinical severity score based on the CEAP classification can help clinicians guide the assessment of patients with CVI and monitor changes over time. For example, the highest CEAP score is applied to patients with active, large ulcers. In a study of 1422 patients with chronic CVI, the overall score for symptoms severity was significantly correlated with CEAP clinical class, after controlling for age, sex, BMI, comorbidities, and the duration of chronic venous disease.[14] Association between CEAP class and the quality of life has been found with the use of this disease-specific questionnaire.[14]

In addition to CEAP, another way to objectively characterize the severity of clinical manifestations of CVI is based on the VCSS Table 5.5.[10] The VCSS consists of 10 clinical descriptors (pain, varicose veins, edema, pigmentation, induration, inflammation, number of active ulcers, duration of active ulcerations, size of the largest active ulcer, and compressive therapy), each graded on the scale from 0 to 3. This scoring system should be used in addition to CEAP as it allows an objective assessment of change over time in response to treatments.[20] A VCSS greater than 8 indicates a patient with severe disease who is at risk for progression and warrants additional diagnostic evaluation or treatment.[12]

Table 5.5 Venous clinical severity scoring (VCSS) system.

	0 (none)	1 (mild)	2 (moderate)	3 (severe)
Pain		Occasional pain	Daily pain (interfering but not preventing daily activities)	Daily pain (limits daily activities)
Varicose veins		Few, scattered including ankle flare	Confined to calf or thigh	Involved calf and thigh
Edema		Limited to foot and ankle	Above ankle but below knee	Extends above knee
Skin pigmentation		Limited to perimalleolar area	Diffuse over lower third of calf	Above lower third of calf
Inflammation		Limited to perimalleolar area	Diffuse over lower third of calf	Above lower third of calf
Induration		Limited to perimalleolar area	Diffuse over lower third of calf	Above lower third of calf
Active ulcer number		1	2	≥ 3

Table 5.5 Venous clinical severity scoring (VCSS) system.—cont'd

	0 (none)	1 (mild)	2 (moderate)	3 (severe)
Active ulcer duration (longest)		<3 mo	>3 mo but <1 y	>1 y
Active ulcer size (largest diameter)		<2 cm	2–6 cm	>6 cm
Compression therapy		Intermittent use of stockings	Use stockings most days	Use stockings every day

Summary

Chronic venous insufficiency is a common etiology of lower extremity ulcerations. The diagnosis is based primarily on clinical evaluation which aims to differentiate venous ulcerations from other pathologies and requires a thorough understanding of the etiologies and pathophysiology of the disease. Signs and symptoms of CVI vary widely and should be thoroughly assessed. Additional diagnostic studies may be useful if symptoms and clinical signs do not correlate. CEAP classification as well as the VCSS system should be used to objectively characterize the severity of clinical manifestations of CVI, appropriately select treatments and interventions, and monitor outcomes and progression or improvement of the VLU.

Figs. 5.2–5.7 provided by Dr. Yana Etkin.

References

1. Nicolaides AN, Allegra C, Bergan J, et al. Management of chronic venous disorders of the lower limbs: guidelines according to scientific evidence. *Int Angiol.* 2008;27(1):1–59.
2. Nicolaides AN, Labropoulos N. Burden and suffering in chronic venous disease. *In Adv Ther.* 2019: 1–4.
3. Collins L, Seraj S. Diagnosis and treatment of venous ulcers. *Am Fam Physician.* 2010;81(8):989–996.
4. O'Meara S, Al-Kurdi D, Ologun Y, Ovington LG, Martin-St James M, Richardson R. Antibiotics and antiseptics for venous leg ulcers. *Cochrane Database Syst Rev.* 2014;(1):Cd003557.
5. Bergan JJ, Schmid-Schonbein GW, Coleridge Smith PD, Nicolaides AN, Boisseau MR, Eklof B. Chronic venous disease. *Minerva Cardioangiol.* 2007;55(4):459–476.
6. Abbade LP, Lastoria S, Rollo HA, Stolf HO. A sociodemographic, clinical study of patients with venous ulcer. *Int J Dermatol.* 2005;44(12):989–992.
7. Abbade LP, Lastória S, Rollo Hde A. Venous ulcer: clinical characteristics and risk factors. *Int J Dermatol.* 2011;50(4):405–411.
8. Abbade LP, Lastória S. Venous ulcer: epidemiology, physiopathology, diagnosis and treatment. *Int J Dermatol.* 2005;44(6):449–456.
9. Lurie F, Passman M, Meisner M, et al. The 2020 update of the CEAP classification system and reporting standards. *J Vasc Surg Venous Lymphat Disord.* 2020;8(3):342–352.

10. Passman MA, McLafferty RB, Lentz MF, et al. Validation of venous clinical severity score (VCSS) with other venous severity assessment tools from the American venous Forum, national venous screening program. *J Vasc Surg*. 2011;54(6 Suppl):2s–9s.
11. Bergqvist D, Lindholm C, Nelzén O. Chronic leg ulcers: the impact of venous disease. *J Vasc Surg*. 1999;29(4):752–755.
12. O'Donnell Jr TF, Passman MA, Marston WA, et al. Management of venous leg ulcers: clinical practice guidelines of the Society for Vascular Surgery ® and the American Venous Forum. *J Vasc Surg*. 2014; 60(2 Suppl):3s–59s.
13. Chiesa R, Marone EM, Limono C, Volonte M, Schaefer E, Petrini O. Effect of chronic venous insufficiency on activities of daily living and quality of life: correlation of demographic factors with duplex ultrasonography findings. *Angiology*. 2007;58(4):440–449.
14. Bergan JJ, Schmid-Schonbein GW, Coleridge Smith PD, Nicolaides AN, Boisseau MR, Eklof B. Chronic venous disease. *N Engl J Med*. 2006;355(5):488–498.
15. Herrick SE, Treharne LJ, deGiorgio-Miller AM. Dermal changes in the lower leg skin of patients with venous hypertension. *Int J Low Extrem Wounds*. 2002;1(2):80–86.
16. Stacey MC, Burnand KG, Bhogal BS, Black MM. Pericapillary fibrin deposits and skin hypoxia precede the changes of lipodermatosclerosis in limbs at increased risk of developing a venous ulcer. *Cardiovasc Surg*. 2000;8(5):372–380.
17. Chiesa R, Marone EM, Limoni C, Volonte M, Petrini O. Chronic venous disorders: correlation between visible signs, symptoms, and presence of functional disease. *J Vasc Surg*. 2007;46(2):322–330.
18. Musil D, Kaletova M, Herman J. Age, body mass index and severity of primary chronic venous disease. *Biomed Pap Med Fac Univ Palacky Olomouc Czech Repub*. 2011;155(4):367–371.
19. Deol ZK, Lakhanpal S, Franzon G, Pappas PJ. Effect of obesity on chronic venous insufficiency treatment outcomes. *J Vasc Surg Venous Lymphat Disord*. 2020;8(4):617–628.e1.
20. Krishnan S, Nicholls SC. Chronic venous insufficiency: clinical assessment and patient selection. *Semin Intervent Radiol*. 2005;22(3):169–177.

CHAPTER 6

Ultrasound evaluation of lower extremity chronic venous disease

Raudel Garcia[1] and Nicos Labropoulos[2]

[1]ChenMed, Miami, FL, United States; [2]Department of Surgery and Radiology, Vascular Laboratory, Division of Vascular and Endovascular Surgery, Stony Brook University Medical Center, Stony Brook, NY, United States

Key points

- All clinical forms of chronic venous insufficiency (CVI) are caused by venous hypertension due to valve incompetence (reflux) and/or venous outflow obstruction in the superficial, deep, and perforating venous systems. Primary venous reflux is the most common cause of CVI.
- Duplex ultrasound (DUS) is the goal standard imaging technique for the diagnosis of valve incompetence (reflux) and obstruction affecting any of the three lower extremity venous systems in patients with venous leg ulcers (VLU).
- Retrograde blood flow with a duration of >1000 ms in the common femoral vein (CFV), femoral vein (FV), and popliteal vein (POPV) and >500 ms in the superficial, perforating, and deep calf veins is considered pathological.
- Most patients with VLU demonstrate combined multisystem reflux patterns. Nearly half of all patients with VLU show reflux in the superficial venous system alone. Isolated venous reflux in the perforating or the deep veins is uncommon. Over two-thirds of VLU have multiple, enlarged refluxing veins beneath the base and the perimeter of the ulcer.
- Postthrombotic changes within the deep veins are detected in about one-third of limbs with VLU, usually at multiple venous sites. Gray scale imaging may reveal contracted, noncompliant, and partially compressible veins with thicker walls and hyperechoic filling defects (synechiae). Additionally, information about vein patency and blood flow patterns within the postthrombotic vein segments can be obtained by means of color flow and spectral Doppler imaging.
- Asymmetric flow patterns within the common femoral veins and reversal of flow within the superficial epigastric or pudendal veins suggest suprainguinal outflow venous obstruction and may require further investigation.

Introduction

Venous leg ulcer (VLU) is a complex clinical scenario of a highly prevalent, gradual, and progressive venous disease. Chronic venous disease (CVD) in its entire clinical spectrum from CEAP Class C0 to C6 is caused by ambulatory venous hypertension due to venous valve incompetence (reflux), venous outflow obstruction, and mechanical dysfunction of the calf muscle pump.[1,2] Chronic venous insufficiency (CVI) is reserved for advanced CVD associated with skin changes (classes C3 to C6).[3] Primary insufficiency is the most common etiology of CVI and venous ulcers.[4]

Patients with VLU have a distinctive and more complex pattern of venous reflux and obstruction. Most ulcerated limbs demonstrate combined or multisystem reflux patterns.[5] Reflux within the superficial veins alone or in combination with the deep and/or the perforating veins is the dominant pattern.[6,7] Over two-thirds of VLU have a cluster of enlarged refluxing veins within and around the ulcerated area connecting to an incompetent superficial venous trunk and/or the perforating venous system.[8,9] Deep venous obstruction is most commonly of postthrombotic origin, and it is nearly always associated with reflux. Postthrombotic changes within the deep veins are detected in about one-third of limbs with VLU, usually at multiple venous sites.[7,8,10,11] Over one-third of CEAP Class five and six patients demonstrate pelvic outflow venous obstruction of at least 50%, and one-fourth reveal obstruction of >80%.[12] Risk factors associated with a higher degree of obstruction are female gender, history of deep vein thrombosis (DVT), and reflux in the deep venous system.[12] About one-third of patients with VLU and pelvic outflow obstruction have a lesion of nonthrombotic etiology (nonthrombotic iliac vein lesion, NIVL).[13] The rate of ulcer healing with venoplasty and stent placement for the management of chronic venous obstruction is significantly higher in NIVL than in postthrombotic iliofemoral veins.[14] Routine evaluation of the outflow suprainguinal veins has been advocated in all patients with nonhealing VLU.[13,15]

The strategies for the procedural management of patients with CVI and VLU may differ according to the location, distribution, and extension of the underlying reflux and obstruction. Duplex ultrasound (DUS) is the imaging modality of choice for detecting and mapping the venous abnormalities of the lower limbs associated with venous hypertension. Additionally, blood flow disturbances within the common femoral (CFV) veins (i.e., nonphasic flow in the proximal CFV or asymmetrical flow pattern) and the superficial epigastric (SEV) veins (i.e., retrograde blood flow in the SEV), detected by means of color flow and pulsed-wave Doppler, may identify venous obstruction in the ilio-caval segment.[16–19] DUS has demonstrated a high sensitivity and specificity for diagnosing reflux and obstruction in the infrainguinal veins, it is noninvasive, very cost-effective, and reproducible. Moreover, DUS is an essential instrument for proper management, procedure guidance, and follow-up.

The accuracy of the DUS results can be limited by factors such as an uncooperative patient, individual body habitus, limited mobility, leg swelling, pain at the ulcer site, presence of bandages or casts, dystrophic calcifications, poor image optimization, examination protocols, and image interpretation. It is worth emphasizing that the DUS examination is highly operator dependent, and its diagnostic accuracy relies on the proper training of the ultrasonographer.

Basic principles

The correlation between the clinical features of an ulcer, such as location, size, and depth, and the underlying anatomical and functional abnormalities detected by means of DUS are relevant for determining the specific etiology. The examiner should be mindful about the importance of this clinical-sonographic association. Before performing the ultrasound study, the examiner should review the patient's medical record, and assess for comorbidities (i.e., diabetes mellitus, obesity, peripheral arterial disease, peripheral neuropathy, malignancies, etc.), ulcer duration and recurrences, previous venous thromboembolic events, past superficial or deep venous interventions (i.e., high ligation and stripping, inferior vena cava filters, stents, ablations, etc.), and other surgeries (i.e., hip and/or knee replacement). Also, it is imperative to perform a focused physical exam.

The exam should be performed with the patient standing for proper measurement of vein diameter and detection of reflux. For patients who are unable to stand, it may be appropriate to scan them in the reverse-Trendelenburg position. The popliteal and all calf veins may be interrogated with the patient sitting at the edge of the examination table with both legs dangling. The room and gel should be comfortably warm to prevent vein spasms. Reflux measurements in the superficial veins are more repeatable when performed in the morning, however, a longer duration of reflux is usually obtained when patients are scanned in the afternoon.[18] Flow augmentation and reflux assessment are obtained using the Valsalva maneuver for the interrogation of the proximal lower extremity veins, and manual or mechanical calf squeezing for the evaluation of the distal veins. Similarly, active dorsiflexion of the ankle may be used for the assessment of the distal veins.[20]

Multilinear high-resolution array transducers are used to image superficial veins and any deep structure up to 6 cm depth. Lower frequency (1–5 MHz) curvilinear transducers are preferred for imaging deep veins, obese or edematous patients because of better penetration. A probe cover is used for scanning the veins in the ulcer area. B-mode images are obtained in the transverse axis for assessment of vein patency and diameter measurement. Color flow and spectral Doppler images are obtained in the longitudinal axis for vein patency, augmentation (characterized as normal, reduced, absent), flow pattern (characterized as phasic, decreased, pulsatile, continuous, regurgitant), flow direction (characterized as antegrade, retrograde, absent), and spontaneity (characterized as spontaneous or nonspontaneous).[21] The machine settings should be adjusted to acquire high-quality images and reliable blood flow velocities.[22,23] Parameters to be adjusted for B-mode include mostly the location and number of focal zones, depth, and time-gain compensation (TGC). The focal zone is typically positioned at the vein lumen or the deeper wall to maximize lateral resolution. Multiple focal zones can be selected; however, this can slowdown the frame rate, and therefore degrades temporal resolution. Similarly, sampling deeper segments can negatively affect the frame rate. The TGC should be properly set to overcome ultrasound attenuation so that the lumen of the vein appears dark in the absence of stasis and thrombosis. B-mode optimization is essential for the subsequent Doppler components of the study. "Low flow" settings are recommended for duplex (color flow) and triplex (spectral waveform) Doppler ultrasound. The color gain should be adjusted as high as possible for adequate vessel lumen filling and wall definition, but without causing bruit. Conventionally, the color bar is preset to demonstrate positive Doppler shifts (toward the transducer) in red, and negative Doppler shifts (away from transducer) in blue. The red color means retrograde venous flow, while the blue color represents anterograde or normal venous flow. The wall filter, and the scale or pulse repetition frequency (PRF) should be set low. The size and depth of the color box inversely affect the frame rate and image resolution. Reliable waveforms may be obtained by adjusting the spectral gain, the scale or PRF, baseline, sweep velocity, sample volume, and the steering angle. The most accurate velocities are recorded when the Doppler angle (angle of insonation) is set at 0 degrees or parallel to the flow. However, because most veins run parallel to the skin an angle of <60 degrees should be used to optimize results. Steering the angle may be difficult when scanning tortuous veins. Velocities may be underestimated when the cursor is no positioned parallel to the vessel axis. In general, when recording the duration of reflux or waveform patterns an angle of 0 degrees can be used as it only affects the speed of the blood. The velocities registered below and above the baseline represent antegrade flow and retrograde flow (reflux), respectively. Retrograde venous flow is defined as physiological or pathological depending on its duration.

Diagnosis of reflux

Venous reflux parameters frequently assessed with DUS include reflux duration, peak refluxing velocity, flow volume, and vein diameter. However, only reflux duration with cutoff values >500 ms for the superficial, perforating, deep femoral vein (DFV) and deep calf veins and >1000 ms for the CFV, FV, and POPVs are considered pathological.[24] Typically, the apposition of valve leaflets is lengthier in larger veins with fewer valves; therefore, different cutoff points are required. Reflux duration is a qualitative variable, and it should not be used to define severity. It is inappropriate to classify reflux as mild, moderate, severe, etc. based on valve closure time.[25]

Venous valve incompetence alone or associated with obstruction is the most common pathology in patients with VLU. Obstruction alone is rare. The combination of reflux and obstruction has the highest morbidity in patients with CVI.[26,27] Venous reflux of primary etiology from varicose veins is the most common finding and results from localized or multisegmental venous wall dilatation and valve dysfunction.[28] Risk factors associated with primary venous insufficiency and VLU are advanced age, genetic predisposition, family history, obesity, and orthostatism.[28] A review article revealed the presence of reflux in the superficial system, either isolated or in conjunction with perforating and/or deep systems, in about 88% of VLU. Isolated superficial vein incompetence was detected in 45% of limbs, whereas isolated deep reflux was very infrequent (12%).[7] Similar findings have been reported by many others.[4,9,29–34] Prolonged venous distention, weak vessel walls or leaflets, injuries, or superficial phlebitis are the main causes of valvular incompetence in the superficial systems.[35]

Incompetent perforating veins (IPV) are associated with superficial and/or deep-vein reflux in nearly 60% of patients with ulceration but are rarely found in the absence of reflux in the other two systems.[36–38] The prevalence, diameter, volume flow, and velocity of IPVs increase with the clinical severity of CVD irrespective of the presence or absence of coexisting deep venous incompetence.[39–45] About 86% of venous ulcers have some degree of reflux in the local area, the pattern of which may differ from the axial vein disease (Fig. 6.1). Deep venous reflux is usually postthrombotic.[46,47] Only one-third of patients with ulcers show evidence of a past episode of thrombosis based on history and noninvasive investigations.[4,48,49] In postthrombotic limbs, the most important predictors for skin damage and VLU are ipsilateral recurrent DVT, popliteal vein reflux, multisegmental involvement, and a combination of reflux and obstruction.[50] Traumatic venous injuries (mechanical, thermal, or chemical) and arteriovenous fistulas may also lead to valvular damage and reflux. Venous reflux due to congenital valve agenesis or aplasia in the deep system is rare.[51–53]

The differentiation between primary and secondary venous reflux is important, and DUS is excellent for this. It allows a clear visualization of the wall and luminal changes and the pattern of reflux. Many patients have asymptomatic thrombosis that is discovered

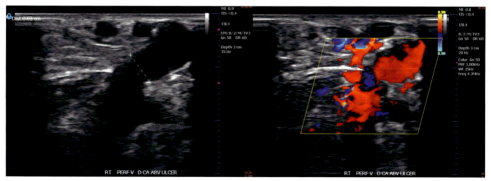

Figure. 6.1 A cluster of incompetent tributary and varicose veins is seen within the ulcer area in connection to a large pathological perforator vein.

only during the DUS examination. Proper correlation between the location of the ulcer, and the anatomical distribution and extension of the reflux is crucial for treatment purposes. As mentioned previously, patients with VLU have a higher prevalence of concomitant incompetence in the perforating and deep veins.[50,54]

Superficial, deep, and perforator veins

Superficial reflux patterns correlate with the location of the ulcer. Typical VLUs are found distally in the calf either medially (56%), laterally (30%), or bilaterally (14%).[8] Bilateral VLUs (both medial and lateral calf) can present as multiple ulcers, or as a single large ulcer affecting the entire leg circumference or a section of it. Medial ulcers are mainly associated with reflux in the great saphenous vein (GSV) and lateral ulcers with reflux in the small saphenous vein (SSV).[8]

Obermayer et al. prospectively evaluated 169 consecutive patients with VLUs for the source of superficial reflux using DUS. The source of the reflux was classified as either axial or crossover type. The axial type described medially located ulcers in connection with an incompetent GSV or medially located IPVs, and laterally located ulcers associated with an incompetent SSV or laterally located IPVs. The crossover type referred to laterally located ulcers in association with an incompetent GSV or medially located IPVs, and laterally located ulcers with an incompetent SSV. The longer reflux distance was used for ulcers with multiple sources of reflux. The location of the first ulcer based on history was used for bilateral (medial and lateral) ulcers. Seventy-nine percent of VLU demonstrated axial reflux mainly associated with medially located ulcers. Eighty-nine percent of medial ulcers revealed axial reflux, 62% in the GSV, and 27% in the medial perforating veins. Only 11% of medial ulcers showed crossover reflux, all within the SSV. Conversely, 54% of lateral ulcers revealed axial reflux, 24% in the SSV, and 30% in the lateral perforating veins. Forty-six percent of lateral ulcers had crossover reflux in the GSV. Twenty-one percent of all ulcers in this study were "atypically" located, concerning the source of reflux.[9]

Most patients with VLU have refluxing tributary and varicose veins. However, because varicose veins are highly prevalent in the general adult population, they can be incidentally found with ulcers of nonvenous etiology.[55] Interestingly, Pittaluga et al. noticed that limbs with saphenous vein reflux and not clinical evidence of varicosities are associated with more advanced clinical classes.[56] The competency status of venous microvalves in the smaller intradermal tributaries may explain the above observations and the reason why many CVD patients do not progress to CEAP class 6.[56–59]

Small saphenous vein incompetence directly correlates with advanced CEAP clinical classes. Lin et al. reported a prevalence of SSV incompetence of 25.8% for patients with C1–C3 disease, and a statistically significant higher percentage (36.1%, $P = .006$) for patients with skin damage, healed or active ulcers (C4–C6). In this study, SSV reflux was found to be highly associated with deep venous reflux.[60] Labropoulos et al. reported an

SSV reflux incidence of 10% in limbs with CVD classes 0–6 and of 5.5% in classes 5 and 6.[61] In this study, isolated SSV reflux was defined as incompetence within the SSV and associated perforating veins. Both the extent of reflux within the SSV and the number of associated IPVs correlated with CEAP clinical class severity.[5,61,62]

Foot and toe ulcers are usually caused by the simultaneous action of multiple contributing factors such as peripheral vascular disease, diabetes mellitus, neuropathy, biomechanical structural deformities, and soft tissue changes.[63] Foot ulcers of mere venous etiology are uncommon. They are usually associated with reflux in the marginal veins, the superficial venous arch, and the plantar and perforating veins. The presence of typical clinical features with surrounding skin hyperpigmentation from hemosiderin deposition warranted detailed interrogation of the pedal veins.[64,65]

The prevalence of deep venous reflux is higher in patients with VLU compared to milder forms of CVD. Overall, more than half of patients with VLU demonstrate deep venous reflux, usually associated with reflux in the superficial and/or perforating veins.[4,7] Deep venous incompetence is secondary (postthrombotic) in two-thirds of all CVD patients.[66] The most common site of deep venous reflux is the popliteal vein; about one-fourth of patients with deep venous reflux demonstrate isolated popliteal incompetence.[6] The importance of popliteal venous reflux in the pathogenesis of VLU is not fully understood; therefore, it is a continuous reason for debate. It has been named by many as the gatekeeper of the leg veins and cited as an important risk factor for disease severity and poor treatment outcomes.[67–70] Brittenden et al. recorded a baseline pattern of venous reflux by means of DUS in 155 patients with active VLU. All patients were managed with compression therapy only. At 24 weeks, more than two-thirds of ulcers healed. They found no significance difference in patterns of either deep or superficial venous reflux between healed and nonhealed ulcers, except with respect to the popliteal vein which adversely affected healing.[68] Neglen et al. evaluated the correlation between the presence of popliteal venous incompetence in 103 limbs with different sonographic hemodynamic parameters and clinical severity. Contrarily, they found no meaningful correlation between these variables.[25]

In respect to deep venous incompetence, many studies have correlated different other ultrasound-derived parameters, besides reflux duration, with disease severity and management outcomes. Neglen et al. observed a better association between higher CEAP clinical class and peak refluxing velocity and rate of reflux, than with reflux duration. The findings were more significant in patients with axial deep reflux to below knee veins, with or without superficial venous incompetence.[25] Similarly, Danielsson et al. reported a significantly higher peak refluxing velocity in patients with skin damage and active ulcers. However, reflux duration did not indicate a significant correlation with disease severity. They suggested that peak refluxing velocity seems to reflect venous malfunction more appropriately.[71] A peak reflux velocity greater than 25.4 cm/s in the popliteal vein and greater than 24.5 cm/s in the femoral vein were found to be strong predictors of disease severity in postthrombotic limbs with deep venous incompetence.[72] Marston et al. evaluated the correlation of deep venous peak refluxing velocity and clinical outcomes after saphenous ablation. Twenty-five limbs in the study had a healed or active ulcer associated with both, deep and superficial venous reflux. The study revealed significantly better clinical outcomes after saphenous ablation in the limbs with a peak refluxing velocity of less than 10 cm/s.[73] Meanwhile, the current guideline recommends reflux time or duration as a sole qualitative parameter for diagnostic purposes.[74] Researchers have not yet reached a consensus as to whether include other DUS-derived variables for the assessment of the magnitude of the underlying hemodynamic venous disturbances of the limb.

There is also controversy about the clinical significance of axial versus segmental deep venous reflux, but for most uninterrupted reflux throughout the entire deep venous system from groin to the calf correlates better with disease severity and delayed- or nonhealing of an ulcer.[25,70,72,75] (Fig. 6.2) If spectral-Doppler parameters, other than reflux time, were to be assessed in limbs with axial deep venous reflux, the quantitative variables registered at the popliteal vein only should be used.[25]

The identification, distribution, and extension of reflux within the deep vein are important for CVI disease management. Venous leg ulcers associated with deep venous insufficiency have a higher recurrence rate.[76] Additionally, patients with concomitant deep and superficial venous reflux have a higher rate of thrombotic complications related to superficial venous interventions and should be counseled appropriately preoperatively.[77] Deep venous reflux resolves after saphenous vein ablation procedures in about one-third of patients with concurrent deep and superficial incompetence.[73,78] Patients with persistent deep venous reflux and recalcitrant venous ulcers may benefit from deep venous reconstruction, including valvuloplasty, valve transposition or transplantation, or valve substitutes.[73,74]

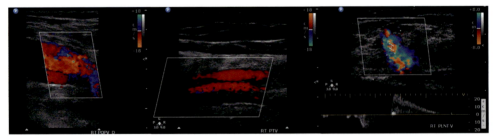

Figure. 6.2 Uninterrupted reflux from the popliteal vein through the posterior tibial and plantar veins as seen in the pictures has high prevalence in postthrombotic limbs of patients with venous ulceration. This is even more common in patients with recurrent ulceration and in those having ulcers that are difficult to heal.

The role IPVs have in the pathogenesis of CVD and VLU is not fully understood. Theoretically, in patients with primary VLUs, calf perforator veins enlarge and become incompetent from increasing draining blood volume from the associated refluxing superficial veins in a descending manner (reentry perforators) and from progressive vein wall disease extension from the superficial venous system. There is no evidence that reflux starts in a perforator and then extends to the superficial veins.[79] In fact, in patients with active VLU, incompetent calf perforators are always associated with superficial, deep, or combined venous reflux.[36–38] The eradication of the associated superficial venous reflux corrects in most cases the reflux in the perforator veins as well.[80,81] Persistent perforator vein reflux after superficial venous interventions is usually caused by the presence of deep venous incompetence, failed ablation, or connections to untreated superficial refluxing segments.[44,81] Anticipatory treatment of the incompetent saphenous trunk and its tributaries along with any pathological perforator veins during one or more procedure sessions may be a reasonable plan in patients with deep venous incompetence.[82]

Identifying and characterizing pathological perforator veins with DUS is fundamental. As previously mentioned, the prevalence and the morphological and functional abnormalities of IPVs adversely correlate with CVD severity.[39–45,83,84] This methodology is not just helpful for initial treatment planning, but also for postprocedural follow up. Pathological perforator veins associated with recalcitrant or recurrent VLU, whether superficial or deep venous reflux is present, should be treated as well. However, the evidence for treating them is weak.[79] The closure rate of IPVs is lower compared with saphenous ablation regardless of the type of procedure (thermal vs. nonthermal). Also, ablated IPVs have a higher rate of recanalization, and de novo IPVs can develop.[85] Ongoing DUS assessment for IPVs is warranted for the management of patients with nonhealing or recurrent VLUs.

Examination protocol

The Vein Glossary,[86] in its most updated version, defines axial reflux as uninterrupted retrograde venous flow from the groin to the calf. Axial reflux could be confined to the superficial (superficial reflux) or the deep (deep reflux) venous systems, or also be present as a combination of all three venous systems (combined reflux). Segmental reflux describes the incompetence of any venous anatomical segment, including femoral, popliteal, and crural for deep veins and above and below the knee for the GSV. For many CVD patients, venous valve incompetence is a progressive phenomenon, from a single segment, to multisegmental, and to axial reflux. This phenomenon directly correlates with disease severity.

Axial reflux whether superficial, deep, or combined, directly correlates with CVD progression and severity. Axial reflux is a typical ultrasound finding in patients with C6 disease (Fig. 6.3). Danielsson et al. scanned the legs of 83 patients with active

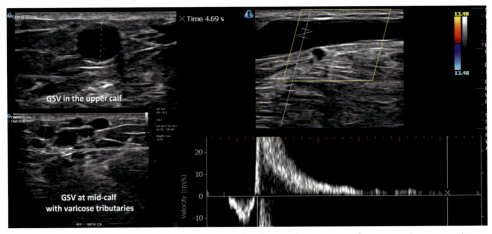

Figure. 6.3 High velocity and long duration axial reflux in the great saphenous vein connecting to multiple refluxing tributary and varicose veins in the mid-calf.

VLU. Axial reflux was detected in most of them (79%). More than half of the legs demonstrated combined reflux.[87] Occasionally, reflux can be detected within a segment of an axial vein with competent valves, both proximally and distally. This can happen to competent superficial segmental trunks with connections to refluxing tributary veins. Current ultrasound protocols interrogate the veins in a segmental fashion, and the final report is just a mere reconstruction of segmental information.[88] Therefore, it is very important that clinicians correlate their patients' clinical picture with the ultrasound findings when formulating a treatment plan.

Because of the variability and multisystem reflux pattern in VLU patients, a comprehensive ultrasound exam of all three venous systems from groin to ankle is recommended.[6] The veins at the ulcer site should be interrogated in detail, and the refluxing segments mapped to their source of origin.[9] An ultrasound protocol for the examination of the infrainguinal veins has been published by the authors recently and is summarized next.[89]

The above knee veins are interrogated with the patient standing in front of the examiner bearing most of the weight on the contralateral leg, and the limb being scanned externally rotated. The below knee veins could be evaluated with the patient sitting on the examination table facing the examiner with legs in a dependent position. The deep, superficial, and perforating veins are examined in this sequence. Compressibility and vein diameter are evaluated using grayscale with the transducer oriented perpendicularly to the vessel. Luminal changes, augmentation, flow direction, and reflux duration are assessed using color and spectral Doppler with the transducer oriented parallelly to the vessel. Manual or automatic squeeze/release maneuver distally to the interrogated vein section is performed by segments from cephalad to caudad. The veins at the groin and the upper thigh can be evaluated by the Valsalva maneuver.

In patients with VLU, all the deep veins from the groin to the calf should be interrogated for obstruction and reflux. The CFV is identified just below the inguinal ligament. Spectral Doppler images of the bilateral CFV on a single split window may help the interpreter physician not to overlook an asymmetric flow pattern indicative of pelvic outflow obstruction. In the presence of severe femoropopliteal postthrombotic venous changes, images of the DFV usually reveal axial transformation with enlargement and reflux. Retrograde flow within the anatomical boundaries of the "saphenous-deep" and "perforator-deep" vein junctions may just reflect a recruitment type of reflux and not a true deep venous valve malfunctioning. This type of deep venous reflux is usually corrected by treating the associated incompetent superficial vein.

The evaluation of the superficial veins starts at the saphenofemoral junction (SFJ). At the groin, the SFJ is seen medially to the CFV and closer to the skin. The depictive sonographic image in the short axis of the SFJ, CFV, and the common femoral artery resembles "Mickey Mouse" shape. Patterns of reflux within the SFJ in continuity with the GSV, accessory saphenous veins, and other common tributaries found at the groin need to be documented. The presence of retrograde flow within the SEV and/or the pudendal veins may suggest suprainguinal outflow venous obstruction.

Numerical and morphological variations of the veins need to be recorded as well. True duplication of GSV is seen in less than 2% of CVD patients.[90] At the proximal thigh, the GSV crosses over the adductor magnus muscle, 3—5 cm below the SFJ. The GSV is identified with high-resolution B-mode DUS imaging using the superficial fascia as an anatomic landmark since in the transverse plane deep and superficial fascia reflects ultrasound intensely, yielding a characteristic appearing of the GSV termed the "saphenous eye." This ultrasonographic anatomic marker is most clearly visualized in the transverse section of the medial thigh and can be utilized to differentiate the GSV from varicose tributaries and other elongated superficial veins that can incorrectly be identified as the GSV. The anterior accessory great saphenous vein (AAGSV) and posterior accessory great saphenous vein have a parallel course to the GSV, laterally and medially, respectively, but within separate compartments. The AAGSV can be distinguished from the GSV by its alignment with the FV and the superficial femoral artery (alignment sign). Any saphenous tributary veins with a diameter of at least 50% of the saphenous trunk should be interrogated for the presence of reflux. The saphenous veins and their accessories below knee are evaluated for vein diameter and reflux time in the same fashion.

The SSV is identified at the popliteal fossa and interrogated from proximal to distal. A small saphenous vein diameter is obtained 3 cm below the saphenopopliteal junction (SPJ) and at the mid-calf. The SSV follows a straight course within its own fascial compartment posteriorly and slightly laterally on the calf. Multiple anatomical variations of the SSV and its termination have been described. One-fourth of patients do have not a true SPJ. The SSV usually derives from duplicated veins seen at the mid-distal calf level. Enlarged saphenous tributary and intersaphenous veins need to be similarly evaluated.

Reflux can be detected in superficial nonsaphenous veins in about 10% of patients with CVD, these are infrequently the main source of ambulatory venous hypertension in patients with active VLU. The scanning protocol for detecting reflux in these veins varies somewhat. Many of these veins lie deeply, therefore a curvilinear probe may be required. Some of these veins include the gluteal, vulvar, and sciatic nerve veins, and the vein of the popliteal fossa.

Below the knee perforator veins larger than 3.5 mm should be interrogated for reflux in all patients with skin damage and active VLU. Escape or exit perforator veins are usually seen at the thigh, whereas terminal or reentry perforators are mostly found one-half distally in the calf (Fig. 6.4). The diameter is measured at the deep fascia level, with the caliper positioned perpendicular to the vein walls. Frequently, the diameter of the fascia defect is wrongly measured instead.

Besides the above examination protocol, a detailed interrogation of the veins around the ulcer bed is highly recommended. A sterile technique was suggested by Hanrahan et al. in 1991 and it is still in use.[34] Manual compression and release of the midfoot is

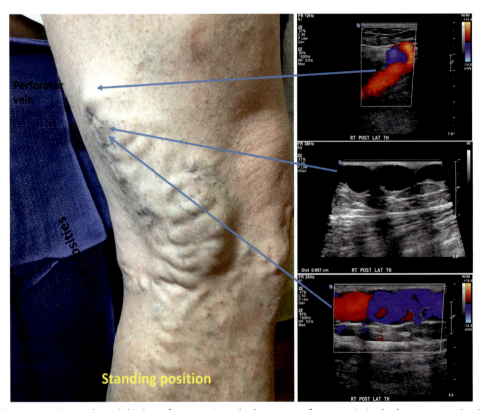

Figure. 6.4 Posterolateral thigh perforator vein is the longest perforator vein in the lower extremity. It was the source of reflux in this patient's very large varicosities that are seen along the lateral limb.

a better option for identifying the source of reflux in the ulcers located on the distal calf. Any refluxing veins passing through the ulcer bed or within 2 cm of the border of the ulcer should be interrogated and followed until its source is found. This seems crucial for procedure planning in C6 disease patients.[8–10] Labropoulos et al. evaluated the local venous reflux pattern of 43 individual VLU. All ulcerated limbs demonstrated either isolated or combined reflux in the "axial" (defined as lower limb veins outside the ulcer bed) superficial and deep veins. Only six (14%) ulcers showed no evidence of local reflux (defined as lower limb veins passing through the ulcerated area). The most common refluxing "axial" calf veins seen in VLU adjacent to the medial malleolus were the GSV, the posterior saphenous arch, the posterior tibial, and the peroneal veins. For ulcers adjacent to the lateral malleolus the SSV with its branches and the peroneal veins were the main source of "axial" calf reflux. Two ulcers located on the anterior calf demonstrated reflux in the anterior saphenous arch and the anterior tibial veins.[8]

Diagnosis of obstruction

Acute DVT is the most common cause of lower limb venous obstruction. Infrainguinal DVT is occasionally associated with pelvic outflow chronic venous obstruction due to extrinsic compression. Chronic venous obstruction may cause stasis, increased blood viscosity, and thrombi formation. The correlation between acute lower extremity DVT and pelvic outflow venous obstruction does not necessarily imply causation. Iliac vein stenosis of 50% or greater is present in about one-quarter to one-third of the general population.[91] Infrainguinal obstruction after DVT is caused by failed or partial recanalization of the thrombosed vein segments. As mentioned before, one-third of patients with VLU demonstrate postthrombotic lower extremity venous changes. Morphological changes in the veins and its valves after acute DVT may result in luminal narrowing and valve incompetence. If fact, deep venous reflux is detected in most patients with VLU and imaging findings compatible with obstruction.[7,8,11,92]

The importance of identifying reflux and obstruction is crucial for the management of patients with VLU. A study by Lawrence et al. revealed venous stenosis of one or more lower extremity veins in 134 out of 832 patients with VLU (approximately 16%). Patients who received procedures to eradicate both, deep venous obstruction, and superficial/perforator veins reflux, achieved better outcomes in terms of ulcer healing and recurrence rate.[93]

Examination protocol

Initial DUS examination for patients with VLU is performed with the aim of identifying the patterns of reflux and obstruction. The DUS examination protocol for this purpose was previously explained in this chapter. When DUS is performed with the sole intention of ruling out obstruction, a simpler methodology is followed. The patient is examined in a supine position with flexion, abduction, and external rotation of the leg that is being examined. The deep veins are interrogated from the cephalad to caudad direction for compressibility, phasicity, spontaneity, augmentation, and flow direction. When assessing for augmentation, the Valsalva maneuver is used for the evaluation of the groin veins, and the calf squeeze/release sequence is for the interrogation of the thigh and calf veins.

Using B-mode with the probe in the short axis, multiple segments (at 2—3 cm intervals) of the veins are compressed until fully obliterated. Color flow and spectral Doppler images are taken on a longitudinal view. Only the affected leg is examined unless the bilateral disease is suspected. It is important to scan the contralateral CFV for the presence of asymmetrical flow patterns, commonly seen in patients with pelvic outflow venous obstruction. High body mass index can make the exam more difficult for the suprainguinal veins due to depth. In experienced hands in most cases this is not an issue. Placing the patient in the left and right decubitus position reduces the distance from the inferior vena cava and iliac veins significantly. Furthermore, in these positions, more pressure can be applied by the transducer without causing pain or discomfort to the patient. In most occasions, simply by doing this, imaging of the central veins is adequate. We perform this examination routinely in patients with high body mass index without having any problems. The lack of training and experience in imaging these veins remains the biggest challenge.

Acute venous thrombosis

Proximal (affects any deep vein segment from CVF to the POPV) acute lower extremity DVT has a worse prognosis than distal limb (affects any deep vein segment from below the popliteal vein: the crural and muscular veins) DVT because of higher risk for pulmonary embolism and severe postthrombotic syndrome (PTS). Acute DVT may be excluded using the combination of clinical probability scores and testing for D-Dimer. DUS is the best diagnostic imaging modality for the diagnosis of DVT in patients with mid-, to high clinical probability or those who cannot be tested for D-dimer.[94]

Concerning DUS findings for DVT include partial or noncompressibility of the vein, filling defects or absence of color flow, decreased or absence of augmentation, nonspontaneity, and decreased or absence of respiratory phasicity. The degree of the above findings may indicate partial or complete obstruction of the vein. Additionally, the acutely thrombosed veins are typically enlarged, demonstrate preserve vein walls morphology, and the lumen stays hypo-, or anechoic. Occasionally, a slightly hyperechoic thrombus can be seen attached to the vein wall or freely floating in the lumen. Free-floating thrombi have always an acute onset. In this circumstance, Doppler may reveal the absence of color.

Chronic venous obstruction

Total occlusion or incomplete recanalization of the vein after acute DVT occurs often. Lower extremity chronic venous obstruction from failed recanalization, and deep venous incompetence from valvular damage may cause ambulatory venous hypertension, and the main symptoms of PTS: venous claudication and edema. One-fourth to one-half of all patients with acute symptomatic DVT may develop PTS. The prevalence and severity of PTS are higher in patients with proximal DVT and worse clot burden.[94] One-fourth to one-third of patients with PTS experience severe symptoms, including VLU.[95] About 10% of patients with PTS will develop VLU within 1–2 years.[96] Postthrombotic venous changes are well identified with DUS. B-mode can reveal vein wall thickening and irregularity, contracted and partially compressible vein, and lumen narrowing with the presence of hyperechoic fibrotic bands attached to the intima. Color flow may demonstrate the absence of flow or filling defects in the form of multiple narrowed and erratic flow channels. There is usually decreased augmentation, respiratory phasicity, and retrograde flow seen with spectral Doppler. Occasionally, multiple collateral vessels can be seen bypassing the obstruction.

Venous obstructions due to extrinsic compression are usually found in the suprainguinal veins. The protocol used for lower extremity DUS can be of great value for diagnosing venous obstruction in the pelvic veins. Findings within the CFV compatible with iliofemoral venous obstruction include reduced or absent augmentation, decreased or absent (continuous) respiratory phasicity, and nonspontaneous flow. Asymmetrical flow patterns within the CFV are highly suggestive for outflow venous obstruction (Fig. 6.5). These are just indirect findings of obstruction that require additional imaging for confirmation (i.e., multiplanar venography and intravenous ultrasound). The presence of collateral blood vessels may obscure the abovementioned sonographic findings. Therefore, normal, and symmetrical flow patterns in the CFV should not completely exclude a more proximal venous obstruction.

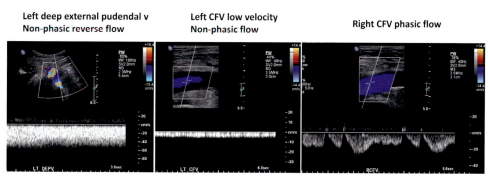

Figure. 6.5 Nonphasic flow and flow asymmetry in a patient with left iliac vein obstruction. The patient had a past iliofemoral DVT and presented with swelling and venous claudication. Nonphasic reversed flow is seen in a collateral (deep external pudendal vein). The left common femoral vein is partially recanalized with the nonphasic low flow velocity. The right common femoral vein has phasic flow. Both the nonphasic flow and the asymmetry to the normal right side indicated obstruction in the left iliac vein. This patient had chronic occlusion of the left common and external iliac veins.

A study by Hui et al.[97] evaluated the diagnostic value of lower extremity DUS for detecting iliocaval obstruction. Abnormal DUS findings in 313 limbs (192 patients) were compared to later performed cross-sectional imaging studies or venography. Intravascular venous ultrasound (IVUS) which has a higher sensitivity was not performed. Nonphasic flow demonstrated a sensitivity of 69.2%, specificity of 82.8%, negative predictive value (NPV) of 78.4%, and positive predictive value (PPV) of 74.8% for diagnosing iliocaval obstruction. Nonresponsive flow to Valsalva demonstrated a sensitivity of 13.6%, specificity of 97.6%, NPV of 61.6%, and PPV of 80.0% for diagnosing iliocaval obstruction. And the combination of nonphasic and nonresponsive flow to the Valsalva maneuver demonstrated a sensitivity of 68.2%, specificity of 87.2%, NPV of 79.6%, and PPV of 78.9%. Kayilioğlu et al.[16] compared DUS flow patterns recorded in 86 patients with the results of venography and IVUS. Among the different used DUS parameters in this study, the combination of CFV monophasic flow at rest and continuous unceasing flow during the Valsalva maneuver had the greatest diagnostic value for iliocaval venous obstructions with a sensitivity of 38.1%, specificity of 100%, NPV of 55.8%, and PPV of 100%. The sensitivity and negative predictive value of these diagnostic parameters increased as the degree of obstruction increased. The diagnostic value of abnormal DUS parameters in the CFV directly correlates with the degree of pelvic outflow venous obstruction. The reported sensitivity, specificity, NPV, and PPV of DUS for the diagnosis of outflow obstruction in a group of patients with >80% stenosis were 77%, 100%, 95%, and 100%, respectively.[18]

Reversal of flow within the SEV is a very specific finding in patients with iliocaval occlusion. Fifteen patients diagnosed with iliocaval obstruction, based on the presence of SEV flow reversal by DUS, underwent CT venography or conventional contrast venography with IVUS. Iliocaval occlusion was confirmed in all 15 patients. Furthermore, SEV flow normalized in all patients who were scanned after completing a successful procedure for recanalization of the occluded veins.[17] Uncommon causes of chronic venous obstruction of the lower extremity veins include popliteal entrapment syndrome, bone or soft tissue masses, lymphadenopathies, arterial aneurysms, abscesses, and hematoma.[89]

Recurrent venous thrombosis

Recurrence DVT is defined as venous thrombosis of a site that was either previously uninvolved or had interval documentation of incident DVT resolution.[98] One-fifth to one-third of patients may experience a recurrent venous thromboembolic event in their lifetime.[94,99] Ipsilateral DVT recurrence has been identified as a risk factor for developing severe PTS and VLU.[100,101] About 6% of patients with a history of DVT may progress to more severe forms of PTS and develop VLU.[99,102] Patients with PTS have a higher risk of experiencing DVT recurrence, making the two events a vicious cycle.[103] Decreasing the risk of ipsilateral DVT recurrence may reduce the occurrence of PTS and VLU.[104]

Ipsilateral recurrent DVT can develop in a new healthy vein and/or postthrombotic venous segments with luminal changes and varies degrees of residual venous obstruction. Postthrombotic luminal changes can be detected in one-third to more than one-half of patients after a single DVT episode.[105–108] The prevalence is higher after two or more ipsilateral DVT events.[105,108] Other factors associated with residual luminal changes include larger clot burden[105] and precipitating factors (provoked vs. unprovoked).[109]

Persistent postthrombotic luminal changes may be a risk factor for DVT recurrence, therefore higher risk for PTS and VLU. Some studies have shown an association between residual venous obstruction and VTE recurrence,[105–107,110] whereas other studies have not.[109,111,112] The presence of residual venous obstruction on surveillance DUS at 3 months follow-up, and a higher degree of scarring and luminal narrowing have been identified as main factors for DVT recurrence by some researchers.[105–107]

New baseline DUS morphological and functional abnormalities of the veins after a lower extremity DVT episode may be also clinically relevant for future evaluation of recurrent episodes. Many DUS criteria for the diagnosis of recurrent DVT within a vein segment with residual postthrombotic changes have been proposed. Some of the proposed sonographic criteria for these circumstances include an increase in the length of the thrombus by >5–9 cm, an increase in vein diameter under maximum transducer compression (thrombus thickness) >2 mm to ≥4 mm, and de novo lack of compressibility of the vein.[113–116]

Imaging during and after intervention

Ultrasound is the most necessary "tool" for performing superficial venous interventions in patients with VLU. Ultrasound is used for percutaneous vein access, injection of sclerosants, catheter insertion, advancing, and positioning of the tip, administration of tumescent anesthesia, and to validate satisfactory closure of the treated vein. The vein can be accessed with the probe oriented perpendicular or parallel to the vessel based on the physician's preference. After insertion, the catheter tip is followed up using ultrasound guidance from the access site to its final position distally to the deep-saphenous junction. The ultrasound is very useful for navigating the catheter through difficult vein segments in the presence of tortuosity, synechiae, or webs. Tumescent anesthesia is administered during endovenous thermal ablation techniques for anesthesia, dissipation of heat without affecting the skin or adjacent nerves, and obliteration of the saphenous vein and tributaries. The anesthetic solution is injected into the perivenous tissue along the entire course of the saphenous vein under direct ultrasound visualization using a transverse or longitudinal axis. Direct ultrasound visualization assures homogenous distribution of the tumescent fluid around the adventitia and prevents the inadvertent intravascular injection of the solution. At the end of the procedure, the treated veins are reassessed with DUS for technical success.[117]

Follow-up DUS after superficial vein treatment is still considered standard of care in the United States. It is performed to determine ablation success and to rule out complications (Fig. 6.6). Some of the treated incompetent veins may never occlude, a few of the occluded veins may later recanalize, and occasionally new trunks (i.e., accessory saphenous veins) may become incompetent (Fig. 6.7). Complete abolition of venous reflux is essential for VLU healing and to lower the risk of recurrence.[71,118]

The most common complications related to superficial venous interventions that can be seen by DUS include acute DVT, endovenous heat-induced thrombosis (EHIT), arteriovenous fistula, superficial thrombophlebitis, hematoma, lymphocele, seroma, and abscess. The incidence of acute DVT after saphenous ablation procedures is about 1%.[119–121] Propagation of a thrombus from the ablated superficial venous trunk into the saphenous-deep junction can be detected after thermal ablation (EHIT) or chemical ablation (postablation superficial thrombus extension, PASTE). Endovenous heat-induced thrombosis and PASTE have very distinct DUS findings and are not considered DVT. This type of thrombus is very stable and often retracts shortly after diagnosis. Only the most severe forms require the use of anticoagulation. Patients with a history of venous thromboembolic events are at a higher risk.[122,123] Other risk factors include the larger diameter of the saphenous vein, concurrent phlebectomy, and male sex.[124,125] The American Venous Forum and the Society of Vascular Surgery have recently reviewed the topic and provided recommendations for prevention and management.[126] For prevention of EHIT, the guideline recommends mechanical (compression) or chemical (anticoagulation) prophylaxis based on individual assessment of the risks, benefits, and alternatives. The tip of the catheter should be positioned >2.5 cm from the deep-saphenous junction. Additionally, the authors suggest positioning patients with large saphenous veins in extreme Trendelenburg position and infiltrating excess amount of tumescent anesthesia at the deep-saphenous junction, particularly the SFJ. The

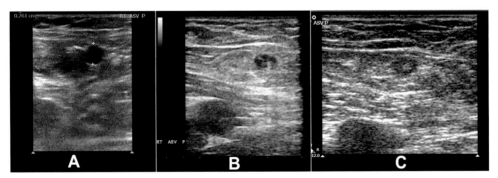

Figure. 6.6 Right anterior accessory great saphenous vein had reflux with a diameter of 7.6 mm. One-week postablation, the vein was occluded from junction to mid-thigh. One-month postablation, the vein was still occluded.

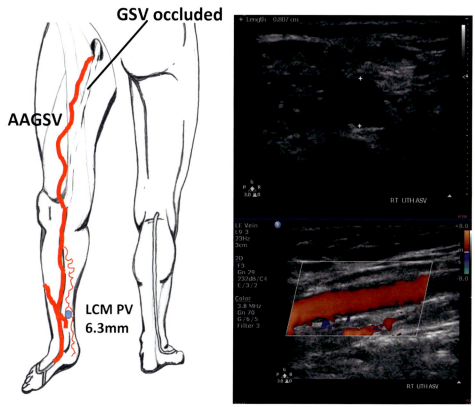

Figure. 6.7 A 63 year old male with long-standing history of bilateral lower extremity venous ulcers presented with new ulcer just above the right medial malleolus. Duplex ultrasound demonstrates reflux in the right anterior accessory great saphenous vein in continuity with nonoccluded below-knee great saphenous vein and multiple varicose veins extending to the ulcer.

recommendations for management suggest no treatment or surveillance for EHIT I, and no treatment with weekly surveillance DUS until resolution for EHIT II. Alternatively, EHIT II patients with a high risk for thrombus extension or pulmonary embolism may be offered antiplatelet therapy versus prophylactic or therapeutic anticoagulation until thrombus retraction or resolution. Patients with EHIT III should be treated with therapeutic anticoagulation and followed with weekly DUS. Anticoagulation should be suspended after thrombus retraction or resolution. Patients with EHIT IV should be managed following the recommendations for provoked proximal DVT as written in the latest Chest guidelines.[127]

An arteriovenous fistula may be occasionally detected with DUS after superficial venous interventions. It could happen because of concomitant thermal injury of the vein and the adjacent artery, or by traumatic needle injury during the administration of tumescent anesthesia. This is usually of nonclinical relevance, but occasionally may be associated with recanalization of the ablated venous segment. Spectral Doppler may reveal continuous blood flow within the venous segment of the fistula with high velocity, and low resistance waveforms. Pulsatile flow can also be seen.[128,129]

The incidence of thrombophlebitis after endovenous thermal ablation is about 5%.[130] Most superficial thrombophlebitis can be managed conservatively with compression, warm compresses, and antiinflammatory medications. Highly symptomatic superficial thrombophlebitis can be treated with incision and drainage (thrombectomy) using ultrasound guidance. Few cases of septic thrombophlebitis requiring surgical resection have been reported.[131,132] Other complications such as abscess, seroma, and fragmentation of the catheter occur less frequently ($<1\%$).[133] These complications can be easily identified with DUS.

Disclosure statement

The authors received no financial support for the research, authorship, and/or publication of this article. The authors declared no potential conflicts of interest with respect to the research, authorship, and/or publication of this article.

References

1. Raju S, Knepper J, May C, Knight A, Pace N, Jayaraj A. Ambulatory venous pressure, air plethysmography, and the role of calf venous pump in chronic venous disease. *J Vasc Surg Venous Lymphat Disord*. May 2019;73:428–440.
2. Raffetto JD, Ligi D, Maniscalco R, Khalil RA, Mannello F. Why venous leg ulcers have difficulty healing: overview on pathophysiology, clinical consequences, and treatment. *J Clin Med*. December 24, 2020;10(1):29.
3. Meissner MH, Gloviczki P, Bergan J, et al. Primary chronic venous disorders. *J Vasc Surg*. December 2007;46(Suppl S):54S–67S.
4. Kanth AM, Khan SU, Gasparis A, Labropoulos N. The distribution and extent of reflux and obstruction in patients with active venous ulceration. *Phlebology*. June 2015;30(5):350–356.
5. Labropoulos N, Leon M, Nicolaides AN, Giannoukas AD, Volteas N, Chan P. Superficial venous insufficiency: correlation of anatomic extent of reflux with clinical symptoms and signs. *J Vasc Surg*. December 1994;20(6):953–958.
6. Labropoulos N, Leon M, Geroulakos AD, Volteas N, Chan P, Nicolaides AN. Venous hemodynamic abnormalities in patients with leg ulceration. *Am J Surg*. June 1995;169(6):572–574.
7. Tassiopoulos AK, Golts E, Oh DS, Labropoulos N. Current concepts in chronic venous ulceration. *Eur J Vasc Endovasc Surg*. September 2000;20(3):227–232.
8. Labropoulos N, Giannoukas AD, Nicolaides AN, Ramaswami G, Leon M, Burke P. New insights into the pathophysiologic condition of venous ulceration with color-flow duplex imaging: implications for treatment? *J Vasc Surg*. July 1995;22(1):45–50.
9. Obermayer A, Garzon K. Identifying the source of superficial reflux in venous leg ulcers using duplex ultrasound. *J Vasc Surg*. November 2010;52(5):1255–1261.
10. van Rij AM, Solomon C, Christie R. Anatomic and physiologic characteristics of venous ulceration. *J Vasc Surg*. November 1994;20(5):759–764.
11. Nicolaides AN. Cardiovascular Disease Educational and Research Trust; European Society of Vascular Surgery; The International Angiology Scientific Activity Congress Organization; International Union of Angiology; Union Internationale de Phlebologie at the Abbaye des Vaux de Cernay. Investigation

of chronic venous insufficiency: a consensus statement (France, March 5-9, 1997). *Circulation*. November 14, 2000;102(20):E126—E163.
12. Marston W, Fish D, Unger J, Keagy B. Incidence of and risk factors for iliocaval venous obstruction in patients with active or healed venous leg ulcers. *J Vasc Surg*. May 2011;53(5):1303—1308.
13. Raju S, Kirk OK, Jones TL. Endovenous management of venous leg ulcers. *J Vasc Surg Venous Lymphat Disord*. April 2013;1(2):165—172.
14. Wen-da W, Yu Z, Yue-Xin C. Stenting for chronic obstructive venous disease: a current comprehensive meta-analysis and systematic review. *Phlebology*. July 2016;31(6):376—389.
15. Kokkosis AA, Labropoulos N, Gasparis AP. Investigation of venous ulcers. *Semin Vasc Surg*. March 2015;28(1):15—20.
16. Kayılıoğlu SI, Köksoy C, Alaçayır I. Diagnostic value of the femoral vein flow pattern for the detection of an iliocaval venous obstruction. *J Vasc Surg Venous Lymphat Disord*. January 2016;4(1):2—8.
17. Kolluri R, Fowler B, Ansel G, Silver M. A novel duplex finding of superficial epigastric vein flow reversal to diagnose iliocaval occlusion. *J Vasc Surg Venous Lymphat Disord*. May 2017;5(3):358—362.
18. Sloves J, Almeida JI. Venous duplex ultrasound protocol for iliocaval disease. *J Vasc Surg Venous Lymphat Disord*. November 2018;6(6):748—757.
19. Lurie F, Comerota A, Eklof B, et al. Multicenter assessment of venous reflux by duplex ultrasound. *J Vasc Surg*. February 2012;55(2):437—445.
20. Gianesini S, Obi A, Onida S, et al. Global guidelines trends and controversies in lower limb venous and lymphatic disease: narrative literature revision and experts' opinions following the vWINter international meeting in Phlebology, Lymphology & Aesthetics, 23-25 January 2019. *Phlebology*. September 2019;34(1 Suppl):4—66.
21. Kim ES, Sharma AM, Scissons R, et al. Interpretation of peripheral arterial and venous Doppler waveforms: a consensus statement from the society for vascular medicine and society for vascular ultrasound. *Vasc Med*. October 2020;25(5):484—506.
22. Coleridge-Smith P, Labropoulos N, Partsch H, Myer K, Nicolaides A, Cavezzi A, et al. Duplex ultrasound investigation of the veins in chronic venous disease of the lower limbs: UIP consensus document. Part 1. Basic principles. *Eur J Vasc Endovasc Surg*. 2006;31:83—92.
23. Leon LR, Labropoulos N. Vascular laboratory: venous duplex scanning. In: Cronenwett JL, Johnston W, eds. *Rutherford's Vascular Surgery*. 8th ed. London: Saunders Elsevier; 2014.
24. Labropoulos N, Tiongson J, Pryor L, et al. Definition of venous reflux in lower-extremity veins. *J Vasc Surg*. October 2003;38(4):793—798.
25. Neglén P, Egger 3rd JF, Olivier J, Raju S. Hemodynamic and clinical impact of ultrasound-derived venous reflux parameters. *J Vasc Surg*. August 2004;40(2):303—310.
26. Labropoulos N, Waggoner T, Sammis W, Samali S, Pappas PJ. The effect of venous thrombus location and extent on the development of post-thrombotic signs and symptoms. *J Vasc Surg*. August 2008;48(2):407—412.
27. Johnson BF, Manzo RA, Bergelin RO, Strandness Jr DE. Relationship between changes in the deep venous system and the development of the postthrombotic syndrome after an acute episode of lower limb deep vein thrombosis: a one- to six-year follow-up. *J Vasc Surg*. February 1995;21(2):307—312. discussion 313.
28. Alavi A, Sibbald RG, Phillips TJ, et al. What's new: management of venous leg ulcers: approach to venous leg ulcers. *J Am Acad Dermatol*. April 2016;74(4):627—640. quiz 641-2.
29. Magnusson MB, Nelzén O, Risberg B, Sivertsson R. A colour Doppler ultrasound study of venous reflux in patients with chronic leg ulcers. *Eur J Vasc Endovasc Surg*. April 2001;21(4):353—360.
30. Grabs AJ, Wakely MC, Nyamekye I, Ghauri AS, Poskitt KR. Colour duplex ultrasonography in the rational management of chronic venous leg ulcers. *Br J Surg*. October 1996;83(10):1380—1382.
31. Adam DJ, Naik J, Hartshorne T, Bello M, London NJ. The diagnosis and management of 689 chronic leg ulcers in a single-visit assessment clinic. *Eur J Vasc Endovasc Surg*. May 2003;25(5):462—468.
32. Ibegbuna V, Delis KT, Nicolaides AN. Haemodynamic and clinical impact of superficial, deep and perforator vein incompetence. *Eur J Vasc Endovasc Surg*. 2006;31:535—541.
33. Ioannou CV, Giannoukas AD, Kostas T, et al. Patterns of venous reflux in limbs with venous ulcers. Implications for treatment. *Int Angiol*. June 2003;22(2):182—187.

34. Hanrahan LM, Araki CT, Rodriguez AA, Kechejian GJ, LaMorte WW, Menzoian JO, et al. Distribution of valvular incompetence in patients with venous stasis ulceration. *J Vasc Surg*. 1991;13: 805–811.
35. Eberhardt RT, Raffetto JD. Chronic venous insufficiency. *Circulation*. July 22, 2014;130(4):333–346.
36. Darke SG, Penfold C. Venous ulceration and saphenous ligation. *Eur J Vasc Surg*. 1992;6:4–9.
37. Lees TA, Lambert D. Patterns of venous reflux in limbs with skin changes associated with chronic venous insufficiency. *Br J Surg*. 1993;80:725–728.
38. Myers KA, Ziegenbein RW, Zeng GH, Matthews PG. Duplex ultrasonography scanning for chronic venous disease: patterns of venous reflux. *J Vasc Surg*. 1995;21:605–612.
39. Bjordal R. Simultaneous pressure and flow recordings in varicose veins of the lower extremity. A haemodynamic study of venous dysfunction. *Acta Chir Scand*. 1970;136:309–317.
40. Christopoulos D, Nicolaides AN, Szendro G. Venous reflux: quantification and correlation with the clinical severity of chronic venous disease. *Br J Surg*. 1988;75:352–356.
41. Delis KT, Ibegbuna V, Nicolaides AN, Lauro A, Hafez H. Prevalence and distribution of incompetent perforating veins in chronic venous insufficiency. *J Vasc Surg*. 1998;28:815–825.
42. Zukowski AJ, Nicolaides AN, Szendro G, et al. Haemodynamic significance of incompetent calf perforating veins. *Br J Surg*. 1991;78:625–629.
43. Stuart WP, Adam DJ, Allan PL, Ruckley CV, Bradbury AW. The relationship between the number, competence, and diameter of medial calf perforating veins and the clinical status in healthy subjects and patients with lower-limb venous disease. *J Vasc Surg*. 2000;32:138–143.
44. Stuart WP, Lee AJ, Allan PL, Ruckley CV, Bradbury AW. Most incompetent calf perforating veins are found in association with superficial venous reflux. *J Vasc Surg*. November 2001;34(5):774–778.
45. Delis KT, Husmann M, Kalodiki E, Wolfe JH, Nicolaides AN. In situ hemodynamics of perforating veins in chronic venous insufficiency. *J Vasc Surg*. 2001;33:773–782.
46. Labropoulos N, Meissner M, Nicolaides AN, Sowade O, Volteas N, Chan P, et al. Venous reflux in patients with previous deep venous thrombosis: correlation with ulceration and other symptoms. *J Vasc Surg*. July 1994;20(1):20–26.
47. Labropoulos N, Gasparis AP, Tassiopoulos AK. Prospective evaluation of the clinical deterioration in post-thrombotic limbs. *J Vasc Surg*. October 2009;50(4):826–830.
48. Labropoulos N, Patel PJ, Tiongson JE, Pryor L, Leon Jr LR, Tassiopoulos AK, et al. Patterns of venous reflux and obstruction in patients with skin damage due to chronic venous disease. *Vasc Endovasc Surg*. 2007 Feb-Mar;41(1):33–40.
49. Meissner MH, Moneta G, Burnand K, et al. The hemodynamics and diagnosis of venous disease. *J Vasc Surg*. 2007;46:4S–24S.
50. Labropoulos N, Leon L, Rodriguez H, Kang SS, Mansour AM, Littooy FN, et al. Deep venous reflux and incompetent perforators: significance and implications for therapy. *Phlebology*. 2004;19(1):22–27.
51. Kistner RL, Eklof B, Masuda EM. Diagnosis of chronic venous disease of the lower extremities: the "CEAP" classification. *Mayo Clin Proc*. April 1996;71(4):338–345.
52. Labropoulos N. CEAP in clinical practice. *Vasc Surg*. 1997;31:224–225.
53. Plate G, Brudin L, Eklof B, Jensen R, Ohlin P. Congenital vein valve aplasia. *Wold J Surg*. 1986;10(6): 929–934.
54. Parker CN, Finlayson KJ, Shuter P, Edwards HE. Risk factors for delayed healing in venous leg ulcers: a review of the literature. *Int J Clin Pract*. September 2015;69(9):967–977.
55. Pannier F, Rabe E. Differential diagnosis of leg ulcers. *Phlebology*. March 2013;28(Suppl 1):55–60.
56. Pittaluga P, Chastane S, Rea B, Barbe R. Classification of saphenous refluxes: implications for treatment. *Phlebology*. 2008;23(1):2–9.
57. Curri SB, Annoni F, Montorsi W. Les microvalvules dans les microveinules [Microvalves in microvenules]. *Phlebologie*. 1987 Jul-Sep;40(3):795–801.
58. Caggiati A, Phillips M, Lametschwandtner A, Allegra C. Valves in small veins and venules. *Eur J Vasc Endovasc Surg*. October 2006;32(4):447–452.
59. Vincent JR, Jones GT, Hill GB, van Rij AM. Failure of microvenous valves in small superficial veins is a key to the skin changes of venous insufficiency. *J Vasc Surg*. December 2011;54(6 suppl l), 62S–9S.c1–3.

60. Lin JC, Iafrati MD, O'Donnell Jr TF, Estes JM, Mackey WC. Correlation of duplex ultrasound scanning-derived valve closure time and clinical classification in patients with small saphenous vein reflux: is lesser saphenous vein truly lesser? *J Vasc Surg*. May 2004;39(5):1053–1058.
61. Labropoulos N, Giannoukas AD, Delis K, et al. The impact of isolated lesser saphenous vein system incompetence on clinical signs and symptoms of chronic venous disease. *J Vasc Surg*. November 2000; 32(5):954–960.
62. Qureshi MI, Lane TR, Moore HM, Franklin IJ, Davies AH. Patterns of short saphenous vein incompetence. *Phlebology*. March 2013;28(Suppl 1):47–50.
63. Boulton AJ, Armstrong DG, Albert SF, et al. American Diabetes Association. American Association of Clinical Endocrinologists. Comprehensive foot examination and risk assessment: a report of the task force of the foot care interest group of the American Diabetes Association, with endorsement by the American Association of Clinical Endocrinologists. *Diabetes Care*. August 2008;31(8):1679–1685.
64. van Bemmelen PS, Spivack D, Kelly P. Reflux in foot veins is associated with venous toe and forefoot ulceration. *J Vasc Surg*. February 2011;53(2):394–398.
65. Stelzner C, Schellong S, Wollina U, Machetanz J, Unger L. Fußläsionen [foot lesions]. *Internist*. November 2013;54(11):1330–1336.
66. Kalra M, Gloviczki P. Surgical treatment of venous ulcers: role of subfascial endoscopic perforator vein ligation. *Surg Clin*. 2003;83:671–705.
67. Shull KC, Nicolaides AN, Fernandes é Fernandes J, et al. Significance of popliteal reflux in relation to ambulatory venous pressure and ulceration. *Arch Surg*. November 1979;114(11):1304–1306.
68. Brittenden J, Bradbury AW, Allan PL, Prescott RJ, Harper DR, Ruckley CV, et al. Popliteal vein reflux reduces the healing of chronic venous ulcer. *Br J Surg*. January 1998;85(1):60–62.
69. Hjerppe A, Saarinen JP, Venermo MA, Huhtala HS, Vaalasti A. Prolonged healing of venous leg ulcers: the role of venous reflux, ulcer characteristics and mobility. *J Wound Care*. November 2010; 19(11):474–478.
70. Kjaer ML, Jorgensen B, Karlsmark T, Holstein P, Simonsen L, Gottrup F, et al. Does the pattern of venous insufficiency influence healing of venous leg ulcers after skin transplantation? *Eur J Vasc Endovasc Surg*. June 2003;25(6):562–567.
71. Danielsson G, Eklof B, Grandinetti A, Lurie F, Kistner RL. Deep axial reflux, an important contributor to skin changes or ulcer in chronic venous disease. *J Vasc Surg*. December 2003;38(6):1336–1341.
72. Yamaki T, Nozaki M, Sakurai H, Takeuchi M, Soejima K, Kono T, et al. High peak reflux velocity in the proximal deep veins is a strong predictor of advanced post-thrombotic sequelae. *J Thromb Haemost*. February 2007;5(2):305–312.
73. Marston WA, Brabham VW, Mendes R, Berndt D, Weiner M, Keagy B, et al. The importance of deep venous reflux velocity as a determinant of outcome in patients with combined superficial and deep venous reflux treated with endovenous saphenous ablation. *J Vasc Surg*. August 2008;48(2): 400–405.
74. O'Donnell Jr TF, Passman MA, Marston WA, Ennis WJ, Dalsing M, O'Donnell Jr TF, et al, Society for Vascular Surgery, American Venous Forum. Management of venous leg ulcers: clinical practice guidelines of the Society for Vascular Surgery ® and the American Venous Forum. *J Vasc Surg*. August 2014;60(2 Suppl):3S–59S.
75. Welch HJ, Young CM, Semegran AB, Iafrati MD, Mackey WC, et al. Duplex assessment of venous reflux and chronic venous insufficiency: the significance of deep venous reflux. *J Vasc Surg*. November 1996;24(5):755–775.
76. Marston WA, Crowner J, Kouri A, Kalbaugh CA. Incidence of venous leg ulcer healing and recurrence after treatment with endovenous laser ablation. *J Vasc Surg Venous Lymphat Disord*. July 2017; 5(4):525–532.
77. Brown CS, Osborne NH, Kim GY, et al. Effect of concomitant deep venous reflux on truncal endovenous ablation outcomes in the Vascular Quality Initiative. *J Vasc Surg Venous Lymphat Disord*. March 2021;9(2):361–368. e3.
78. Puggioni A, Lurie F, Kistner RL, Eklof B. How often is deep venous reflux eliminated after saphenous vein ablation? *J Vasc Surg*. September 2003;38(3):517–521.

79. Labropoulos N. Current views on the management of incompetent perforator veins. *Ann Phlebology.* 2020;18(1):1—3.
80. Al-Mulhim AS, El-Hoseiny H, Al-Mulhim FM, et al. Surgical correction of main stem reflux in the superficial venous system: does it improve the blood flow of incompetent perforating veins? *World J Surg.* July 2003;27(7):793—796.
81. Stuart WP, Adam DJ, Allan PL, Ruckley CV, Bradbury AW. Saphenous surgery does not correct perforator incompetence in the presence of deep venous reflux. *J Vasc Surg.* November 1998;28(5):834—838.
82. Goldschmidt E, Schafer K, Lurie F. A systematic review on the treatment of nonhealing venous ulcers following successful elimination of superficial venous reflux. *J Vasc Surg Venous Lymphat Disord.* July 2021;9(4):1071—1076.
83. Delis KT. Leg perforator vein incompetence: functional anatomy. *Radiology.* April 2005;235(1):327—334.
84. Liu X, Zheng G, Ye B, Chen W, Xie H, Zhang T, et al. Factors related to the size of venous leg ulcers: a cross-sectional study. *Medicine (Baltim).* February 2019;98(5):e14389.
85. Dillavou ED, Harlander-Locke M, Labropoulos N, Elias S, Ozsvath KJ. Current state of the treatment of perforating veins. *J Vasc Surg Venous Lymphat Disord.* January 2016;4(1):131—135.
86. Perrin M, Eklof B, Maleti O. *The Vein Glossary.* 2020.
87. Danielsson G, Arfvidsson B, Eklof B, Kistner RL, Masuda EM, Satoc DT, et al. Reflux from thigh to calf, the major pathology in chronic venous ulcer disease: surgery indicated in the majority of patients. *Vasc Endovasc Surg.* 2004 May-Jun;38(3):209—219.
88. Lurie F. Anatomical extent of venous reflux. *Cardiol Ther.* 2020;9:215—218.
89. Garcia R, Labropoulos N. Duplex ultrasound for the diagnosis of acute and chronic venous diseases. *Surg Clin.* April 2018;98(2):201—218.
90. Kockaert M, de Roos KP, van Dijk L, Nijsten T, Neumann M. Duplication of the great saphenous vein: a definition problem and implications for therapy. *Dermatol Surg.* January 2012;38(1):77—82.
91. Meissner MH, Khilnani NM, Labropoulos N, et al. The symptoms-varices-pathophysiology classification of pelvic venous disorders: a report of the American vein & lymphatic society international working group on pelvic venous disorders. *J Vasc Surg Venous Lymphat Disord.* May 2021;9(3):568—584.
92. Neglén P, Thrasher TL, Raju S. Venous outflow obstruction: an underestimated contributor to chronic venous disease. *J Vasc Surg.* November 2003;38(5):879—885.
93. Lawrence PF, Hager ES, Harlander-Locke MP, et al. Treatment of superficial and perforator reflux and deep venous stenosis improves healing of chronic venous leg ulcers. *J Vasc Surg Venous Lymphat Disord.* July 2020;8(4):601—609.
94. Chopard R, Albertsen IE, Piazza G. Diagnosis and treatment of lower extremity venous thromboembolism: a review. *JAMA.* November 3, 2020;324(17):1765—1776.
95. Kahn SR, Ginsberg JS. Relationship between deep venous thrombosis and the postthrombotic syndrome. *Arch Intern Med.* January 12, 2004;164(1):17—26.
96. Ashrani AA, Heit JA. Incidence and cost burden of post-thrombotic syndrome. *J Thromb Thrombolysis.* November 2009;28(4):465—476.
97. Hui JZ, Goldman RE, Mabud TS, Arendt VA, Kuo WT, Hofmann LB, et al. Diagnostic performance of lower extremity Doppler ultrasound in detecting iliocaval obstruction. *J Vasc Surg Venous Lymphat Disord.* September 2020;8(5):821—830.
98. Heit JA, Lahr BD, Petterson TM, Bailey KR, Ashrani AA, Melton 3rd LJ, et al. Heparin and warfarin anticoagulation intensity as predictors of recurrence after deep vein thrombosis or pulmonary embolism: a population-based cohort study. *Blood.* November 3, 2011;118(18):4992—4999.
99. Schulman S, Lindmarker P, Holmström M, et al. Post-thrombotic syndrome, recurrence, and death 10 years after the first episode of venous thromboembolism treated with warfarin for 6 weeks or 6 months. *J Thromb Haemost.* April 2006;4(4):734—742.
100. Ziegler S, Schillinger M, Maca TH, Minar E. Post-thrombotic syndrome after primary event of deep venous thrombosis 10 to 20 years ago. *Thromb Res.* January 15, 2001;101(2):23—33.

101. Kahn SR. Frequency and determinants of the postthrombotic syndrome after venous thromboembolism. *Curr Opin Pulm Med*. September 2006;12(5):299−303.
102. Polak MW, Siudut J, Plens K, Undas A. Prothrombotic clot properties can predict venous ulcers in patients following deep vein thrombosis: a cohort study. *J Thromb Thrombolysis*. November 2019; 48(4):603−609.
103. Stain M, Schönauer V, Minar E, et al. The post-thrombotic syndrome: risk factors and impact on the course of thrombotic disease. *J Thromb Haemost*. December 2005;3(12):2671−2676.
104. Kahn SR. The post-thrombotic syndrome: the forgotten morbidity of deep venous thrombosis. *J Thromb Thrombolysis*. February 2006;21(1):41−48.
105. Yoo T, Aggarwal R, Wang TF, Satiani B, Haurani MJ. Presence and degree of residual venous obstruction on serial duplex imaging is associated with increased risk of recurrence and progression of infrainguinal lower extremity deep venous thrombosis. *J Vasc Surg Venous Lymphat Disord*. September 2018;6(5):575−583.
106. Stephenson EJ, Liem TK. Duplex imaging of residual venous obstruction to guide duration of therapy for lower extremity deep venous thrombosis. *J Vasc Surg Venous Lymphat Disord*. July 2015;3(3): 326−332.
107. Donadini MP, Ageno W, Antonucci E, et al. Prognostic significance of residual venous obstruction in patients with treated unprovoked deep vein thrombosis: a patient-level meta-analysis. *Thromb Haemost*. January 2014;111(1):172−179.
108. Galli M, Ageno W, Squizzato A, et al. Residual venous obstruction in patients with a single episode of deep vein thrombosis and in patients with recurrent deep vein thrombosis. *Thromb Haemost*. July 2005; 94(1):93−95.
109. Carrier M, Rodger MA, Wells PS, Righini M, LE Gal G. Residual vein obstruction to predict the risk of recurrent venous thromboembolism in patients with deep vein thrombosis: a systematic review and meta-analysis. *J Thromb Haemost*. June 2011;9(6):1119−1125.
110. Prandoni P, Lensing AW, Prins MH, et al. Residual venous thrombosis as a predictive factor of recurrent venous thromboembolism. *Ann Intern Med*. December 17, 2002;137(12):955−960.
111. Watson HG. RVO−real value obscure. *J Thromb Haemost*. June 2011;9(6):1116−1118.
112. LE Gal G, Carrier M, Kovacs MJ, et al. Residual vein obstruction as a predictor for recurrent thromboembolic events after a first unprovoked episode: data from the REVERSE cohort study. *J Thromb Haemost*. June 2011;9(6):1126−1132.
113. Prandoni P, Cogo A, Bernardi E, et al. A simple ultrasound approach for detection of recurrent proximal-vein thrombosis. *Circulation*. October 1993;88(4 Pt 1):1730−1735.
114. Gibbs H. The diagnosis of recurrent deep venous thrombosis. *Aust Prescr*. 2007;30:38−40.
115. Linkins LA, Pasquale P, Paterson S, Kearon C. Change in thrombus length on venous ultrasound and recurrent deep vein thrombosis. *Arch Intern Med*. September 13, 2004;164(16):1793−1796.
116. Linkins LA, Stretton R, Probyn L, Kearon C. Interobserver agreement on ultrasound measurements of residual vein diameter, thrombus echogenicity and Doppler venous flow in patients with previous venous thrombosis. *Thromb Res*. 2006;117(3):241−247.
117. Nyamekye IK. A practical approach to tumescent local anaesthesia in ambulatory endovenous thermal ablation. *Phlebology*. May 2019;34(4):238−245.
118. Magnusson MB, Nelzén O, Volkmann R. Leg ulcer recurrence and its risk factors: a duplex ultrasound study before and after vein surgery. *Eur J Vasc Endovasc Surg*. October 2006;32(4):453−461.
119. Healy DA, Kimura S, Power D, et al. A systematic review and meta-analysis of thrombotic events following endovenous thermal ablation of the great saphenous vein. *Eur J Vasc Endovasc Surg*. September 2018;56(3):410−424.
120. Aurshina A, Ascher E, Victory J, et al. Clinical correlation of success and acute thrombotic complications of lower extremity endovenous thermal ablation. *J Vasc Surg Venous Lymphat Disord*. January 2018;6(1):25−30.
121. Merchant Jr R, Kistner RL, Kabnick LS. Is there an increased risk for DVT with the VNUS closure procedure? *J Vasc Surg*. September 2003;38(3):628.
122. Harlander-Locke M, Jimenez JC, Lawrence PF, Derubertis BG, Rigberg DA, Gelabert HA, et al. Endovenous ablation with concomitant phlebectomy is a safe and effective method of treatment

for symptomatic patients with axial reflux and large incompetent tributaries. *J Vasc Surg.* 2013;58: 166−172.
123. Puggioni A, Marks N, Hingorani A, Shiferson A, Alhalbouni S, Ascher E, et al. The safety of radio-frequency ablation of the great saphenous vein in patients with previous venous thrombosis. *J Vasc Surg.* 2009;49:1248−1255.
124. Sufian S, Arnez A, Labropoulos N, Lakhanpal S. Incidence, progression, and risk factors for endovenous heat-induced thrombosis after radiofrequency ablation. *J Vasc Surg Venous Lymphat Disord.* April 2013;1(2):159−164.
125. Sufian S, Arnez A, Labropoulos N, Lakhanpal S. Endovenous heat-induced thrombosis after ablation with 1470 nm laser: incidence, progression, and risk factors. *Phlebology.* June 2015;30(5):325−330.
126. Kabnick LS, Sadek M, Bjarnason H, et al. Classification and treatment of endothermal heat-induced thrombosis: recommendations from the American venous Forum and the society for vascular surgery. *Phlebology.* February 2021;36(1):8−25.
127. Kearon C, Akl EA, Ornelas J, et al. Antithrombotic therapy for VTE disease: CHEST guideline and expert panel report. *Chest.* 2016;149:315−352.
128. Theivacumar NS, Gough MJ. Arterio-venous fistula following endovenous laser ablation for varicose veins. *Eur J Vasc Endovasc Surg.* August 2009;38(2):234−236.
129. Rudarakanchana N, Berland TL, Chasin C, Sadek M, Kabnick LS. Arteriovenous fistula after endovenous ablation for varicose veins. *J Vasc Surg.* May 2012;55(5):1492−1494.
130. Dermody M, O'Donnell TF, Balk EM. Complications of endovenous ablation in randomized controlled trials. *J Vasc Surg Venous Lymphat Disord.* October 2013;1(4):427−436.
131. Grady Z, Aizpuru M, Farley KX, Benarroch-Gampel J, Crawford RS. Surgical resection for suppurative thrombophlebitis of the great saphenous vein after radiofrequency ablation. *J Vasc Surg Cases Innov Tech.* November 22, 2019;5(4):532−534.
132. Gupta S, Mansuri N, Kowdley G. Group G streptococcus leading to necrotizing soft tissue infection after left lower extremity radiofrequency venous ablation. *J Vasc Surg Cases Innov Tech.* April 28, 2019; 5(2):110−112.
133. Mazayshvili K, Akimov S. Early complications of endovenous laser ablation. *Int Angiol.* April 2019; 38(2):96−101.

CHAPTER 7

The diagnosis of major venous outflow obstruction in chronic venous insufficiency

Jovan N. Markovic[1], Martin V. Taormina[1] and Ellen D. Dillavou[2]

[1]Department of Surgery, Division of Vascular Surgery, Duke University School of Medicine, Durham, NC, United States;
[2]Vascular Surgery, WakeMed Hospital System, Raleigh, NC, United States

Introduction

Chronic venous disease (CVD) can present in many different stages as the development and progression of this prevalent venous disorder usually occur over the course of many years. Typical early symptoms include leg swelling, pain, heaviness, burning, aching, and skin discoloration with pigmentation. As the disease progresses, patients can develop skin changes and ulcerations. Determining the underlying etiology of venous hypertension is critical for diagnosis and treatment planning. Defining whether venous hypertension results from venous reflux or venous outflow obstruction is a critical component in selecting the appropriate intervention. Valvular dysfunction and venous reflux are the major underlying causes of chronic venous insufficiency (CVI). However, iliac outflow obstruction, from either postthrombotic changes or external compression, is becoming increasingly recognized as an important factor in the pathogenesis of venous insufficiency. The combination of reflux and outflow obstruction results in the highest levels of venous hypertension and the most severe clinical symptoms. Therefore, timely and accurate diagnosis of venous obstruction is critical in the management of these patients. This chapter will examine a diagnostic algorithm and diagnostic modalities used for the assessment of venous hypertension caused by venous outflow obstruction in patients with venous ulceration.

Iliac and pelvic veins anatomy and hemodynamics

Given the continuity of the venous system, especially pertinent to the relationship between abdominal, pelvic, and infrainguinal venous anatomy and hemodynamics, combined with the increasing knowledge we are gaining regarding pelvic venous congestion syndrome, it is imperative that a comprehensive approach that evaluates venous system as a continuum rather than an approach that evaluates isolated venous segments is undertaken. Understanding the nominal pelvic venous anatomy and normal venous hemodynamics is the first and fundamental step in understanding the pathophysiology of venous outflow obstruction, its diagnosis, and the selection of the most adequate treatment modality. The most caudal infrainguinal deep vein is the common femoral vein (CFV) which is located between the confluence of the femoral, deep femoral, and great saphenous vein inferior to the inguinal ligament. As it ascends posterior to the inguinal ligament, the CFV becomes the external iliac vein (EIV). On its cephaloid course the EIV confluences with the internal iliac vein (IIV) to form the common iliac vein (CIV). The EIV also represents an outflow for the inferior epigastric vein, deep circumflex iliac vein, and pelvic veins. The inferior epigastric vein communicates superiorly with the superior epigastric vein. The left and right CIV join on the right side of the L5 vertebrae and form the inferior vena cava (IVC). The right side of the spine at the L5 level represents an important anatomic landmark during venous stenting, as visualization of hardware used during stenting on the left side or aligned with the spine is indicative either of anatomic variation (i.e., left-sided IVC) or misplacement of the hardware during the procedure. Angulation of the confluence of the left CIV with the IVC is more prominent when compared to angulation between the right CIV and the IVC. In addition, the left CIV is located posterior to the right common iliac artery, and this anatomic proximity has an important role in the pathogenies of the May—Thurner syndrome when the left CIV is compressed between the right iliac artery and the convexity of the sacroiliac spine. In the pelvis, the venous system comprises a complex network of interconnected venous plexuses. The IIV represents an outflow for superior and inferior gluteal, sacral, sciatic, lumbar, obturator, and internal pudendal veins as well as for the visceral veins including the gonadal, hemorrhoidal, vesico-prostatic (in males) and uterine and vesicovaginal veins (in females). The right gonadal vein joins the IVC directly and the left gonadal vein joins the left renal, which drains into the IVC.[1–4] It is important to note that the above-mentioned anatomy varies significantly, emphasizing the importance of taking an individualized approach to each patient and each extremity.

Diagnosis of chronic venous insufficiency and venous hypertension
History and physical

The history taking and physical examination are critical to initiate diagnosis and therapy of the venous ulcer patient. However, these findings are often vague and may not differentiate the exact pathophysiology and anatomic location of the venous dysfunction. The differential diagnosis of leg pain can be complex. The differential diagnosis includes arterial occlusive disease, venous disease, neuropathic pain, musculoskeletal disorders, spinal stenosis or compression, trauma, and many others. When patients present with ulcerations, a multitude of other differentials should be considered such as vasculitis, pyoderma gangrenosum, ulcerated malignancies (such as basal cell or squamous cell carcinomas), fungi, or traumatic injuries just to name a few.

A detailed history should be performed to aid in the appropriate pathway for diagnostic evaluation. There are several etiologies of CVD, which include postthrombotic syndrome (PTS) resulting from previous deep vein thrombosis (DVT) and nonthrombotic iliac vein lesions (NIVLs). The postthrombotic syndrome produces scarring directly related to venous thrombosis, which may result in reflux and valvular incompetence, creating chronic venous hypertension. PTS is the most common long-term complication of DVT and is seen in approximately 20%—50% of patients with prior DVT.[1] Most frequent cause of severe PTS is seen with iliofemoral DVT.[2,3] Inquiring about previous thrombotic events is one of the key aspects of the history, but many patients do not know the exact location or extent of the DVT. Thus, pelvic imaging in addition to detailed imaging of the entire lower extremity should be considered for all patients who report a history, or have clinical presentation suggestive of DVT.

Patients who do not have a history of previous thrombosis and still have symptoms of venous hypertension should be screened for NIVL. The most commonly seen NIVL is left iliac vein compression between the spine and right iliac artery, also known as May—Thurner syndrome, as mentioned earlier.[5] There are other anatomic variants less frequently seen compressing the right iliac vein or the vena cava. The resultant compression produces venous hypertension from outflow obstruction.

Physical findings of CVI include leg swelling, hemosiderin deposition resulting in skin pigmentation, dry, scaly skin, and healed or active ulcerations. Medical causes of leg swelling should be ruled out such as liver disease, congestive heart failure, pulmonary hypertension, lymphedema, nephrotic syndrome, acute and chronic kidney disease, pregnancy, and medication-induced swelling (see Chapter 9).

Diagnostic modalities
Arterial testing

Every patient presenting with leg ulceration should have a complete arterial workup, and all patients need to have a documented pulse exam, which is critical in the decision pathway. Adequate perfusion is paramount for wound healing. Evaluation for risk factors of peripheral arterial disease is critical. These include male gender, age, smoking, hypertension, dyslipidemia, diabetes, obesity, coexisting cardiovascular disease, stroke, renal insufficiency, and family history. Concomitant arterial and venous insufficiency, seen in approximately 20% of ulceration patients, should be evaluated in any patient with CVD. If the patient has a palpable pulse, no further testing may be required. However, in the absence of a palpable pulse or weak pulses, further testing is mandatory. Measuring an ankle-brachial index (ABI) is a safe and reliable method of quantifying appropriate flow to the extremity, which is essential, especially if conservative measures such as compression therapy may be initiated. When measuring ankle pressure, the lower extremity cuff should be placed on the distal-most portion of the leg. This is compared to the brachial pressure which should reveal the overall perfusion to the lower extremity. Based on this result, further evaluation may be needed. Any patient with a venous ulcer and an ABI lower than 0.5 should be considered for revascularization. This is to ensure adequate perfusion to heal and to allow for adequate compression. Compression can result in new wounds and is generally avoided if the ABI is less than 0.5. Above normal ABI values are indicative of a noncompressible calcified artery, and don't exclude the presence of peripheral artery disease.

Correction of arterial insufficiency of slowly healing ulcers is essential to improve wound healing. Minimally invasive procedures such as angiography may be helpful in the diagnosis of the level of disease and intervention planning to increase distal perfusion. A surgical bypass may be warranted depending on the clinical scenario.

Duplex ultrasonography

Duplex ultrasonography imaging is the most critical diagnostic modality for evaluating the lower extremity venous system. It is affordable, safe, and effective in measuring many anatomic and hemodynamic components of the venous system to better delineate underlying pathophysiology. It must be emphasized that duplex examination is highly operator dependent, and each patient should undergo a complete venous duplex evaluation by a qualified registered vascular technologist in an accredited vascular lab. Duplex scans should be obtained in the supine position for assessment of DVT and in a standing position for the venous reflux exam. Compression and flow patterns with augmentation should be obtained to identify both acute and chronic venous obstruction or incompetent valves.[4]

The common femoral, deep femoral, superficial femoral, and popliteal veins are evaluated for reflux and thrombosis. The saphenous veins and perforators are evaluated as well. The reflux scan is done either standing or in the reversed Trendelenburg position if the patient cannot safely stand. The common femoral vein and the saphenofemoral junction are assessed with the Valsalva maneuver and with distal compression and release. The veins are augmented, and valve closure times are recorded. Reflux is defined at valvular closure times greater than 1.0 s for femoral and popliteal veins and greater than 0.5 s for the great, small, and anterior saphenous veins as well as perforating veins (see Chapter 6).

Venous outflow obstruction has been historically underappreciated as a cause of venous hypertension and associated venous disease. Existing guidelines for the management of lower extremity CVD in general do not adequately account for manifestations of pelvic and/or abdominal venous disorders. One cause is a lack of a noninvasive screening study that reliably identifies the problem.[6–8] Duplex may provide a role in the diagnosis of venous obstruction. It is worth noting that formal ultrasonography training for the detection of venous outflow obstruction is lacking, despite the fact that extensive experience and training are required for accurate assessment. The exam of venous outflow begins at the CFV union and is completed with the IVC evaluation, and it should include the abovementioned suprainguinal venous territories. The CFV waveform is interrogated for signs of indirect evidence of obstruction. Signals exhibiting monophasic waveforms, loss of respiratory variation, and poor augmentation of distal limb compression are characteristics of proximal outflow obstruction. Direct visualization of the abdominal and pelvic venous system may improve diagnostic accuracy. However, high BMI, body habitus, intestinal gas, and operator variability may limit good visualization and assessment of these structures. Strict criteria defining the degree of stenosis have not been well defined, limiting the overall effectiveness of duplex for direct imaging. Traditionally duplex ultrasound assessment was based on direct and indirect criteria discussed below.

Direct duplex ultrasonography criteria include planimetric evaluation, venous luminal changes, and poststenotic to prestenotic venous blood velocity ratio.[6,7] Planimetric evaluation and luminal changes allow direct visualization of the obstruction making in some cases possible to differentiate stenosis from occlusion. Velocity ratio assessment requires careful interpretation, as its absence does not exclude obstruction because the vein can be completely occluded, be partially recanalized, or have long segment stenosis. Relatively recently Metzger et al. evaluated 102 lower extremities in 51 patients with CVI to correlate duplex ultrasound findings with intravascular ultrasound (IVUS). All the patients underwent duplex ultrasonography, IVUS, and venography. The study showed that when present, a velocity ratio >2.5 indicates stenosis of >50%.[7] During the ultrasonographic evaluation, it is important not to apply excessive pressure by the ultrasound transducer to the skin (and subsequently deeper anatomic structures), as the pressure can affect the measurements and overestimate obstruction.

Indirect duplex ultrasonography criteria include nonphasic flow in the proximal CFV, asymmetrical CFV flow, nonphasic flow with Valsalva maneuver, presence of collaterals, no velocity or low-velocity augmentation in the CFV during thigh compression, the reverse of flow in the ipsilateral internal iliac vein and in the deep external pudendal vein, cephalad flow in the inferior epigastric vein, and difficulty compressing the CFV. Obstruction can be present despite phasic flow patterns and good flow augmentation in the CFV.[5,8–10] When present, aforementioned indirect signs suggest obstruction but they cannot be used for definitive diagnosis or to differentiate between extrinsic compression, stenosis, and occlusion, or luminal changes. If such criteria are identified, further investigation is warranted. Although useful, duplex ultrasonography lacks diagnostic criteria for defining obstruction in a great number of patients. It is also worth mentioning that one of the major limitations of the duplex exam is operator dependency.

CT venography

CT venography (CTV) is the most used imaging modality for the investigation of iliofemoral and inferior vena cava pathology.[9] CTV provides useful information pertinent to DVT and anatomic factors such as vein compression from surrounding structures. It can be divided into indirect and direct venography. Indirect CTV is performed with the peripheral administration of contrast while direct CTV is performed with direct contrast delivery to the affected extremity. The indirect exam is most commonly performed. It is performed by injecting weight-adjusted (75–100 mL) of nonionic contrast over 50 s via the antecubital vein access (most commonly) using an 18 or 20-gauge needle. This is followed by a second scan delay.

Direct CTV is performed by injection of contrast into the affected limb, typically into the foot.[9] This is done to achieve a high concentration of contrast into the affected limb and to obtain essentially a venogram of the extremity. This technique is most frequently reserved for planning endovascular reconstructions.

Lin et al. evaluated 2963 Doppler exams and found abnormal monophasic waveforms in the CFV in 124 of cases.[5] In this subgroup, 87% (108) of cases were evaluated with abdominal/pelvic CT scans resulting in the detection of DVT and diffuse stenosis or hypoplasia of the iliac vein in 42% and 6% of cases, respectively. However, the study also showed that in 45 cases CT showed no iliac outflow obstruction despite the abnormal monophasic CFV waveforms detected on ultrasonography. Nevertheless, given the high morbidity associated with DVT, a CT scan evaluation of iliac veins is warranted in patients with abnormal monophasic CFV waveforms.

Marston et al. analyzed 78 cases with either healed or active venous ulcers to determine the diagnostic accuracy of duplex ultrasonography for detecting the iliocaval venous outflow obstruction, compared to CTV or magnetic resonance venography (MRV).[11] Study showed respective rates for specificity, sensitivity, positive predictive value and negative predictive value of 100%, 77%, 100%, and 95%. Authors used the absence of abnormal CFV waveform with respiratory variation and flow augmentation with compression as criteria for duplex scan evidence of outflow obstruction. As the results suggested, the absence of abnormal waveforms in the CFV did not rule out iliocaval obstruction. Falsely negative duplex evaluation of the CFV can be seen in patients with severe obstruction of the CFV or in patients with more proximal obstruction.

The major drawback to CTV in general is the radiation dose, iodinated contrast (making it relatively contraindicated in renal insufficiency and in patients with contrast allergy) and artifacts from implantable devices such as orthopedic hardware.

MRV

Another valuable diagnostic tool is contrast-enhanced or nonenhanced magnetic resonance imaging, which is especially useful in patients with contraindications to CT angiography such as contrast-induced nephropathy or allergic reactions. It is also helpful in pregnant patients. It offers significant advantages with no need for iodinated contrast administration and no radiation exposure. MR venography is highly accurate when compared with conventional venography. It is particularly accurate for imaging of pelvic veins.[12,13] It also has advantages for central occlusions as well as etiologies for pulmonary emboli, such as a central mass or other lesions not readily identified or seen by ultrasound. The technique has improved over the years; however, it is expensive and requires dedicated radiologists and well-trained radiographic staff to achieve acceptable diagnostic imaging accuracy.[9] In 2022, Saleem at al. compared IVUS and MRV in 505 patients who were evaluated for chronic iliofemoral venous obstruction.[12] Of these, 78 (15%) underwent MRV evaluation. This study demonstrated that MRV had a high sensitivity but low specificity when compared to IVUS examination. It also showed that MRV overestimated the severity of the stenosis in EIV and CIV. Based on these findings the authors suggested that MRV should not be used for the definitive diagnosis of chronic iliofemoral venous obstruction patients.

Venography

Venography for many years was the gold standard in the identification of iliofemoral venous pathology.[9,14,15] Conventional catheter-based diagnostic venography is typically used if an intervention is planned, such as a filter placement, thrombolysis, thrombectomy, and stent placement. A needle is inserted into the venous system and iodinated contrast is injected to visualize the venous system. Acute and chronic lesions can be seen, and appropriate interventions are selected. Acute DVT can be identified, and a selection of therapies can be determined. In addition, nonthrombotic lesions can be evaluated and repaired as needed. IVUS may be required to get confirmation of stenosis or to aid in selecting the appropriate size stents. This will be discussed in more detail in the following sections. In 2017 Cagne et al. published The Venogram versus IVUS for Diagnosing Iliac vein Obstruction (VIDIO) trial[16] which compared the diagnostic efficacy of multiplanar venography with IVUS in 100 patients with clinically suspected iliofemoral vein obstruction evaluated at 11 sites in the United States and three in Europe. Venograms were used to measure vein diameter and IVUS was used for diameter and area measurements. All multiplanar venograms were performed using 30 degrees right and left anterior oblique views and anteroposterior views. A 50% diameter stenosis and a 50% cross-sectional area reduction by venography and IVUS, respectively, were considered significant. The reference vein used for calculation of the reduction of diameter and the cross-sectional area was defined either as the most normal appearing proximal or distal venous segment to the target lesion, or when no normal reference vein was identified, a precedent from the literature was used as a proxy reference. The study showed superior sensitivity of IVUS compared with multiplanar venography for identifying and quantifying iliofemoral vein obstruction and showed that IVUS changed the treatment plan in 57% of patients. Additionally, significant obstructive lesions were not detected with multiplanar venography in 26.3% of patients. In the subsequent analysis of the same trial, the same group of authors evaluated IVUS findings as a predictor for significant symptom improvement defined as greater than a four-point reduction in the revised Venous Clinical Severity Score at a 6-month follow-up.[17] Study showed a threshold of >54% area stenosis by IVUS may be used to better predict clinical improvement. In patients with nonthrombotic lesions, IVUS-measured diameter rather than the area of stenosis was a predictor of clinical improvement with a threshold of >64% diameter stenosis. However, it is worth emphasizing that the aforementioned IVUS parameters should be used as general guidance rather than absolute indications for treatment. A decision to intervene should be based on the judgment of an experienced practitioner in the context of radiologic findings and the severity of presenting symptoms. For example, a patient with IVUS identified stenosis of <50% and active ulcer despite conservative treatment should be considered for correction of venous outflow obstruction.

IVUS

Intravascular ultrasound is the current gold standard to identify stenotic lesions in the venous system, especially the vena cava and the iliac veins as well as the surrounding structures. It is an essential tool for diagnosing deep venous disease and guiding safe and effective endovascular treatments.[18] It provides accurate real-time information and the assessment of surrounding anatomic structures with no radiation and no contrast agents. It is also independent of body habitus making it helpful in treating morbidly obese patients. It is an invasive procedure that is utilized for a preballoon assessment and to assess the amount of stenosis, aid in stent sizing, identify landing zones and evaluate the outcomes of interventions. Intravascular ultrasound avoids false results as it evaluates the entire wall of the vessel in real time. A relatively recent systematic review that compared IVUS with multidimensional contrast imaging modalities for the characterization of chronic occlusive iliofemoral venous disease showed high sensitivity and specificity associated with IVUS for guiding venous interventions in patients with chronic iliac venous obstruction.[19] The study also showed that venography often underestimated the severity as well as the existence of venous stenosis. It is worth emphasizing that in patients with IVUS identified stenosis collateral vessels were not detected on all venograms. Notably, in patients with compression of right iliac vein (>50% stenosis on IVUS) no collateral veins were detected on venography. In patients with contralateral vein compression (>50% compression of right iliac vein on IVUS) venography identified associated collateral vessels in 85% of cases. Shammas et al. showed that venous stenosis percentage on venography was underestimated by 15.2% when compared with IVUS.[20]

Diagnostic algorithm

All patients presenting with symptoms of CVI should undergo a thorough history and physical examination. Symptoms of leg swelling, heaviness, aching, throbbing, skin discoloration, and ulcers direct a differential diagnosis toward venous insufficiency. The arterial system should be evaluated, and the workup should proceed based on physical findings. If the patient has clearly palpable pulses, then a venous workup should ensue. However, thickened, scarred skin or swelling can make physical exam difficult, so if there is any doubt, ABIs need to be obtained.

A venous duplex is then obtained to determine if there is reflux or evidence of venous outflow obstruction. In some instances, duplex alone may find direct evidence of venous obstruction. If there are subtle signs of obstruction and no confirmatory duplex findings then obtaining a CTV or an MRV is helpful. As we mentioned previously, MRV accuracy is institution dependent so it is recommended to coordinate with a radiologist to confirm an MRV can provide the needed information. The findings may suggest obstructive pathology requiring venography and IVUS to confirm the diagnosis and intervene to alleviate symptoms. Obstructive components of venous insufficiency should be alleviated before addressing any reflux identified in the lower extremity to reduce the probability of recurrent reflux. We have summarized the algorithm in Fig. 7.1.

In summary, a significant limitation that characterizes the management of major venous outflow obstruction is the lack of a noninvasive screening and diagnostic modality to identify and characterize the major venous outflow obstruction. Historically CTV and MRV were used for venous outflow imaging in patients with CVI signs and symptoms that were out of proportion with the ultrasound findings or in patients with ultrasonographic changes suggestive of iliocaval obstruction.[11,21] However, the lack of acquisition of useful hemodynamic data by CTV or MRV and the notion that these imaging modalities are associated with the risk of not identifying intraluminal webs and other chronic findings that may contribute to pathologic venous return function represent major limitations for using them as a definitive imaging tool for diagnosis and treatment planning. Currently, IVUS and venography are considered the gold standard for diagnosis; however, they are invasive and expensive and can be performed at highly specialized centers, therefore they cannot be used to screen large numbers of patients. An increasing number

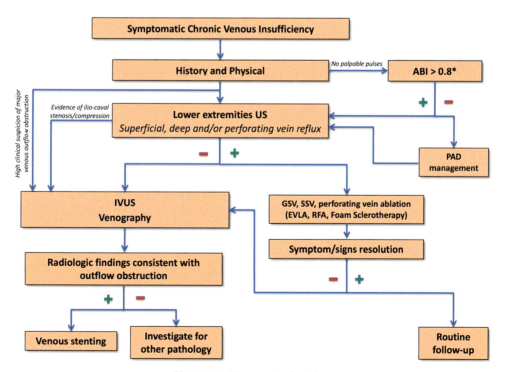

Figure 7.1 Diagnostic algorithm.

of studies demonstrated IVUS to be more sensitive than venography and this diagnostic modality should be considered the main imaging for patients with clinically suspected major venous outflow obstruction. In addition for evaluating iliocaval obstruction increasing number of evidence is becoming available pertinent to the role of abdominal and pelvic collateral venous network and the role of pelvic venous congestion as an underlying etiology in patients with signs and symptoms indicative of CVI.[22–25]

References

1. Meissner MH. Lower extremity venous anatomy. *Semin Intervent Radiol*. 2005;22:147–156.
2. Notowitz LB. Normal venous anatomy and physiology of the lower extremity. *J Vasc Nurs*. 1993;11:39–42.
3. Ouriel K, Green RM, Greenberg RK, Clair DG. The anatomy of deep venous thrombosis of the lower extremity. *J Vasc Surg*. 2000;31:895–900.
4. Seligman B. Venous anatomy of the lower extremity. *Clin Med*. 1961;8:1119–1124.
5. Lin EP, Bhatt S, Rubens D, Dogra VS. The importance of monophasic Doppler waveforms in the common femoral vein: a retrospective study. *J Ultrasound Med*. 2007;26:885–891.
6. Labropoulos N, Borge M, Pierce K, Pappas PJ. Criteria for defining significant central vein stenosis with duplex ultrasound. *J Vasc Surg*. 2007;46:101–107.
7. Metzger PB, Rossi FH, Kambara AM, et al. Criteria for detecting significant chronic iliac venous obstructions with duplex ultrasound. *J Vasc Surg Venous Lymphat Disord*. 2016;4:18–27.
8. Bach AM, Hann LE. When the common femoral vein is revealed as flattened on spectral Doppler sonography: is it a reliable sign for diagnosis of proximal venous obstruction? *AJR Am J Roentgenol*. 1997;168:733–736.
9. Kayilioglu SI, Koksoy C, Alacayir I. Diagnostic value of the femoral vein flow pattern for the detection of an iliocaval venous obstruction. *J Vasc Surg Venous Lymphat Disord*. 2016;4:2–8.
10. Sermsathanasawadi N, Pruekprasert K, Pitaksantayothin W, et al. Prevalence, risk factors, and evaluation of iliocaval obstruction in advanced chronic venous insufficiency. *J Vasc Surg Venous Lymphat Disord*. 2019;7:441–447.
11. Marston W, Fish D, Unger J, Keagy B. Incidence of and risk factors for iliocaval venous obstruction in patients with active or healed venous leg ulcers. *J Vasc Surg*. 2011;53:1303–1308.
12. Saleem T, Lucas M, Raju S. Comparison of intravascular ultrasound and magnetic resonance venography in the diagnosis of chronic iliac venous disease. *J Vasc Surg Venous Lymphat Disord*. 2022;10:1066–1071 e2.
13. How GY, Quek LHH, Huang IKH, et al. Intravascular ultrasound correlation of unenhanced magnetic resonance venography in the context of pelvic deep venous disease. *J Vasc Surg Venous Lymphat Disord*. 2022;10:1087–1094.
14. Lau I, Png CYM, Eswarappa M, et al. Defining the utility of anteroposterior venography in the diagnosis of venous iliofemoral obstruction. *J Vasc Surg Venous Lymphat Disord*. 2019;7:514–521 e4.
15. Rollo JC, Farley SM, Oskowitz AZ, Woo K, DeRubertis BG. Contemporary outcomes after venography-guided treatment of patients with May-Thurner syndrome. *J Vasc Surg Venous Lymphat Disord*. 2017;5:667–676 e1.
16. Gagne PJ, Tahara RW, Fastabend CP, et al. Venography versus intravascular ultrasound for diagnosing and treating iliofemoral vein obstruction. *J Vasc Surg Venous Lymphat Disord*. 2017;5:678–687.
17. Gagne PJ, Gasparis A, Black S, et al. Analysis of threshold stenosis by multiplanar venogram and intravascular ultrasound examination for predicting clinical improvement after iliofemoral vein stenting in the VIDIO trial. *J Vasc Surg Venous Lymphat Disord*. 2018;6:48–56.e1.
18. Raju S, Martin A, Davis M. The importance of IVUS assessment in venous thrombolytic regimens. *J Vasc Surg Venous Lymphat Disord*. 2013;1:108.
19. Saleem T, Raju S. Comparison of intravascular ultrasound and multidimensional contrast imaging modalities for characterization of chronic occlusive iliofemoral venous disease: a systematic review. *J Vasc Surg Venous Lymphat Disord*. 2021;9:1545–1556 e2.

20. Shammas NW, Shammas GA, Jones-Miller S, et al. Predicting iliac vein compression with computed tomography angiography and venography: correlation with intravascular ultrasound. *J Invasive Cardiol*. 2018;30:452–455.
21. Wolpert LM, Rahmani O, Stein B, Gallagher JJ, Drezner AD. Magnetic resonance venography in the diagnosis and management of May-Thurner syndrome. *Vasc Endovasc Surg*. 2002;36:51–57.
22. Daugherty SF, Gillespie DL. Venous angioplasty and stenting improve pelvic congestion syndrome caused by venous outflow obstruction. *J Vasc Surg Venous Lymphat Disord*. 2015;3:283–289.
23. Mahmoud O, Vikatmaa P, Aho P, et al. Efficacy of endovascular treatment for pelvic congestion syndrome. *J Vasc Surg Venous Lymphat Disord*. 2016;4:355–370.
24. Antignani PL, Lazarashvili Z, Monedero JL, et al. Diagnosis and treatment of pelvic congestion syndrome: UIP consensus document. *Int Angiol*. 2019;38:265–283.
25. Gibson K, Minjarez R. Vascular disease patient information page: pelvic venous reflux (pelvic congestion syndrome). *Vasc Med*. 2019;24:467–471.

CHAPTER 8

Hypercoagulable states associated with chronic venous insufficiency

Samuel Anthony Galea and Emma Wilton
Oxford University Hospitals, NHS Foundation Trust and Buckinghamshire Healthcare NHS Trust, Oxford, United Kingdom

Venous disease is the most common cause of leg ulceration in the United Kingdom accounting for 60%—80% of cases.[1] Its prevalence is an estimated 0.1%—0.3% in the United Kingdom.[1] Approximately 5%—10% of patients with venous ulcers have diseased deep-venous systems only. An estimated 1% of total health costs in the Western world result from the therapy costs of chronic leg ulcers.[2] Venous leg ulceration, therefore, places a significant burden on healthcare expenditure. Prompt and appropriate prevention, investigation, and treatment of venous leg ulcers are paramount.

Venous leg ulcers are the most severe manifestation of chronic venous disease (CVD) and the result of elevated ambulatory venous pressure within the lower limb and venous hypertension. This can be caused by either deep, superficial, and/or perforating vein reflux or venous outflow obstruction. In the deep veins, thrombosis may cause valvular damage (resulting in reflux) or scarring and chronic obstruction. Other causes of chronic venous obstruction include congenital abnormalities, May—Thurner syndrome (compression of the left iliac vein by the left common iliac artery), and external compression by a pelvic mass. When compared to primary venous disease, a higher rate of ulceration and progression of CVD is observed in patients who suffer from obstruction from secondary venous disease and venous reflux including postthrombotic syndrome (PTS).[3]

Abnormal thrombus formation is central to the acute pathophysiology of venous disease. Deep vein thrombosis (DVT) and potentially subsequent pulmonary embolism (PE) both result from abnormal thrombus formation in the lower extremity venous circulation that is deep to the muscular (deep) fascia.

Thrombus in the deep veins impedes venous outflow from the lower limb due to a decrease in the cross-sectional luminal diameter of the vein. This decrease in the area causes an increase in outflow resistance, which in turn leads to a decrease in the outflow volume and an associated increase in venous pressure. These changes in the hemodynamics, along with perivascular inflammation lead to the signs and symptoms of DVT, which characteristically results in "calor" (warmth), "dolor" (pain), "tumor" (swelling), and "rubor" (redness). As the inflammatory process starts to subside, so do the symptoms that may disappear altogether should the vein fully recanalize especially if the DVT did not provoke structural damage to the integrity of the wall and/or valve. The rate of recanalization decreases the more proximal (cephalad) the DVT is, with recanalization rates of 80% for calf veins, decreasing to 20% for iliac veins.[4] Patients who have failed to recanalize completely and have residual thrombus in the vein have an increase in the venous outflow pressure, which may result in secondary valvular damage leading to reflux.[4] The presence of reflux and/or venous obstruction may be risk factors for the development of PTS. The thrombotic process leads to increased outflow resistance and decreased outflow volume with increased venous pressure, which, together with perivascular inflammation, is responsible for the characteristic symptoms and signs of DVT. Patients present with swelling, pain, and tenderness, usually in the calf, but symptoms may also involve the thigh in the case of an iliofemoral DVT. This is a chronic debilitating condition that is a long-term complication of DVT. Around 20%–50%[4] of people who have had a DVT in the leg will develop some postthrombotic symptoms within 5 years. PTS is characterized by chronic pain, swelling, edema, venous ectasia, skin induration, venous claudication, and ulceration,[5] with the latter developing in 5%–10% of cases.[6] Other symptoms associated with PTS include cramps, heaviness, paraesthesia, and pruritus in the affected limb. It has a significant impact on quality of life. Most patients who develop PTS become symptomatic between 6 months and 2 years from the acute episode[7] and up to 10% of patients with PTS will develop venous ulceration over a 10-year period.[8] The risk of developing PTS is increased if there is incomplete resolution of symptoms at 1 month, a proximal (iliofemoral) DVT, previous ipsilateral DVT, high BMI,[9] compliance to oral anticoagulation and the presence of residual DVT identified on Duplex ultrasound after anticoagulation treatment has been initiated.[10] Understanding the conditions that may predispose to abnormal thrombus formation will therefore help inform clinicians on the management of the patient.

The hypercoagulable state refers to an increase in a patient's propensity to develop thrombosis due to a change in the blood physiology as a consequence of an inherited and/or acquired condition(s). This forms just one of the contributing factors toward the development of venous thrombosis. The triad of features originally coined in 1856 by Rudolf Virchow[11] has changed little since its inception. The Virchow triad states that a thrombus is generated in view of disturbances in the blood flow, endothelial lining, and/or blood composition.

The hypercoagulable state is a medical conundrum with little in the way of hard evidence to guide clinical practice. Data from relatively recent studies[12–15] shape the most recent clinical practice and provide an opportunity for continued research in the field. Although thrombophilia screening is a useful tool, it has certain limitations. These include the fact that literature pertaining to the use of thrombophilia testing has traditionally been based on studies that did not control for ethnic diversity and that the absence of positive results does not completely rebut the possibility that the patient has a known, or indeed inherited, condition.

There are two ways of classifying the causes of the hypercoagulable state. It can be subdivided based on a gain of hypercoagulable function or a loss of antithrombotic function[16] (Table 8.1) or based on the pattern of inheritance into acquired or inherited (Table 8.2).

Protein C and protein S

Johan Stenflo in 1976 was the first scientist to isolate protein C from bovine plasma,[17] while in 1982 Griffin et al. were among the first to associate low protein C plasma concentrations with clinical venous thrombosis.[18] Both Protein C and Protein S are vitamin K-dependent plasma glycoproteins that are synthesized in the liver by the hepatocytes. Both are inherited in an autosomal dominant fashion and thus 50% of the next generation are affected. The prevalence of Protein C deficiency stands between 0.2% and 0.4% in the general population[19] and accounts for 10%–15% of all familial thrombophilia.[4] A Type I deficiency of Protein C involves a reduction in both the antigenic concentration and biochemical activity. A Type II deficiency of Protein C involves a reduction in the biochemical activity, yet the antigen concentration remains normal.[19,20] The prevalence of Protein S is uncertain but is estimated between 0.3% and 0.13%.[21] Protein S is bound to CP4 protein in circulation and is only active in the free form. PS deficiency is classified as type I (low total and free antigen, reduced activity), type II (normal total and free antigen, reduced activity), and type III (normal total antigen, reduced free antigen and activity).

Protein C is a precursor protein and is activated when transformed into the active serine protease-activated protein C (APC). This is done by thrombin which is potentiated by the binding of thrombin to thrombomodulin on the surface of platelets. Protein S serves as a cofactor to enhance the APC functionality, and after combining in the presence of calcium and phospholipids they together inhibit thrombin formation directly and indirectly by inactivating factors Va and VIIIa, respectively, by cleavage at arginine residues.[20,22] Patients who are homozygotes or compound heterozygotes have Protein

Table 8.1 Examples of conditions that are either prothrombotic or antithrombotic.

Gain of prothrombotic function	Loss of antithrombotic function
Factor V Leiden	Protein C deficiency
Prothrombin G20210A variant	Protein S deficiency
Elevation of procoagulant factors such as factor VIII, von Willebrand factor	Antithrombin deficiency
Elevation of factors V, VII, IX, and XI	

Table 8.2 Examples of hereditary and acquired hypercoagulable conditions.[4]

Hereditary	Acquired
Protein C deficiency	Antiphospholipid antibodies on two occasions more than 12 weeks apart. Three assays are performed: 1. Lupus anticoagulant 2. Anticardiolipin antibodies 3. Anti-β-2 glycoprotein I antibodies
Protein S deficiency	Paroxysmal nocturnal hemoglobinuria
Antithrombin deficiency	Myeloproliferative syndromes with JAK2V617F mutation
Factor V Leiden	Hemolytic states, e.g., sickle cell crises
Prothrombin G20210A variants	Any inflammatory disease such as infections, e.g., pneumonia, rheumatoid arthritis, inflammatory bowel disease, systemic lupus erythematosus, Adamantiades-Behçet disease.
Dysfibrinogenaemia	Nephrotic syndrome (loss of antithrombin in the urine)
Factor XIII 34val Fibrinogen (G) 10034T A and/or B alleles of the ABO blood group Prothrombin Yukuhashi (II R596L)	

C levels just compatible with life (<0.01 U/mL) and typically present in the neonatal period with spontaneous skin necrosis (neonatal purpura fulminans), or later in life with 7–10-fold increased risk of venous thromboembolic events (VTE) if less severely affected (Protein C levels 0.4–0.6 U/mL) (Table 8.3).[4]

Antithrombin

Antithrombin (formerly known as antithrombin III or heparin cofactor I) is the most infrequent of hereditary thrombophilia and has an incidence of 0.02%–0.05% in the general population.[24] Antithrombin was first discovered by Brinkhous et al. in 1939,[25] and antithrombin deficiency was first described clinically in a Swedish family with multiple VTE by Olav Egeberg in 1965.[26] It is inherited in an autosomal dominant Mendelian

Table 8.3 Main characteristics of protein C (PC), protein S (PS), and antithrombin (AT).[20,23]

	Protein C	Protein S	Antithrombin
History	Isolated from bovine plasma by Johan Stenflo in 1976 and named protein C because it was the third protein to elute from DEAE-Sepharose	Described by Di Scipio in 1977, named protein S in reference to its isolation and characterization in Seattle Gene characterization	Discovered by Brinkhous in 1939 and described clinically by Olav Egeberg in 1965
Gene	PROC located on chromosome 2, position 2q13-q14	PROS1, PROS2 both located on chromosome 3, position 3p11.1-3q11.2	SERPINC1 chromosome 1q23-25
Inheritance	Autosomal dominant	Autosomal dominant	Autosomal dominant
Prevalence (% of the general population)	0.2%–0.4%	0.2%	0.002%
Synthesis	Hepatocytes, endothelial cells,	Hepatocytes, endothelial cells, human testis Leydig cells, vascular smooth muscle cells, and megakaryocytes	Hepatocytes
Molecular weight	62 kDa	71 kDa	58.2 kDa
Protein	The zymogenic form of plasma PC is activated by thrombin in the presence of thrombomodulin and endothelial protein C receptor leading to activated PC (APC) generation	60% bound to C4bBP-β chain (inactive), 40% free PS (physiologically active), together total PS	α_2 globulin found as 2isomers α globulin (90%–95%) β globulin (5%–10%)
Concentration in plasma	3–5 μg/mL	20–25 μg/mL	112–140 μg/mL
Half life	6–8 h	42 h	57 h
Assays	Amidolytic PC assays for routine	Immunoassays for free and total PS,	Amidolytic AT functional assays,

Continued

Table 8.3 Main characteristics of protein C (PC), protein S (PS), and antithrombin (AT).[20,23]—cont'd

	Protein C	Protein S	Antithrombin
	screening for PC deficiency, more specific than coagulation assays; immunoassays (turbidimetric, nephelometric, ELISA)	and clotting assays for PS activity	antigenic assays for quantitative measure of the total amount of AT

fashion. Antithrombin can be found as an $\alpha 2$-[27] or β-globulin[23] single-chain glycoprotein. This antiproteinase enzyme works by inactivating factor Xa and factor II (thrombin) and to a lesser extent factors IXa, XIa, and XIIa. Its function is potentiated by heparin that binds to antithrombin and as a cofactor accelerates the inactivation of aforementioned coagulation cascade factors, thus executing heparin's anticoagulation function.[28] Antithrombin deficiency can be either quantitative (type I) characterized by an overall reduction in the production of antithrombin, or qualitative (type II) characterized by a functional reduction in antithrombin activity. Type II deficiency can be further subdivided into Type IIa, due to a mutation in the thrombin binding site and Type IIb, due to a defect in the heparin-binding site. Type IIb is less thrombogenic than type IIb.[23] A type III deficiency is also recognized as producing a pleiotropic effect. It is estimated that 50% of the patients who have this thrombophilia will develop a DVT.[24] Others demonstrated that patients with antithrombin deficiency conferred a risk ratio of 8.1 of developing thrombosis when compared to patients without thrombophilia.[29] Apart from being inherited, antithrombin deficiency can be acquired.[23,30] Underlying etiology can be broadly divided into the following:

- Decreased antithrombin production, for example, liver cirrhosis
- Increase antithrombin loss, for example, protein-losing enteropathy, nephrotic syndrome
- Consumptive/inactivation of antithrombin, for example, cancer, polytrauma, sepsis hepatic veno-occlusive disease
- Drug induced, for example, heparin therapy, L-asparaginase therapy.

Factor V Leiden

Factor V Leiden was discovered by Dahlback in 1993[31] and is the most common genetic disorder that causes hypercoagulability.[32] Factor V Leiden accounts for 20%—25% of genetic causes of VTE and 50% of familial thrombophilia.[33] Factor V is an enzyme encoded by chromosome 1 locus q24.2.[34] The most commonly inherited form of improper factor V function was studied at Leiden University by Dutch investigators, thus giving it its name. Factor V Leiden is characterized by a resistance to APC. Activated Protein C is an enzyme that breaks down Factor Va at amino acid positions R (arginine) 306, R506, R679 and thus inactivating the Factor Va in the process. Factor V Leiden refers to the point substitution mutation of guanine for adenine at nucleotide position 1691, thus resulting in the substitution of glutamine for arginine at position 506 of the Factor V protein. This causes a conformational change to the R506 cleavage site, rendering the anticoagulant APC much slower at inactivating the procoagulant Factor Va. The cleaved Factor V also serves as a cofactor for APC inactivation of factor VIIIa.[35] Patients who are heterozygotes for Factor V Leiden carry a three- to eight-fold increased risk for VTE, which increases to 10–80-fold increase for homozygotes.[22] Factor V Leiden can be tested for using APC resistance assays or DNA molecular testing.

Activated protein C resistance

Apart from Factor V Leiden, there are other causes that cause resistance to the function of APC. Factor V Cambridge and Factor V Hong Kong affect cleavage site R306 with the former substituting the arginine with threonine and the latter with a glycine. Factor V Liverpool is another form of APC with isoleucine being substituted for threonine at position 359. Other forms of APC resistance included Factor V Glu666Asp, Factor V HR2, and Factor V Ala485Lys.[36] Individually these genetic disorders are certainly rare and have uncertain thrombogenic and clinical value. Such disorders may coexist and thus contribute to other genetic or acquired risk factors.[37] Additionally, APC resistance may be an acquired condition, most commonly associated with antiphospholipid syndrome.

Prothrombin G20210A gene mutations

Roots et al. in 1996 described a point mutation at position 20,210 of the gene encoding for factor II. The prothrombin gene is found on chromosome 11 (locus p11-q12). The group of investigators noted that patients who had a substitution of guanine with adenosine in position 20,210, resulted in an increase in the incidence of VTE.[38] This point mutation leads to a gain in function as the gene locus plays a regulatory role in gene transcription, resulting in an increased amount of mRNA (that can be subjected to posttranscriptional modification) and increased production of prothrombin. Typical normal (G20210G) plasma levels of factor II are at 1.05 U/mL (range, 0.55–1.56 U/mL), heterozygotes (G20210A/G20210G) 1.32 U/mL (range 0.95–1.78 U/mL) while homozygous (G20210A) 1.70 U/mL. There are various ways of laboratory diagnosing prothrombin G20210A mutations. Direct measurement of the plasma prothrombin levels may have poor sensitivity due to the physiological variations in plasma prothrombin levels. The diagnosis of this thrombophilia is based on genetic detection of the mutation by PCR techniques.[39] In the Leiden Thrombophilia Study, the prothrombin G20210A mutation was found in 6.2% of affected patients and 2.3% of the healthy individuals used as the control branch of the study. In this study, elevated plasma levels of the prothrombin lead to a 2.8-fold increased VTE risk.[37]

Dysfibrinogenemia

Fibrinogen is a 340kDA protein composed of a hexameric arrangement of α-, β-, and γ-peptide chains. It is encoded by *FGA*, *FGB*, and *FGG* genes on chromosome 4.[40] Congenital dysfibrinogenemia is often caused by frameshift mutations caused by single base deletion or insertion, single base substitutions, and mutation in the regulatory region. Mutations in the FGA and FGG account for 99.3% of dysfibrinogenemia.[41] Fibrinogen is among the most abundant proteins in the body with a plasma concentration of 2–4 g/L and is produced by the liver. Fibrinogen is a substrate for the formation of thrombin that develops due to the polymerization of the peptide chains. Apart from the genetic disorders described earlier, dysfibrinogenemia can also be due to acquired causes. Acquired causes of dysfibrinogenemia can be due to assay interference with, for example, direct thrombin inhibitors, a posttranslational modification due to abnormal sialylation in liver disease, autoantibody formation as happens in monoclonal gammopathy of undetermined significance, myeloma or autoimmune disease.[42,43] Dysfibrinogenemia leads to both bleeding and thrombotic tendencies.[44] The diagnosis is typically made by assessing fibrinogen assays and measurement of the prothrombin time (PT), activated partial thromboplastin time (aPTT), thrombin time, and reptilase time and by performing genetic analyses.

Antiphospholipid syndrome

Antiphospholipid syndrome encompasses a family of acquired causes that result in an autoantibody-induced thrombophilia,[45] that is, the production of immunoglobulins that bind to antigens on the phospholipid surfaces such as β-2 glycoprotein1 predisposing the patient to VTE.[4] The diagnosis of antiphospholipid syndrome relies on the Sapporo clinical criteria proposed in 1998 which were later modified in the Sydney criteria in 2006. The criteria include one of two clinical and one of two laboratory findings. Clinical findings include diagnosing a vascular thrombosis, for example, arterial/venous/small vessel thrombus excluding superficial venous thrombosis, and/or pregnancy morbidity.[46] Pregnancy morbidity was defined as one or more unexplained death of a morphologically normal fetus at/or beyond 10-week gestation, one or more premature birth before 34 weeks due to eclampsia/preeclampsia or placental insufficiency, or three or more spontaneous abortions before 10 weeks gestation with no maternal hormonal or anatomic abnormalities and parental chromosomal causes. Laboratory criteria included the presence of lupus anticoagulant, anticardiolipin IgG/IgM by standard ELISA, and anti-β-2 glycoprotein-1 antibody IGG/IgM on standard ELISA.[47] Laboratory findings need to be persistently high on two or more occasions spaced at least 12 weeks apart.[4] This is because these findings may be temporarily elevated due to other causes. The antiphospholipid syndrome may be found in isolation (primary) or found in combination with an auto-immune condition such as systemic lupus erythematosus or rheumatoid arthritis.[48] Antibodies may also be elevated due to medications or infections. The causes include lupus anticoagulant syndrome and anticardiolipin syndrome, but can also be drug induced, for example, hydralazine, cocaine, quinidine, interferon, procainamide hydrochloride, phenytoin, phenothiazines, quinine, and pyrimethamine/sulfadoxine. Finally, the most severe form of antiphospholipid syndrome, catastrophic antiphospholipid syndrome, involves a rapid onset multifocal thrombosis with evidence of systemic inflammatory response syndrome, causing unusual organ involvement and carries a high mortality rate.[49]

Testing for antiphospholipid syndrome may include prolongation of the PT, PTT, and Russell viper venom time as a result of lupus anticoagulant's affinity to surface phospholipids.[32] Lupus anticoagulant activity may be suspected based on an unexplained PTT prolongation that is not reversed when the patient's plasma is diluted at a 1:1 ratio with normal plasma. Apart from clot-based functional assays, ELISA can also be used to detect various antiphospholipid antibodies. Thrombosis occurs due to the increased production of thromboxane B_2 induced by the binding of antiphospholipid antibodies to the β-2 glycoprotein1receptors on platelets. Production of thromboxane A_2 leads to platelet activation, aggregation, and adhesion to collagen resulting in VTE.[45] The thrombotic risk for patients with antiphospholipid syndrome stands at 5.5% for symptomatic patients while that for anticardiolipin syndrome stands at 1% risk for patients with low titers and 6% for patients with high titters.[32]

Paroxysmal nocturnal hemoglobinuria

Paroxysmal nocturnal hemoglobinuria (PNH) was first described by Dr. Paul Strübing in 1882.[50] It is a hypercoagulable state in which there is a bone marrows failure manifested a hemolytic anemia, thrombosis, and peripheral blood cytopenia.[51] It involves a mutation in the X-linked phosphatidylinositol glycan class gene in hematopoietic stem cells. This gene encodes for the glycosylphosphatidylinositol (GPI) protein that helps anchor other protein moieties to the surface of erythrocytes. Deficiency of this protein leads to an inability of proteins which usually regulate the production of complement decay-accelerating (CD) factor 55 and CD59, to attach to hematopoietic stem cells leading to a loss of inhibition. The consequence of this disinhibition results in a complement-mediated hemolysis of PNH cells. This hemolysis is exacerbated by other conditions that activate the complement system, for example, sepsis, surgery, and trauma. The hypercoagulable state in PNH is multifactorial. One proposed mechanism is secondary to the absence of CD59 and CD99 on platelets. This leads to the production of prothrombotic microparticles containing phosphatidylserine which when externalized serves as a binding site for prothrombinase and tenase complexes.[52] Hemolysis leads to free hemoglobin that consumes the nitric oxide, in turn, contributes to platelet activation and aggregation. Complement activation, specifically C5a, results in a proinflammatory condition and prothrombotic state with the release of interleukin 6, interleukin 8, and tumor necrosis factor α. Aalternatively, another cause of the hypercoaguable state may be attributable to the dysregulation of fibrinolysis due to defective GPI linked proteins such as heparin sulfate, urokinase-type plasminogen activator receptor and tissue factor pathway inhibitor.[51] The incidence of VTE in PNH varies from 36% to 73% depending on the ethnicity of the patient.[53] Treatment for PNH involves the use of eculizumab, which is a humanized anti-C5 monoclonal antibody.[54]

Myeloproliferative neoplasms

2016 World Health Organization classification of Breakpoint Cluster Protein-Abelson gene (BCR-ABL) negative myeloproliferative neoplasms encompasses polycythemia vera (PV), essential thrombocythemia (ET), primary myelofibrosis (PMF), chronic myeloproliferative neoplasms (MPN), chronic neutrophilic leukemia, chronic eosinophilic leukemia and mastocytosis, and prefibrotic myelofibrosis. Of these PV, ET, myelofibrosis, and prefibrotic myelofibrosis are associated with thrombosis.[55] The incidence of thrombosis is 30%—41% in PV, 19%—32% in ET, and 7.2%—15% in PMF.[56] The *JAK2* mutation has been identified as one of the genetic causes of MPN. The JAK2 is a gene located on chromosome 9p24, with the V617F being the most common mutation. This gain-of-function mutation replaces guanine for thymine on the 1849th nucleotide producing a protein that substitutes valine for phenylalanine. This gene, when transcripted, encodes for nonreceptor tyrosine kinase involved in cellular growth, differentiation, development, or histone modification. The pathogenesis of thrombus formation in MPN is multifactorial with quantitative and qualitative abnormalities in cell lineages due to clonal proliferation of the stem cell, as well as expression of procoagulant and proteolytic properties, secretion of inflammatory mediators, and production of adhesion signals. The presence of the JAK2V617F is itself a risk factor for the development of thrombosis. When present in PV, this carries a 7.1 relative risk, ET patients with JAK2V617F have an odds ratio of 1.92 while in PMF it carries a hazard ratio of 3.13. Different forms of MPN have different treatment modalities which include the use of aspirin, hydroxyurea, phlebotomy, anagrelide, and interferon α.[57]

Other hypercoagulable states that increase the risk of developing VTE and potential subsequent venous ulceration

Age

Age has been identified as an independent risk factor for the development of venous thromboembolism. There is a tendency to present with more atypical symptoms and a higher incidence of PE as the presenting complaint.[58] Patients having underlying comorbidities including congestive cardiac failure, pulmonary edema, acute myocardial infarction, atrial tachyarrhythmia, urinary/respiratory infections, sepsis, cancer, lung fibrosis, arthritis, and hyperthyroidism and having had a hysterectomy are at increased risk.[59] With an increase in the patient's age, there is an increase in the risk of VTE with a steep increase in incidence above the age of 60–65, with a higher rate for males.[58]

Combined oral contraceptive pill and hormone replacement therapy

WM Jordan was the first to describe thromboembolic complications of the combined oral contraceptive pill (COCP) following its inception in the 1960s.[60] The relative risk of thromboembolism in women taking the COCP is 3.5. The risk decreased slightly between the first- and second-generation COCPs from 3.2 to 2.8, respectively; however, it increased to 3.8 with the third-generation COCPs. The thrombotic nature of the COCP is related to the loss of the anticoagulant effect of antithrombin, Protein S, and tissue factor pathway inhibitor and the simultaneous increase in the levels of factors II, VI, VIII, and Protein C.[61] The progestin-only contraceptives do not carry an increased risk of VTE compared to the COCP and is used in patients who are higher risk of developing VTE.[62]

For menopausal female patients who are taking hormone replacement therapy, the relative risk increase varies between 1.43 and 1.63.[63] In this case, the risk is greatest in the first year of use and is related to the dose of estrogen used and the method of delivery favoring the transdermal over the oral route of administration.[64]

Puerperium

The pregnant female is at an increased risk of developing venous thromboembolism due to a combination of factors. These include hormonal changes as described above as well as mechanical compression causing venous outflow obstruction, reduced mobility, and vascular injury. The risk of developing a DVT is fourfold compared to the general population in the antenatal period, increasing to 21.5–84-fold increased risk in the first 6 weeks postpartum.[65]

Cancer

The incidence of VTE in cancer is 4—7 times more common while 20% of patients who present with a VTE have an underlying malignancy. The physiology related to the development of a VTE remains elusive and multifaceted. One of the mechanisms is due to a mass effect, causing compression of venous structures and venous outflow obstruction. Caner-related therapies also increase the risk of thrombosis, including surgery, the use of central venous catheters, and chemotherapeutic agents. The development of a VTE increases cancer-related morbidity and mortality.[66]

When and who to investigate?

The 2021 European Society of Vascular Surgery (ESVS) Clinical Practice Guidelines on the Management of Venous Thrombosis[4] provide some insights into whether we should be investigating patients for thrombophilia; however, they have a weak level of evidence to back these guidelines. ESVS recommends that thrombophilia testing should not be pursued for patients with a provoked DVT (Class III, Level C recommendation). For patients with an unprovoked DVT and a positive family history of DVT in a first-degree relative, thrombophilia screening should be considered (Class IIa, Level C). This is even more pertinent for younger patients, that is, less than 45 years old, where patients have a longer period of time over which to develop recurrence. In this clinical context, the clinician may consider extending the anticoagulation treatment period, and so testing would alter clinical management. On a similar note, Kakkos et al. also recommend (Class IIa, Level C) testing for antiphospholipid syndrome in patients who experienced an unprovoked DVT as a form of risk assessment when considering anticoagulation cessation. Other indications for thrombophilia testing are those young patients, that is, less than 45 years old, who have unprovoked DVTs at unusual sites and have a strong first-degree family history of VTE. Thrombophilia testing can be also considered for antiphospholipid antibodies for females who had a DVT and a history of intrauterine fetal death and recurrent miscarriages, intrauterine fetal growth restriction, and preeclampsia. This clinical practice guideline also details information on when to take the thrombophilia screen. It is important for the test not to be taken in the acute period of a VTE. Should the patient be on heparin, vitamin K antagonist, or a direct oral anticoagulant (DOAC), plasma level assays should be deferred until at least 2 weeks after cessation. Genetic testing and thrombophilia test can be done despite the use of vitamin K antagonists and DOACs.

These clinical practice guidelines are confirmed by the National Institute for Health and Care Excellence (NICE) guidelines who also add that thrombophilia screening should not be conducted for the first-degree relatives of patients who sustained a DVT and have been found to have thrombophilia condition.[67] The American Society of Hematology,[68] in their most recent clinical practice guidelines, does not provide any recommendation on which cohort of patients would benefit from thrombophilia testing. The British Society of Hematology last published clinical practice guidelines in 2010, yet its message remains pertinent to today's practice. Similar to more recent clinical practice guidelines, it does not recommend thrombophilia testing for patients presenting for the first time with a VTE. Testing should be considered in patients with a strong family history of unprovoked recurrent thrombosis (Level C) however do not recommend how to select such patients. It recommends the testing for Protein C and S for patients who develop skin necrosis after starting a vitamin K antagonist (Level 2B) as well as for neonates with purpura fulminans (Level 1B). It also provides clinical guidance on whether to test asymptomatic first-degree relatives, which it does not recommend. For pregnant women, these guidelines recommend thrombophilia testing for the asymptomatic pregnant women who have a positive family history of a first-degree relative who had an unprovoked VTE or a VTE related to pregnancy/use of the COCP or a secondary to a minor risk factor.

Some evidence that supports the notion of not to test for thrombophilia conditions comes from the post hoc analysis of patients enrolled in the Multiple Environmental and Genetic Assessment of risk factors for venous thrombosis (MEGA) study published by Coppens et al. in 2008. They demonstrated that the odds ratio for recurrence was 1.2 for tested for thrombophilia versus nontested patients even after correction for various possible confounding factors.[69] The same author in another publication described real-world data with a true reflection of daily practice in a retrospective analysis of thrombophilia requests done in the Netherlands. It noted that the results only influenced management in 23% of patients pointing out the futility of the request.[70] This was reiterated by Kwon et al. in 2013, who in their study noted that only 18% out of 2081 patients tested positive for thrombophilia. The study also found that 6%, 13%, and 4.5% had appropriate confirmatory testing for Protein C, Protein S, and antithrombin deficiency, respectively. It also noted that only 12% of patients with abnormal results were followed up.[71]

A study published by Cohn et al. in 2008 attempted to address the psychological impact of the thrombophilia test. It managed to retrieve six studies that had addressed this topic however the lack of uniformity of reporting led to an inability to pool the results into a meta-analysis. While a majority of patients were satisfied with the news of being a carrier, there was a large percentage of other participants who were worried about the information.[72] The psychological impact of thrombophilia screening was also addressed in the ESVS clinical practice guidelines. It argues that unnecessary testing may cause undue anxiety and medicalization.

Thrombophilia testing does not come without additional cost and thus cost-effectiveness models are used by service planners in clinical situations where finances are finite. The TREATS study found that thrombophilia screening was only cost-effective for women being started on hormone replacement therapy with a net cost of

£6824 per adverse clinical complication prevented. Even more cost-effective, in this subset of the population, would be to test women who have a personal or family history, reducing the cost to £2446 per adverse clinical complication prevented.[73] Marchetti et al. noted that extending the anticoagulation program from 6 months to 2 years to prevent recurrent VTE based on positive results of thrombophilia screening for factor V Leiden cost $12,833 per quality-adjusted life year (QALY).[74] Similar prolongation of anticoagulation programs for thrombophilia screen-detected factor V Leiden patients was found to be effective by Auerbach et al. at the cost of $3804/QALY. Prolongation of anticoagulation for patients with thrombophilia screen-detected antiphospholipid syndrome was also found to be cost-effective at $2928/QALY.[75]

One of the major drawbacks of the current research is the fact that most of the studies we currently have were conducted on a population of similar ethnicity. Various ethnicities may depict different incidences of genetic hereditary mutations. An example of this is how rare factor V Leiden is associated with African-American demographics.[76] Moreover, the pathological ranges for which one is considered positive may vary with the age of the patient, making a thrombophilia diagnosis difficult to make at a young age.[77]

Thrombophilia and post-thrombotic syndrome

Various studies have tried to elucidate any link between thrombophilia and PTS. A meta-analysis published in 2013 demonstrated that there was no significant association between the two. In this study, there was some thrombophilia that could not be included in the meta-analysis; however, even in this subset of conditions, there appears to be no association between PTS and thrombophilia. This study also showed a protective effect that factor V Leiden and prothrombin mutation had for the development of PTS.[78] Other authors base the association between PTS and thrombophilia solely on the risk of DVT recurrence associated with thrombophilia.[79] Another retrospective study directly assessed the development of PTS for patients who tested positive for thrombophilia. Thrombophilia patients had an adjusted hazard ratio of 1.23 for developing PTS as compared to patient who tested negative. This was further subdivided according to the thrombophilia; factor V Leiden hazard ratio 0.42, Protein S deficiency hazard ratio 1.08, Protein C hazard ratio 0.96, lupus anticoagulant hazard ratio 0.81, prothrombin gene mutation hazard ratio 1.33.[80] Another review article could not establish any association between thrombophilia and the development of PTS or chronic venous ulcers in the context of severe PTS.[81]

Thrombophilia and venous ulceration

Based on currently available evidence, the Society for Vascular Surgery and the American Venous Forum in their Clinical Practice Guidelines for Management of Venous Leg Ulcers recommend selective thrombophilia laboratory evaluation for all patients with a history of recurrent venous thrombosis and chronic recurrent venous leg ulcers.[82] It is worth emphasizing that these recommendations were based on a weak level of evidence (Grade 2C) derived from observational studies or case reports that indicate the need for additional, high-quality data pertinent to the impact of hypercoagulable states on pathogenesis and management of venous ulcers. The association between thrombophilia and venous ulceration has been the subject of a number of studies. One of the studies found that multiple (>3) thrombophilia and homozygous mutation were statistically significantly more common in patients who developed venous ulcers before the age of 50 years.[83] Statistical significance was not found when two factors were taken into consideration. This study had the highest prevalence of thrombophilia with a 100% of the subject being found to test positive. An explanation for this could be the fact that the authors chose to test for PAI-1 polymorphism and MTHFR gene mutations that are not commonly tested for in other published literature. Indeed, there were no cases of antithrombin, Protein C, and Protein S in their study.[83] Other authors report 41% of patients with chronic venous ulcers to have at least one form of thrombophilia including APC resistance, anticardiolipin antibodies, Factor V Leiden, lupus anticoagulant, Protein S deficiency, Protein C deficiency, antithrombin, and prothrombin 20210A mutation, with respective rates of 16%, 14%, 13%, 9%, 7%, 6%, 5%, and 4%.[84] The thrombophilia was not related to a history of DVT, deep reflux, or disease severity. The diagnosis of thrombophilia was neither associated with an ulcer duration/recurrence, ulcer surface area, or ulcer-related pain.[84] Interestingly, a patient with five of above-mentioned underlying hypercoagulable states had no DVT history and no evidence of deep venous disease on clinical and ultrasonographic evaluation. Another study postulated that patients with thrombophilia had an increase in the incidence of clinical and subclinical thrombosis leading to the development of superficial and deep venous reflux which in turn is a risk factor for the development of varicose veins, venous hypertension secondary to venous reflux and obstruction and ultimately chronic venous ulcer. This was concluded as the prevalence of thrombophilia in patients who had active or healed ulcers and/or varicose veins approached that observed in post-DVT patients, whereas the thrombophilia rate in the control population was comparable to prevalence in the general population.[85] This message was reiterated by Bradbury et al. They pointed out that relying on patients to self-report a history of DVT is highly unreliable as these may be both clinical and subclinical DVTs. They speculate that the ulcers may be caused by isolated infra popliteal postthrombotic venous pathology, which is difficult to report by duplex ultrasonography, with a more distal location of DVT being a feature of some forms of thrombophilia.[86] Gaber et al. evaluated 100 consecutive patients with leg ulcers and identified the Factor V Leiden mutation as a specific thrombophilia state associated with chronic venous ulceration in 36% of patients.[87]

Venous leg ulceration has a significant impact on a patient's quality of life. It also has a significant cost implication for both patients and health services. Hypercoagulable states contribute to this formation by the manifestation of PTS following an acute lower limb DVT. Therefore, understanding the conditions that may predispose to DVT formation will help to guide the management of the patient, both in the acute and chronic phases. The detection of combined hereditary thrombophilia defects may significantly influence decisions on the type and duration of anticoagulation to try to prevent further venous thromboembolism and the associated increased risk of developing PTS and subsequent venous ulceration.

References

1. *SIGN Guideline No. 120. Management of Chronic Venous Leg Ulcers: A National Clinical Guideline*. 2010.
2. Nelzen O. Leg ulcers: economic aspects. *Phlebology*. 2000;15. s.l.
3. Labropoulos N, Grasparis AP, Pefanis D, Leon Jr LR, Tassiopoulos AK. Secondary chronic venous disease progresses faster than primary. *J Vasc Surg*. 2009:704−710.
4. Stavros K. European society for vascular surgery (ESVS) 2021 clinical practice guidelines on the management of venous thrombosis. *Eur J Vasc Endovasc Surg*. 2020;61(1):9−82. https://doi.org/10.1016/j.ejvs.2020.09.023.
5. Labropoulos N, Leon M, Nicolaides AN, et al. Venous reflux in patients with previous deep venous thrombosis: correlation with ulceration and other symptoms. *J Vasc Surg*. 1994:20−26.
6. Galanaud JP, Monreal M, Kahn SR. Epidemiology of the post-thrombotic syndrome. *Thromb Res*. 2018:100−109.
7. Prandoni P, Kahn SR. Post-thrombotic syndrome: prevalence, prognostication and need for progress. *Br J Haematol*. 2009:286−295.
8. Schulman S, Lindmarker P, Holmström M, et al. Post-thrombotic syndrome, recurrence, and death 10 years after the first episode of venous thromboembolism treated with warfarin for 6 weeks or 6 months. *J Thromb Haemostasis*. 2006:734−742.
9. Ageno W, Piantanida E, Dentali F, et al. Body mass index is associated with the development of the post-thrombotic syndrome. *Thromb Haemostasis*. 2003:305−309.
10. Prandoni P, Lensing AWA, Prins MH, et al. Residual venous thrombosis as a predictive factor of recurrent venous thromboembolism. *Ann Intern Med*. 2002:955−960.
11. Virchow RLK. Thrombosis und Emboli. s.l.: *Gesammelte Abhandlungen zur Wissenschaftlichen Medicine*. Vols. 449−454. Meidinger Sohn & Co.; 1856:33.
12. Connors JM. Thrombophilia testing and venous thrombosis. *N Engl J Med*. 2017:1177−1187.
13. Stevens SM, Woller SM, Bauer KA, et al. Guidance for the evaluation and treatment of hereditary and acquired thrombophilia. *J Thromb Thrombolysis*. 2016:154−164.
14. Garcia-Horton A, Kovacs MJ, Abdulrehman J, Taylor JE, Sharma S, Lazo-Langner A. Impact of thrombophilia screening on venous thromboembolism management practices. *Thromb Res*. 2017: 76−80.
15. Moll S. Thrombophilia: clinical−practical aspects. *J Thromb Thrombolysis*. 2015:367−378.
16. Crowther MA, Kelton JG. Congenital thrombophilic states associated with venous thrombosis: a qualitative overview and proposed classification system. *Ann Intern Med*. 2003;128−134:138 [PubMed: 12529095].
17. Stenflo J. A new vitamin K-dependent protein: purification from bovine plasma and preliminary characterization. *J Biol Chem*. 1976:355−363.
18. Griffin JH, Evatt B, Zimmerman TS, Kleiss AJ, Wideman C. Deficiency of protein C in congenital thrombotic disease. *J Clin Investig*. 1981:1370−1373.
19. Folsom AR, Aleksic N, Wang L, Cushman M, Wu KK, White RH. Protein C, antithrombin, and venous thromboembolism incidence A prospective population-based study. *Arterioscler Thromb Vasc Biol*. 2002:1018−1022.

20. Wypasek E, Undas A. Protein C and protein S deficiency—practical diagnostic issues. *Adv Clin Exp Med*. 2013:459—467.
21. Dykes AC, Walker ID, McMahon AD, Islam SIAM, Tait RC. A study of Protein S antigen levels in 3788 healthy volunteers: influence of age, sex and hormone use, and estimate for prevalence of deficiency state. *Br J Haematol*. 2001:636—641.
22. Fitridge R, Thompson M. *Mechanisms of Vascular Disease*. Adelaide: The University of Adelaide Press; 2011.
23. Patnaik M, Moll S. Inherited antithrombin deficiency: a review. *Haemophilia*. 2008:1229—1239.
24. Lipe B, Ornstein D. Deficiencies of natural anticoagulants, protein C, protein S, and antithrombin. *Circulation*. 2011:365—e368.
25. Brinkhous KM, Smith HP, Warner ED, Seegers WH. Inhibition of blood clotting and unidentified substances which acts in conjunction with heparin to prevent the conversion of prothrombin to thrombin. s.l. *Am J Physiol*. 1939;Seminars in Trhombosis and Haemostasis, Vols 125:683—687.
26. Egeberg O. Inherited antithrombin deficiency causing thrombophilia. *Thrombosis Diath Haemorrh*. 1965;13:516—530. s.l.
27. Lechner E, Thaler K. Antithrombin III deficiency and thromboembolism. *Clin Haematol*. 1981;10(2): 369—390. https://doi.org/10.1016/S0308-2261(21)00229-0.
28. Marciniak E, Farley C, DeSimone P. Familial trhombosis due to antithrombin III deficiency. *Blood*. 1974;43(2):219—231. https://doi.org/10.1182/blood.V43.2.219.219.
29. Martinelli I, Mannucci PM, De Stefano V, et al. Different risks of thrombosis in four coagulation defects associated with inherited thrombophilia: a study of 150 families. *Blood*. 1998;92:2353—2358. s.l.
30. Gaman AM, Gaman GD. Deficiency of antithrombin III (AT III)—case report and review of the literature. *Curr Health Sci J*. 2014:141—143.
31. Dahlback CB, Carlsson M, Svensson PJ. Familial thrombophilia due to a previously unrecognized mechanism characterized by poor anticoagulant response to activated protein C: prediction of a cofactor to activated protein. In: *Proceedings of the National Academy of Sciences of the United States of America*Vol 90.
32. Thomas RH. Hypercoagulability syndromes. *Arch Intern Med*. 2001;161(20):2433—2439. s.l.
33. Ridker PM, Hennekens CH, Lindpaintner K, et al. Mutation in the gene coding for coagulation factor V and the risk of myocardial infarction, stroke, and venous thrombosis in apparently healthy men. *N Engl J Med*. 1995;332:912—917. s.l.
34. Kujovich J. *GeneReviews*. Seattle: Univeristy of Washington; 2018.
35. Kujovich JL. Factor V Leiden thrombophilia. *Genet Med*. 2011;13(1):1—16. https://doi.org/10.1097/GIM.0b013e3181faa0f2.
36. Sharma A, Bhakuni T, Biswas A, et al. Prevalence of factor V genetic variants associated with Indian APCR contributing to thrombotic risk. *Clin Appl Thromb Hemost*. 2017:596—600.
37. van der Meer FÜM, Koster T, Vandenbroucke JP, Briet E, Rosendaa FR. The Leiden thrombophilia study (LETS). *Thromb Haemostasis*. 1997:631—635.
38. Poort SR, Rosendaal FR, Reitsma P, Bertina RM. A common genetic variation in the 3′-untranslated region of the prothrombin gene is associated with elevated plasma prothrombin levels and an increase in venous thrombosis. *Blood*. 1996:3698—3703.
39. McGlennen RC, Key NS. Clinical and laboratory management of the prothrombin G20210A mutation. *Arch Pathol Lab Med*. 2002;126(11):1319—1325. https://doi.org/10.5858/2002-126-1319-CALMOT.
40. Casini A, Blondon M, Lebreton A, et al. Natural history of patients with congenital dysfibrinogenemia. *Thromb Haemost*. 2015;125(3):553—561. https://doi.org/10.1182/blood-2014-06-582866.
41. Wei A, Wu Y, Xiang L, Yan J, Cheng P, Deng D. Congenital dysfibrinogenemia caused by γAlαa327Val mutation: structural abnormality of D region. *Hematology*. 2021:305—311.
42. MacDonald BMW, Stephen G. Acquired hypofibrinogenemia: current perspectives. *Hematol Res Rev*. 2016:217—225.
43. Tiscia GL, Mrgaglione M. Human fibrinogen: molecular and genetic aspects of congenital disorders. *Int J Mol Sci*. 2018:1597.

44. Shapiro SE. Diagnosis and Management of dysfibrinogenemia. *Clin Adv Hematol Oncol.* 2018;16(9): 602–605.
45. Corban MT, Duarte-Garcia A, McBane RD, Matteson EL, Lerman LO, Lerman A. Antiphospholipid syndrome; role of vascular endothelial cells and implications for risk stratification and targeted therapeutics. *J Am Coll Cardiol.* 2017:2317–2330.
46. Devreese KMJ, Ortel TL, Pengo V, Delaat B. Laboratory criteria for antiphospholipid syndrome: communication from the SSC of the ISTH. *Int Soc Thromb Haemost.* 2018:809–813.
47. Miyakis S, Lockshin MD, Atsumi T, et al. International consensus statement on an update of the classification criteria for definite antiphospholipid syndrome (APS). *Int Soc Thromb Haemost.* 2006: 295–306.
48. Ames PRJ, Merashli M, Bucci T, et al. Antiphospholipid antibodies and autoimmune haemolytic anaemia: a systematic review and meta-analysis. *Int J Mol Sci.* 2020;21(11):4120. https://doi.org/10.3390/ijms21114120.
49. Aguiar CL, Erkan D. Catastrophic antiphospholipid syndrome: how to diagnose a rare but highly fatal disease. *Therapeutic Advances in Musculoskeletal Disease.* 2013:305–314.
50. Strübing P. Paroxysmale Haemoglobinurie. *DMW (Dtsch Med Wochenschr).* 1882;8(1):17. s.l.
51. Brodsky RA. Paroxysmal nocturnal hemoglobinuria. *Blood.* 2015:2804–2811.
52. Hill A, Kelly RJ, Hillmen P. Thrombosis in paroxysmal nocturnal hemoglobinuria. *Blood.* 2013: 4985–4996.
53. Araten DJ, Thaler HT, Luzzatto L. High incidence of thrombosis in African-American and Latin-American patients with paroxysmal nocturnal haemoglobinuria. *Thromb Haemostasis.* 2005:88–91.
54. Luzzatto L, Risitano AM, Notaro R. Paroxysmal nocturnal hemoglobinuria and eculizumab. *Haematologica.* 2010:523–526.
55. Sekhar M. Prevention and management of thrombosis in myeloproliferative neoplasms. *Clin Adv Hematol Oncol.* 2017;15(3):178–181.
56. Cai Chia Y, Ramli M, Woon PY, Johan MF, Hassan R, Asiful Islam M. Molecular genetics of thrombotic myeloproliferative neoplasms: implications in precision oncology. *Genes Dis.* 2021.
57. Barbui T, Finazzi G, Falanga A. Myeloproliferative neoplasms and thrombosis. *Blood.* 2013: 2176–2184.
58. Silverstein MD, Heit JA, Mohr DN, Petterson TM, O'Fallon WM, Melton LJ. Trends in the incidence of deep vein thrombosis and pulmonary embolism; a 25-year population-based study. *JAMA Intern Med.* 1998;158(6):585–593. https://doi.org/10.1001/archinte.158.6.585.
59. Yayan J, Bals R. Relative risk of deep vein thrombosis in very elderly patients compared with elderly patients. *Clin Appl Thromb/Haemost.* 2016:77–84.
60. Jordan WM. Pulmonary embolism. *Lancet.* 1961:1146–1147.
61. Stegeman BH, De Bastos M, Rosendaal FR, et al. Different combined oral contraceptives and the risk of venous thrombosis: systematic review and network meta-analysis. *Br Med J.* 2013;347(f5298). https://doi.org/10.1136/bmj.f5298.
62. Tepper NK, Whiteman MK, Marchbanks PA, James AH, Curtis KM. Progestin-only contraception and thromboembolism: a systematic review. *Contraception.* 2016:678–700.
63. Vinogradova Y, Coupland C, Hippisley-Cox J. Use of hormone replacement therapy and risk of venous thromboembolism: nested case-control studies using the QResearch and CPRD databases. *Br Med J.* 2019;364(k4810). https://doi.org/10.1136/bmj.k4810.
64. Wu O. Postmenopausal hormone replacement therapy and venous thromboembolism. *Gend Med.* 2005:518–527.
65. Ghaji N, Boulet SL, Tepper N, Hooper W. Trends in venous thromboembolism among pregnancy-related hospitalizations, United States, 1994–2009. *Am J Obstet Gynecol.* 2013;209:5, 433.e1-133.e4338. https://doi.org/10.1016/j.ajog.2013.06.039
66. Agnelli G, Verso M. Management of venous thromboembolism in patients with cancer. *J Thromb Haemostasis.* 2011:316–324.
67. National Institute for Health and Care Excellence. *Venous Thromboembolic Disease: Diagnosis, Management and Thrombophilia Testing.* NICE. NICE Guideline; 2020.

68. Ortel TL, Neumann I, Walter A, et al. American Society of Hematology 2020 guidelines for management of venous thromboembolism: treatment of deep vein thrombosis and pulmonary embolism. *Blood Advances*. 2020;4.
69. Coppens M, Reijnders JH, Middeldorp S, Doggen CJM, Rosendaal FR. Testing for inherited thrombophilia does not reduce the recurrence of venous thrombosis. *J Thromb Haemostasis*. 2008:1474−1477.
70. Coppens M, Van Mourik JA, Eckmann CM, Buller HR, Middeldorp S. Current practise of testing for inherited thrombophilia. *J Thromb Haemost*. 2007:1979−1981.
71. Kwon A, Roshal M, De Sancho MT. Evaluating adherence to clinical guidelines for thrombophilia screening. *Blood*. 2013;14(5):982−986. https://doi.org/10.1111/jth.13284.
72. Cohn DM, Vansenne F, Kaptein AA, De Borgie CAJM, Middeldorp S. The psychological impact of testing for thrombophilia: asystematic review. *J Thromb Haemostasis*. 2008:1099−1104.
73. Wu O, Robertson L, Twaddle S, et al. Creening for thrombophilia in high-risk situations: a meta-analysis and cost-effectiveness analysis. *Br J Haematol*. 2005:80−90.
74. Marchetti M, Pistorio A, Barosi G. Extended anticoagulation for prevention of recurrent venous thromboembolism in carriers of factor V Leiden-cost-effectiveness analysis. *Thromb Haemost*. 2000: 752−757.
75. Auerbach AD, Sanders GD, Hambleton J. Cost-effectiveness of testing for hypercoagulability and effects on treatment strategies in patients with deep vein thrombosis. *Am J Med*. 2004;116(12):816−828. https://doi.org/10.1016/j.amjmed.2004.01.017.
76. Cushman M. Epidemiology and risk factors for venous thrombosis. *Semin Hematol*. 2008:62−69.
77. Bruce A, Massicotte MP. Thrombophilia screening: whom to test? *Thromb Hemost*. 2012;120(7): 1353−1355. https://doi.org/10.1182/blood-2012-06-430678.
78. Rabinovich A, Cohen JM, Prandoni P, Kahn SR. Association between thrombophilia and the post-thrombotic syndrome: a systematic review and meta-analysis. *J Thromb Haemostasis*. 2014:14−23.
79. Kreidy R. Pathophysiology of post-thrombotic syndrome: the effect of recurrent venous thrombosis and inherited thrombophilia. *Int Sch Res Netw Vasc Med*. 2011;2011(513503). https://doi.org/10.5402/2011/513503.
80. Spieza L, Campello E, Giolo E, Villalta S, Prandoni P. Thrombophilia and the risk of post-thrombotic syndrome: retrospective cohort observation. *J Thromb Haemost*. 2010:211−213.
81. Rabinovich A, Cohen JM, Prandoni P, Kahn SR. Association between thrombophilia and the post-thrombotic syndrome. *Int J Vasc Med*. 2013;12(1):14−23. https://doi.org/10.1111/jth.12447.
82. O'Donnell Jr TF, Passman MA, Marston WA, et al. Society for vascular surgery; American venous forum. Management of venous leg ulcers: clinical practice guidelines of the society for vascular surgery and the American venous forum. *J Vasc Surg*. 2014 Aug;60(2):3S−59S. https://doi.org/10.1016/j.jvs.2014.04.049.
83. Calistru AM, Baudrier T, Gonçalves L, Azevedo F. Thrombophilia in venous leg ulcers: a comparative study in early and later onset. *Indian J Dermatol Venereol Leprol*. 2012;78(3):406. https://doi.org/10.4103/0378-6323.95477.
84. MacKenzie RK, Ludlam CA, Ruckley V, Allan PL, Burns P, Bradbury AW. The prevalence of thrombophilia in patients with chronic venous leg ulceration. *J Vasc Surg*. 2002:718−722.
85. Darvall KAL, Sam RC, Adam DJ, Silverman SH, Fegan CD, Bradbury AW. Higher prevalence of thrombophilia in patients with varicose veins and venous ulcers than controls. *J Vasc Surg*. 2009: 1235−1241.
86. Bradbury AW, MacKenzie RK, Burns P, Fegan C. Thrombophilia and chronic venous ulceration. *Eur J Vasc Endovasc Surg*. 2002:97−104.
87. Gaber Y, Siemens HJ, Schmeller W. Resistance to activated protein C due to factor V Leiden mutation: high prevalence in patients with post-thrombotic leg ulcers. *Br J Dermatol*. March 2001;144(3): 546−548. https://doi.org/10.1046/j.1365-2133.2001.04081.x.

CHAPTER 9

The chronically swollen leg with ulcers—finding the cause: theory and practice

Pier Luigi Antignani[1], Luca Costanzo[2], Giacomo Failla[2] and Francesco Paolo Palumbo[3]

[1]Vascular Center, Nuova Villa Claudia, Rome, Italy; [2]Angiology Unit, San Marco Hospital, Department of Cardiovascular Disease, A.O.U. "G. Rodolico-San Marco", University of Catania, Catania, Italy; [3]Surgery Unit, Villa Fiorita Clinic, Prato, Italy

Introduction and epidemiology

In the setting of leg ulcer and edema, a common challenge for primary care physicians is to determine the cause and find an effective treatment.

It has been estimated that within the general population of the United Kingdom, approximately 3.99 people in every 1000 have chronic edema, which is thought to correlate with the increasingly aging population and associated polymorbidity. In people aged over 85 years, the prevalence of chronic edema increases to 12 people in every 1000.[1] Development of chronic edema is also related to lifestyle choices, such as immobility and obesity, both of which are expected to rise exponentially for the next 10–15 years.

The prevalence of chronic edema has a significant association with the presence of a wound. Moffatt et al. reported that between 52% and 69% of patients cared for by community nurses had chronic edema, and of these, 73% also had a leg ulcer.[2]

Edema is defined as a palpable swelling caused by an increase in interstitial fluid volume producing noticeable clinical signs and symptoms.

The fluid between the interstitial and intravascular spaces is regulated by the capillary hydrostatic pressure gradient and the oncotic pressure gradient across the capillary.[3] The accumulation of fluid occurs when local or systemic conditions disrupt this equilibrium, leading to increased capillary hydrostatic pressure, increased plasma volume, decreased plasma oncotic pressure (hypoalbuminemia), increased capillary permeability, or lymphatic obstruction.

In an article published in 2006, Ely and coworkers tried to suggest the most common causes of edema on the basis of data provided by evidence-based medicine, recognizing venous insufficiency as the most frequent, followed by idiopathic or cyclic edema (edema that develops in women in the fertile period) and venous hypertension associated with sleep apnea.[4]

Edema can be observed in one leg or both: some pathologies such as obesity, endocrinopathies, congestive heart disease, kidney, and liver disease, as well as severe states of malnutrition, affect two legs. More frequently, edema affects one of the lower limbs, and its genesis is related to the presence of a deep vein thrombosis or superficial thrombophlebitis. Other causes are the presence of a popliteal Baker's cyst, cellulite, or a muscle tear. Also, swelling may affect mostly one limb in venous hypertension linked to chronic venous insufficiency, postthrombotic syndrome and postural alterations, and lymphedema.[5]

It should be emphasized that not all edematous legs develop ulcerative lesions of the skin and in some subjects the ulcerative lesions are not accompanied by underlying venous pathologies.

Therefore, can ulcer pathophysiology be attributed only to the presence of edema? Furthermore, why do some subjects with severe edema do not develop an ulcer?

Literature data have indicated in the formation and development of edema a factor favoring hypoxia at the level of ulcerative lesion, but at the microcirculation level an alteration of the endothelial state has never been thoroughly investigated.[6]

A particularly important aspect was underlined by the studies of Raffetto and coll,[7] which highlight the involvement of the endothelium in relation to changes in the inflammatory state triggered by venous disease.

In subjects suffering from gravitational or occupational edema, the slowing or abolition of venous flow due to lack of contraction of the calf can dilate the time of contact of the blood with the venous wall allowing the creation of a chronic inflammatory state and of a marked hypoxemia in the cutaneous microcirculation.[8–10]

In general, the long immobility allows the accumulation of fluids in the tissues and prevents their physiological removal, but at the same time it alters the integrity of the glycocalyx and allows the triggering of chronic localized inflammation with the recall of leukocytes and macrophage cells that release reactive oxygen species (ROS), reactive nitrogen species (RNS), and matrix metalloproteinases.

Many patients suffer from lower limb ulcers limbs in absence of a diagnosed vascular or systemic disease. In a review published in 1949 by Anning,[11] the importance of gravity in the genesis of ulcerations has been emphasized: in previous work by Brodie[12] in 1846 and Gay in 1868,[13] they excluded the possibility that the varices were responsable for ulcerations in a study where was highlighted the presence of thrombosis of the deep venous vessels after careful dissections.

Despite this publication, the idea that the appearance of ulcers was exclusively related to varicose pathology was revived and amplified by the works of Dickson Wright[14] and Homans.[15]

The data reported by Birger in 1941[16] highlighted the role of thrombosis as the main reason for the genesis of the ulcer, as well as the works of Nilzèn (1945),[17] Bauer (1946),[18] and Birger (1947)[19] reinforced the idea of a postthrombotic genesis of leg ulcers.

Allen, Barker, and Hines in 1946[20] gave particular importance to venous stasis resulting from previous thrombotic events in the iliofemoral site as the main cause of local hypoxia in the skin of the leg. The possibility that a minor trauma could cause the onset of a chronic ulcerative lesion was explained by the reduced repair capacity of the tissues following the hypoxic state.

Already in the drafting of this article, the Author notes that, however, not all patients with leg ulcerations have a history of varicose veins and/or thrombosis and hypothesizes the role of muscle inactivity as a cause of venous stasis.[21]

In recent years, epidemiological data indicate that venous ulcer has a larger impact than those from other origins, although their actual incidence amounted to around 50%–55%.

At the same time, interest has focused on the role of chronic inflammation that triggers changes in the skin that lead to ulcer formation. The data emerging from the observations have allowed to hypothesize new activation pathways of inflammation, which assume an important role in cell populations in wound tissue, and to propose therapeutic alternatives for the treatment of the disease.[22]

Despite the introduction of Clinical, Etiological, Anatomical, and Pathophysiological, data relating to the real incidence of venous ulcers are extremely variable. There are several reasons that can explain the heterogeneity of the reports, but it seems clear that the percentages suggested in the 1990s (up to 75% of ulcers) appear to be resized from the latest studies. Large-scale studies and other studies on restricted populations showed that the diagnosis of venous ulcers varies from 45% to 57%.[5,23]

Because of the lack of standardized diagnostic criteria and the variable professional figures involved, several lesions are misdiagnosed as "venous."

On 31,619 patients examined, Korber[24] in 20.3% of ulcerative lesions of the leg was not identified as a cause. Also, Failla et al.[25] report that about 17% of leg ulcers are associated with venous edema. This study showed that a combination of factors such as postural and walking alterations can contribute to the onset of leg ulceration that was originally attributed to a venous genesis.

Assessment of edema

In the first evaluation, it is extremely important to assess the history of edema: duration of the edema, whether it is painful or it changes with position, if it improves overnight, if it is unilateral or bilateral, medication history, and presence of comorbidities. Typically, we can differentiate acute onset edema (e.g., <72 h) from chronic. Acute swelling of a limb is more characteristic of deep venous thrombosis, cellulitis, ruptured popliteal cyst, and acute compartment syndrome from trauma.[26] The chronic accumulation of more generalized edema is due to the onset or exacerbation of chronic systemic conditions, such as congestive heart failure, renal disease, or hepatic disease.[27]

Unilateral edema in absence of trauma can be attributable to compromise of venous or lymphatic drainage that can result from thrombosis, venous insufficiency, venous or lymphatic obstruction by extrinsic compression or lymphatic destruction (e.g., congenital vs. secondary from a tumor, radiation, or filariasis). Bilateral or generalized swelling suggests a systemic cause, such as chronic heart failure, pulmonary hypertension, chronic renal or hepatic disease (causing hypoalbuminemia), protein-losing enteropathies, or severe malnutrition. Also, some drugs can be responsible for edema onset, such as antihypertensive drugs (e.g., calcium channel blockers, alfa-blockers), antidepressants, antivirals, chemotherapeutics, cytokines, hormones (e.g., corticosteroids), and nonsteroidal antiinflammatory drugs. The mechanism often includes the retention of salt and water with increased capillary hydrostatic pressure. Among comorbidities, some are frequently associated with edema such as cardiac, renal, thyroid, or hepatic disease.[27] Another disease that is emerging as a cause of edema is obstructive sleep apnea.

Quite often the clinical scenario is confusing due to the simultaneous involvement of several factors such as the presence of comorbidities, venous disease, obesity, and hypomobility.

For a differential diagnosis, it is necessary to establish whether there are diseases capable of causing alterations in the protein component of the serum (kidney disease, liver disease, or heart failure) and which are characterized by the involvement of both limbs, by an edema affecting predominantly a limb. In this case, the causes are to be found in the presence of a venous obstruction or in the possible presence of a Baker's cyst or a muscle tear. Other times the picture is very confusing due to the simultaneous involvement of a venous axis (postthrombotic syndrome, for example) and obesity or lymphedema.

However, not all edematous legs subsequently develop ulcerative skin lesions, just as in some patients' ulcerative lesions are not accompanied by underlying venous pathologies.

Excluding the presence of an organ disease and serious involvement of the venous axis, can the pathophysiology of these lesions be attributed only to the presence of edema? And why do some subjects, despite having a significant increase in volume, do not develop the appearance of an ulcer?

The meticulous anamnestic investigation of the lifestyle can help in identifying subjects who for various reasons (age, osteoarticular pathologies, sedentary lifestyle) spend hours in a sitting position without mobilizing the leg muscles and who sometimes develop ulcerative lesions of the leg.

Unfortunately, this does not allow to identify the factors triggering the onset of the skin lesion.

Literature data have indicated in the formation and development of edema a factor favoring hypoxia at the level of ulcerative lesion (see works by Palumbo and Failla), but at the microcirculation level, an alteration of the endothelial state has never been thoroughly investigated.

A particularly important aspect was underlined by the studies of Raffetto and coll., which highlight the involvement of the endothelium in relation to changes in the inflammatory state triggered by venous disease.[7]

In particular in subjects suffering from gravitational or occupational edema, the slowing or abolition of venous flow due to lack of contraction of the calf can dilate the time of contact of the blood with the venous wall and determine in some patients the creation of a chronic phlogistic state, favoring a marked hypoxemia of the cutaneous microcirculation (terminal type) with the recruitment of neutrophils and macrophages capable of altering the structure of the glycocalyx and allowing the accumulation of liquids in extravasal tissues.[28] In general, the long immobility, on the one hand, allows the accumulation of fluids in the tissues and prevents their physiological removal; on the other hand, it alters the integrity of the glycocalyx and allows the triggering of a chronic localized inflammation with the recall of leukocytes and macrophage cells to release free radicals and **metalloproteinases**.[29]

Physical examination

Pitting, tenderness, and skin changes have to be evaluated in the edema. Pitting is an indentation that remains in the edematous area after pressure is applied. This occurs when fluid in the interstitial space has a low concentration of protein, which is associated with decreased plasma oncotic pressure and disorders caused by increased capillary pressure (e.g., chronic heart failure, nephrotic syndrome, obstruction to venous or lymphatic deflux); conversely, in nonpitting edema, the indentation does not persist and it is mainly associated to mid-to early stage lymphedema, myxedema, and lipedema. The resulting fluid build-up can be composed of a variety of substances, including proteins, salts, and water. Furthermore, nonpitting edema is the expression of a chronic disease when fibrosis of the subcutaneous tissue occurs.[6]

While assessing the skin, the following signs have to be noted: dryness, signs of cellulitis, color of the skin, and pigmentation.

Examination has to start from the foot and between the toes to determine if there are any skin breaks or infection present. Notably, in lymphedema, there is an inability to tent the skin of the dorsum of the second toe, which is called the Kaposi—Stemmer sign.[30]

Diagnostic studies
Laboratory test

In presence of edema, some laboratory tests are useful to rule out the systemic cause of edema. Generally, complete blood count, urinalysis, electrolytes, creatinine, hepatic enzyme, blood sugar, thyroid-stimulating hormone, and proteinogram should be performed. Notably, a serum albumin below 2 g/dL often leads to edema and can be caused by liver disease, nephrotic syndrome, or protein-losing enteropathy. Additional tests may be indicated depending on the clinical presentation, such as brain natriuretic peptide to rule out chronic heart failure or D-dimer to exclude vein thrombosis. However, while false negative D-dimers are rare, an elevated D-dimer is common in several conditions such as inflammatory states.

Imaging

Venous ultrasonography is the imaging modality of choice in the evaluation of suspected deep venous thrombosis. Also, leg ultrasound can be used for the diagnosis of chronic venous insufficiency, can rule out peripheral arteriopathy, and can evaluate subcutaneous tissue (presence of fluid or fibrotic tissue), presence of lymphadenopathy or cysts (such as popliteal cyst). However, lymph flow cannot be assessed with ultrasonography; therefore, indirect radionuclide lymphoscintigraphy, which shows absent or delayed filling of lymphatic channels, is the method of choice for evaluating lymphedema in those cases that are challenging to diagnose by clinical examination.

Additional imaging such as computed tomography or magnetic resonance imaging may be required in some specific cases such as clinical suspicion of compression of the left iliac vein by the right iliac artery (May–Thurner syndrome) or diagnosis of musculoskeletal etiologies (such as a gastrocnemius tear). Also, in case of suspicion of cardiac origin, echocardiography should be performed.[31]

Goals

The main objective of the diagnosis of the edematous leg is aimed not only at the systemic treatment of the pathology that caused the edema, but above all at the use of compression therapy. The distinction between a leg carried. Once it is excluded that there may be systemic contraindications to the bandage (nephrotic syndrome, severe heart failure, and systemic diseases with general impairment), if the patient is able to walk, the main indication for treatment is compression therapy.[32]

Knowing how to distinguish a bandable ulcer from a nonbandable one is essential, regardless of the cause, especially if the patient is treated at home. Placing the right indication for compression treatment allows you not to compromise and accelerate the healing of the ulcer itself.

In this regard, the use of Continuous Wave (CW) doppler or alternative ankle-brachial index (ABI) measurement systems is essential. The cut-off reported in the literature for safety margins is the ABI value between 0.8 and 1.3.

Experienced personnel under medical supervision can also bandage limbs with ABI below 0.8 with moderate compression and under close medical supervision.

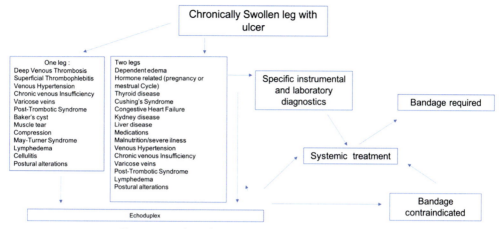

Figure. 9.1 Flow chart on diagnostics and bandaging.

Arteriopathic patients with critical ischemia therapy can also undergo compression therapy using intermittent pneumatic pressure as reported in the Fig. 9.1.[33]

References

1. Moffatt CJ, Keeley V, Franks PJ, Rich A, Pinnington LL. Chronic oedema: a prevalent problem for UK health services. *Int J Wounds*. 2017;14:772−781.
2. Moffatt CJ, Gaskin R, Sykorova M, et al. Prevalence and risk factors for chronic oedema in UK community nursing services. *Lymphatic Res Biol*. 2019;17(2):147−154.
3. Trayes KP, Studdiford JS, Pickle S, Tully AS. Edema: diagnosis and management. *Am Fam Physician*. July 15, 2013;88(2):102−110. PMID: 23939641.
4. Ely JW, Osheroff JA, Lee Chambliss M, Ebell MH. Approach to leg edema of unclear etiology. *J Am Board Fam Med*. 2006;19(2):148−160.
5. Failla G. Epidemiologia ed inquadramento clinico Atti del VII Congresso Nazionale AIUC Firenze 2009. *Acta Vulnol*. 2009;7(Supp 1 al n, 3):42.
6. Gasparis AP, Kim PS, Khilnani VM, Labropoulos N. Diagnostic approach to lower limb edema. *Phlebology*. 2020;35(9):650−655.
7. Raffetto JD, Ligi D, Maniscalco R, Khalil RA, Mannello F. Why venous leg ulcers have difficulty healing: overview on pathophysiology, clinical consequences, and treatment. *J Clin Med*. 2021;10:29. https://doi.org/10.3390/jcm10010029.
8. Failla G, Palumbo FP. The modern approach to venous ulcers. In: Allegra C, Antignani PL, Kalodiki E, eds. *Tips and Tricks in Angiology Torino*. Minerva Medica Ed.; 2016:182−184.
9. Palumbo FP, Failla G. Use of laser-speckle contrast analysis in the study of "non healing" leg ulcers—a preliminary study. *J Dermatol Cosmetol*. 2018;2:00028.
10. Palumbo FP, Failla G, Adamo G, Antignani PL. Laser-speckle evaluation of microcirculation in leg ulcers. *Acta Phlebologica*. 2016;17:86−90.
11. Anning ST. The aetiology of gravitational ulcers of the leg. *BMJ Aug*. 1949;27:458.
12. Brodie SBC. *Lectures Illustrative of Various Subjects in Pathology and Surgery*. Longmans, London cit. in (8); 1846.
13. Gay J. *On Varicose Disease of the Lower Extremities. The Lettsomian Lectures of 1867 (cit in 8)*. 1868.
14. Wright AD. *Lancet,1,457 cit in (8)*. 1931.
15. Homans J. *Circulatory Diseases of the Extremities*. New York: Macmillan; 1939. cit. in (8).

16. Birger I. *Nord.Med.,12,3542. cit. in (8)*. 1941.
17. Nilzen A. *Acta Chir Scand.,92,285. cit. in (8)*. 1945.
18. Bauer G. H. *Lancet 1, 447 cit. in (8)*. 1946.
19. Birger I. *Acta Chir Scand*. 1947;95(suppl. 129 (8)).
20. Allen EV, Barker NW, Hines EA. *Peripheral Vascular Diseases. Saunders, Philadelphia cit. in (8)*. 1946.
21. Robertson F. Appraisal of causes and treatment of venous ulcers of the leg. *Can Med Assoc J*. 1956;75:42.
22. Franks PJ, Barker J, Collier M, et al. Management of patients with venous leg ulcers: challenges and current best practice. *J Wound Care*. 2016;25(Suppl 6):S1–S67.
23. Moffatt CJ, Franks PJ, Doherty DC, Martin R, Blewett R, Ross F. Prevalence of leg ulceration in a London population. *QJM*. July 2004;97(7):431–437.
24. Körber A, Klode J, Al-Benna S, et al. Etiology of chronic leg ulcers in 31,619 patients in Germany analyzed by an expert survey. *J Dtsch Dermatol Ges*. February 2011;9(2):116–121.
25. Failla G, Palumbo FP, Serantoni S, et al. *Prevalence posture and leg ulcers. Epidemiological study of 220 ambulatory patients observed in the district of eastern Sicily in 2012*. In: *Paper presented at: VI Congresso CO.R.T.E Rome March 9, 11*. 2016.
26. Young JR. The swollen leg. Clinical significance and differential diagnosis. *Cardiol Clin*. 1991;9:443–456.
27. Yale SH, Mazza JJ. Approach to diagnosing lower extremity edema. *Compr Ther*. 2001;27:242–252.
28. Braverman I. The cutaneous microcirculation. The journal of investigative dermatology. In: *Symposium Proceedings/the Society for Investigative Dermatology, Inc. [and] European Society for Dermatological Research. 5. 3-9. 10.1046*. 2001.
29. Ali MM, Mahmoud AM, Le Master E, Levitan I, Phillips SA. Role of matrix metalloproteinases and histone deacetylase in oxidative stress-induced degradation of the endothelial glycocalyx. *Am J Physiol Heart Circ Physiol*. 2019;316:H647–H663.
30. Szuba A, Rockson SG. Lymphedema: classification, diagnosis, and therapy. *Vasc Med*. 1998;3:145–146.
31. Antignani PL, Benedetti-Valentini F, Aluigi L, et al. Diagnosis of vascular disease ultrasound investigation guidelines. *Int Angiol*. October 31, 2012. Suppl 1 to Issue 5.
32. World Union of Wound Healing Societies (WUWHS). *Principles of Best Practice: Compression in Venous Leg Ulcers. A Consensus Document*. London, UK: MEP Ltd; 2008.
33. Mosti G, Mattaliano V, Polignano R, Masina M. La terapia compressive nel trattamento delle ulcere cutanee. *Acta Vulnol*. 2009:113–135.

CHAPTER 10

Lower extremity wounds associated with mixed venous and arterial insufficiency and relevant differential diagnosis

Enjae Jung and Robert B. McLafferty
Division of Vascular Surgery, Oregon Health and Science University, Portland, OR, United States

Introduction

Leg ulcers are a very common clinical finding and continue to be an increasing problem worldwide. The most common underlying etiologies encountered in the outpatient clinic are chronic venous disease (CVD), peripheral neuropathy caused by diabetes mellitus, and peripheral artery disease (PAD) with approximately 70% of leg ulcers being caused by CVD and roughly 20% caused by PAD or mixed arteriovenous disease.[1,2] In patients with venous leg ulcers (VLU), the reported incidence of concomitant PAD has ranged from 15% to 30%.[3–5] The presence of PAD complicates the treatment of patients with VLU as standard protocols cannot be applied without evaluation and treatment of the underlying PAD. This chapter will review the approach to patients with combined arterial and venous insufficiency including clinical presentation, diagnostic evaluation, and treatment. We will also discuss and compare less common etiologies in the differential diagnosis that can mimic VLU with or without a PAD component.

Evaluation

History and physical examination

When a healthcare provider evaluates a patient with an ulcer in the gaiter area of the leg that is presumed to be a VLU, several important parts of the history and physical examination can provide important clues to the presence of coexisting CVD and PAD. The presence or history of the typical risk factors for PAD such as smoking, hypercholesterolemia, hypertension, diabetes mellitus, and renal insufficiency or failure in association with advanced age should immediately trigger a high suspicion. Additionally, concomitant coronary artery disease, cerebrovascular disease, or remote or current history of claudication are important factors in the history. More severe manifestations for PAD such as ischemic rest pain or ulceration can be gleaned by asking about pain in the forefoot, particularly at night while in bed when the foot is elevated and whether the patient has any concomitant ulcers on the feet or toes. A family history of any of the atherosclerotic occlusive diseases, particularly in patients less than 60 years of age, is equally important in determining a predilection to having concomitant PAD.

The physical examination remains equally important in determining whether or not a VLU is complicated by PAD. Inspection should note the location, size, and depth of the ulcer. Peripheral artery disease is more frequently identified in patients with a severe form of CVD[6] and in VLU complicated by PAD, the ulcers are often multiple, larger, deeper, and concomitantly infected with minimal evidence of granulation tissue (Fig. 10.1).[7,8] Further scrutiny by inspection in patients with concomitant PAD may reveal chronic nail changes, a paucity of hair growth, toe ulcers often lurking between web spaces and dependent rubor with elevation pallor. Peripheral pulse exam may reveal absent pulses in the femoral, popliteal, dorsal pedal, or posterior tibial arteries. The absence of

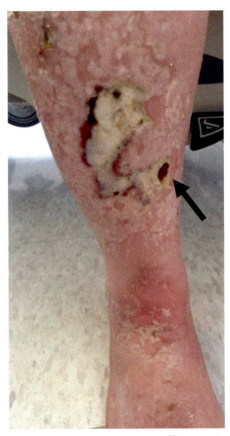

Figure 10.1 Patient with combined arterial and venous insufficiency. On history and physical exam, he was noted to have severe rest pain and dependent rubor. Note the pale fibrous ulcer base with minimal granulation tissue and a separate deeper ulcer with a punched out appearance (*black arrow*).

pulses in the foot should never be solely ascribed to the presence of CVD. The reason being that often in patients with CVD, there are other confounding factors that can worsen edema in the foot, such as obesity, chronic lower extremity dependency, lymphedema, chronic renal insufficiency, and/or congestive heart failure. The presence of edema from these diagnoses should never lead to the presumption that pulses are probably present but cannot be palpated. In contrast, patients with significant edema and with palpable pulses may still have mild-to-moderate PAD. Therefore, because of variability in pulse examination and the relatively common occurrence of PAD in the presence of VLU, it is recommended that all patients with a presumed VLU have ankle-brachial index (ABI) testing performed. In those patients where a reliable ABI cannot be performed, other forms of PAD evaluation are necessary to assure freedom from PAD.

Vascular laboratory testing

Measurement of the ABI is the primary method for establishing the diagnosis of PAD with a ratio of less than 0.9 having a high sensitivity and specificity compared to the criterion standard of contrast arteriography.[9] While this is most commonly performed in the vascular laboratory, this test can easily be performed in the outpatient setting or bedside by dividing the ankle blood pressure (as determined from the higher of the dorsal pedal or posterior tibial artery using a hand-held portable Doppler ultrasound) by the higher of the two systolic brachial blood pressures. Most vascular laboratories objectively define the presence of PAD with an ABI of less than 0.9, and normal ranging from 0.9 to 1.3. Conditions such as diabetes mellitus, renal failure, and morbid obesity may artificially elevate the ratio to greater than 1.3 due to the inability to adequately compress the tibial arteries due to calcification of the vessels or woody edema. In addition, Doppler-derived arterial waveforms are often assessed at the time of the ankle pressure measurements. A waveform is triphasic when normal, then becoming biphasic to monophasic with loss of amplitude as the PAD severity progresses. A review of the waveform morphology along with a toe-brachial index (TBI) can prevent being misled by a falsely elevated ABI. Toe-brachial index is commonly performed by vascular laboratories, and an index of less than 0.7 is indicative of hemodynamically significant arterial insufficiency. Not uncommonly, VLU location on the lower aspect of the leg may not allow for an ABI measurement, and a TBI is preferred. The presence of PAD can be further defined by duplex arterial mapping. A skilled technologist can image from the infrarenal aorta to the tibial arteries to obtain a preliminary map and hemodynamically significant stenoses—typically reported as greater than 50%. Most often, history, physical examination, and ankle- or toe-brachial indices are sufficient to determine whether other more detailed imaging may be needed to guide further PAD treatment with endovascular or open procedures.

Arteriography

If ABI demonstrates moderate-to-severe PAD, typically ABI less than 0.5 or 0.7, additional arterial imaging is most often indicated to determine whether arterial revascularization can be performed. Arteriography includes computed tomographic arteriography (CTA), magnetic resonance arteriography (MRA), and contrast arteriography. Contrast arteriography has traditionally been the criterion standard, but with the improved image quality of CTA and MRA, these forms of imaging can be less invasive than contrast arteriography and still allow for optimal planning for either endovascular and/or open surgical procedures.

Treatment

Treatment of VLU requires multimodal therapy with compression therapy, wound care, and treatment of underlying venous and/or arterial pathology when indicated. There are multiple forms of compression therapy, a wide array of wound care products and skin substitutes, various techniques of percutaneous ablation of superficial venous reflux, and relief of venous obstruction with angioplasty and stenting, which are all discussed in depth in other chapters. First and foremost, when significant PAD is identified with concomitant CVD and VLU, optimizing arterial flow by medical, endovascular, and/or open surgical treatment remains paramount to healing a VLU despite the other critical steps in the VLU treatment algorithm. In essence, it should be the first or primary step on the road to healing.

Compression remains the mainstay treatment in patients with VLU and without it, the large majority of VLUs will not heal despite other adjuvant procedures to treat CVD. The challenge in patients with concomitant PAD is that compression can reduce skin and peri-wound perfusion pressures leading to worsening of VLU or even frank necrosis and gangrene. Clinical practice guidelines recommend limiting the use of compression to patients with ABI less than 0.5 or to those patients with absolute ankle pressure less than 60 mmHg.[9,10] In patients with ABI between 0.5 and 0.8, compression therapy should be initiated after consultation with a vascular specialist and should be modified with either inelastic wraps or by reducing the number of layers of compression and close surveillance.[11,12] In a limb with an ABI less than 0.5, with or without ipsilateral ischemic rest pain, ulcer, or gangrene of the foot, aggressive arterial revascularization should be performed if feasible.

While patients with VLU and reliable ABIs between 0.5 and 0.9 may tolerate compression, we recommend having a low threshold for the treatment of PAD in addition to those venous treatments such as venous ablation and stenting for obstruction. Often, VLUs with concomitant PAD can be large, long-standing, recalcitrant, and/or recurrent. This is in combination with other significant comorbidities such as advanced age, immobility, obesity, renal insufficiency, and diabetes. Therefore, to optimize the rate and permanence of healing, aggressive treatment of PAD when feasible is warranted. The decision as to whether medical treatment, endovascular and/or open surgical revascularization is the best option often remains unique to each individual patient's constellation of symptoms and comorbidities. That said, the use of advanced endovascular procedures with less durability substituting for a typically preferred open surgical revascularization may be a better option. Open arterial bypass procedures to below the knee arteries in patients with VLU can be fraught with serious wound complications that can be limb-threatening. Additionally, finding adequate venous autogenous conduits may be challenging in this population.

A small subset of patients with VLU who have advanced PAD will not be candidates for endovascular or open revascularization. Other medical modalities that may be of assistance in treating PAD include intermittent pneumatic compression and hyperbaric oxygen therapy (HBO). Intermittent pneumatic compression has been shown to increase collateral flow and while data is lacking in the combined VLU/PAD population, adding this treatment can be beneficial—barring no limitations from pain or compliance. Intermittent pneumatic compression can be put over multilayer bandages in patients with less virulent PAD and may help with edema control as well.[13,14] Hyperbaric oxygen therapy has been shown to increase oxygen delivery in patients with moderate-to-severe chronic limb ischemia and prevent limb loss.[15] In these cases, it is important to remember that while the underlying cause is chronic venous hypertension, the ulcer is also exacerbated by the arterial ischemia component. Again, barring no contraindications to HBO, this treatment may improve the probability of healing the ulcer.

Differential diagnosis

While the more common causes of nonhealing lower extremity ulcers include CVD, PAD, and diabetic foot ulcers, up to 5% of lower extremity wounds are atypical with less common etiologies.[1,17] The incidence is higher in wounds that are refractory to vascular interventions, compression, and optimal wound care. A history and physical exam whereby the provider gives careful attention to the associated signs and symptoms of VLU (with or without PAD) and more importantly when they are absent—should immediately alert to a possible atypical cause. Categories of chronic wounds from atypical etiologies include those caused by inflammation, infection, malignancy, and microvascular occlusion. Tissue or punch biopsy that includes the peri-ulcer skin border is an important diagnostic tool and should be performed when an atypical cause is suspected. Depending on the ulcer size and characteristics on the exam, two to three different areas should be biopsied.

Inflammatory ulcers

Inflammation can affect blood vessels of various sizes and sites depending on the underlying pathology. Inflammatory ulcers most often are caused by ischemia from vasculitis of the small and medium-sized arteries and arterioles. This results in evolving necrosis of the skin and subcutaneous tissue of the lower extremity resulting in petechiae, erythematous nonblanching macules or nodules, purpura, and subsequent ulceration. Diagnoses include rheumatoid arthritis, systemic lupus erythematosus, polyarteritis nodosa, hypersensitivity vasculitis, granulomatosis with polyangiitis, Sjögren's syndrome, cryoglobulinemia, scleroderma, and dermatomyositis.[17,18] Patient may manifest systemic features of fever, myalgia, arthralgia, and malaise. Biopsy and blood chemistry tests (erythrocyte sedimentation rate, C-reactive protein) checking for systemic inflammation are critical to the diagnosis. In partnership with a rheumatology consultation, treatment focuses on control of the systemic disease and often involves antihistamines, corticosteroids, and immunosuppressive agents. Ulcer treatment should include debridement of necrotic tissue, control of infection and biofilm formation, absorption of excess exudate, protection against further trauma, and providing a moist wound healing environment.

Infectious ulcers

While all leg ulcers have bacterial colonization, some ulcers can be primarily due to infection of the skin from bacterial, fungal, spirochete, or protozoal organisms, which then cause varying degrees of skin necrosis. The history can reveal a minor trauma to the area involved. Examples include ecthyma gangrenosum from pseudomonas, ecthyma from streptococcus, and gas gangrene from clostridium species. Infectious cutaneous ulcers from other fungal or protozoal infections are most commonly present in immunosuppressed patients. Presentation varies depending on the underlying organism and diagnosis is made by identifying the causative organism either from a swab culture or tissue biopsy that is sent for culture. Treatment requires immediate initiation of antibiotics because of the tendency for rapid spread with further skin involvement and the increased risk of systemic involvement. Gas gangrene caused by clostridium species is an emergency and most often requires surgical debridement. Patients often have had recent surgery in the surrounding area and present with malaise, fever, tachycardia, pain out proportion to physical findings, and high white blood cell count. Exam of the affected area can include blistering and pale skin that can turn bronze, gray, dark red, purple or black. Crepitus may or may not be present.

Malignant ulcers

Skin cancers may arise on the lower legs, especially in patients who have been habitually sun-exposed. However, chronic ulceration from other causes may also predispose to malignancy. In a prospective observational trial, authors found skin cancer in 10% of patients with chronic lower extremity wounds attributed to venous and/or arterial disease.[19] Most frequent primary skin cancer is basal cell carcinoma followed by squamous cell carcinoma (Fig. 10.2). Additional nonepithelial skin cancers that can cause leg ulceration include melanoma, sarcoma, and cutaneous lymphomas while lung, breast, and head and neck cancers develop most frequently as cutaneous metastasis (Fig. 10.3). Neoplastic

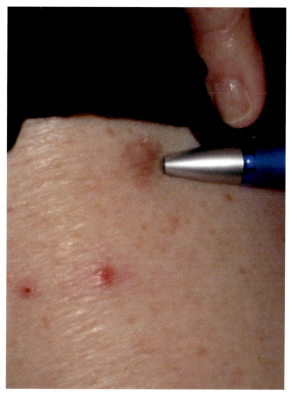

Figure 10.2 (A) The most common skin cancer is basal cell carcinoma which can present as open wounds, pearly bumps, scars with rolled edges with or without central indentation, or red patches as seen here. (B) The second most common skin cancer is squamous cell carcinoma which too can present as scaly red patches, open wounds, or thickened wart-like skin.

ulcers can vary in presentation from nodular ulcerated lesions, ulcerated plaques, to chronic ulcers with exuberant granulation tissue and pseudoepithelium.[20] One or more biopsies of the ulcer margin are indicated if there is doubt about the diagnosis or in cases where there is evidence of stalled healing with standard treatment of VLU.

Microvascular occlusive ulcers

Examples of ulcers due to microvascular occlusion include cryoglobulinemia, antiphospholipid syndrome, cholesterol emboli, calciphylaxis, warfarin-induced skin necrosis, and sickle cell disease (Fig. 10.4). While the underlying mechanism for occlusion of the small blood vessel varies (e.g., cryoagglutination, coagulopathies, platelet aggregation, embolus), the end result is skin ulceration and subcutaneous necrosis from microvascular occlusion. The clinical presentation may vary with the underlying etiology, but in general these ulcers are usually acute in onset, extremely painful, and often have retiform purpura as a common associated finding.[21,22] Again, the history is key to the diagnosis that may or may not require confirmatory blood laboratory testing and a biopsy of the involved skin.

Figure 10.2, cont'd.

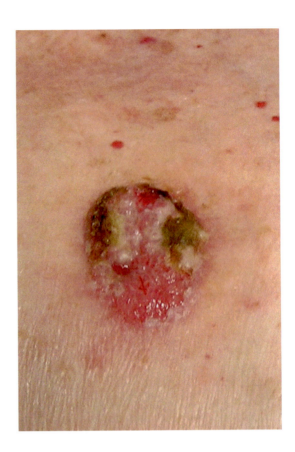

Pyoderma gangrenosum

Pyoderma gangrenosum (PG) is a rare noninfectious neutrophilic dermatosis that classically presents with a very painful, undermined, violaceous border and wound bed that is often necrotic and fibrinous (Fig. 10.5). It can occur in association with an underlying systemic disease such as inflammatory bowel disease, rheumatoid arthritis, hematologic disease, or malignancies, but is often idiopathic and a diagnosis of exclusion.[23] Pyoderma gangrenosum is frequently misdiagnosed due to the rarity of the condition, lack of available testing, and overlapping presentations of more common wound etiologies. Pyoderma gangrenosum may also develop at sites of other wounds including VLU or arterial ulcers[24,25] and conversely, chronic PG can cause necrosis and scarring that may exacerbate venous and arterial ulcers. Treatment of PG is immunosuppression with systemic corticosteroids and cyclosporine, or TNF-α inhibitors as effective second-line treatments.[26] Ulcer treatment is supportive with emphasis on trauma prevention and cautious debridement of only frankly necrotic tissue due to the high risk of further pathergy.[27] Negative pressure wound therapy may be helpful only after inflammation has been adequately controlled.

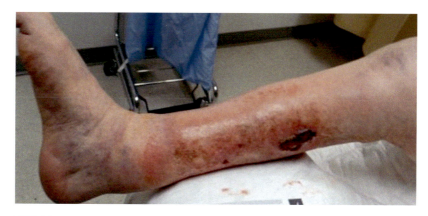
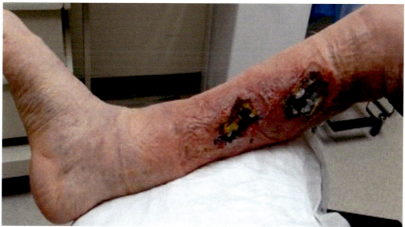

Figure 10.3 (A) Patient with recurrent sarcoma of the extremity on initial presentation. (B) Same patient a month later.

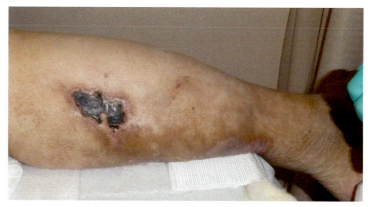

Figure 10.4 Patient with calciphylaxis wound on the posteromedial calf. Early lesions can present with violaceous, plaque-like subcutaneous nodules, indurations, or livedo reticularis and is quite painful on exam. These lesions then progress to ischemic or necrotic ulcers with eschar as seen here.

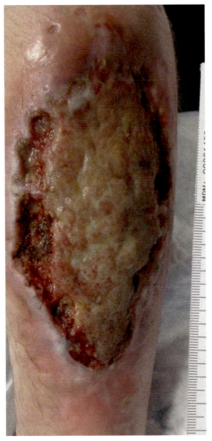

Figure 10.5 Patient with large pyoderma gangrenosum wound on the medial calf. The wound is characterized by deep ulceration with a violaceous border that overhangs the ulcer bed.

References

1. Singer AJ, Tassiopoulos A, Kirsner RS. Evaluation and management of lower extremity ulcers. *N Engl J Med*. 2017;377(16):1559−1567.
2. Marston W. Mixed arterial and venous ulcers. *Wounds*. 2011;23(12):351−356.
3. Humphreys ML, Stewart AH, Gohel MS, et al. Management of mixed arterial and venous leg ulcers. *Br J Surg*. 2007;94:1104−1107.
4. Callam MJ, Harper D, Dale JJ, et al. Arterial disease in chronic leg ulceration: an underestimate hazard? Lothian and forth valley leg ulcer study. *Br Med J*. 1987;294:929−931.
5. Marston WA, Carlin RE, Passman MA, et al. Healing rates and cost efficacy of outpatient compression treatment for leg ulcers associated with venous insufficiency. *J Vasc Surg*. 1999;30:491−498.
6. Matic M, Matic A, Duran V, et al. Frequency of peripheral arterial disease in patients with chronic venous insufficiency. *Iran Red Crescent Med J*. 2016;18(1):e20781.
7. Lantis J, Boone D, Lee L, et al. The effect of percutaneous intervention on wound healing in patients with mixed arterial venous disease. *Ann Vasc Surg*. January 2011;25(1):79−86.
8. Georgopoulos S, Kouvelos GN, Koutsoumpelis A, et al. The effect of revascularization procedures on healing of mixed arterial and venous leg ulcers. *Int Angiol*. August 2013;32(4):368−374.

9. Norgren L, Hiatt WR, Dormandy JA, et al. Inter-society consensus for the management of peripheral arterial disease (TASC II). *J Vasc Surg*. 2007;45:S5–S67.
10. Lurie L, Brajesh KL, Antignani PL, et al. Compression therapy after invasive treatment of superficial veins of the lower extremities: clinical practice guidelines of the American venous forum, society for vascular surgery, American college of phlebology, society for vascular medicine, and international union of phlebology. *J Vasc Surg Venous Lymphat Disord*. 2019;7(1):17–28.
11. Robson MC, Cooper DM, Aslam R, et al. Guidelines for treatment of venous ulcers. *Wound Repair Regen*. 2006;14:649–662.
12. O'Donnell TF, Passman MA, Marston WA, et al. Management of venous leg ulcers: clinical practice guidelines from the society for vascular surgery and the American venous forum. *J Vasc Surg*. 2014;60:3S–59S.
13. Moran PS, Teliguer C, Harriington P, et al. A systematic review of intermittent pneumatic compression for critical limb ischemia. *Vasc Med*. 2015;20(1):41–51.
14. Kavros SJ, Delis KT, Turner NS, et al. Improving limb salvage in critical ischemia with intermittent pneumatic compression: a controlled study with 18-month follow-up. *J Vasc Surg*. 2008;47(3):543–549.
15. Shields RC. Hyperbaric oxygen therapy for critical limb ischemia. In: Dieter RS, Dieter RA, Dieter RA, Nanjundappa A, eds. *Critical Limb Ischemia*. Switzerland: Springer; 2017:483–489.
16. Agale SV. *Chronic Leg Ulcers: Epidemiology, Aetiopathogenesis, and Management*. Ulcers; 2013. article ID 413604.
17. Rubano J, Kerstein M. Arterial insufficiency and vasculitides. *J Wound Ostomy Cont Nurs*. 1996;28:147–152.
18. Shah JB. Approach to commonly misdiagnosed wounds and unusual leg ulcers. In: Sheffield PJ, Fife CE, eds. *Wounds Care Practice*. 2nd ed. Flagstaff: Best Publishing; 2006:590–591.
19. Senet P, Combemale P, Debure C. Malignancy and chronic leg ulcers the value of systematic wound biopsies: a prospective, multicenter, cross-sectional study. *Arch Ddermatol*. 2012;148(6):704–708.
20. Janowska A, Dini V, Organes T, et al. Atypical ulcers: diagnosis and management. *Clin Interv Aging*. 2019;14:2137–2143.
21. Shanmugam VK, Angra D, Rahimi H, et al. Vasculitic and autoimmune wounds. *J Vasc Surg Venous Lymphat Disord*. 2017;5(20):280–292.
22. Piette WW. Cutaneous manifestations of microvascular occlusion syndrome. In: Bolognia J, Jorizzo J, Schaffer J, eds. *Dermatology*. 3rd ed. Vol. 1. Philadelphia, London: Elsevier Saunders; 2012:369.
23. Gottrup F, Karlsmark T. Leg ulcers: uncommon presentations. *Clin Dermatol*. 2005;23:601–611.
24. Lloret P, Redondo P, Sierra A, Cabrera J. Mixed skin ulcers misdiagnosed as pyoderma gangrenosum and rheumatoid ulcer: successful treatment with ultrasound-guided injection of polidocanol microfoam. *Dermatol Surg*. 2006;32:749–752.
25. Rosina P, Cunego S, Franz CZ, et al. Pathergic pyoderma gangrenosum in a venous ulcer. *Int J Dermatol*. 2002;41:166–167.
26. Alavi A, French LE, Davis MD, et al. Pyoderma gangrenosum: an update on pathophysiology, diagnosis and treatment. *Am J Clin Dermatol*. 2017;18:355–372.
27. Wong LL, Borda LJ, Liem TK, et al. Atypical pyoderma gangrenosum in the setting of venous and arterial insufficiency. *Int J Low Extrem Wounds*. 2001. https://doi.org/10.1177/15347346211002334.

CHAPTER 11

Assessment tools and wound documentation for patients with chronic venous insufficiency

Michael Palmer[1], Rick Mathews[1], Gregory L. Moneta[1] and Khanh P. Nguyen[1,2,3]

[1]Oregon Health & Science University, Department of Surgery, Division of Vascular & Endovascular Surgery, Portland, OR, United States; [2]Oregon Health & Science University, Department of Biomedical Engineering, Portland, OR, United States; [3]Portland VA Health Care System, Research & Development, Portland, OR, United States

Wound assessment tools

Chronic venous disease (CVD) includes all manifestations of venous hypertension such as telangiectasias, reticular veins, varicose veins, dermatitis, edema, pigmentation, fibrosis, and venous leg ulcers (VLU) and is caused by venous reflux and/or obstruction. Currently, an estimated 23% of adults in the United States (US) have varicose veins while up to 5% have a more severe disease (characterized by skin changes, ulcers, etc.). Approximately 1%—3% of patients with chronic venous insufficiency have active or healed ulcers. The direct medical cost in the United States is estimated to be between 150 million and one billion US dollars annually.[1,2] The clinical presentation of CVD is complex, varying drastically even between individuals with similar underlying pathology. The need for a standardized classification system to better document, communicate, follow patients, and assess treatment was realized early on and the first venous classification system was developed in the 1970s. Multiple others followed but the first to gain widespread acceptance was the CEAP classification. Today, clinical, etiological, anatomical, and pathophysiology (CEAP) is the widely used classification system for venous disease with additional adjuncts for severity scoring: Venous Severity Score (VSS), Venous Clinical Severity Score (VCSS), Venous Segmental Disease Score (VSDS), Venous Disability Score (VDS), and for quality of life (QoL) assessment (Short Form-36 (SF-36), Venous Insufficiency Epidemiological and Economic Study (VEINES)-QoL/SYM, Aberdeen Varicose Vein Questionnaire (AVVQ), Chronic Venous Insufficiency Questionnaire (CIVIQ), Charing Cross Venous Ulceration Questionnaire (CXVUQ).

CEAP classification

Historically, CVD scoring systems focused on the patient's clinical appearance leading to a lack of reproducibility and diagnostic accuracy. This was realized by the American Venous Forum prompting the development and publication of CEAP classification in 1994 and1996, respectively.[3]

CEAP is primarily intended as a classification system for CVD and is not intended to measure disease severity or impact on quality of life. Additional scoring systems to measure disease severity and quality of life that both antedate and postdate CEAP serve in those regards as adjuncts to CEAP and are also discussed later. CEAP classification is based on the clinical manifestations (C), etiologic factors (E), primary, secondary, congenital, anatomic distribution (A), deep, superficial, perforator veins and specific venous segments), and underlying pathophysiology (P), reflux, obstruction, compression) of venous disease. The CEAP classification system has been incorporated into CVD reporting standards and is strongly recommended for classifying CVD in clinical publications of CVD including diagnosis and treatment helping it gain widespread use. As our understanding of the venous disease has increased and the use of CEAP increased the need for revisions was recognized and these occurred in the years 2004 and 2020.[3,4]

In the most 2020 recent update, the clinical, C, component of CEAP remains determined by physician designation of signs of CVD. No apparent disease remains denoted as C0; C1 remains telangiectasias/reticular veins C3 remains edema and C5 a healed venous ulcer. However, C2 is now divided into varicose veins and recurrent varicose veins; and C4 (skin changes secondary to CVD) was further divided into C4a; pigmentation or eczema, C4b; lipodermatosclerosis or atrophie blanche and C4c; corona phlebectatia while C6 representing active venous ulcers also has an additional designation for recurrent venous ulcer, C6r (Table 11.1). Each is further characterized by a subscript for symptomatic (s) or asymptomatic (a). Basic CEAP reports the single highest C classification in a

Table 11.1 CEAP-clinical classification.

C class	Description
C0	No visible or palpable signs of disease
C1	Telangiectasias or reticular veins
C2	Varicose veins
C2r	Recurrent varicose veins
C3	Edema
C4	Changes in skin and subcutaneous tissue secondary to CVD
C4a	Pigmentation or eczema
C4b	Lipodermatosclerosis or atrophie blanche
C4c	Corona phlebectatica
C5	Healed venous ulcer
C6	Active venous ulcer
C6r	Recurrent venous ulcer[a]

[a]Each C class should be subcharacterized by the subscript s (symptomatic) or a (asymptomatic).
Adapted from the 2020 update of the CEAP classification system Lurie F, Passman M, Meisner M, et al. The 2020 update of the CEAP classification system and reporting standards [published correction appears in *J Vasc Surg Venous Lymphat Disord.* 2021 Jan;9(1):288]. *J Vasc Surg Venous Lymphat Disord.* 2020;8(3):342–352. doi: 10.1016/j.jvsv.2019.12.075.

limb while advanced CEAP reports all components of CEAP, including E, A, and P. The etiologic classification now consists of congenital, primary, secondary-intravenous, secondary-extravenous, and no cause identified (Table 11.2). The anatomic classification subdivides the veins involved as either superficial, deep, perforating, or no venous anatomic location identified. Any combination of the three may be used and the limb involved should be specified: right (R) or left (L). The different subdivisions are further classified using specific abbreviations for the anatomic segments involved (Table 11.3). The pathophysiologic classification has both basic and advanced designations. The basic designations are r (reflux), o (obstruction), r/o (reflux and obstruction), and n (no venous pathology). The advanced designation includes the addition of the specific abbreviated anatomic (A) segments (Table 11.4).[3]

CEAP was designed to be a discriminative instrument classifying CVD in one or both limbs. As such, it was meant to categorize the different CVD domains at a single point in time, allowing them to be described efficiently, compared between patients, and standardized for research. It was not designed to measure disease severity, change over time, therapy outcomes, or the burden of venous disease on patient's lives. These issues are addressed by the VSS, as well as, several different QoL scores.[3,5,6]

Venous Severity Score

CEAP initially was accompanied by a simple severity score based on ability to work and response to the use of support devices. It was subjective, poorly discriminatory, and thus rightly never gained traction. Clearly, evaluative adjuncts to complement CEAP were needed to grade severity and assess outcomes. In response to this need, the American Venous Forum developed and in the year 2000 disseminated the VSS based on the elements and concepts of CEAP. VSS consists of three components including the VCSS, VSDS, and the VDS.

Revised Venous Clinical Severity Score

The VCSS is based on the clinical (C) aspects of CEAP and after revision to include compliance with compression therapy in 2010 is now made up of 10 clinical parameters:

Table 11.2 CEAP-etiologic classification.

E class	Description
Ep	Primary
Es	Secondary
Esi	Secondary intravenous
Ese	Secondary extravenous
Ec	Congenital
En	No cause identified

Adapted from the 2020 update of the CEAP classification system Lurie F, Passman M, Meisner M, et al. The 2020 update of the CEAP classification system and reporting standards [published correction appears in *J Vasc Surg Venous Lymphat Disord.* 2021 Jan;9(1):288]. *J Vasc Surg Venous Lymphat Disord.* 2020;8(3):342−352. doi: 10.1016/j.jvsv.2019.12.075.

Table 11.3 CEAP-anatomic classification.

A class			Description
As	Superficial	Tel	Telangiectasia
		Ret	Reticular veins
		GSVa	Great saphenous vein above knee
		GSVb	Great saphenous vein below knee
		SSV	Small saphenous vein
		AASV	Anterior accessory saphenous vein
		NSV	Nonsaphenous vein
Ad	Deep	IVC	Inferior vena cava
		CIV	Common iliac vein
		IIV	Internal iliac vein
		EIV	External iliac vein
		PELV	Pelvic veins
		CFV	Common femoral vein
		DFV	Deep femoral vein
		FV	Femoral vein
		POPV	Popliteal vein
		TIBV	Crural (tibial) vein
		PRV	Peroneal vein
		ATV	Anterior tibial vein
		PTV	Posterior tibial vein
		MUSV	Muscular veins
		GAV	Gastrocnemius vein
		SOV	Soleal vein
Ap	Perforator	TPV	Thigh perforator vein
		CPV	Calf perforator vein[a]
An	No venous anatomic location identified		

[a]Anatomic locations also reported under each pathophysiologic (P) classification.
Adapted from the 2020 update of the CEAP classification system Lurie F, Passman M, Meisner M, et al. The 2020 update of the CEAP classification system and reporting standards [published correction appears in *J Vasc Surg Venous Lymphat Disord*. 2021 Jan;9(1):288]. *J Vasc Surg Venous Lymphat Disord*. 2020;8(3):342−352. doi: 10.1016/j.jvsv.2019.12.075.

Table 11.4 CEAP-pathophysiologic classification.

P Class	Description
Pr	Reflux
Po	Obstruction
Pr,o	Reflux and obstruction[a]

[a]Advanced abbreviations from anatomic (A) location(s) should be reported under each pathophysiologic (P) class.
Adapted from the 2020 update of the CEAP classification system Lurie F, Passman M, Meisner M, et al. The 2020 update of the CEAP classification system and reporting standards [published correction appears in *J Vasc Surg Venous Lymphat Disord*. 2021 Jan;9(1):288]. *J Vasc Surg Venous Lymphat Disord*. 2020;8(3):342−352. doi: 10.1016/j.jvsv.2019.12.075.

pain, varicose veins, venous edema, skin pigmentation, inflammation, induration, number of ulcers, duration of ulcers, size of ulcers, and compliance with compression therapy. Each item is assigned a score of 0—3 based on severity, and categories are scored individually to generate a dynamic score (Table 11.5). The score is performed for both legs and is dynamic over time. This allows for comparison of severity, assessing change over time,

Table 11.5 Revised venous clinical severity score (rVCSS).

Parameters	Score			
Pain	0-none	1-occasional pain or discomfort	2-daily pain or discomfort that interferes with but does not prevent regular activities	3-daily pain or discomfort that limits most daily activities
Varicose veins	0-none	1-few: scattered (isolated branch varicosities or clusters), corona phlebectatica	2-confined to the calf or thigh	3-involves calf and thigh
Venous edema	0-none	1-limited to foot and ankle	2-extends above ankle but below knee	3-extends to or above the knee
Skin pigmentation	0-none or focal	1-limited to perimalleolar area	2-diffuse over lower third of calf	3-wider distribution above lower third of calf
Inflammation	0-none	1-limited to perimalleolar area	2-diffuse over lower third of calf	3-wider distribution above lower third of calf
Induration	0-none	1-Limited to perimalleolar area	2-diffuse over lower third of calf	3-wider distribution above lower third of calf
Active ulcer number	0	1	2	>/= 3
Active ulcer duration	n/a	<3 months	>3 months but <1 year	Not healed for >1 year
Active ulcer size	n/a	Diameter <2 cm	Diameter 2—6 cm	Diameter >6 cm
Use of compression therapy	1-not used	1-Intermittent use of stockings	2-wears stockings most days	3-full compliance

Adapted from CEAP & Venous Severity Scoring CEAP & Venous Severity Scoring. https://www.veinforum.org/medical-allied-health-professionals/avf-initiatives/ceap-venous-severity-scoring/. Accessed June 16, 2021.

and assessing treatment outcomes. The VCSS has been extensively studied and revised (rVCSS), showing good validity, reliability, and correlation with CEAP. It is well accepted due to its ease of use and its ability to perform longitudinal surveillance[7–12].

Venous Segmental Disease Score

This score combines the anatomic (A) and pathophysiologic (P) components of CEAP. Of the 18 venous segments described by CEAP, the 11 major venous segments were chosen and are graded according to the presence of reflux and/or obstruction on venous duplex ultrasonography or phlebology. As different segments have major or minor roles in reflux and obstruction, the scoring system was designed by giving a baseline score of one to each segment and lower scores (1/2) for segments with minor significance, while those with major significance were assigned a score of two. Both reflux and obstruction are given a maximum score of 10 representing the most severe disease'.[5,9] The current VSDS is presented in Table 11.6.

Venous disability score

The VDS is a modification of the original CEAP disability score. Although the original score attempted to quantify disability, it was not all inclusive and contained ambiguous language. Disability was assessed based on the ability to perform an eight-hour workday. Patients with venous disease are a diverse population and many did not meet this inclusion criterion. Additionally, the previous disability score discussed the ability to work with or without a supportive device, which was not further described. The VDS

Table 11.6 Venous segmental disease score.

Reflux		Obstruction
1/2	Small saphenous	
1	Great saphenous	1
1/2	Perforators, thigh	
1	Perforators, calf	
2	Calf veins, multiple (PT alone = 1)	1
2	Popliteal vein	2
1	Superficial femoral vein	1
1	Profunda femoral vein	1
1	Common femoral vein	2
	Iliac vein	1
	IVC	1
10	Maximum score	10[a,b,c]

[a]Must be visualized on imaging (duplex ultrasonography or phlebography).
[b]Reflux: all the valves in the segment are incompetent.
[c]Obstruction: total occlusion or >50% stenosis of at least half of the segment.
Adapted from Venous severity scoring: An adjunct to venous outcome assessment.

addressed these limitations by substituting the ability to return to prior normal activity levels for the 8 h workday and replacing supportive devices with compression and/or elevation (Table 11.7).[5,8]

Quality of life measures

Quality of life measurements are a necessary adjunct for evaluating the effects of disease on patients, as well as the effects of treatments. In health care, QoL can be broken down into the patient's emotional, physical, financial, and social well-being. Subjective in nature, these measures vary based on the patient's own perception of their life and disease state. It has been shown that assessing QoL allows for better patient-healthcare provider communication and provides insights into what matters to patients.[3,7,13]

Quality-of-life assessments are constructed in multiple forms including generic and disease specific. Of the generic assessments, the 36 Item Short Form Health Survey (SF-36) is most commonly used for venous disease. Disease-specific assessments focus on specific details related to the disease process and the effects of its treatment. Examples include the AVVQ, the CIVIQ, the Venous Insufficiency Epidemiological and Economic Study (VEINES-QoL/SYM), and the Charing Cross Venous Ulcerations Questionnaire (CCVQ).

QoL: SF-36

The 36-item Short Form Health Survey (SF-36) is a generic QoL survey developed by the Rand Corporation and the Medical Outcomes Study that has good reliability and validity in CVD. This general QoL assessment consists of questions related to both physical and mental health, breaking the two categories down further into eight categories: physical functioning, role limitations due to physical problems (role-physical), bodily pain, general health perceptions, vitality, social functioning, role limitations due to emotional problems (role-emotional), and mental health. A score is generated from 0 to 100 with higher scores associated with better general health perception[9,14–17].

Table 11.7 Venous disability score.

Score	Description
0	Asymptomatic
1	Symptomatic but able to carry out usual activities[a] without compressive therapy
2	Can carry out usual activities[a] only with compression and/or limb elevation
3	Unable to carry out usual activities[a] even with compression and/or limb elevation

[a]Usual activities = patients activities before onset of disability from venous disease.
Adapted from Venous severity scoring: An adjunct to venous outcome assessment.

QoL: Aberdeen Varicose Vein Questionnaire

The AVVQ is a thirteen-part, disease-specific QoL measurement developed by Garratt et al. in 1993 for specific use in patients with varicose veins.[4] It takes into consideration both the physical symptoms and the social issues, with scores calculated from 0 (no effect) to 100 (severe effect). Studies comparing the AVVQ to other QoL assessments showed a good correlation.[18]

QoL: Chronic Venous Insufficiency Questionnaire

The CIVIQ is a disease disease-specific self-questionnaire that has undergone several revisions, and the most current version is CIVIQ-20. It consists of 20 questions related to the presence/intensity of impairment experienced and the level of importance assigned to the impairment by the patient. The questions are broken down into four categories: psychological (nine items), pain (four items), physical (four items), and social (three items). Each item is scored from one to five with a minimum score of 20 and a maximum score of 100. A global index score is calculated from these numbers with higher numbers equating to greater QoL.[9,19,20]

QoL: VEINES-QoL/SYM

Venous Insufficiency Epidemiological and Economic Study is a disease-specific instrument that has been validated in CVD and deep venous thrombosis. This instrument consists of 35 items classified into two categories, QoL and symptoms, generating separate summary scores. VEINES-QoL includes 26 items, 25 of which are used to quantify the effect of disease on QoL while VEINES SYM consists of 10 items measuring physical symptoms. Each item is scored on a two- to seven-point scale with higher scores representing better QoL[9,17,20–23].

QoL: Charing Cross Venous Ulcerations Questionnaire

The Charing Cross Venous Ulcerations Questionnaire (CCVUQ) is a QoL measure developed by Smith et al. in 2000 to assess patients with venous ulcer disease.[24] Up to that point no specific tool had been developed for this specific pathology. The items addressed by the questionnaire were selected from patient interviews, literature review, and expert opinion and were administered along with the SF-36. When administered with the SF-36 the CCVUQ showed good reliability, validity, and responsiveness, correlating with the eight domains of the SF-36.[9,24]

Wound documentation

As stated earlier, a common sequela of severe and chronic lower extremity venous insufficiency secondary to reflux or obstruction are VLUs. Detailed documentation of VLU allows for thorough assessment and tracking of wound healing or lack thereof keeping in mind there may be more than one VLU per lower extremity.

It is important to clearly indicate the location of the ulcer(s), particularly if there is more than one to accurately assess the rate of wound healing over time. While most VLUs are located at the medial ankle or perimalleolar area, some can extend onto the dorsum of the foot or to mid or even to the proximal calf. Thus, another common location of VLU is in the gaiter area between the ankle and calf. To allow for longitudinal tracking of VLU, a description of the anatomical location of the VLU can specify the location on the foot referencing its relative location to bony landmarks if applicable (medial malleolus, tarsal bones including the medial calcaneus, talus, navicular, and cuneiforms and lateral cuboid, metatarsals, and phalanges) and specifying its overall position on the calf or ankle (medial, lateral, superior, inferior, anterior, posterior), or foot (also proximal or distal, dorsal or plantar aspects of the foot). Ulcer dimensions including maximal width, length, and depth are noted to calculate wound volumes that can be used to track progression or regression of the VLU (Fig. 11.1). The depth of the wound and extent of penetration through the skin, subcutaneous or fascial (rare) layers should be noted. This designates the ulcer as a partial-thickness wound involving only layers of the skin (epidermis or partial dermis) or a full-thickness wound through the dermis, subcutaneous tissue, and adipose layers. Primary venous ulcers complicated by infection, vasculitis, pressure necrosis, or arterial insufficiency can also extend to muscle or bone. In such cases, involvement of key structural components such as cartilage, joint capsule, ligament, or tendon should also be noted.

Documentation of the characteristics of the ulcer is helpful. A description of the wound bed should indicate the presence or absence of granulation tissue and if present, the quality of the granulation tissue. In some cases, VLUs will demonstrate the hypergranulation of tissue, and this should be noted as its presence can impair healing by preventing the migration of epithelial cells across the ulcer surface. Slough, eschar, or fibrin in the ulcer bed should also be noted as they impede wound healing as well as epithelialization that indicates wound healing. New epithelialization is encouraging as it indicates the ulcer is making progress toward closure.

A description of the wound edge should indicate if epibole or tunneling is present. This occurs when upper epidermal cells roll down over lower epidermal cells and migrate down

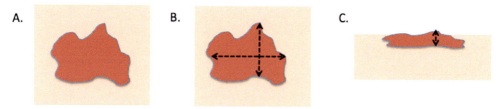

Figure 11.1 Measurement of maximal wound width, length ("birds-eye" view, A and B), and depth (lateral view, C).

the sides of an ulcer instead of across the VLU. Documentation of undermining, tunneling, or sinus tracks should measure the depth and direction of penetration. Typically, an area of tunneling is documented with reference to a clock face; for example, tunneling can be described as being "present from 1 o'clock to 3 o'clock" (Fig. 11.2). Although less common in VLUs, documentation of these features indicates a more difficult wound that may require aggressive wound debridement or skin excision to promote healing.

Presence or absence of exudate should be documented. The quality of the exudate can reflect the possibility of bacterial infection or colonization and can be described as follows: sanguineous, serosanguinous, purulent, and foul purulent. In addition, the amount of drainage can be semiquantified as follows: none, scant, minimal, moderate, or large. Another technique to semiquantify drainage is by the number of dressings used over a period of time and can be used to track either a decrease or increase in drainage.

A description of the peri-VLU tissue and skin is important. Description of the color of the skin and tissues, tissue consistency (induration or bogginess), lipodermatosclerosis, and temperature should be noted. Also, the presence or absence of edema in the skin, phlebolymphedema around the wound and in the leg should be noted. The skin around the wound should be evaluated and noted for irritation, maceration, excoriation, ecchymoses, or other abnormalities. Uncommon in VLUs, but the wound should be assessed for callous or periwound fibrosis that would indicate hyperkeratosis at the wound edge. If present, callous or hyperkeratosis should be excised to promote VLU healing.

Presence or absence of odor should be documented. Any odors from the ulcer should be noted including malodor or other odors that may indicate infection or colonization of particular bacteria or fungus. Some bacteria or fungi may have distinctive odors (for example, "a sweet grape-like smell" may indicate the presence of *pseudomonas aeruginosa*).

Evaluation for indicators of possible infection is a part of the routine assessment of VLUs. Presence or absence of erythema in the periwound skin or tissue or leg is noted. Increased drainage, odor, warmth, edema, induration, and particularly pain is suggestive of infection. In addition, fever or hypothermia or other constitutional symptoms can suggest progression to severe or systemic infection.

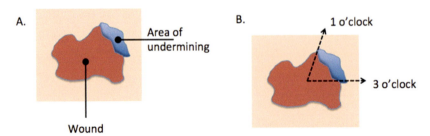

Figure 11.2 Example of how to document the presence of undermining.

When present, documentation of the location, intensity quality, and duration of pain is important. Whether it is spontaneous or elicited pain with movement, on examination, or other manipulation is recorded and as with pain anywhere alleviating or exacerbating factors are also noted. While many venous ulcers are not that tender many are exquisitely tender to touch or pressure and/or especially with sharp debridement.

Debridement routinely requires topical or injectable lidocaine and gentle debridement. For large or extensive VLUs, debridement under sedation or anesthesia may be necessary. Numerous beside and operative tools are available for debridement and tools used for sharp debridement should be documented, which can include scalpel blade (#15, #10 or #11), forceps, scissors, nippers, curettes, and even ronguers.

Figs. 11.3—11.8 show examples of VLU with varying degrees of severity. Individual wounds are measured from wound edge to wound edge. However, in situations where there are numerous, small individual wounds clustered in one area, the maximal dimensions of the complex of wounds are measured for simplicity, which may include small areas of intervening epidermis and dermis as shown in Fig. 11.7.

VLU measurements

VLU measurements are a standard means to document the course of healing. VLUs notoriously deviate from healing patterns of other chronic wounds and accurate dimension measurement represents the most fundamental form of tracking healing.

Historically, visual, and linear measurements were sufficient but with the advent of both invasive and noninvasive imaging modalities, wound characterization has advanced aiding understanding of progression. These adjunctive modalities aim to reduce human error and improve the reproducibility of wound measurements and assessments over time. They have also helped standardize care by avoiding errors inevitably associated with visual assessment and characterization alone. While often not used in the routine clinical care of VLUs, magnetic resonance imaging (MRI), ultrasound, and biopsy permit

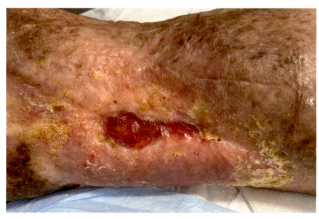

Figure 11.3 VLU with some granulation tissue, minimal biofilm, and serous/serosanguinous drainage.

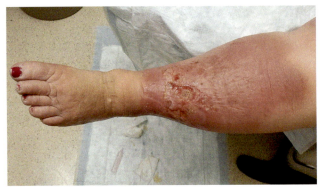

Figure 11.4 Full thickness VLU. Note exposed fascia and minimal granulation.

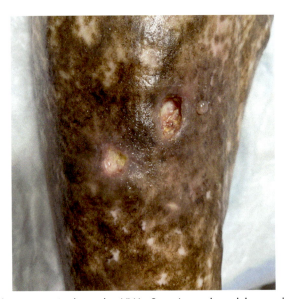

Figure 11.5 Full thickness, anterior lower leg VLUs. Superior and caudal wound with undermining at 12 o'clock to 2 o'clock.

accurate description of ulcer extension into different dermal layers.[25] They also permit volumetric measurements (see below).

Linear measurements are often documented with photography and digital planimetry of the evolving wound and can be reviewed and sequentially tracked by those involved in the management of the ulcer.

In the Apligraf and Integra trials for extracellular-matrix components containing dermal patches, complete wound closure along with reepithelialization was required to determine complete wound healing.[26] The Apligraf trial measured wound duration, area, and depth in the context of a 1998 staging system developed by the International

Assessment tools and wound documentation for patients with chronic venous insufficiency 191

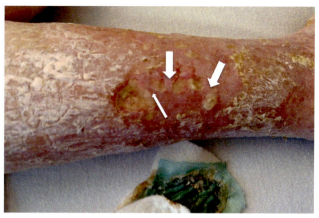

Figure 11.6 VLU with fibrinous exudate, debris, and periwound maceration. Cluster of multiple, small wounds tracked by measuring maximal length and width of the leg involved for ease of documentation.

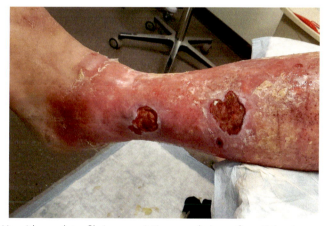

Figure 11.7 VLUs with exudate, fibrin, excoriations, and circumferential cutaneous inflammation.

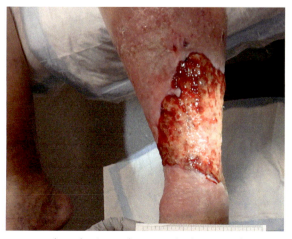

Figure 11.8 Severe postthrombotic syndrome and a long-standing circumferential VLU.

Association of Enterostomal Therapists, currently known as the Wound, Ostomy, Continence Nurses Society. These parameters along with the ankle-brachial index (ABI) determined the progress of wound healing in recruited patients.[26]

However, a severe limitation of physical dimension characterization via linear measurement, even with the aid of photography and digital planimetry, is the inability to predict wound healing or determine ulcer depth or stage of healing. This is particularly problematic for VLUs given their often unpredictable course. Thus, characterization of VLUs by epithelialization, perfusion, physical dimensions, and presence of infection, all are required for accurate VLU monitoring.

Imaging modalities for wound assessment and measurement

Given the limitations of bedside wound measurements, imaging modalities can be used in instances where the assessment of wound dimensions is especially challenging. The depth of the wound allows for the characterization of the levels of the tissue involved. Computed tomography (CT), MRI, and ultrasound imaging are all modalities used to characterize wound depth and underlying pathology. CT imaging is an accessible imaging modality with high resolution for the assessment of VLUs. MRI is often the most sensitive for measuring wound depth and can also characterize moisture. Ultrasound is most often a more convenient cost-effective option with utilization at the point-of-care without radiation.[27]

Time-of-flight camera imaging measures the phase shift of light reflected from the ulcer surface to determine the depth of the wound.[28] Similarly, hyperspectral imaging can capture diffusely reflected light to reconstruct a three-dimensional image of the VLU, giving depth measurements. Optical coherence tomography can utilize low coherence interferometry[29] to generate images of the internal structure of the tissue underlying the ulcer and can be utilized to determine ulcer depth. Spatial frequency domain imaging and near-infrared imaging spectroscopy measure frequencies or wavelengths of absorbed light to facilitate ulcer imaging.[30]

Apps and cameras used for point-of-care wound measurements

Linear measurement of VLUs using a ruler or a Kundin gauge is associated with variability of measurements. Additionally, time considerations for measuring ulcers manually without the use of mobile applications are not trivial when compared to providers using mobile applications to document and track wounds. Finally, the ease of integration with Electronic Medical Records software and sharing of wound documentation helps interprofessional teams in coordinating care for VLU patients. Standardization, streamlining, faster characterization, and ease of documentation and sharing of measurements are the focus of a number of applications now entering the realm of venous ulcer care measurement, documentation and monitoring (Table 11.8).

Table 11.8 Applications and cameras for assessing wound dimensions at the point of care.

Application	Year launched	Features	Cost (USD)
"+WoundDesk"	2016	Semiautomated measurements	Free
"Parable"	2016	Linear measurements, convenient documentation	Free
"Swift skin and wound" applications suite	2017	Linear measurements, progress tracking, enterprise software	Free
"Silhouettelite"	2017	Linear measurements, convenient documentation	Free
"Pointclickcare skin and wound" Applications suite	2018	Automated linear measurements	Free
"Wound measurement"	2018	Linear measurements, convenient documentation	$0.99
"WoundwiseIQ"	2018	Linear measurements, convenient documentation	Free
"Woundmatrix patient"	2019	Linear measurements, convenient documentation	Free
"Imito" applications suite	2016–20	Linear measurements, chronological documentation, report generation, multiple applications from the same developer	Free
"Wound capture"	2020	Streamlined wound tracking process	Free

Point-of-care assessment of wound perfusion
ABI, TCPO2, and SPY angiography

Increasingly, the predicton of wound healing has become a focus of wound clinics.[29] Monitoring the vascularization of VLUs to assess adequate perfusion and predict wound healing can be an integral part of VLU care. Current methods to evaluate perfusion of the wound region include but are not limited to the ABI, color Doppler, transcutaneous oxygen pressure ($TCPO_2$), and SPY angiography (Novadaq Technologies Inc.). ABI, first described in 1950,[31] provides an index of large vessel perfusion to the ankle. $TCPO_2$ measures oxygen tension of dermal capillaries quantifying skin oxygenation and is a measure of wound perfusion.

However, ABI, Doppler, and $TCPO_2$ lack the capability to track regional perfusion changes limiting their use in the care of chronic wounds like VLUs. SPY angiography is based on indocyanine green (ICG) angiography. A 0.1% solution of 0.1 mg/kg ICG is administrated intravenously in a setting with dim light. ICG angiography provides real-time, quantitative, and qualitative data on regional vascularization of the wound

Table 11.9 Commonly used VLU diagnosis (ICD-9 and ICD-10), wound care management, and debridement codes (CPT).

Diagnosis codes: ICD-9 codes	
* Varicose veins of lower extremities	454
With ulcer	454
With inflammation (any part or unspecified site)	454.11
With ulcer and inflammation of LE (any part or unspecified)	454.12
With other complications (edema, pain, swelling)	454.8
Asymptomatic varicose veins	454.9

Diagnosis codes: ICD-10 codes

Varicose Vein Classification (I83: _ _ _)

Varicosities with:	Laterality	Location
0-ulcer	1-right	1-thigh
1-inflammation	2-left	2-calf
2-both ulcer and inflammation	0-unspecified	3-ankle
8-pain		4-heel/mid foot
		5-foot other
		8-leg other
		9-unspecified

Chronic Venous Hypertension (Idiopathic) Classification (I87: 3 _ _)

	Venous hypertension with:	Laterality
	1-ulcer	1-right
	3-both ulcer and inflammation	2-left
		3-bilateral
		9-unspecified

Post Thrombotic Syndrome (PTS) with Ulcer Classification (I87: 0 _ _)

	PTS with:	Laterality
	1-ulcer	1-right
	3-ulcer and inflammation	2-left
		3-bilateral
		9-unspecified

Non-Pressure Chronic Ulcer Classification (L97: _ _ _)

Location	Laterality	Severity/Depth
1-thigh	1-right	1-skin
2-calf	2-left	2-fat
3-ankle	0-unspecified	3-muscle/tendon
4-heel/midfoot		4-bone/joint
5-foot other		5-muscle without evidence of necrosis
8 leg other		6-bone without evidence of necrosis
9-Unspecified		8-muscle with other specified severity
		9-unspecified severity

Table 11.9 Commonly used VLU diagnosis (ICD-9 and ICD-10), wound care management, and debridement codes (CPT).—cont'd

Nonpressure chronic ulcer of skin, not elsewhere classified: L98.4

Wound care management codes

Debridement (e.g., high-pressure water jet with/without suction, sharp selective debridement with scissors, scalpel, and forceps); open wound (e.g., fibrin, devitalized epidermis and/or dermis, exudate, debris, biofilm); including topical application, wound assessment, use of a whirlpool, when performed and instructions for ongoing care, per session, total wound surface area; first 20 cm^2 or less	97597
Each additional 20 cm^2 or part thereof (list separately in addition to code for primary procedure)	97598
Removal of devitalized tissue from wound(s), nonselective debridement, without anesthesia (e.g., wet-to-moist dressings, enzymatic, abrasion, larval therapy) including topical application, wound assessment and instructions for ongoing care, per session	97602

Debridement codes

Debridement: skin, muscle, bone	11012
Debridement, subcutaneous tissue (includes epidermis and dermis, if performed); first 20 cm^2 or less	11042
Each additional 20 cm^2 or part thereof (list separately in addition to code for primary procedure) (use 11045 in conjunction with 11042)	10045
Debridement, muscle and/or fascia (includes epidermis, dermis, and subcutaneous tissue, if performed), first 20 cm^2 or less	11043
Each additional 20 cm^2 or part thereof (list separately in addition to code for primary procedure) (use 11046 in conjunction with 11043)	11046
Debridement, bone (includes epidermis, dermis, subcutaneous tissue, muscle, and/or fascia, if performed); first 20 cm^2 or less	11044
Each additional 20 cm^2 or part thereof (list separately in addition to code for primary procedure) (use 11047 in conjunction with 11044)	11047

*Common ICD-9 and ICD-10 codes used to classify VLU are listed with ICD-10 codes indicating laterality of the lower extremity involved. In addition, wound care management and debridement codes depend on the amount of wound surface area treated.

for monitoring the perfusion of wounds. Patel et al. used ICG angiography data to accurately diagnose nonhealing patients who were initially misdiagnosed using ABI and TCPO$_2$ modalities.[32]

Diagnosis and procedure codes

The Table 11.9 lists the diagnosis (ICD-9 and ICD-10), wound care management, and debridement codes Current Procedural Terminatology (CPT) commonly used for VLU management and procedures.[33,34]

Appropriate ICD-9 or ICD-10 codes need to be associated and in accordance with wound care management and procedure codes. Wound care management codes include debridement procedures and should not be used simultaneously with debridement codes.

References

1. Piazza G. Varicose veins. *Circulation*. 2014;130(7):582−587. https://doi.org/10.1161/CIRCULATIONAHA.113.008331.
2. Kabnick LS, Scovell S. Overview of Lower Extremity Chronic Venous Disease. In: Eidt JF, Mills JL, Collins K, eds. *UpToDate. Waltham, Mass.* UpToDate; 2021.
3. Lurie F, Passman M, Meisner M, et al. The 2020 update of the CEAP classification system and reporting standards. *J Vasc Surg Venous Lymphat Disord*. 2020;8(3):342−352. https://doi.org/10.1016/j.jvsv.2019.12.075. published correction appears in J Vasc Surg Venous Lymphat Disord. 2021 Jan; 9(1):288.
4. Garratt AM, Macdonald LM, Ruta DA, Russell IT, Buckingham JK, Krukowski ZH. Towards measurement of outcome for patients with varicose veins. *Qual Health Care*. 1993;2(1):5−10. https://doi.org/10.1136/qshc.2.1.5.
5. CEAP & Venous Severity Scoring. https://www.veinforum.org/medical-allied-health-professionals/avf-initiatives/ceap-venous-severity-scoring/. Accessed June 16, 2021.
6. Khilnani NM, Davies AH. CEAP: a review of the 2020 revision. *Phlebol*. 2020;35(10):745−748. https://doi.org/10.1177/0268355520961239.
7. Meissner MH, Natiello C, Nicholls SC. Performance characteristics of the venous clinical severity score. *J Vasc Surg*. 2002;36(5):889−895. https://doi.org/10.1067/mva.2002.128637.
8. Rutherford RB, Padberg Jr FT, Comerota AJ, Kistner RL, Meissner MH, Moneta GL. Venous severity scoring: an adjunct to venous outcome assessment. *J Vasc Surg*. 2000;31(6):1307−1312. https://doi.org/10.1067/mva.2000.107094.
9. Vasquez MA, Munschauer CE. Venous Clinical Severity Score and quality-of-life assessment tools: application to vein practice. *Phlebol*. 2008;23(6):259−275. https://doi.org/10.1258/phleb.2008.008018.
10. Gil Vasquez MA, Wang J, Mahathanaruk M, Buczkowski G, Sprehe E, Dosluoglu HH. The utility of the Venous Clinical Severity Score in 682 limbs treated by radiofrequency saphenous vein ablation. *J Vasc Surg*. 2007;45(5):1008−1015. https://doi.org/10.1016/j.jvs.2006.12.061.
11. let JL, Perrin MR, Allaert FA. Clinical presentation and venous severity scoring of patients with extended deep axial venous reflux. *J Vasc Surg*. 2006;44(3):588−594. https://doi.org/10.1016/j.jvs.2006.04.056.
12. Ricci MA, Emmerich J, Callas PW, et al. Evaluating chronic venous disease with a new venous severity scoring system. *J Vasc Surg*. 2003;38(5):909−915. https://doi.org/10.1016/s0741-5214(03)00930-3.
13. Jenkinson C. Quality of life. Encyclopædia Britannica. https://www.britannica.com/topic/quality-of-life. Published May 6, 2020. Accessed June 16, 2021.
14. Kosinski M, Keller SD, Hatoum HT, Kong SX, Ware Jr JE. The SF-36 Health Survey as a generic outcome measure in clinical trials of patients with osteoarthritis and rheumatoid arthritis: tests of

data quality, scaling assumptions and score reliability. *Med Care*. 1999;37(5 Suppl):MS10–MS22. https://doi.org/10.1097/00005650-199905001-00002.
15. Ware Jr JE, Gandek B. Overview of the SF-36 health survey and the International quality of life assessment (IQOLA) Project. *J Clin Epidemiol*. 1998;51(11):903–912. https://doi.org/10.1016/s0895-4356(98)00081-x.
16. Ware Jr JE, Kosinski M, Bayliss MS, McHorney CA, Rogers WH, Raczek A. Comparison of methods for the scoring and statistical analysis of SF-36 health profile and summary measures: summary of results from the Medical Outcomes Study. *Med Care*. 1995;33(4 Suppl):AS264–AS279.
17. Ware Jr JE, Kosinski M, Gandek B, et al. The factor structure of the SF-36 health survey in 10 countries: results from the IQOLA Project. International quality of life assessment. *J Clin Epidemiol*. 1998;51(11):1159–1165. https://doi.org/10.1016/s0895-4356(98)00107-3.
18. Kuet ML, Lane TR, Anwar MA, Davies AH. Comparison of disease-specific quality of life tools in patients with chronic venous disease. *Phlebol*. 2014;29(10):648–653. https://doi.org/10.1177/0268355513501302.
19. Launois R, Mansilha A, Jantet G. International psychometric validation of the chronic venous disease quality of life questionnaire (CIVIQ-20). *Eur J Vasc Endovasc Surg*. 2010;40(6):783–789. https://doi.org/10.1016/j.ejvs.2010.03.034.
20. Launois R, Reboul-Marty J, Henry B. Construction and validation of a quality of life questionnaire in chronic lower limb venous insufficiency (CIVIQ). *Qual Life Res*. 1996;5(6):539–554. https://doi.org/10.1007/BF00439228.
21. Bland JM, Dumville JC, Ashby RL, et al. Validation of the VEINES-QOL quality of life instrument in venous leg ulcers: repeatability and validity study embedded in a randomised clinical trial. *BMC Cardiovasc Disord*. 2015;15:85. https://doi.org/10.1186/s12872-015-0080-7. Published 2015 Aug 11.
22. Kahn SR, Lamping DL, Ducruet T, et al. VEINES-QOL/Sym questionnaire was a reliable and valid disease-specific quality of life measure for deep venous thrombosis. *J Clin Epidemiol*. 2006;59(10):1049–1056. https://doi.org/10.1016/j.jclinepi.2005.10.016. published correction appears in J Clin Epidemiol. 2006 Dec;59(12):1334.
23. Lamping DL, Schroter S, Kurz X, Kahn SR, Abenhaim L. Evaluation of outcomes in chronic venous disorders of the leg: development of a scientifically rigorous, patient-reported measure of symptoms and quality of life. *J Vasc Surg*. 2003;37(2):410–419. https://doi.org/10.1067/mva.2003.152.
24. Smith JJ, Guest MG, Greenhalgh RM, Davies AH. Measuring the quality of life in patients with venous ulcers. *J Vasc Surg*. 2000;31(4):642–649. https://doi.org/10.1067/mva.2000.104103.
25. Lazarus GS, Cooper DM, Knighton DR, et al. Definitions and guidelines for assessment of wounds and evaluation of healing. *Arch Dermatol*. 1994;130(4):489–493.
26. Curran MP, Plosker GL. Bilayered Bioengineered skin Substitute (Apligraf). *BioDrugs*. 2002;16:439–455.
27. Frykberg RG, Banks J. Challenges in the treatment of chronic wounds. *Adv Wound Care*. 2015;4:560–582.
28. Gaur A, Sunkara R, Raj ANJ, Celik T. Efficient wound measurements using RGB and depth images. *Int J Biomed Eng Technol*. 2015;18:333–358.
29. Izatt JA, Choma MA, Dhalla A-H. Theory of Optical Coherence Tomography. In: Drexler W, Fujimoto JG, eds. *Optical Coherence Tomography*. Cham, Switzerland: Springer; 2015:65–94.
30. Li S, Mohamedi AH, Senkowsky J, Nair A, Tang L. Imaging in chronic wound diagnostics. *Adv Wound Care*. 2020;9(5):245–263.
31. Winsor T. Influence of arterial disease on the systolic blood pressure gradients of the extremity. *Am J Med Sci*. 1950;220:117–126.
32. Patel HM, Bulsara SS, Banerjee S, et al. Indocyanine green angiography to Prognosticate healing of foot ulcer in Critical limb Ischemia: a Novel technique. *Ann Vasc Surg*. 2018;51:86–94. https://doi.org/10.1016/j.avsg.2018.02.021.
33. American Medical Association. *CPT Professional*. Chicago, IL: American Medical Association; 2018.
34. ICD List. https://icdlist.com/icd-10/index.

SECTION 3

Nonoperative management of chronic venous insufficiency and wound care

CHAPTER 12

Compression therapy in venous leg ulcers

Hugo Partsch
Department of Dermatology, University of Vienna, Vienna, Austria

Introduction

Compression therapy is the basic treatment modality in venous leg ulcer management and has been shown to be effective for healing ulcers and maintaining healing.[1,2] In every individual case, the underlying venous pathology should be determined, preferably by duplex ultrasonographic investigation; moreover, methods to correct the underlying pathophysiology (such as surgery, endovenous ablation, or sclerotherapy) should be considered. Severe arterial occlusive disease and other possible causes of nonhealing wounds should be excluded as part of the treatment plan. Before compression is applied, important recommendations concerning the underlying skin condition, the presence of varicose veins, and the status of arterial flow should be considered.

1. Local therapy: Skin moisturizing cremes should be applied to the leg when the stockings are taken off in the evening or when a bandage is removed, but not before putting on the stocking in the morning, to avoid stocking damage. Compression stockings with integrated skin care have been developed to prevent exsiccation, redness, and dryness; these stockings may improve skin hydration and patient compliance.[3] After taking care of the ulcer and its surroundings by adequate, not adhering to local therapy keeping the ulcer moist, sufficient layers of gauze pads should be placed over the ulcer to avoid secretion penetration into the outer stocking.
2. Correction of venous reflux: Although optimal local therapy is important,[3] it should always be combined with ample compression of varicose veins around a venous ulcer. The eradication of such veins can result in improved ulcer healing[4]; furthermore, correction of superficial venous reflux represents a paradigm shift in venous ulcer therapy.[4] The easiest way to correct superficial venous reflux is by performing foam sclerotherapy under duplex ultrasound guidance, in addition to compression.[4]
3. Assessment of arterial circulation: Before strong compression is applied, arterial flow status should be evaluated, preferably using a hand-held Doppler or by measuring the ankle-brachial index (ABI) (see Chapter 10. Compression therapy is contraindicated in patients with the severe arterial occlusive disease (ABI < 0.5).[5] Whenever peripheral arterial disease is suspected by weak or absent pulses, Doppler investigation and ABIs should be performed. An arteriogram should be performed in inconclusive cases and for treatment planning.

Compression devices

Different devices can be used for venous ulcer compression therapy (Table 12.1).

Medical compression stockings

The use of compression stockings may be considered for treating venous ulcers if the ulcers are not too large (less than 5 cm^2) and not too long-standing (evolving for less than 3 months).[6] Even inexperienced patients can self-administer this kind of treatment, which generates a constant degree of pressure and allows patients to change the dressing, as well as clean and wash the ulcer. One of the disadvantages of these stockings is that the exudates may contaminate the stockings, which must be frequently washed; this weakens the stocking fibers. Randomized controlled trials have shown favorable ulcer healing results using different types of compression stockings to reduce leg edema which is one of the obstructing factors for ulcer healing.[6-8]

Light compression stockings may be used to keep the local ulcer dressing in place. A Class II compression stocking placed on top of the light stocking will not only add pressure to the underlying light stocking, but also increase the stiffness of the entire kit.[9] Several two-layered ulcer stockings have been introduced (Venosan kit, Venotrain, Bauerfeind, Medi Ulcertec, Ulcer kit Gloria, Saphenamed UCV, Hartmann). The basic layer that keeps the ulcer dressing in place stays on the leg overnight, whereas the second stocking is placed over the basic layer during the daytime. This regimen allows the patient to clean the ulcer and change the dressing, as needed. In addition, this treatment modality is cost-effective, as no medical personnel is required to perform dressing changes.

Table 12.1 Types of compression devices.

- Graduated compression stockings (ready-made or custom made)
    - Custom made
    - Standard size
        - Knee length
        - Thigh length
        - Compression tights
- Bandages
    - Single component—multiple components
    - Inelastic—elastic
- Adjustable compression wraps
- Intermittent pneumatic compression
    - Single chamber
    - Sequential chambers
        - Foot-pump
        - Lower leg
        - Full leg, trunk

Stockings with a zipper have been introduced in the United States (Ulcer-care, Jobst-Beiersdorf). A ready-to-use tubular device, which can be washed and reused, was introduced in some European countries under the name Tubulcus or Rosidal mobil. Below-knee stockings are usually prescribed, and strong compression stockings that generate a pressure of more than 30 mmHg at the ankle level are recommended.[2,8]

After the ulcer heals, the use of compression stockings is essential to prevent recurrence.[2,8] Daily compliance may be a challenge, especially in elderly patients who have difficulty applying elastic stockings. To facilitate the handling of compression stockings, aids in taking them off and putting them on have been developed (e.g., Butler, nylon or silk socks, Slim slide, Easy slide); these aids are designed to help slide the stocking over the heel. Wearing rubber gloves can also facilitate the procedure, as they provide a better grip when handling the stockings.

Currently, compression stockings that follow national guidelines are graduated[10]; hence, they exert a higher pressure distally than proximally. This principle is reasonable for thromboprophylactic stockings worn by patients who are in the supine position, but not for treating a mobile patient. Our approach has been revolutionized by the introduction of reverse pressure gradient stockings, which are not only easier to put on and take off, but also superior to the abovementioned stockings in terms of hemodynamic outcomes.[11,12]

Since problems associated with putting on and taking off the stockings are the most important reasons for compression therapy-related noncompliance, this approach will hopefully have a positive impact on compliance in the future. Recently, compression stockings without pressure-relieving parts over the foot and heel have been developed and have shown excellent application facilitation.[13]

There are several classes of compression stockings based on the pressure they exert on the leg. Unfortunately, there are considerable discrepancies between the national standards of stocking classification, as shown in Table 12.2. For instance, Class II stocking is defined by a pressure range at the ankle level of 15–20 mmHg, 18–24 mmHg, 20–30 mmHg, and 23–32 mmHg in France, the United Kingdom, the United States, and Germany and other European countries, respectively.[14] In addition, the descriptive terms vary considerably; a stocking labeled "medium or moderate" in European countries may be labeled "strong" in the United States.

Therefore, for international standardization of terminology, it is recommended to use the pressure range in mmHg rather than the compression class, and to specify the description of "mild," "light," "medium, moderate," and "strong, firm, or extra firm" by adding the pressure range.[14] In conclusion, compression stockings are highly effective in reducing edema[15]; however, they do not play a primary role in treating leg ulcers because of insufficient hemodynamic efficiency.[16,17]

Table 12.2 Pressure ranges of medical compression stockings according to different national regulations[a].

Compression class	European union (CEN)	United States	United Kingdom	France	Germany
I (mild)	15–20				
(Moderate)		14–17			
(Light)	10–15	18–21			
II	23–32 (moderate)	20–30 (firm)	18–24 (medium)	15–20	23–32 (medium)
III	34–46 (strong)	30–40 (extra firm)	25–35 (strong)	20–36	34–46 (strong)
IV	>49 (very strong)	40+		>36	>49 (very strong)

CEN: The European Committee for Standardization.
[a]The values indicate the pressure (mmHg) that should be exerted by the hosiery at a hypothetical cylindrical ankle shape above the ankle.

Compression bandages

For the routine management of venous ulcers, stockings cannot replace compression bandages, which may exert much higher pressure. Table 12.3 shows the values of interface pressures of bandages measured on the distal leg according to the British standards, the only existing standard for bandages,[18] and compares these pressure values with newly proposed pressure values from a consensus conference on compression bandaging.[19]

The pressure values indicated by the consensus group are clearly higher than those from the British standard. Emphatically, all pressure values from Table 12.3 are mainly based on assumptions and therefore need to be substantiated by clinical research involving more pressure measurements on the human leg.

Table 12.3 Interface pressure ranges (mmHg) exerted by compression bandages on the distal leg in the supine position[a].

	Consensus group	BS 7505
Light compression	<20	<20
Moderate compression	20–40	21–30
Strong compression	40–60	31–40
Very strong compression	>60	41–60

[a]All values are in mmHg. The definitions proposed by a consensus group[19] re compared with those of the British Standard (BS) 7505.[18]

Interface pressure and stiffness measured on the leg

The pressure on the leg is an important parameter that is used to determine the efficacy of a bandage. Another important parameter is stiffness, which characterizes the elastic property of the compression material, defined by an increase in pressure due to an increase in the circumference of the leg during movement.[20]

Both parameters can be assessed on the leg using simple, battery-powered, portable transducers such as Picopress (Microlab, Italy) or the Kikuhime tester (MediTrade, Soro, Denmark). Such measurements will likely be indispensable in future trials comparing different compression products, and for training purposes.

A preferred measuring point is an area on the leg that shows the most pronounced changes in curvature and circumference during active standing and walking, which is approximately 8–12 cm above the inner malleolus.[20,21] A comparison between the in vivo and in vitro measurements of the pressure and stiffness of compression stockings revealed a good correlation.[9]

Elastic and inelastic bandages

Table 12.4 summarizes practical differences in compression materials based on the elastic properties of single layers.

o **Elastic compression bandages:** These are bandages that incorporate materials that exert pressure when stretched.
o **Inelastic compression bandages:** These are bandages that, when tightly applied, exert pressure, which increases when movement causes the calf muscles to contract.

Elastic long-stretch materials are relatively easy to handle and can be used by patients. In contrast to inelastic materials, these elastic bandages produce an active force owing to the elastic constriction of their fibers. For elastic bandages, the pressure drop with time is minimal; therefore, such bandages may cause pain and discomfort when the patient sits or lies down, especially when they are applied too tightly. Single-component elastic bandages or compression stockings are applied in the morning, preferably before getting

Table 12.4 Elastic property of single-layer compression material.

	Inelastic/No stretch	Inelastic/short stretch	Elastic/long stretch
Maximal stretch (%)	5	<100	>100
Stiffness	Very high	High	Low
Applied by	Trained expert	Trained staff	Patient or relatives
Stays on leg	Up to several days	Up to several days	Removed before night time

up and removed before going to bed at night. During walking, the peak pressure waves of elastic materials are lower than those of inelastic materials (Fig. 12.1).

Inelastic materials produce a much higher pressure increase when a patient performs dorsiflexion in an upright position, compared to elastic materials. The pressure increase during standing and the pressure amplitudes during ankle movement are useful parameters for characterizing stiffness.[20]

For a practical differentiation between elastic and inelastic materials, it has been proposed to define the difference between standing and supine pressures measured using a small pressure transducer at the medial gaiter region as the static stiffness index.[20] Static stiffness index values greater than and less than 10 are indicative of inelastic and elastic materials, respectively. This parameter signifies the range between the effective working pressure and tolerable resting pressure.[21]

A good compression bandage is characterized by well-tolerated resting pressures and high-pressure peaks during walking. An inelastic bandage achieving pressure peaks of 80 mmHg will compress the superficial and deep veins intermittently (Fig. 12.1). This can be demonstrated by compressing the leg with blood pressure cuffs containing an ultrasound-permeable window. Using a duplex evaluation, it can be demonstrated that, in the upright position, a pressure of 40–60 mmHg will reduce the diameter of the leg veins.[22]

It has been shown that adequate external compression can reduce venous reflux and venous hypertension.[23] Basically, a compression stocking pressure of 20 mmHg will be too weak to achieve hemodynamic improvement in an ambulatory patient (Fig. 12.1).

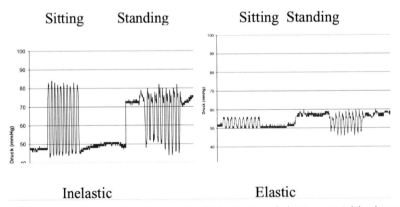

Figure 12.1 Subbandage pressure (mmHg y-axis) with inelastic and elastic material, both applied with a resting pressure of 50 mmHg on the distal lower leg in the sitting position. During up and down movement of the feet, the pressure gradient between muscle systole (peak values) and muscle diastole (lowest values) is much higher for the inelastic than for the elastic material. When the patient is standing up, the pressure rises with the inelastic bandage by 22 mmHg and only by 8 mmHg with the elastic.

However, this low pressure may be sufficient to prevent edema formation and accelerate venous blood flow in a horizontal position.

Two main disadvantages of inelastic bandages must be considered: first, bandage pressure reduction starts immediately after bandage application. After only 1 h following bandage application, the initial resting pressure drops by approximately 25%, mainly due to an immediate limb volume decrease. However, the hemodynamic benefits of the bandage are maintained. Second, an adequate application of an inelastic compression bandage is not easy, and therefore requires training. Owing to the rapid pressure reduction after bandage application, inelastic bandages should be applied with a much higher initial tension than elastic bandages. An inadequate inelastic bandaging technique is likely the main reason for the poor clinical outcomes described in some studies.[1] Basic differences between elastic and inelastic compression are summarized in Table 12.4.

"Single-layer" and multilayer bandages

Usually, "single-layer bandages" are applied with an overlap of approximately 50%. The use of "single-layer bandages" (e.g., One Ace bandage) is insufficient to treat a venous leg ulcer. Because they are always applied with some overlap, it may be argued that "single-layer bandages" do not exist.

Randomized controlled trials have clearly shown that multi-layered compression is more effective in healing venous ulcers than "single-layer compression."[1] Multilayer bandages may consist of one or several components of different compression materials. One of our preferred inelastic multicomponent bandages is an Unna boot bandage wrapped over by a textile short-stretch bandage, as shown in Fig. 12.2.

The main problem with inelastic bandages is the fact that they are applied too loosely and in most cases by the staff,[2] and lose pressure very quickly due to the regression of edema (with the subsequent reduction of the leg diameter). Frank Shaw, an engineer and inventor from California, whose wife suffered from leg lymphedema, made an invention that was globally accepted by compression experts. He cut the leather boots of his wife in the lengthwise direction and adjusted the loose edges using Velcro strips, thereby inventing a new class of adjustable inelastic compression wraps under the trade name of Circaid.[24] Moreover, there are several related products on the market, with an increasing number of publications demonstrating the value of such adjustable inelastic wraps in lymphedema and leg ulcer therapy[25–28] (Fig. 12.3). Circaid products have a simple built-in pressure system guide card, allowing the patient to assess the pressure range during application (Fig. 12.4) (for venous ulcers without an arterial component, we recommend a range of 40–60 mmHg). The first study showing a greater healing rate of leg ulcers using the abovementioned wraps compared to conventional four-layer bandages was a randomized clinical trial published by Villavicencio and his group in 2005.[29]

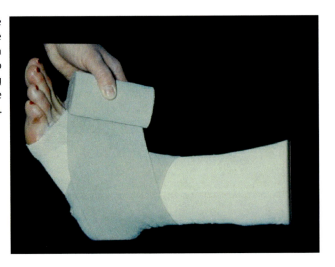

Figure 12.2 Unna boot bandage (10 m applied under considerable tension) and wrapped over by an inelastic bandage (Rosidal K). To avoid wrinkles and folds during application, the zinc paste bandage must be cut and molded to the leg.

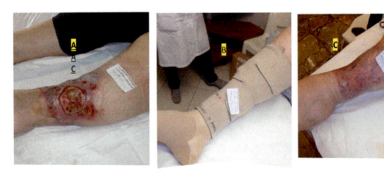

Figure 12.3 (A) 55-year-old patient with a posttraumatic ulcer and postthrombotic syndrome. (B) Performed self-management with Circaid. (C) After 20 weeks of self-treatment, the ulcer was healed. Note: No compression on the foot. *(Courtesy Dr. Mosti.)*

The main advantage of these wraps is that they can be applied by trained patients. A recently published study showing better results regarding the efficacy of these wraps on ulcer healing demonstrated the hemodynamic advantages of short-stretch bandages, but without the disadvantage of pressure loss because patients can adjust the material once they detect pressure decrease.[30] This study also showed that the material of the adjustable compression wraps (ACW) devices used was cheaper than that of the conventional comparator, which had to be replaced with every bandage change. This cost-effectiveness is even more significant when considering the costs of using professional healthcare staff for wraps application versus the self-application of the wraps by a patient.[2]

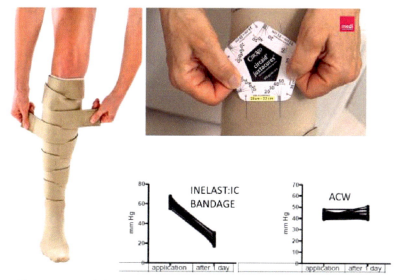

Figure 12.4 The pressure of the device can be assessed by a calibration card (upper right), which shows unchanged values after 1 week due to self-adaptations (bottom right) in contrast to an inelastic bandage changed at weekly intervals (bottom left).

Intermittent pneumatic compression

Intermittent pulsatile external compression produces beneficial physiological changes, including hematologic, hemodynamic, and endothelial effects, thereby promoting ulcer healing.[31] The combined application of intermittent and sustained graduated compressions can improve ulcer healing. This could be achieved even in patients with coexisting arterial occlusive disease.[32]

Adjunctive treatment may be extremely helpful in patients who have edema and are unable to walk, or who suffer from a stiff ankle, especially when the concomitant arterial disease is also present. In such cases, there is usually a vicious circle starting with pain and inability to walk (which results in sitting for many hours), and progressing to edema, which reduces the arterial skin perfusion, and therefore worsens the pain. Bed rest with leg elevation is not recommended for patients in this situation; intermittent pneumatic compression (IPC) can reduce edema and enhance arterial blood flow.[33]

Intermittent pneumatic compression can be beneficial in patients with ischemia, due to intermittent periods without compression which may result in short intervals of reactive hyperemia in the tissues. This is opposite from the principle associated with bandages and stockings which is characterized by sustained, continuous compression.

The positive effect of IPC on ulcer healing is difficult to verify because this therapy is always applied in combination with conventional compression. "Hybrid devices" are currently on the market and present a combination of sustained and intermittent compression using inelastic, inflatable boots; these devices show promising preliminary results in patients with ulcers.[34]

Pelottes and pads

Venous ulcers are frequently localized posterior to the inner malleolus or in flat areas of the medial lower leg. The pressure exerted by a bandage or stocking is low in these areas as explained by Laplace's law, which defines the inverse relationship between the radius of a patient's limb and the pressure applied.[35] A local increase in pressure can be achieved by applying rubber foam pads over the ulcer region, thereby decreasing the radius of the leg segment (Fig. 12.5). Care should be taken to flatten the edges of such pelottes to avoid creating sharp impressions on the skin by these devices (Fig. 12.6A and B).

Prevention of ulcer recurrence

The management of leg ulcers consists of two phases:
1. The healing phase, which is concluded by complete epithelialization.
2. The maintenance phase after ulcer healing, in which a frequently occurring recurrence should be prevented.

Generally, it is easier to heal a venous ulcer than to keep it healed. To keep the ulcer healed, continuous compression is essential ("maintenance phase"). Medical compression using below-knee stockings (30–45 mmHg) constitutes the preferred compression method. Patients who are unable to put on the stockings may use elastic bandages or ACW instead. The eradication of venous reflux using surgery, endovenous ablation, or sclerotherapy should be considered in every patient.

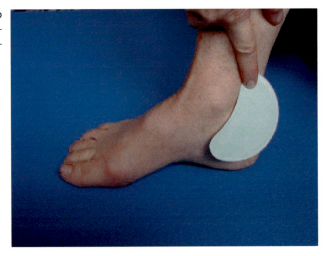

Figure 12.5 Rubber foam pad to increase the local bandage pressure over an ulcer in the retro malleolar region.

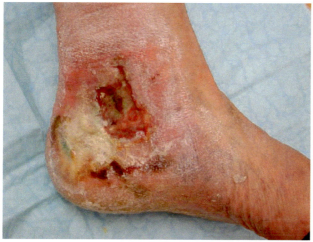

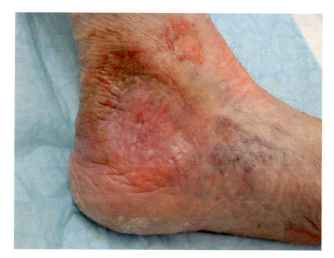

Figure 12.6 (A) Venous ulcer due to a postthrombotic syndrome behind the inner malleolus. High local pressure promotes debridement. (B) After 16 weeks of compression therapy using unna boot bandages and a rubber foam pad, the ulcer is healed (the impression of the pad can still be seen). Consideration to venous refluxes by surgery or by sclerotherapy and wearing of compression stockings is essential to keep the ulcer closed.

Compression techniques—practical guidelines

Many different bandage types[36] and compression bandaging techniques have been described. The general rules are as follows:

o Elastic bandages are easier to handle than inelastic bandages; moreover, they may be applied by staff who are not specifically trained in managing venous diseases, or by the patients themselves. This is also true for compression stockings.

- Inelastic materials (such as zinc paste) should be applied to generate much higher resting pressure. This is achieved by carefully pressing the bandage roll toward the leg surface during the application to obtain a maximum contact between the leg and the zinc paste. To avoid irregular folds and gaps the zinc paste roll can be tailored by cutting wrinkles to assure it is molded to follow the shape of the leg.
- To obtain an equal pressure distribution without constricting bands or folds, it is recommended to stop the zinc paste application after a sufficient amount has been applied to mold the zinc material on the leg following the cone-shaped leg surface. A 10 m long bandage is recommended for each lower leg. After the lower leg is covered with several layers, a 5 m-long, short-stretch bandage is wrapped over the covered leg, and the patient is encouraged to start walking immediately for a minimum of 30 min.
- The abovementioned short-stretch bandage can be washed and reused with each bandage change.
- To prevent the occurrence of iatrogenic skin injury using scissors, it is important to carefully unwrap the zinc paste layers and cut them only when the leg circumference has decreased, and when there is enough space to accommodate the scissors.
- After several minutes, the pressure decreases to approximately 40 mmHg due to an immediate, but considerable, edema reduction.
- In the edematous phase, the bandage loosens after a few days; hence, the bandage should be reapplied or a short-stretch bandage should be wrapped over the previously applied bandage. The same is recommended when exudates from the ulceration penetrate the bandage. This may occur, especially during the initial treatment phase, and the patient should be instructed to return to the healthcare facility if this happens. Thereafter, the bandage is changed every 7 days on average.
- Bandaging should cover the foot and extend cephalically to the capitulum fibulae. The initial turn may be placed around the ankle or between the heel and dorsal tendon to fix the bandage. Thereafter, the bandage is taken down to the foot and to the base of the toes. The ankle joint is always bandaged with the foot in maximal dorsal extension.
- Taking the bandage up to the leg and layer overlapping are performed in a spiral fashion or with a "figure of eights."
- Graduated compression is achieved by exerting higher pressures on the distal lower leg than on the proximal calf. Local pressure over ulcers or firm lipodermatosclerotic areas can be increased using pads and pelottes. Tendons and shins should always be protected by cotton wool.
- There should be an overlapping of the layers between 30% and 50%.
- The proximal end of the bandage should be applied with reduced pressure to cover the capitulum fibulae.
- Bandaging of the lower leg is sufficient for most patients. Compression bandages, starting below the flexed knee and reaching the inguinal fold, are only recommended for

patients with extensive swelling or phlebitis of the thigh. The flexor tendons in the popliteal fossa are protected using cotton wool.
o Thigh bandages can best be applied using adhesive material starting from the proximal lower leg and extending cephalad to the proximal thigh. To compress the veins, a sub-bandage pressure of at least 40 mmHg should be applied at the mid-thigh level.
o Highly exudative ulcers may require frequent dressing changes during the initial phase. However, exudation will likely subside after several days of firm compression.
o Walking exercises are essential to optimize the effect of compression therapy.

Summary

Compression therapy is an essential component in the management of patients with venous leg ulcers, as it counteracts the underlying pathophysiologic factors responsible for their development. In these patients, venous outflow obstruction, and especially valvular incompetence in large and small veins, lead to venous reflux and ambulatory venous hypertension, which is the trigger for further changes in the microcirculation. Underlying venous hypertension can be reduced by correcting venous reflux using endovenous and surgical procedures and/or external venous compression.

To achieve hemodynamically efficient leg vein compression, higher pressures are needed (corresponding to resting values of 50 mmHg); these pressures are not bearable in a lying position. Such compression may be achieved by using stiff, inelastic bandage materials (such as Unna boot), which exert pressures of more than 100 mmHg during muscle contraction. Intermittent pressure peaks during muscle systole while walking and intermittently occlude the veins by blocking venous reflux, thereby reducing venous hypertension. Such bandages should be applied by well-trained and experienced staff. The following three principles showed clinically proven benefits regarding the use of compression regimens in patients with leg ulcers:

1. Concomitant arterial occlusive disease-causing "mixed ulcers" is not a contraindication for compression therapy; a systolic ankle pressure >60 mmHg is a good indication for the use of inelastic material, as long as the well-trained staff applies a bandage with a resting pressure not exceeding 40 mmHg.[5]
2. Reflux correction should not be delayed until the ulcer is healed; it should be performed in the acute phase of ulceration.[4]
3. Due to the introduction of ACW, effective self-management is currently possible.[28]

It is reasonable to anticipate that the principle of adequate compression therapy will be understood and applied in the future, not only in treating medical centers, but also in the homes of adequately trained and compliant patients with ulcers.

References

1. O'Meara S, Cullum N, Nelson EA, Dumville JC. Compression for venous leg ulcers. *Cochrane Database Syst Rev*. November 14, 2012;11(11):CD000265. https://doi.org/10.1002/14651858.CD000265.pub3. PMID: 23152202; PMCID: PMC7068175.
2. Nelson EA, Bell-Syer SE. Compression for preventing recurrence of venous ulcers. *Cochrane Database Syst Rev*. September 9, 2014;2014(9):CD002303. https://doi.org/10.1002/14651858.CD002303.pub3. PMID: 325203307; PMCID: PMC7138196.
3. Harries RL, Bosanquet DC, Harding KG. Wound bed preparation: TIME for an update. *Int Wound J*. September 2016;13(Suppl 3):8–14. https://doi.org/10.1111/iwj.12662. PMID: 27547958.
4. Gohel MS, Mora MSc J, Szigeti M, et al. Early venous reflux ablation trial group. Long-Term clinical and cost-effectiveness of early endovenous ablation in venous ulceration: a randomized clinical trial. *JAMA Surg*. September 23, 2020;155(12):1113–1121. https://doi.org/10.1001/jamasurg.2020.3845. Epub ahead of print. PMID: 32965493; PMCID: PMC7512122.
5. Rabe E, Partsch H, Morrison N, et al. Risks and contraindications of medical compression treatment - a critical reappraisal. An international consensus statement. *Phlebology*. August 2020;35(7):447–460. https://doi.org/10.1177/0268355520909066. Epub 2020 Mar 2. PMID: 32122269; PMCID: PMC7383414.
6. Partsch H, Horakova MA. Kompressionsstrümpfe zur Behandlung venöser Unterschenkelgeschwüre [Compression stockings in treatment of lower leg venous ulcer]. *Wien Med Wochenschr*. 1994;144(10–11):242–249. German. PMID: 7856197.
7. Ashby RL, Gabe R, Ali S, et al. VenUS IV (Venous leg Ulcer Study IV) - compression hosiery compared with compression bandaging in the treatment of venous leg ulcers: a randomised controlled trial, mixed-treatment comparison and decision-analytic model. *Health Technol Assess*. September 2014;18(57):1–293. https://doi.org/10.3310/hta18570. PMID: 25242076; PMCID: PMC4781202.
8. Health Quality Ontario. Compression stockings for the prevention of venous leg ulcer recurrence: a health technology assessment. *Ont Health Technol Assess Ser*. February 19, 2019;19(2):1–86. PMID: 30828407; PMCID: PMC6394515.
9. Partsch H, Partsch B, Braun W. Interface pressure and stiffness of ready made compression stockings: comparison of in vivo and in vitro measurements. *J Vasc Surg*. October 2006;44(4):809–814. https://doi.org/10.1016/j.jvs.2006.06.024. PMID: 17012005.
10. Mosti G, Partsch H. Compression stockings with a negative pressure gradient have a more pronounced effect on venous pumping function than graduated elastic compression stockings. *Eur J Vasc Endovasc Surg*. August 2011;42(2):261–266. https://doi.org/10.1016/j.ejvs.2011.04.023. Epub 2011 May 25. PMID: 21612949.
11. Wittens C, Davies AH, Bækgaard N, et al. Editor's choice—management of chronic venous disease: clinical practice guidelines of the European society for vascular surgery (ESVS). *Eur J Vasc Endovasc Surg*. June 2015;49(6):678–737.
12. Couzan S, Leizorovicz A, Laporte S, et al. A randomized double-blind trial of upward progressive versus degressive compressive stockings in patients with moderate to severe chronic venous insufficiency. *J Vasc Surg*. November 2012;56(5):1344–1350.e1. https://doi.org/10.1016/j.jvs.2012.02.060. Epub 2012 May 15. PMID: 22592040.
13. Buset CS, Fleischer J, Kluge R, et al. Compression stocking with 100% donning and doffing success: an open label randomised controlled trial. *Eur J Vasc Endovasc Surg*. January 2021;61(1):137–144. https://doi.org/10.1016/j.ejvs.2020.09.027. Epub 2020 Oct 28. PMID: 33129680.
14. Neumann HA, Partsch H, Mosti G, Flour M. Classification of compression stockings: report of the meeting of the international compression club, copenhagen. *Int Angiol*. April 2016;35(2):122–128. Epub 2015 Feb 12. PMID: 25673312.
15. Mosti G, Picerni P, Partsch H. Compression stockings with moderate pressure are able to reduce chronic leg oedema. *Phlebology*. September 2012;27(6):289–296. https://doi.org/10.1258/phleb.2011.011038. Epub 2011 Nov 16. PMID: 22090466.

16. Nicolaides A, Kakkos S, Baekgaard N, et al. Management of chronic venous disorders of the lower limbs. Guidelines according to scientific evidence. Part I. *Int Angiol.* June 2018;37(3):181—254. https://doi.org/10.23736/S0392-9590.18.03999-8. PMID: 29871479.
17. Mayberry JC, Moneta GL, DeFrang RD, Porter JM. The influence of elastic compression stockings on deep venous hemodynamics. *J Vasc Surg.* January 1991;13(1):91—99. https://doi.org/10.1067/mva.1991.25386. PMID: 1781813.
18. British Standard. *Specifications for the Elastic Properties of Flat, Non-adhesive, Extensible Fabric Bandages.* BS 7505. 1995:1—5.
19. Partsch H, Clark M, Mosti G, et al. Classification of compression bandages: practical aspects. *Dermatol Surg.* May 2008;34(5):600—609. https://doi.org/10.1111/j.1524-4725.2007.34116.x. Epub 2008 Feb 6. PMID: 18261106.
20. Partsch H, Clark M, Bassez S, et al. Measurement of lower leg compression in vivo: recommendations for the performance of measurements of interface pressure and stiffness: consensus statement. *Dermatol Surg.* February 2006;32(2):224—232. https://doi.org/10.1111/j.1524-4725.2006.32039.x. PMID: 16442043.
21. Partsch H, Schuren J, Mosti G, Benigni JP. The static stiffness index: an important parameter to characterise compression therapy in vivo. *J Wound Care.* September 2016;25(Suppl 9):S4—S10. https://doi.org/10.12968/jowc.2016.25.Sup9.S4. PMID: 27608740.
22. Partsch B, Partsch H. Calf compression pressure required to achieve venous closure from supine to standing positions. *J Vasc Surg.* October 2005;42(4):734—738. https://doi.org/10.1016/j.jvs.2005.06.030. PMID: 16242562.
23. Partsch B, Mayer W, Partsch H. Improvement of ambulatory venous hypertension by narrowing of the femoral vein in congenital absence of venous valves. *Phlebology.* 1992;7:101—104, 1992.
24. httpw.veindirectory.org/magazine/article/industry-spotlight/herthas-story.
25. Protz K, Heyer K, Dörler M, Stücker M, Hampel-Kalthoff C, Augustin M. Compression therapy: scientific background and practical applications. *J Dtsch Dermatol Ges.* September 2014;12(9):794—801. https://doi.org/10.1111/ddg.12405. Epub 2014 Aug 18. PMID: 25134422.
26. Williams A. A review of the evidence for adjustable compression wrap devices. *J Wound Care.* May 2016;25(5):242—247. https://doi.org/10.12968/jowc.2016.25.5.242. PMID: 27169339.
27. Stather PW, Petty C, Howard AQ. Review of adjustable velcro wrap devices for venous ulceration. *Int Wound J.* August 2019;16(4):903—908. https://doi.org/10.1111/iwj.13116. Epub 2019 Mar 21. PMID: 30900365.
28. Caprini JA, Partsch H, Simman R. Venous ulcers. *J Am Coll Clin Wound Spec.* December 4, 2013;4(3):54—60. https://doi.org/10.1016/j.jccw.2013.11.001. PMID: 26236636; PMCID: PMC4511547.
29. Blecken SR, Villavicencio JL, Kao TC. Comparison of elastic versus nonelastic compression in bilateral venous ulcers: a randomized trial. *J Vasc Surg.* December 2005;42(6):1150—1155. https://doi.org/10.1016/j.jvs.2005.08.015. PMID: 16376207.
30. Mosti G, Mancini S, Bruni S, et al, MIRACLE Trial investigators. Adjustable compression wrap devices are cheaper and more effective than inelastic bandages for venous leg ulcer healing. A multicentric Italian randomized clinical experience. *Phlebology.* March 2020;35(2):124—133. https://doi.org/10.1177/0268355519858439. Epub 2019 Jun 24. PMID: 31234752.
31. Comerota AJ. Intermittent pneumatic compression: physiologic and clinical basis to improve management of venous leg ulcers. *J Vasc Surg.* April 2011;53(4):1121—1129. https://doi.org/10.1016/j.jvs.2010.08.059. Epub 2010 Nov 3. PMID: 21050701.
32. Oresanya L, Mazzei M, Bashir R, et al. Systematic review and meta-analysis of high-pressure intermittent limb compression for the treatment of intermittent claudication. *J Vasc Surg.* February 2018;67(2):620—628.e2. https://doi.org/10.1016/j.jvs.2017.11.044. PMID: 29389425.
33. Mani R, Vowden K, Nelson EA. Intermittent pneumatic compression for treating venous leg ulcers. *Cochrane Database Syst Rev.* 2001;4:CD001899. https://doi.org/10.1002/14651858.CD001899. Update in: Cochrane Database Syst Rev. 2008;(2):CD001899. PMID: 11687129.
34. Harding KG, Vanscheidt W, Partsch H, Caprini JA, Comerota AJ. Adaptive compression therapy for venous leg ulcers: a clinically effective, patient-centred approach. *Int Wound J.* June 2016;13(3):317—325. https://doi.org/10.1111/iwj.12292. Epub 2014 May 7. PMID: 24802769.

35. Lee BB, Nicolaides AN, Myers K, et al. Venous hemodynamic changes in lower limb venous disease: the UIP consensus according to scientific evidence. *Int Angiol*. June 2016;35(3):236–352. Epub 2016 Mar 24. PMID: 27013029.
36. Mosti G, De Maeseneer M, Cavezzi A, et al. Society for vascular surgery and American venous forum guidelines on the management of venous leg ulcers: the point of view of the international union of phlebology. *Int Angiol*. June 2015;34(3):202–218. Epub 2015 Apr 21. PMID: 25896614.

CHAPTER 13

Wound healing: adjuvant therapy and treatment adherance

Juliet Blakeslee-Carter[1] and Marc A. Passman[2]

[1]University of Alabama at Birmingham, Division of Vascular Surgery and Endovascular Therapy, Birmingham, AL, United States; [2]Department of Surgery, University of Alabama at Birmingham, Division of Vascular Surgery and Endovascular Therapy, Birmingham, AL, United States

Introduction

Venous wound management has evolved significantly within the last decade into a distinct multidisciplinary subspecialty. While once considered a narrow field, modern providers treating venous wounds must now be familiar with a huge range of knowledge including normal and abnormal wound healing pathophysiology, methods for correction of underlying venous disease, general principles of wound healing, a wide range of targeted adjunctive treatments for chronic wounds, and methods for maintaining patient adherence to therapy. In order to be effective and economically viable, wound care techniques must be deployed in a targeted and thoughtful manner, always keeping the broader care of the patient in perspective.

Venous wounds follow a predictable healing progression when responding to appropriate treatment. The pathophysiology of wound healing and failure is reviewed significantly in more detail in the previous chapters (see Chapters 1–3), but is pertinent to review briefly. Studies have demonstrated that wound healing ideally follows a well-organized and predictable exponential course.[1] Wounds should heal at least 10% per week as they progress through the four dynamic stages of healing: hemostasis, inflammation, proliferation, and remodeling.[2] Failure to progress through any one stage results in a chronic wound. Historically, chronic wounds were defined as those that persist past 6–12 weeks[3]; however, more modern definitions classify chronic as those wounds that have deviated significantly from expected healing trajectory.[2] Once a wound has progressed to chronic, it places heavy clinical and socioeconomic burdens on the patient and the healthcare system. For this reason, early and thorough wound care should be cornerstone treatment for venous ulcers.

Chronic venous ulcers are prevalent, costly, and complex; but appropriate wound care does improve outcomes and reduce financial burden. Venous ulcers are the second most common type of wound treated in comprehensive wound care clinics.[4] and the treatment of venous ulcers is resource heavy. Epidemiologic studies have demonstrated that 15% of venous ulcers never fully heal, while recurrence occurs in up to 60% of patients with venous ulcers.[5] Due to complex treatment required and prolonged recovery, it is estimated that venous ulcers cost the US Healthcare system roughly $3 billion yearly, result in the loss of over one million working days,[6] and lead to significantly higher rates of clinical depression.[7] Recent studies estimate that the burden of chronic venous wounds will continue to increase as the prevalence of morbidities in an aging population rises along with increasing healthcare costs.[8] Despite this, wound care providers continue to hold an important role in treatment, as progress in understanding wound biology and therapy continues to be made. Diligent and well-executed wound care remains a cornerstone for treatment of people with venous ulcers.

The objective of this chapter is to address the basic principles of wound care, introduce standard and adjuvant therapies, provide a treatment algorithm for venous wound management, and discuss methods for improving continuity of wound care, all with a specific focus on providing an evidence-based pathway of venous ulcer targeted wound care. Literature regarding wound biology and care is of varying quality. As such, recommendations and data to support the techniques presented in this chapter are derived from high-quality systematic reviews and randomized clinical trials when available. The 2015 Society of Vascular Surgery (SVS) Clinical Best Practice Guidelines from the American Venous Forum (AVF)[9] will be referenced throughout this chapter, with pertinent guidelines to this chapter summarized in Table 13.1 (summary of guideline strengths are shown in Table 13.2). It should be noted that several well-studied and supported topics, including compression and negative pressure wound therapy, that are germane to wound care are covered separately in dedicated chapters (see Chapters 12 and 14).

Patient and wound assessment

The first step in effective wound management is a baseline examination of the wounds and establishing a thorough understanding of the individual patient's health history and potential physical and social barriers to healing. The development of a venous ulcer is a late symptom of a systemic disease process that must be addressed simultaneously with wound care. Diagnostic tools for the assessment of venous ulcers and methods for correction of venous insufficiency are reviewed elsewhere, but they must be among the first steps to wound care in order to prevent ulcer recurrence rates that are as high as 60% in patients without correction of underlying disease.[10]

Identification and preemptive management of potential patient barriers to healing is key to wound care success.[11] Factors that influence the pathophysiology of wound healing include immunosuppression, renal disease, diabetes, heart disease, liver disease, smoking, nutrition, and the presence of infection.[12] Effective wound care is a component of a more comprehensive systemic patient–centered treatment plan. As such, wound care should address the needs of the current wound and preemptively manage potential extrinsic and intrinsic patient healing barriers. Proactive wound care establishes a global holistic perspective on wound healing that is fundamental to success through fewer

Table 13.1 Summary of referenced 2015 society of vascular surgery clinical best practice guidelines from the American Venous Forum.[9]

Guideline		Level of evidence	Details
Best practice			
3.3		Best practice	Recommend serial venous leg ulcer wound measurement and documentation.
3.12		Best practice	Recommend that all patients with venous leg ulcers should be classified based on venous disease. Classification assessment including clinical CEAP.
Wound bed preparation			
4.1	Wound cleansers	Grade 2 level C	Venous leg ulcers should be cleansed initially and at each dressing change, performed with a minimum of chemical or mechanical trauma.
4.2	Debridement indications	Grade 1 level B	Venous leg ulcers should receive thorough debridement at initial evaluation. Health care providers should choose from a number of debridement methods.
4.3	Pain management	Grade 1 level B	Local anesthesia should be administered to minimize discomfort associated with surgical venous leg ulcer debridement. Regional or general anesthesia may be required in select cases.
4.4	Surgical debridement	Grade 1 level B	Surgical debridement should be performed for venous leg ulcers with slough, nonviable tissue, or eschar. Serial wound assessment is important in determining the need for repeat debridement.
4.5	Hydrosurgical debridement	Grade 2 level B	Hydrosurgical debridement is an alternative to standard

Continued

Table 13.1 Summary of referenced 2015 society of vascular surgery clinical best practice guidelines from the American Venous Forum.[9]—cont'd

	Guideline	Level of evidence	Details
4.6	Ultrasonic debridement	Grade 2 level C	surgical debridement of venous leg ulcers. Suggest against ultrasound debridement over surgical debridement.
4.7	Enzymatic debridement	Grade 2 level C	Enzymatic debridement should not be used over surgical debridement, but may be used as an alternative when no clinician trained in surgical debridement is available.
4.8	Biologic debridement	Grade 2 level B	Larval therapy for venous leg ulcers can be used as an alternative to surgical debridement.

Dressings

4.14	Dressing selection	Grade 2 level C	Topical dressings should manage ulcer exudate and maintain a moist wound bed. Selection should absorb wound exudate and protect peri-ulcer skin.
4.15	Dressings containing antimicrobials	Grade 2 level A	Recommend against the routine use of topical antimicrobial containing dressings in the treatment of noninfected ulcers.
4.16	Peri-ulcer skin management	Grade 2 level C	Suggest application of skin lubricants underneath compression.

Adjuvant therapy

4.18	Indications for adjuvant therapies	Grade 1 level B	Recommend adjuvant wound therapy options for venous leg ulcers that fail to demonstrate improvement after a minimum of 4−6 weeks standard wound therapy.
4.19	Skin grafting	Grade 2 level B	Skin graft should not be a primary therapy for leg ulcers.

Table 13.1 Summary of referenced 2015 society of vascular surgery clinical best practice guidelines from the American Venous Forum.[9]—cont'd

	Guideline	Level of evidence	Details
			Grafting may be utilized in conjunction with compression in large ulcers that have failed to show signs of healing with standard care for 4–6 weeks.
4.25	Electrical stimulation	Grade 2 level C	Suggest against electrical stimulation therapy for venous leg ulcers.
4.26	Ultrasound therapy	Grade 2 level B	Suggest against routine ultrasound therapy for venous leg ulcers.

Table 13.2 Summary of grade and level of evidence classifications.

Grade practice recommendations		
	Recommendation	Supporting evidence
Grade A	Strongly recommend	Supported by level 1a or 1b evidence and expertise consensus.
Grade B	Recommend	Supported by level 2 or 3 evidence and findings across research are consistent.
Grade C	Optional	Supported by level 2, 3, or 4 evidence; however, the findings across studies may be inconsistent.
Grade D	Optional	Supported by expert opinion and level 4 or 5 evidence.

Level of evidence

Level 1	1a 1b	Metaanalysis of multiple well-conducted randomized clinical trials. Individual randomized controlled trials.
Level 2	2a 2b	Systematic review of cohort studies with homogeneity of outcomes. Individual cohort studies or low-quality randomized control trials.
Level 3	3a 3b	Systematic review of case control studies. Individual case control studies.
Level 4		Case series or low-quality cohort and case control studies.
Level 5		Expert opinion or basic science research without clinical application verification.

complication, improved clinical and socioeconomic outcomes, and decreased ulcer recurrence rates. Medical management of many patient comorbidities is discussed thoroughly in other chapters of this book (see Chapter 18), but are important to keep in mind when reading this chapter, as wound care is a single component of an extensive treatment plan.

Wound assessment, thorough history and physical, must be completed prior to developing a care plan. Wound duration, and any history of healing or recurrence, is among most pertinent historical details as clinical trials have demonstrated that these factors will predict rate of wound progression along expected trajectory and is predictive of both overall successful wound healing[13,14] and patient morbidity and mortality.[15] Other pertinent clinical questions relate to prior treatment strategies, alleviating factors, and pain control techniques.

Documentation of physical exam wound characteristics is summarized in Table 13.3.[16] Physical assessment of the wound should focus on four broad areas: wound size, tissue quality, exudate, and wound edges.[16] Evaluation should include physical measurements along with photographs, and should initially be performed weekly as the wound progression (SVS-AVF Best Practice Guideline 3.3).[9] Size measurements should be taken as methodically as possible with an easily available standardized measurement tool, such as a ruler or gauge. Appearance of tissue should note any frankly necrotic areas along with eschar and slough that would require debridement. Exudate is very common in venous ulcers, and is typically more abundant in chronic wounds. The quality of the exudate can indicate the presence of excess edema along with infection. Skin adjacent to the wound should be evaluated for signs of underlying varicose veins along with any signs of maceration or infection. Based on SVS-AVF Venous Ulcer Guidelines (Guideline 3.12, Level 1A evidence, Table 13.1), physical exam findings along with patient history should be used to calculate the American Venous Forum Clinical Severity Score and CEAP score (discussed in detail in Chapter 11).[17] The clinical severity score has been utilized for assessment of clinical outcomes,[18] but can also be used to track wound progression across time.

Table 13.3 Wound characteristics to be documented.

Size	Length, width, depth
Tissue appearance	Amount and presence of: necrotic Tissue, slough, eschar, granulation tissue
Exudate	Amount, color, type, smell Serous, serosanguineous, sanguinous, purulent
Wound edge	Maceration, advancing epithelium, cellulitis or erythema

Standard therapies for wound bed management
General principles of wound bed management

Wound bed management is the process by which providers manipulate the wound environment to facilitate endogenous healing and optimize effectiveness of wound care therapies in order to promote successful epithelialization. In 2003, an international consortium of wound care experts drafted consensus documents to provide a conceptual framework when approaching wound bed preparation.[19] The findings can be summarized by the acronym "Time": tissue management, inflammation and infection control, moisture balance, and edge care.[20] In addition to following standard therapy technique, venous ulcers benefit from frequent (at least every other week) observation to allow for early intervention should problems arise.[21] These general principles will be covered in two main sections on treatments (debridement and cleansers) and wound dressings. Several key treatment modalities including compression, infection control, and medical management are covered separately.

Wound bed preparation; debridement and cleansing
Debridement

Debridement is the first and most important step to successful venous ulcer wound care.[20] Debridement is the process by which physical and biologic healing barriers are removed from the wound bed in order to facilitate regrowth of healthy tissue. The presence of nonviable tissue and debris in a wound bed promotes significant inflammation which propagates cellular death and ongoing wound festering; therefore, prompt removal of such tissue is foundational to proper care.[22] Debridement eliminates a chronic wound and resets the pathophysiology so that the wound will behave like an acute wound with the opportunity to follow the expected healing trajectory. Thorough debridement decreases rates of infection, increases healing rate, and increases rates of successful epithelialization.[23]

Debridement should focus on removal of three main barriers: callus/eschar, nonviable/necrotic tissue, and biofilm/bacteria.[24] Eschar and callus are functional barriers to healing that impede penetration of topical treatment agents and prevent migration of healthy cells. Necrotic tissue is both a functional and biologic barrier to healing. Its presence increases risk for infection and places the wound in a chronic proinflammatory state. Biofilm is a carbohydrate-rich polymeric matrix that forms on chronic venous wounds and fosters a fertile environment for bacterial proliferation while preventing penetration of key immune system cells.[25] Debridement establishes an environment where healthy tissue can flourish.

The SVS-AVF Venous Ulcer Guidelines recommend at least one treatment of thorough debridement followed by sequential debridement as needed (Guideline 4.2, Level 1B evidence, Table 13.1).[9] There are many successful methods for debridement. Choice of method will depend on patient's anatomy, condition of the wound, provider's skill with each method, and effectiveness. In many cases, more than one debridement method may be utilized.

Sharp debridement

Sharp debridement, also known as surgical debridement, is widely considered the standard of care for successful debridement. Sharp debridement can be accomplished with a scalpel, curette, scissors, or other surgical instruments depending on specific anatomical needs. Sharp debridement is fast, effective, and with proper technique can be applied to almost any wound.[26] The SVS-AVF Venous Ulcer Guidelines (Guideline 4.4, Level 1B evidence, Table 13.1) recommend sharp debridement for all patients with significant biofilm, slough, necrotic tissue, or eschar as this technique is the most successful for the removal of tough and copious nonviable tissue.[9]

Although widely utilized, sharp debridement is not ideal in all patients. Sharp debridement may not be indicated in patients with acute venous wounds in which there is very little slough, as sharp debridement may remove a significant portion of healthy tissue along with the minimal nonviable tissue. Additionally, pain is often a significantly limiting factor in a patient's ability to tolerate sharp debridement. Local anesthetics can alleviate discomfort in a subset of patients, but for more extensive wounds local blocks or general anesthesia are often required. As such, patients who cannot be placed under general anesthesia are often not ideal candidates for extensive surgical debridement. Additionally, not every clinical practice will have a surgical environment with trained practitioners, and the required instruments.

Hydrosurgical debridement

Hydrosurgery is a relatively new technology that serves as an alternative form of mechanical debridement to surgical debridement.[27] The patented hydrosurgery system relies on the Venturi effect to generate a partial vacuum on the tissue of interest and then utilizes a high-pressure stream of sterile normal saline as a simultaneous cutting and aspiration tool that targets only nonviable tissue within the vacuum.[28] Hydrodebridement can be utilized in both inpatient and outpatient clinical settings,[29] and in select patients can be performed under local anesthesia.[28]

Several clinical trials have established hydrosurgery as a safe alternative to surgical debridement, and highlighted circumstances in which hydrosurgery may offer benefits compared to traditional surgical debridement. The SVS-AVF Venous Ulcer Guidelines (Guideline 4.5, Level 2B evidence, Table 13.1) recognize hydrosurgical debridement as an alternative to surgical debridement when surgical debridement may be limited.[9] In such instances, hydrosurgery is ideal for chronic wounds with an abundance of slough and biofilm. Studies have shown that compared to surgical debridement, hydrosurgery is equally effective for slough removal while preserving a higher amount of underlying healthy tissue.[30] Hydrosurgery is also effective for cavitating wounds that may be difficult to access with traditional surgical technique. However, contrary to surgical debridement, hydrosurgery is less effective for removal of thick and hard eschar.[29] In well-selected patients with appropriate wounds, hydrosurgery offers similar rates of healing compared to surgical debridement, but had shorter operative times and significantly lower rates of intraoperative blood loss.[27,30]

Hydrosurgery is becoming a widely adopted and available technology. Initial barriers to adoption included cost and availability, both of which significantly improved over the last decade.[31] While initial cost of the hydrosurgery system remains moderate, studies have demonstrated long-term costs equal to surgical debridement due to shorted operative times and decreased requirements on individual surgical trays.[27,31,32]

Enzymatic debridement

Enzymatic debridement works through chemical disruption and digestion of the extracellular proteins in a wound bed. Enzymatic debridement is excellent for moist wound beds with significant slough, but when treating wounds with significant eschar, enzymatic debridement is ideal to use in conjunction with other forms of debridement as it is unable to penetrate through eschar.[33] The most common combination is surgical debridement followed by several days of enzymatic debridement.[34] Clinical trials have demonstrated effectiveness and safety of enzymatic debridement, but have not established that enzymatic debridement is more efficacious than surgical debridement.[35] The SVS-AVF Venous Ulcer Guidelines (Guideline 4.7, Level 2C evidence, Table 13.1) recommend enzymatic debridement as a supplement to surgical debridement, or as an alternative to surgical debridement when no surgeons are available, but does not recommend enzymatic debridement over surgical debridement.[9] Enzymatic debridement should not be used in patients with allergies to the product, and is also not ideal for heavily infected wounds as the burden of tissue is often too high.

To achieve optimal effectiveness, enzymatic wound gels must be carefully monitored and maintained in an appropriate environment. Enzymes can be denatured and inactivated in highly acidic or basic environments, at high temperatures, and when improperly utilized with antiseptic agents.[36] Studies have shown that the enzymes are stable within wound beds for up to 48 h and established that daily enzymatic exchange is sufficient to produce significant results.[35]

There are two main enzymes formulas available for use: collagenase and papain urea. Both of these formulas have been shown through in vitro and in vivo studies to liquefy necrotic tissue without harming the underlying healthy granulation tissue.[33] Collagenase is a water-soluble proteinase derived from the *Clostridium histolyticum* bacteria, which functions through targeted hydrolysis of peptide bonds in a triple helix formation. This enzyme is unique in that it will only target triple helix bonds, and will therefore not target keratin, fat, or fibrin. Collagenase is widely used because of its unique ability to break down collagen.[33] Papain-Urea is a proteolytic enzyme derived from the *Carica papaya* plant, which functions through targeted disruption of sulfhydryl groups in tissue. Urea is an activator for the papain, which was shown in studies to double the digestive effects of papain.[33] Papain-urea is unable to break down collagen and can produce significant stinging upon application which may decrease patient adherence.[37]

Biologic debridement

Biologic debridement, also known as maggot debridement therapy (MDT), produces results through placement of larvae directly into the wound bed and allowing digestion of necrotic material. MDT utilizes sterile medical grade lab grown fly larvae (from the *Lucilia sericata* species of fly), which are then placed in a net gauze and put into the wound bed and exchanged every two-three days.[38] MDT allows for direct ingestion of necrotic tissue, but also prevents overgrowth of bacteria by increasing the pH of the wound bed as the maggots release ammonia. Clinical trials have demonstrated MDT to be effective, safe, and expeditious.[39] MDT is ideal for patients with significant barriers to surgical intervention or for patients with poor perfusion and limited options for revascularization and wound salvage.[38] The SVS-AVF Venous Ulcer Guidelines recommend biologic debridement as a safe alternative to surgical debridement (Guideline 4.8, Level 2B evidence, Table 13.1).[9]

Although this technique for debridement has been recognized for centuries, MDT has recently undergone a resurgence. Modern medicine has increased ease of access, simplified application, and decreased costs of maggot therapy through increasing number of labs that produce maggots, utilizing overnight delivery services, and packaging the maggots in a convenient net gauze.[40] "Social disgust" was once considered a barrier to maggot therapy, but recent studies have demonstrated that the stigma associated with maggot therapy is diminishing in both patients and providers as education on this effective wound therapy continues to improve.[41]

Cleansers and irrigation

Wound cleansing should be a consistent component of wound care throughout the process of healing. Although often overlooked compared to debridement, cleansing has significant benefits when implemented correctly and can cause harm when used improperly. While some cleansing techniques may provide a small amount of debridement, cleansing is meant to follow debridement and is performed serially with each dressing change. Cleansers aim to remove contaminants, bacteria, debris, exudate, and dressing remnants that can accumulate on the wound bed throughout the healing process.[42] The exact chemical composition of each cleanser differs, but many contain surfactants to aid in biofilm removal, antimicrobials for bacterial control, and moisturizing solutions to protect healthy tissue. Cleansing promotes an optimal environment for wound healing while also optimizes wound visualization which facilitates accurate wound assessment. The SVS-AVF Venous Ulcer Guidelines (Guideline 4.1, Level 2C evidence, Table 13.1) recommend cleansing during initial wound evaluation and at each subsequent dressing change unless there is a contraindication present.[9]

The effectiveness of any cleanser depends on the chemical composition of the solution and the technique with which the cleanser is utilized. Cleansing can be accomplished through manual application with gauze pads, irritation, or a combination of both[43]: both of these methods must be used carefully as excessive vigorous cleansing can damage healthy soft tissue and can inadvertently push contaminants (such as bacteria) deeper into the wound bed.[44] Irrigation is the generally preferred method, and when performed correctly has been shown to decrease infection rates through removal of wound bed bacterial load and removal of debris that can fester.[43] Effective cleansers should be nontoxic to healthy tissue, chemically stable, effective at microorganism removal, cost effective, and widely available.[45]

Pain management during wound bed preparation

Pain management is an often overlooked component of wound bed preparation; however, inadequately treated pain negatively affects wound healing and patient quality of life.[46] For years it was incorrectly assumed that venous ulcers were nonpainful due to their superficial nature[47]; however, it has been clearly demonstrated that venous ulcers procedure chronic pain that can be significantly increased during wound bed debridement and cleansing procedures.[48] Venous ulcer pain occurs through nociceptive pain, appropriate physiologic response to tissue damage, and neuropathic pain with sensitized nerves and exaggerated stimuli responses. Pain is often worsened by overlying infection, reactions to topical agents, and peri-wound skin maceration.[49] Chronic pain is present in up to 80% of patients with venous ulcers,[50] and 65% of patients rated their baseline pain as at least moderate and 40% of patients reported that pain during dressing changes and cleansing was intolerable[51] and lasted for more than 1 hour following debridement.[52] Not only is chronic pain prevalent, but its presence was associated with high levels of depression and increased rates of disability.[48] As such, providers must be comfortable and efficient with assessing and managing pain associated with venous wounds.

Effective pain assessment should occur prior to any procedures and evaluate baseline pain and potential for procedural pain. Factors related to location, duration, intensity, aggravating activities, and barriers to functioning should be identified.[53] Chronic baseline pain is often managed by a primary care provider or a long-term pain clinic; however, specialized wound care clinics are now incorporating pain management specialists that are able to implement advanced, cognitive, and alternative methods for pain management.[54] Acute pain associated with debridement and cleansing should be managed by the providers performing the procedures through a combination of local and general anesthesia techniques.

The SVS-AVF Venous Ulcer Guidelines (Guideline 4.3, Level 1B evidence, Table 13.1) recommend local anesthetic to be utilized during all procedures, with general anesthetic used as needed for extensive debridement.[9] These guidelines are based on research that demonstrates pain during procedures to be the leading cause of inadequate debridement which can lead to increased rates of infection and decreased rates of healing.[49] Trials have demonstrated topical lidocaine/prilocaine cream to be particularly effective when applied 30–45 min prior to procedure and when used to treat wounds less than 50 cm^2 in total area.[49,55] Topical rather than injected lidocaine is often preferred because it is not ideal to inject lidocaine in already agitated and possibly macerated periwound skin.[56] General anesthesia may be indicated for wounds that are larger, highly infected, complex or cavitating, or in patients with anxiety surrounding procedures. Blocks are an excellent alternative in patients who are not candidates for general anesthesia due to medical comorbidities but who require high pain management than can be provided by topical solutions.

Wound dressings

Wound dressings play a fundamental role in appropriate care of venous ulcers. When used properly and tailored to the specific wound, dressings foster the correct microenvironment that promotes healing. Providers must understand that the needs of each wound are unique, and no single dressing type can be universally employed for all wounds. Selection of an appropriate dressing is a dynamic process in which the wound must be continually evaluated and the dressings adjusted to meet the needs of the wound as it progresses. This section will introduce the theory and review the indications for the three major dressing classifications: passive, interactive, and active (Fig. 13.1). Supplemental therapies to dressings, including compression therapy and active antimicrobial management, are covered separately in Chapters 12 and 15.

The role of a wound dressing has changed significantly over the last several decades, and the role of passive dressings has greatly diminished while interactive (semiocclusive/occlusive) and active dressings have flourished. Historically, dressings were not thought to

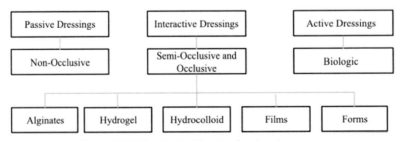

Figure 13.1 Broad classification for dressing types.

play a strong active physiologic role in tissue healing, and were mostly used as a physical barrier against outside elements and to capture exudate.[57] As such, early passive dressings (nonocclusive) were designed to promote a dry environment and not retain any moisture; however, this practice was abandoned after several groundbreaking studies demonstrated significantly increased rates or epithelization in moist wound environments.[57–59] Improved scientific understanding of the pathophysiology allowed the development of a wide variety of dressings, from gels and films to active biologics, that each target a specific wound need.

There are several features unique to venous ulcers that must be taken into account when selecting dressing types. Special consideration must be given to the following areas when selecting a dressing for a venous ulcer: patient experience, amount of wound drainage/exudate, shape and accessibility of wound, health of peri-wound skin, and need to add compression over topical dressing. The patient's ability to adhere to treatment will be heavily influenced by the complexity, frequency, duration, and discomfort associated with dressing changes. Every effort should be made to use the simplest format that will be effective.[58] In addition to patient experience, degree of exudate will heavily impact choice of dressing. Dressings with high absorptive capabilities, such interactive dressings (occlusive and semiocclusive), are often selected early on during venous ulcer healing when the wound is likely to produce the most exudate. Finally, in order to be economically sustainable, an ideal dressing must be widely available, cost effective, and have a reasonably long shelf life.[58] It must be kept in mind, that all venous wound dressings should be used in conjunction with compression therapy, which will be covered in Chapter 12. Within this chapter, each section will address the ways in which specific dressing types do or do not meet the above criteria.

Interactive dressings: occlusive and semi-occlusive

Interactive dressings, semiocclusive and occlusive categories, promote healing through the maintenance of moist and warm environments while successfully managing exudate. The mechanism by which interactive dressings promote healing is twofold: these dressings retain the body's active enzymes required for epithelization while also physically improving the migration of new skin cells. Moist environments retain key proteinases that degrade exudate and promote phagocytosis, growth factors that promote healing, and fibroblasts that stimulate epithelial cell propagation.[60] Additionally, it is hypothesized that hydrated environments actually facilitate easier cellular migration of epidermal cells across moist surfaces.[61] The many types of interactive dressings, semiocclusive and occlusive, vary in how much moisture is retained, and selection can be tailored to account for the level of exudate. Table 13.4 demonstrates the relationship between amount of exudate and choice of interactive dressing type. Attaining appropriate moisture balance through appropriately absorptive dressings is crucial to healing, as excessive moisture

Table 13.4 Dressing choices for degree of ulcer exudate.

Amount of exudate	Dressing types
Minimal	Film, hydrogel
Mild	Hydrocolloid
Moderate	Alginate
Significant	Hydrocellular foam, NPWT

NPWT, negative pressure wound therapy.

will require more frequent dressing changes and can potentially harm the wound and surrounding skin.[58] The indications and pathophysiology of interactive dressings are discussed below.

Wet-to-dry dressings

Wet-to-dry dressings are considered semiocclusive when wet and transition to a debridement modality when dry. This method can promote a moist environment if kept wet with saline; however, the transition to dry actually promotes vapor and heat loss through evaporative cooling. This method is often not well tolerated in venous ulcers that produce a significant amount of exudate and therefore do not allow drying. For this reason, although prevalent across a very wide range of medical specialties, within venous ulcer literature, the wet-to-dry method is considered a debridement modality rather than a true dressing and is not recommended as a primary dressing for venous ulcers.

Hydrogel dressings

Hydrogel dressings are semi-permeable, nonadhesive, films built from cross-linked hydrophilic polymers.[58] The majority of hydrogels are synthetic material, but a select few have been discovered directly in nature.[62] Hydrogels come as either a sheet material, impregnated onto a sponge or gauze, or amorphous free gel. Originally studied by medical engineers in 1894,[63] hydrogels were primarily used in soft contact lenses until tissue engineers discovered several properties of hydrogels that make them ideal for wound dressings.[62] Hydrogels are capable of absorbing and retaining a moderate amount of water given their partially hydrated matrix (80%) of insoluble polymers. This partial hydrated state also enables significant gas exchange across hydrogels, which prevents oversaturation and generates a breathable environment.[58] In addition to their chemical advantages, hydrogels have very low self-adhesion and a soothing cooling effect, making them ideal for fragile wound such as venous ulcers.[62] Hydrogels should be exchanged daily.[64] Hydrogels have extremely stable chemical bonds and low reactivity to other compounds[65]; researchers are currently studying how to harness these properties to produce a system by which the hydrogel matrix is impregnated with bioactive molecules (stem cells, antibiotics) for direct and prolonged delivery onto wound beds.[66]

Hydrocolloid dressings

Hydrocolloids are semiocclusive, self-adhesive, inactive dressings that have quickly become one of the most prevalent venous ulcer dressings since their discovery in the 1960's.[67] Hydrocolloid dressings are comprised of two layers: an inner layer of a hydrophilic polymer matrix (gelatin, pectin, and carboxymethylcellulose) that absorbs excessive moisture and an outer impermeable polyurethane adhesive layer that serves as a physical barrier against bacteria and debris.[68] The exact capacity for exudate absorption varies by brand and thickness of the inner layer. In general, hydrocolloids are less absorptive compared to hydrogels and alginates and are ideal for venous ulcers once they are in a phase of healing with less exudate. Once the hydrocolloid inner layer becomes saturated, it becomes thick and gelatinous. These dressings are impermeable to bacteria and fluid, but semipermeable to gas which allows for gaseous exchange.[64] Hydrocolloid dressings are designed to be worn up to a week, although the appropriate duration will vary depending on the level of exudate. These less frequent dressings changes are often advantageous to the patient's schedule and to the wound bed; however, this does not mean that the wound should not be evaluated on a regular basis as infections can progress quickly if not detected early.[69] Hydrocolloids should never be used on an infected wound bed. Hydrocolloids have inherent gentle adhesion and therefore do not require external tape, and are relatively atraumatic to the surrounding healthy skin during dressing changes. Current hydrocolloid research is focusing on developing new biodegradable hydrocolloids through combination with nanomaterials.[64]

Alginates

Alginates are naturally occurring anionic biopolymers derived from brown seaweed (*Phaeophyceae*).[58] Alginates are rich in calcium, sodium salts, mannuronic, and guluronic acid which all contribute to the semipermeability, immunologic, and naturally hemostatic properties that make alginates so functional in wound dressings.[58] In addition to their physiologic properties, alginates are highly biocompatible[70] and can be cross-linked with varying polymers to form wafers, gels, foams, and films to place directly on wounds.[64] The exact properties of alginate dressings can be altered based on what polymers are added; however, in general, all alginate dressings are highly absorptive (holding 15—20 times their weight in fluid) due to their ability to exchange calcium and sodium ions and are indicated for wounds with moderate-to-heavy exudate.[64] Due to excellent moisture removal, alginates are designed to be changed daily and used in conjunction with a secondary dressing.[58] If not changed appropriately, a thick foul-smelling gelatinous substance can build up within the dressing and be mistaken for an infection. Alginate dressing removal is often accomplished with saline irrigation and is therefore nontraumatic to the wound bed and nonirritating to the patient.[64] Alginate dressings are inherently hemostatic, as calcium ion exchange with the extracellular matrix within the wound bed activates the coagulation cascade. Therefore, alginates are ideal for wounds that have undergone recent debridement and may have a small residual amount of bleeding and irritation within the wound bed.[70] In addition to inherent hemostatic properties, alginates are naturally immunologic. These antibacterial properties can be further enhanced through fusion with silver impregnated polymers. Researchers are currently experimenting with cross-linked alginate—chitosan dressings that may further enhance antimicrobial properties.[64]

hydrocellular foam dressings

Hydrocellular foam dressings are highly absorbent multilayer nontraumatic dressings. Foam dressings are composed of two layers. The inner layer is a spongelike matrix of highly absorbent hydrophilic polyurethane (or less commonly silicone foam) that serves to remove exudate while maintaining a moist wound bed.[71] The outer layer is a waterproof nonpermeable polyethylene glycol that not only serves a physical barrier against bacterial penetration but also enhances the durability of the dressing.[71] Exact capacity for exudate absorption will vary depending on the thickness of the inner layer, but in general, foam dressings are ideal for managing moderate exudate.[72] Foam dressings can be found in both nonadherent and self-adherent varieties and are designed to remain in place for two-three days at a time[58] Nonadherent foam dressings require a secondary dressing, which may not be ideal in patients with especially fragile skin. Nonadherent foam dressings can be particularly useful in deep or cavitating wounds due to their ability to conform to the shape of the wound and reduce "dead" space.[72]

semipermeable film dressings

Film dressings are made from semipermeable transparent self-adherent polyurethane. Films have excellent gas exchange, but have minimal absorptive capacity. If used on wound with significant exudate, the film can trap and pool fluid against the wound bed and healthy skin which will impair granulation tissue formation and lead to maceration of healthy skin. For this reason, films are ideal for wound with minimal exudate that is superficial.[57] Advantages of films include their transparency which allows for easy wound monitoring and their high adherence which allows for placement across joints.[58] Films are designed to be changed every two-three days, and require extra care during removal as they may be traumatic to surrounding skin.

Active (biologic) dressings

The active dressing category is comprised of all dressings that deliver biologically active particles that directly and actively modulate cellular responses. Active dressings encompass growth factors, stem cell therapy, human dermal equivalent therapy, and active antibiotic delivery. While active dressings have been on the market since 1990's, many potential techniques remain in the clinical trial phase of development due to the complexities of these active ingredients in both mechanism of action and delivery system requirements.[73] Despite this, their potential has been demonstrated in both animal and human trials, and they represent burgeoning category within wound healing research that is expected to generate up to $5 billion in sales within the next several years.[73] This section will focus on growth factor therapy, while antibiotic therapy, stem cell therapy, and human dermal equivalent will be covered separately in Chapter 19.

Growth factor therapy

Growth factors are a broad category of endogenous polypeptides responsible for generating and controlling cellular function on a molecular level. On a large scale, growth factors are responsible for cellular migration, proliferation, and differentiation. Specific to wound healing, growth factors trigger the physiologic processes for mitigating immune responses, forming granulation tissue, and producing angiogenesis. A single growth factor may be responsible for thousands of processes, and the exact cellular response depends on the cell type and microenvironment. Harnessing the full potential of growth factors is a delicate process because growth factors are extremely sensitive to, and dependent upon, both molecular configuration and temporal relationship to other cellular processes.[74]

There are several large families of growth factors. The most commonly used factors within wound healing are discussed below. Platelet-derived growth factors, including vascular endothelial growth factor, on a board level specific to wound healing, work to attract fibroblasts, activate macrophages, and stimulate endothelial cell migration and proliferation.[75] Epidermal Growth Factors, including Transforming Growth Factor-*a*, stimulate keratinocytes and fibroblasts.[75] Fibroblast Growth Factors play a significant role in activating endothelial cells and promoting extracellular matrix remodeling.[74] Each of these has demonstrated significant benefits in wound healing within both animal models and human clinical trials.[76]

Harnessing the full potential of growth factors is exceedingly dependent upon achieving the ideal delivery system. Growth factors have very low chemical stability which is further limited by short half-lives.[77] Therefore, delivery systems must be able to achieve rapid initial molecular release followed by a slow sustained release.[75] Currently, growth factors are prepared in topical gels, creams, and ointments through a variety of techniques including mixing and cross-linking, encapsulation (micro- and nanospheres), and ion exchange.[74,75] Despite significant research, growth factor delivery systems still require application at high doses and high frequency to overcome the low in vivo molecular stability, poor drug penetration pasty exudate, and restricted tissue absorption.[78] These high doses can potentially increase risk for side effects and costs of therapy.[78] Research is currently investigating bioactive gel platelet-rich plasma, cells directly harvested and isolated from the patient, as a means to overcome delivery and side effect barriers because these cells are specific to each patient. Results from these studies have been promising, but largely remain in the clinical trial stage.[77,79]

Peri-wound management

The importance of appropriate peri-wound skin management cannot be overstated. Preventative measures must be taken to avoid damage to healthy skin surrounding the wound, as damage to this tissue will increase patient's discomfort, risk for infection, and ultimately delay wound closure.[58] Studies have shown that up to 25% of patients with venous ulcers experience some peri-wound skin damage from either excessive moisture, moisture-associated skin damage, contact dermatitis secondary to adhesive tape, or adhesive trauma (shearing and friction forces).[80] Patients at highest risk for peri-wound skin injury are those with preexisting neuropathy, underlying peripheral arterial disease, and elderly patients with fragile skin.[58] Patients with peri-wound skin damage experience pain, pruritus, and potential new wounds due to breakdown of the *stratum corneum*.[81] Loss of the *stratum corneum* leaves skin vulnerable to further damage, precipitating a cycle of skin loss and wound progression, and impairs keratinocyte migration from the wound edge to the wound bed which leads to overall delayed wound healing.[82] Factors that can exacerbate damage to peri-wound skin include abnormal pH due to wound bed exudate, overlying infection, contact dermatitis secondary to medication application in wound bed, and mechanical stress.[83] Prevention is the basic tenement to management of peri-wound skin.

The first basic principle is to maintain clean peri-wound skin. Skin should be cleansed regularly, with a cleanser that maintains a slightly acidic environment (pH 5—5.5), and kept free of gross contaminants. Following appropriate cleansing, an appropriate moisture balance must be maintained through matching the wound dressing to the level of exudate. Excessive moisture retention will lead to maceration, but similarly, excessively dry environments will lead to skin cracking and fissures which can allow irritants to penetrate the skin and increase risk for skin damage. In the setting of a dressing that is achieving a moist environment without allowing for retention of excess fluid, the next step to protective skin management is the use of atraumatic adhesives. Atraumatic adhesives are especially significant in elderly patients with fragile skin. In this particular population, wrapping the bandage rather than using adhesives may be appropriate, but special care must be taken to not make the wrap excessively tight. In addition to atraumatic adhesion, a protective skin barrier should be employed.[82] The SVS-AVF Venous Ulcer Guidelines (Guideline 4.16, Level 2C evidence) recommend the use of protective barriers in all venous ulcer wound dressings.[9] A comprehensive overview of skin barriers is shown in Table 13.5.[82] Research has not demonstrated superiority of one barrier over another, but use of any barrier has shown significant benefits compared to lack of any barrier.

Adjuvant therapy for venous ulcer wound healing

There are several adjuvant therapies available to aid in the healing of a recalcitrant chronic venous ulcer. The SVS-AVF Venous Ulcer Guidelines (Guideline 4.18, Level 1B evidence) recommend introduction of adjuvant therapies into the treatment regimen for

Table 13.5 Overview of skin protection products.[80]

Product category	Mechanism	Advantages	Disadvantages
Paraffin wax ointment	Physical and chemical barrier. Semipermeable, reduces transepidermal water loss. Paraffin wax prevents shearing damage	Maintains optimal moisture balance. Adheres well to slightly damp skin. Easy to apply and remove without trauma. Low cost, widely available.	May interfere with primary dressing adherence. Tacky texture may adhere to debris, dust, and bacteria. Melts at high temperatures.
Zinc oxide paraffin wax ointment	Zinc oxide thickens paraffin wax: improves physical barrier. Antiinflammatory. Zinc oxide has natural antiinflammatory properties.	Less permeable than traditional paraffin wax, improved barrier against fluids such as urine and sweat.	Thick texture may be difficult to apply and remove. May interfere with primary dressing absorption.
Silicone ointments	Physical but not chemical barrier.	Easy application, no greasy residue. Very comfortable.	Silicone allergies arc relatively common. Silicone not approved for use in open wounds, care must be taken to avoid inadvertent application directly onto wound bed.
Film-forming polymers in solvents	Sprayed onto skin, forms a film once dried. Prevents friction and shearing damage.	Nonmessy. No excess bulk added to dressings. Docs not interfere with primary wound dressing adherence.	Flammable. Ineffective if applied too thinly. Removal requires organic solvent, may cause stinging or irritation.
Cyanoacrylate liquids	Liquid form is applied to skin, quickly dries, and directly polymerizes with the healthy skin forming a thin but impenetrable layer.	No solvent required for application, minimizes risk for irritation. Highly resistant to fluids. Allows gaseous exchange.	Resistant to removal—if attempting to remove early, may cause trauma. Allergies to cyanoacrylates are relatively common. Relatively expensive.

wounds that have failed to progress within four-six weeks of starting appropriate wound care, even in patients with uncorrected underlying venous insufficiency.[9] Studies show that wounds should heal in a well-organized predictable pattern with significant granulation being achieved within at least 2 weeks.[84] Failure to progress by at least 30% wound size reduction in 4 weeks is associated with a less than 20% chance of healing without escalating interventions.[85] Prior to starting adjuvant therapies, the etiology of the wound should be again confirmed along with assessment of patient's compliance with standard therapies. Further, the SVS-AVF Venous Ulcer Guidelines (Guideline 3.5, Level 1C evidence) recommend performing a wound biopsy in patients with failure to progress within 6 weeks and in all ulcers with atypical features, as malignancy may be an underlying barrier to healing.[9] The list of adjuvant therapies is extensive, and many common adjuvant therapies, such as negative pressure therapy and biologic/cellular therapy, will be discussed in separate dedicated chapters (see Chapters 14 and 19). The following section will cover split thickness skin grafting and tissue equivalents, ultrasound therapy, and electrical stimulation therapy.

Skin grafting

Skin grafting for treatment of venous ulcers is most likely to benefit more chronic wounds of larger sizes. Grafting options include autografting (split thickness skin grafting—harvesting skin), allografting (donor cells, lab grown skin), xenografting (skin harvested from animals), and human skin equivalents (bioengineered skin).[86] The most recent SVS-AVF Venous Ulcer Guidelines (Guideline 4.19, Level 2B evidence) recommend against skin grafting as a primary treatment for venous ulcers, but do recommend grafting for ulcers that have failed to respond to standard care for four-six weeks.[9]

Skin grafting requires a moist, but not exudative, wound bed with healthy underlying tissue free from significant inflammation and infection.[87] Given these requirements, researchers were initially pessimistic toward skin grafting over venous ulcers given their chronic inflammation, significant edema, and compromised microcirculation.[87] Despite this, research has demonstrated significant success with all forms of grafting compared to standard wound care.[88]; however, bioengineered skin (human skin equivalent) appears to confer the most significant benefit. No clinical trials have demonstrated one form of grafting to be superior to another form of grafting.[89]

Recent studies suggest that the most benefit from split thickness skin grafting is seen in patients with massive venous ulcers (area >100 cm^2) given that ulcers of this magnitude demonstrate significantly slower healing and significantly lower rates of total epithelialization.[57,90] In massive venous ulcers, research has shown split thickness skin grafting to be economically equivalent to traditional wound care, despite the added expenses of hospitalization and anesthesia for skin harvesting, making this a reasonable option in patients with otherwise very prolonged treatment courses.[90]

The benefits of skin grafting appear to be enhanced when employed in conjunction with negative pressure wound therapy. Negative pressure therapy is discussed separately in detail in Chapter 14. The benefits that negative pressure therapy confer are reduced edema and decreased bacterial load, both of which have been demonstrated to be key in the success of graft success. Studies have demonstrated improved overall rates of success and decreased duration of treatment when skin grafting is followed by a course of negative pressure therapy.[91]

Despite the significant benefits, skin grafting is not ideal for use in all patients. Absolute contraindications include active malignancy, overlying infection, and coagulopathy. Relative contraindications include active smoking, anticoagulation medications, chronic corticosteroids, and significant malnutrition.[92] Skin grafts also perform poorly when placed over joints or on weight-bearing surfaces.[93]

Ultrasound therapy

Ultrasound therapy for the treatment of venous ulcers is a relatively new technology. Ultrasound for wound management was first explored in the early 2000's in diabetic foot ulcers, with recent expansion into venous ulcers.[94] At this time, the evidence for ultrasound therapy is still being explored and the SVS-AVF Venous Ulcer Guidelines (Guideline 4.26, Level 2B evidence) do not yet recommend the use of ultrasound therapy for venous ulcers.

In animal models, low frequency (20–60 Hz) ultrasound therapy increases macrophage activity, increases leukocyte adhesion, promotes collagen production, and stimulates angiogenesis through processes called cavitation, microstreaming, and frequency resonance.[95] Microstreaming is the process by which the physical movement of the ultrasound probe across the wound bed leads to a therapeutic movement of fluid from out of the wound.[96] Ultrasound waves produce micron-sized bubbles within the wound fluid that vibrate vigorously and transfer mechanical energy to proteins within the wound bed that alters their conformation and leads to physiologic changes. Cavitation and resonance result from molecular vibrations in resonance with the ultrasound waves, leading to temporary conformational changes in key proteins that can result in increased cellular activity.[97,98]

Despite the observed benefits in in vitro studies, the evidence to support ultrasound therapy in venous ulcers is still being explored and the benefits have not yet been clearly established. One reason for the lack of conclusive evidence is the lack of standardization within experimental protocols. Frequency of application ranges from one to three times per week with duration ranges from four to 12 weeks of therapy and frequency of ultrasound therapy varies widely by institution.[95,96,99,100] Despite the current research limitations, ultrasound therapy is an exciting area of research and may become an integral part of wound care in the future.

Electrical stimulation

Electrical stimulation therapy is an adjunctive procedure in which electric current is directly applied across the venous wound bed. Electric therapy within wound beds has been studied for many decades, with studies going back hundreds' of years in which gold flakes were placed directly into wounds and then electrically stimulated.[101] While the earliest studies of electric therapy within venous ulcers in the 1960's were not promising,[102] recent work in vivo and in vitro is encouraging. As an evolving technology, there is not yet enough evidence for the SVS-AVF Venous Ulcer Guidelines (Guideline 4.25, Level 2C evidence) to recommend electrical therapy for the treatment of venous ulcers, but these guidelines are continually updated and the next iteration is expected to continue to evaluate the newest clinical evidence.

Every human cell can be considered to be an electrical unit that generates an electric field through electrochemical gradient generated and maintained by ion exchange across the cellular membrane. The skin functions as one large electrical field that becomes disrupted with the presence of a wound.[103] Early in vitro studies demonstrated that cultured epithelial cells, under the influence of direct current (DC), travel along the lines of the electric field, and migration of these cells can be directly influenced through polarization of the field.[104] Since then, studies in vitro have demonstrated electrical therapy to significantly increase the proliferation of cultured epithelial cells and their cellular signaling activity which resulted in significantly faster epithelialization and increased recruitment of macrophages and fibroblasts.[105] Additionally, in vitro studies demonstrate bacterial inhibition seen with varying levels of electric stimulation[106] and improved cutaneous microcirculation seen in healthy patients.[107]

When researched within in vivo venous ulcers, the results have been mixed. Studies demonstrate significant benefit through decreasing pain in the wound bed and surrounding skin, which did significantly improve quality of life.[108] Double-blind clinical trials have shown electrical stimulation provides significant benefits in venous ulcers in patients who have not undergone surgical correction of their venous disease,[109] but have not demonstrated these benefits in patients who have surgical venous intervention.[110] No standard format for delivery of electric current has been established, likely contributing to the lack of conclusive results, with studies often including DC, alternating current, and high-voltage pulsed current at varying voltages.[103,111] Studies have identified several contraindications to electrical stimulation therapy, which include skin malignancy, wound infection, ion residues in the wound (previous use of topical iodine or silver dressings), and cardiac pacing devices.[111]

Optimizing care delivery

Wound care clinics

Wound care clinics are becoming standard of care for the treatment of venous ulcers. Forces driving this evolution in healthcare delivery are social, clinical, and financial. Since the 1980's, comprehensive wound care clinics have opened at an exponential rate in order to meet the demand of a growing population with complex wound needs.[112] The models for building a comprehensive wound care team, clinical benefits, and financial considerations are discussed in this section.

There are three main models for wound care clinics. Private practice free standing centers, community hospital cents, and academic hospital systems. While private practice clinics have the benefit of being relatively easy to access with close proximity to patient residential areas, they are often limited in their ability to offer on-site ancillary services and are associated with the highest costs. Community hospital clinics benefit from improved access to ancillary services, such as operating rooms, but may not have the full diversity of providers. Academic centers benefit from a large number of specialist providers, access to ancillary services, and a centralized location for all healthcare needs, but can often be hampered by long wait-times due to large patient volumes.[113] Regardless of setting, comprehensive clinics should aim to centralize providers available for wound care. Common providers include vascular surgeons, podiatry, infectious disease, pain management specialists, endocrinology, nutrition, physical therapy, and social work.[114] The importance of social work to the success of a wound care clinic cannot be overstated. Studies have shown that the presence of social workers within a clinic increases patient's adherence to treatment plans through coordination of transportation to appointments, facilitating at home nursing visits, and working with insurance to limit direct costs to the patients.[113]

Wound care clinic have improved clinical outcomes compared to standard practices for delivery of care. Studies have demonstrated that wounds heal in shorter durations.[21] with higher rates of success when treated in a comprehensive wound center.[115] Additionally, patients treated at wound centers experience significantly lower rates of sepsis and osteomyelitis due to earlier intervention,[116] fewer hospitalizations,[117] and significantly lower rates of lower extremity amputations.[118] Due to improved outcomes, patient-reported quality of life metrics are significantly higher in patients treated within a wound care clinic.[113]

Despite the significant clinical benefits, many providers express reluctance to start or participate in wound care clinics due to financial barriers. Wound care clinics are associated with significant start-up and maintenance costs that do not always result is clinic profitability. Studies have shown that profits are often marginal; however, profits are growing as legislation is enacted to reward outcomes rather than individual services.[114] In general, wound care clinics associated with academic hospital centers demonstrate the largest profit margins due to their low start-up costs, maintenance fees that are built into the system's budget,[113] and ability to provide ancillary care, such as hyperbaric oxygen, that are well reimbursed through insurance.[119] Profitability for wound care clinics is likely to continue to improve as new alternative payment plans for Medicare/Medicaid are developed and reward the excellent clinical outcomes achieved by comprehensive clinics.[114]

Table 13.6 SIMPLE strategies for patient adherence.[122]

Strategy category	Examples
S: Simplify regimens	Reduce frequency, amount, and duration whenever possible. Match regimen to preexisting daily activities of the patient.
I: Impart knowledge	Ensure information is delivery at the literacy level of the patient. Encourage patient to voice any questions or concerns. Employ "repeat back" teaching methods
M: Modify beliefs regarding intervention	Evaluate for cultural or social barriers to treatment. Evaluate patient's opinion on barriers and benefits of treatment.
P: Patient communication	Active listening with in-person meetings. Provide information in many forms (mail, online, pamphlets, voicemails). Encourage family participation for additional understanding and open communication.
L: Leave behind the bias	Evaluate patient and provider bias.
E: Evaluation of adherence	Employ methods to accurately gauge adherence. Patient, self-reported. Journal recording activities (dressing changes). Measuring size of wound. Rate of supply utilization.

Adherence to therapy

Improving adherence to medical plans should be at the forefront of provider's concerns. Difficulty with adherence, previously called noncompliance,[120] to venous ulcer treatment plans, is understandable given the long and difficult recovery process. A recent metaanalysis demonstrates nonadherence ranges up to 40% with regards to appointments, dressing changes, and compression therapy.[121] Poor adherence has been linked to significantly worse clinical outcomes, which often perpetuate worse adherence due to feelings of hopelessness.[121] Difficulty with adherence cannot be attributed to a single factor, but studies have shown that socioeconomic status, marital status, culture, and social support may have a larger influence compared to disease or therapy-specific factors.[122,123]

The provider is responsible for initiating interventions to increase adherence to treatment. Although no clinical trials exist comparing methods within venous ulcer treatments, the wound care center at the Cleveland Clinic has described the "SIMPLE" method that can be applied to any patient and any treatment method.[124] This method focuses on **s**implifying routines, **i**mparting new knowledge, **m**odifying patient beliefs, **p**atient communication, **l**eaving the bias, and **e**valuating adherence. Examples within each category can be seen in Table 13.6.[124] It is fundamental to remember that the burden of increasing adherence is placed on the provider rather than the patient, and that removing judgment toward patients that struggle with adherence will significantly improve provider–patient relationships.

Treatment algorithm

The following wound treatment algorithm for management of patients with venous ulcers is shown in Fig. 13.2. Although the treatment plan for each individual patient will vary, this algorithm can provide a starting point to any clinical protocol used based on accepted standards of care. From these general categories, specific wound dressing following the details covered in this chapter can be selected based on the directed therapy needed for clinical needs of the wound. It is important when applying this algorithm to reassess patients following these strategies at each clinical evaluation of the venous ulcer wound.

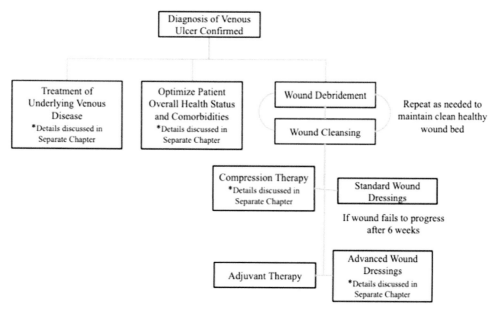

Figure 13.2 Treatment algorithm.

References

1. Robson MC, Hill DP, Woodske ME, Steed DL. Wound healing trajectories as predictors of effectiveness of therapeutic agents. *Arch Surg.* 2000;137(7):773–777. https://doi.org/10.1001/archsurg.135.7.773.
2. Robson MC, Steed DL, Franz MG. Wound healing: biologic features and approaches to maximize healing trajectories. *Curr Probl Surg.* 2001;38(2):72–140. https://doi.org/10.1067/msg.2001.111167.
3. Lazarus GS, Cooper DM, Knighton DR, et al. Definitions and guidelines for assessment of wounds and evaluation of healing. *Wound Repair Regen.* 1994;130(4):489–493. https://doi.org/10.1046/j.1524-475X.1994.20305.x.
4. Kisner RL, Shafritz R, Karl R, Stark RAW. Emerging treatment options for venous ulceration in today's wound care practice. *Ostomy/Wound Manag.* 2010;56(4):1–11.
5. Lamping DL, Schroter S, Kurz X, Kahn SR, Abenhaim L. Evaluation of outcomes in chronic venous disorders of the leg: development of a scientifically rigorous, patient-reported measure of symptoms and quality of life. *J Vasc Surg.* 2003;37(2):410–419. https://doi.org/10.1067/mva.2003.152.
6. Fernandes Abbade LP, Lastória S. Venous ulcer: epidemiology, physiopathology, diagnosis and treatment. *Int J Dermatol.* 2005;44(6):449–456. https://doi.org/10.1111/j.1365-4632.2004.02456.x.
7. Green J, Jester R, McKinley R, Pooler A. The impact of chronic venous leg ulcers: a systematic review. *J Wound Care.* 2014;23(12):601–612. https://doi.org/10.12968/jowc.2014.23.12.601.
8. Sen CK. Human wounds and its burden: updated 2020 compendium of estimates. *Adv Wound Care.* 2021;8(2):39–48. https://doi.org/10.1089/wound.2021.0026.
9. O'Donnel TF, Passman MA, Marston WIA, Ennis WJ, Michael Dalsing RLK. *Society of Vascular Surgery - Managment of Venous Leg Ulcers - Clinical Practice Guildlines of the Society for Vascular Surgery and the American Venous Forum.* 2014.
10. Marston WA, Crowner J, Kouri A, Kalbaugh CA. Incidence of venous leg ulcer healing and recurrence after treatment with endovenous laser ablation. *J Vasc Surg Venous Lymphat Disord.* 2017;5(4):525–532. https://doi.org/10.1016/j.jvsv.2017.02.007.
11. Anderson K, Hamm RL. Factors that impair wound healing. *J Am Coll Clin Wound Spec.* 2012;4(4):84–91. https://doi.org/10.1016/j.jccw.2014.03.001.
12. McDaniel JC, Browning KK. Smoking, chronic wound healing, and implications for evidence-based practice. *J Wound Ostomy Cont Nurs.* 2014;41(5):415–422. https://doi.org/10.1097/WON.0000000000000057.
13. Sheehan P, Jones P, Caselli A, John M, Giurini AV. Percent change in wound area of diabetic foot ulcers over a 4-week period is a robust predictor of complete healing in a 12-week prospective trial. *Diabetes Care.* 2003;26(6):1879–1882.
14. Meaume S, Ourabah Z, Cartier H, et al. Evaluation of a lipidocolloid wound dressing in the local management of leg ulcers. *J Wound Care.* 2005;14(7):329–334. https://doi.org/10.12968/jowc.2005.14.7.26798.
15. Gelfand JM, Hoffstad O, Margolis DJ. Surrogate endpoints for the treatment of venous leg ulcers. *J Invest Dermatol.* 2002;119(6):1420–1425. https://doi.org/10.1046/j.1523-1747.2002.19629.x.
16. Moore WS. In: Houston M, ed. *Vascular and Endovascular Surgery.* 8th ed. Elsevier Saunders; 2013.
17. Vasquez MA, Rabe E, McLafferty RB, et al. Revision of the venous clinical severity score: venous outcomes consensus statement: special communication of the American venous Forum ad hoc outcomes working group. *J Vasc Surg.* 2010;52(5):1387–1396. https://doi.org/10.1016/j.jvs.2010.06.161.
18. Vasquez MA, Wang J, Mahathanaruk M, Buczkowski G, Sprehe E, Dosluoglu HH. The utility of the venous clinical severity score in 682 limbs treated by radiofrequency saphenous vein ablation. *J Vasc Surg.* 2007;45(5):1008–1015. https://doi.org/10.1016/j.jvs.2006.12.061.
19. Schultz GS, Sibbald RG, Falanga V, et al. Wound bed preparation: a systematic approach to wound management. *Wound Repair Regen.* 2003;11(1):S1–S28. https://doi.org/10.1046/j.1524-475X.11.s2.1.x.
20. Schultz GS, Barillo DJ, Mozingo DW, et al. Wound bed preparation and a brief history of TIME. *Int Wound J.* 2004,1(1).19–32. https://doi.org/10.1111/j.1742-481x.2004.00008.x.

21. Warriner RA, Wilcox JR, Carter MJ, Stewart DG. More frequent visits to wound care clinics result in faster times to close diabetic foot and venous leg ulcers. *Adv Skin Wound Care*. 2012;25(11):494–501. https://doi.org/10.1097/01.ASW.0000422629.03053.06.
22. Goren I, Kämpfer H, Podda M, Josef Pfeilschifter SF. Leptin and wound inflammation in diabetic ob/ob mice:differential regula- tion of neutrophil and macrophage influx and a potential role for the scab as a sink for inflammatory cells and mediators. *Diabetes*. 2003;52(11):2821–2832.
23. Cardinal M, Eisenbud DE, Armstrong DG, et al. Serial surgical debridement: a retrospective study on clinical outcomes in chronic lower extremity wounds: original Research - clinical Science. *Wound Repair Regen*. 2009;17(3):306–311. https://doi.org/10.1111/j.1524-475X.2009.00485.x.
24. Knox KR, Datiashvili RO, Granick MS. Surgical wound bed preparation of chronic and acute wounds. *Clin Plast Surg*. 2007. https://doi.org/10.1016/j.cps.2007.07.006.
25. Eming SA, Krieg T, Davidson JM. Inflammation in wound repair: molecular and cellular mechanisms. *J Invest Dermatol*. 2007. https://doi.org/10.1038/sj.jid.5700701.
26. Frykberg RG, Banks J. Challenges in the treatment of chronic wounds. *Adv Wound Care*. 2015;4(9): 560–582. https://doi.org/10.1089/wound.2015.0635.
27. Caputo WJ, Beggs DJ, DeFede JL, Simm L, Dharma H. A prospective randomised controlled clinical trial comparing hydrosurgery debridement with conventional surgical debridement in lower extremity ulcers. *Int Wound J*. 2008;5(2):288–294. https://doi.org/10.1111/j.1742-481X.2007.00490.x.
28. Mosti G, Mattaliano V. The debridement of chronic leg ulcers by means of a new, fluidjet-based device. *Wounds*. 2006;18(8):227–237.
29. Ferrer-Sola M, Sureda-Vidal H, Altimiras-Roset J, et al. Hydrosurgery as a safe and efficient debridement method in a clinical wound unit. *J Wound Care*. 2017;26(10):593–600. https://doi.org/10.12968/jowc.2017.26.10.593.
30. Liu J, Ko JH, Secretov E, et al. Comparing the hydrosurgery system to conventional debridement techniques for the treatment of delayed healing wounds: a prospective, randomised clinical trial to investigate clinical efficacy and cost-effectiveness. *Int Wound J*. 2015;12(4):456–461. https://doi.org/10.1111/iwj.12137.
31. Granick MS, Posnett J, Jacoby M, Noruthun S, Ganchi PA, Datiashvili RO. Efficacy and cost-effectiveness of a high-powered parallel waterjet for wound debridement. *Wound Repair Regen*. 2006;14(4):394–397. https://doi.org/10.1111/j.1743-6109.2006.00136.x.
32. Gurunluoglu R. Experiences with waterjet hydrosurgery system in wound debridement. *World J Emerg Surg*. 2007;2(10):10–22. https://doi.org/10.1186/1749-7922-2-10.
33. Smith RG. Enzymatic debriding agents: an evaluation of the medical literature. *Ostomy Wound Manag*. 2008;54(8):16–34.
34. Ramundo J, Gray M. Enzymatic wound debridement. *J Wound Ostomy Cont Nurs*. 2008;35(3): 273–280. https://doi.org/10.1097/01.WON.0000319125.21854.78.
35. Patry J, Blanchette V. Enzymatic debridement with collagenase in wounds and ulcers: a systematic review and meta-analysis. *Int Wound J*. 2017;14(6):1055–1065. https://doi.org/10.1111/iwj.12760.
36. Avila-Rodríguez MI, Meléndez-Martínez D, Licona-Cassani C, Aguilar-Yañez JM, Benavides J, Sánchez ML. Practical context of enzymatic treatment for wound healing: a secreted protease approach (review). *Biomed Rep*. 2020;13(1):4–14. https://doi.org/10.3892/br.2020.1300.
37. S BK, Pawar PM, M S, S S, V JK, Kunju RD. A prospective study on effectiveness of use of papain urea based preparation in dressings compared with regular conventional dressings in diabetic foot ulcers. *Int Surg J*. 2017;4(6):23–27. https://doi.org/10.18203/2349-2902.isj20172396.
38. Armstrong DG, Salas P, Short B, et al. Maggot therapy in "lower-extremity hospice" wound care: fewer amputations and more antibiotic-free days. *J Am Podiatr Med Assoc*. 2005;95(3):254–257. https://doi.org/10.7547/0950254.
39. Zubir MZM, Holloway S, Noor NM. Maggot therapy in wound healing: a systematic review. *Int J Environ Res Publ Health*. 2020. https://doi.org/10.3390/ijerph17176103.
40. Sherman RA. Maggot therapy takes us back to the future of wound care: new and improved maggot therapy for the 21st century. *J Diabet Sci Technol*. 2009. https://doi.org/10.1177/193229680900300215.

41. King C. Changing attitudes toward maggot debridement therapy in wound treatment: a review and discussion. *J Wound Care*. 2020. https://doi.org/10.12968/jowc.2020.29.Sup2c.S28.
42. Pilcher M. Wound cleansing: a key player in the implementation of the TIME paradigm. *J Wound Care*. 2016;25(3):7–9. https://doi.org/10.12968/jowc.2016.25.Sup3.S7.
43. Atiyeh BS, Dibo SA, Hayek SN. Wound cleansing, topical antiseptics and wound healing. *Int Wound J*. 2009;6(6):420–430. https://doi.org/10.1111/j.1742-481X.2009.00639.x.
44. Edlich RF, Rodeheaver GT, Morgan RF, Berman DE, Thacker JG. Principles of emergency wound management. *Ann Emerg Med*. 1988;17(12):1284–1302. https://doi.org/10.1016/S0196-0644(88)80354-8.
45. Main RC. Should chlorhexidine gluconate be used in wound cleansing? *J Wound Care*. 2008;17(3):112–114. https://doi.org/10.12968/jowc.2008.17.3.28668.
46. Coulling S. Fundamentals of pain management in wound care. *Br J Nurs*. 2007;16(11):S4–S6. https://doi.org/10.12968/bjon.2007.16.sup2.23693.
47. World Union of Wound Healing Societies. *Principles of Best Practice: Minimising Pain at Wound Dressing-Related Procedures. A Consensus Document*. London MEP Ltd; 2004.
48. Bechert K, Abraham SE. Pain management and wound care. *J Am Col Certif Wound Spec*. 2009;1(2):65–71. https://doi.org/10.1016/j.jcws.2008.12.001.
49. Briggs M, Nelson EA, Martyn-St James M. Topical agents or dressings for pain in venous leg ulcers. *Cochrane Database Syst Rev*. 2012;2012(11):11–77. https://doi.org/10.1002/14651858.cd001177.pub3.
50. Leren L, Johansen E, Eide H, Falk RS, Juvet LK, Ljoså TM. Pain in persons with chronic venous leg ulcers: a systematic review and meta-analysis. *Int Wound J*. 2020;17(2):466–484. https://doi.org/10.1111/iwj.13296.
51. Hofman D, Ryan TJ, Arnold F, et al. Pain in venous leg ulcers. *J Wound Care*. 1997;6(5):222–224. https://doi.org/10.12968/jowc.1997.6.5.222.
52. Price PE, Fagervik-Morton H, Mudge EJ, et al. Dressing-related pain in patients with chronic wounds: an international patient perspective. *Int Wound J*. 2008;5(2):159–171. https://doi.org/10.1111/j.1742-481X.2008.00471.x.
53. Price P, Fogh K, Glynn C, Krasner DL, Osterbrink J, Sibbald RG. Managing painful chronic wounds: the wound pain management model. *Int Wound J*. 2007;4(1):4–15. https://doi.org/10.1111/j.1742-481X.2007.00311.x.
54. Caralin Schneider BA, Scott Stratman BS, Daniel Federman MDHL-TM. Dealing with pain: an approach to chronic pain management in wound patients. *Todays Wound Clin*. 2020;14(11):18–21.
55. Gottrup F, Jørgensen B, Karlsmark T, et al. *Reducing Wound Pain in Venous Leg Ulcers with Biatain Ibu: A Randomized, Controlled Double-Blind Clinical Investigation on the Performance and Safety*. Wound Repair Regen; 2008. https://doi.org/10.1111/j.1524-475X.2008.00412.x.
56. Gaufberg SV, Walta MJ, Workman TP. Expanding the use of topical anesthesia in wound management: sequential layered application of topical lidocaine with epinephrine. *Am J Emerg Med*. 2007. https://doi.org/10.1016/j.ajem.2006.11.013.
57. Gloviczki P, Dalsing MC, Lurie TWW F. In: Gloviczki ML, ed. *Handbook of Venous and Lymphatic Disorders*. 4th ed. CRC Press; 2006.
58. Jones V, Joseph E, Grey KGH. Wound dressings. *BMJ*. 2006;332(7544):777–780.
59. Winter GD. Formation of the scab and the rate of epithelization of superficial wounds in the skin of the young domestic pig. *Nature*. 1962. https://doi.org/10.1038/193293a0.
60. Svensjo T, Pomahac B, Yao F, Slama J, Eriksson E. Accelerated healing of full-thickness skin wounds in a wet environment. *Plast Reconstr Surg*. 2000;106(3):602–612. https://doi.org/10.1097/00006534-200009010-00012.
61. Junker JPE, Kamel RA, Caterson EJ, Eriksson E. Clinical impact upon wound healing and inflammation in moist, wet, and dry environments. *Adv Wound Care*. 2013;2(7):348–356. https://doi.org/10.1089/wound.2012.0412.
62. Francesko A, Petkova P, Tzanov T. Hydrogel dressings for advanced wound management. *Curr Med Chem*. 2019;25(41):246–250. https://doi.org/10.2174/0929867324666170920161246.
63. VB J. The hydrogel and the crystalline hydrate of copper oxide. *Anorg Chem*. 1984;5(1):466–483.

64. Stoica AE, Chircov C, Grumezescu AM. Nanomaterials for wound dressings: an Up-to-Date overview. *Molecules*. 2020;25(11):2699. https://doi.org/10.3390/molecules25112699.
65. Talebian S, Mehrali M, Taebnia N, et al. Self-healing hydrogels: the next paradigm shift in tissue engineering? *Adv Sci*. 2019;6(15):180. https://doi.org/10.1002/advs.201801664.
66. Opt Veld RC, Walboomers XF, Jansen JA, Wagener FADTG. Design considerations for hydrogel wound dressings: strategic and molecular advances. *Tissue Eng Part B*. 2020;26(3):230−248. https://doi.org/10.1089/ten.teb.2019.0281.
67. Thomas S. Hydrocolloid dressings in the management of acute wounds: a review of the literature. *Int Wound J*. 2008. https://doi.org/10.1111/j.1742-481X.2008.00541.x.
68. Dhivya S, Padma VV, Santhini E. Wound dressings - a review. *Biomedicine*. 2015;5(4):22. https://doi.org/10.7603/s40681-015-0022-9.
69. Barnes HR. Wound care: fact and fiction about hydrocolloid dressings. *J Gerontol Nurs*. 1993;19(6): 23−26. https://doi.org/10.3928/0098-9134-19930601-08.
70. Aderibigbe BA, Buyana B. Alginate in wound dressings. *Pharmaceutics*. 2018;10:42. https://doi.org/10.3390/pharmaceutics10020042.
71. Yamane T, Nakagami G, Yoshino S, et al. Hydrocellular foam dressings promote wound healing associated with decrease in inflammation in rat periwound skin and granulation tissue, compared with hydrocolloid dressings. *Biosci Biotechnol Biochem*. 2015;79(2):185−189. https://doi.org/10.1080/09168451.2014.968088.
72. Weller CD, Team V, Sussman G. First-line interactive wound dressing update: a comprehensive review of the evidence. *Front Pharmacol*. 2020;11(1):155. https://doi.org/10.3389/fphar.2020.00155.
73. Carver C. *The Future of Bioactive Wound Care Dressings*. 2016.
74. Park JW, Hwang SR, Yoon IS. Advanced growth factor delivery systems in wound management and skin regeneration. *Molecules*. 2017;22(8):1259. https://doi.org/10.3390/molecules22081259.
75. Mandla S, Davenport Huyer L, Radisic M. Review: multimodal bioactive material approaches for wound healing. *APL Bioeng*. 2018;2(2):215. https://doi.org/10.1063/1.5026773.
76. Grose R, Werner S. Wound-healing studies in transgenic and knockout mice. *Appl Biochem Biotechnol Part B Mol Biotechnol*. 2004;28(2):147−166. https://doi.org/10.1385/MB:28:2:147.
77. Laiva AL, O'Brien FJ, Keogh MB. Innovations in gene and growth factor delivery systems for diabetic wound healing. *J Tissue Eng Regen Med*. 2018;12(1):296−312. https://doi.org/10.1002/term.2443.
78. Losi P, Briganti E, Magera A, et al. Tissue response to poly(ether)urethane-polydimethylsiloxane-fibrin composite scaffolds for controlled delivery of pro-angiogenic growth factors. *Biomater*. 2010; 31(10):5336−5344. https://doi.org/10.1016/j.biomaterials.2010.03.033.
79. Long DW, Johnson NR, Jeffries EM, Hara H, Wang Y. Controlled delivery of platelet-derived proteins enhances porcine wound healing. *J Contr Release*. 2017;253(1):73−81. https://doi.org/10.1016/j.jconrel.2017.03.021.
80. Cutting KF, White RJ. Avoidance and management of peri-wound maceration of the skin. *Prof Nurse*. 2002;18(1):35−36.
81. Konya C, Sanada H, Sugama J, Okuwa M, Kamatani Y, Gojiro Nakagami KS. Skin injuries caused by medical adhesive tape in older people and associated actors. *J Clin Nurs*. 2010;19(9):1236−1242.
82. Woo KY, Beeckman D, Chakravarthy D. Management of moisture-associated skin damage: a scoping review. *Adv Ski Wound Care*. 2017;30(11):494−501. https://doi.org/10.1097/01.ASW.0000525627.54569.da.
83. Gray M, Black JM, Baharestani MM, et al. Moisture-associated skin damage: overview and pathophysiology. *J Wound Ostomy Cont Nurs*. 2011;38(3):233−241. https://doi.org/10.1097/WON.0b013e318215f798.
84. Demidova-Rice TN, Hamblin MR, Herman IM. Acute and impaired wound healing: pathophysiology and current methods for drug delivery, part 1: normal and chronic wounds: biology, causes, and approaches to care. *Adv Ski Wound Care*. 2012;25(7):304−314. https://doi.org/10.1097/01.ASW.0000416006.55218.d0.
85. Team V, Chandler PG, Weller CD. Adjuvant therapies in venous leg ulcer management: a scoping review. *Wound Repair Regen*. 2019;27(5):562−590. https://doi.org/10.1111/wrr.12724.

86. Yammine K, Assi C. A meta-analysis of the outcomes of split-thickness skin graft on diabetic leg and foot ulcers. *Int J Low Extrem Wounds*. 2019;18(1):23–30. https://doi.org/10.1177/1534734619832123.
87. Pascarella L, Schönbein GWS, Bergan JJ. Microcirculation and venous ulcers: a review. *Ann Vasc Surg*. 2005;19(6):921–927. https://doi.org/10.1007/s10016-005-7661-3.
88. Azzopardi EA, Boyce DE, Dickson WA, et al. Application of topical negative pressure (vacuum-assisted closure) to split-thickness skin grafts: a structured evidence-based review. *Ann Plast Surg*. 2013;70(1):23–29. https://doi.org/10.1097/SAP.0b013e31826eab9e.
89. Jones JE, Nelson EA, Al-Hity A. Skin grafting for venous leg ulcers. *Cochrane Database Syst Rev*. 2013;(1):2013. https://doi.org/10.1002/14651858.CD001737.pub4.
90. Yang CK, Alcantara S, Goss S, Lantis JC. Cost analysis of negative-pressure wound therapy with instillation for wound bed preparation preceding split-thickness skin grafts for massive (>100 cm^2) chronic venous leg ulcers. *J Vasc Surg*. 2015;61(4):995–999. https://doi.org/10.1016/j.jvs.2014.11.076.
91. Raad W, Lantis JC, Tyrie L, Gendics C, Todd G. Vacuum-assisted closure instill as a method of sterilizing massive venous stasis wounds prior to split thickness skin graft placement. *Int Wound J*. 2010;7(2):81–85. https://doi.org/10.1111/j.1742-481X.2010.00658.x.
92. Adams DC, Ramsey ML. Grafts in dermatologic surgery: review and update on full- and split-thickness skin grafts, free cartilage grafts, and composite grafts. *Dermatol Surg*. 2005;8(2):1055–1067. https://doi.org/10.1111/j.1524-4725.2005.31831.
93. Zilinsky I, Farber N, Weissman O, et al. Defying consensus: correct sizing of full-thickness skin grafts. *J Drugs Dermatol JDD*. 2012;11(4):520–523.
94. Ennis WJ, Foremann P, Mozen N, Massey J, Conner-Kerr T, Meneses P. Ultrasound therapy for recalcitrant diabetic foot ulcers: results of a randomized, double-blind, controlled, multicenter study. *Ostomy Wound Manage*. 2005.
95. Ruby Chang YJ, Perry J, Cross K. Low-frequency ultrasound debridement in chronic wound healing: a systematic review of current evidence. *Plast Surg*. 2017;25(1):21–26. https://doi.org/10.1177/2292550317693813.
96. Ennis WJ, Valdes W, Gainer M, Meneses P. Evaluation of clinical effectiveness of MIST ultrasound therapy for the healing of chronic wounds. *Adv Skin Wound Care*. 2006. https://doi.org/10.1097/00129334-200610000-00011.
97. Webster DF, Pond JB, Dyson M, Harvey W. The role of cavitation in the in vitro stimulation of protein synthesis in human fibroblasts by ultrasound. *Ultrasound Med Biol*. 1978;4(4):343–351. https://doi.org/10.1016/0301-5629(78)90023-6.
98. Johns LD. Nonthermal effects of therapeutic ultrasound: the frequency resonance hypothesis. *J Athl Train*. 2002;37(3):293–299.
99. Voigt J, Wendelken M, Driver V, Alvarez OM. Low-frequency ultrasound (20-40 kHz) as an adjunctive therapy for chronic wound healing: a systematic review of the literature and meta-analysis of eight randomized controlled trials. *Int J Low Extrem Wounds*. 2011;10(4):190–199. https://doi.org/10.1177/1534734611424648.
100. Watson JM, Kang'ombe AR, Soares MO, et al. VenUS III: a randomised controlled trial of therapeutic ultrasound in the management of venous leg ulcers. *Health Technol Assess*. 2011;15(13):1–192. https://doi.org/10.3310/hta15130.
101. Katelaris PM, Fletcher JP, Ltttle JM, Mcentyre RJ, Jeffcoate KW. Electrical stimulation in the treatment of chronic venous ulceration. *Aust N Z J Surg*. 1987;57:605–607. https://doi.org/10.1111/j.1445-2197.1987.tb01434.x.
102. Assimacopoulos D. Low intensity negative electric current in the treatment of ulcers of the leg due to chronic venous insufficiency. Preliminary report of three gases. *Am J Surg*. 1968;115:683–687. https://doi.org/10.1016/0002-9610(68)90101-3.
103. Thakral G, LaFontaine J, Najafi B, Talal TK, Kim P, Lavery LA. Electrical stimulation to accelerate wound healing. *Diabet Foot Ankle*. 2013;4(10):81. https://doi.org/10.3402/dfa.v4i0.22081.
104. Soong HK, Parkinson WC, Bafna S, Sulik GL, Huang SCM. Movements of cultured corneal epithelial cells and stromal fibroblasts in electric fields. *Investig Ophthalmol Vis Sci*. 1990;31(11):2278–2282.

105. Gilbert TL, Griffin N, Moffett J, Ritz MC, George FR. The provant wound closure system induces activation of p44/42 MAP kinase in normal cultured human fibroblasts. *Ann N Y Acad Sci*. 2002;961: 168−171. https://doi.org/10.1111/j.1749-6632.2002.tb03076.x.
106. Asadi MR, Torkaman G. Bacterial inhibition by electrical stimulation. *Adv Wound Care*. 2014;3(2): 91−97. https://doi.org/10.1089/wound.2012.0410.
107. Cramp AFL, Gilsenan C, Lowe AS, Walsh DM. The effect of high- and low-frequency transcutaneous electrical nerve stimulation upon cutaneous blood flow and skin temperature in healthy subjects. *Clin Physiol*. 2000;20(2):150−157. https://doi.org/10.1046/j.1365-2281.2000.00240.x.
108. Jünger M, Arnold A, Zuder D, Stahl HW, Heising S. Local therapy and treatment costs of chronic, venous leg ulcers with electrical stimulation (dermapulse): a prospective, placebo controlled, double blind trial. *Wound Repair Regen*. 2008. https://doi.org/10.1111/j.1524-475X.2008.00393.x.
109. Houghton PE, Kincaid CB, Lovell M, et al. Effect of electrical stimulation on chronic leg ulcer size and appearance. *Phys Ther*. 2003;83(1):17−28. https://doi.org/10.1093/ptj/83.1.17.
110. Franek A, Taradaj J, Polak A, Cierpka L, Blaszczak E. Efficacy of high voltage stimulation for healing of venous leg ulcers in surgically and conservatively treated patients. *Phlebolo*. 2006;35(3):127−133. https://doi.org/10.1055/s-0037-1622138.
111. Houghton PE. Electrical stimulation therapy to promote healing of chronic wounds: a review of reviews. *Chronic Wound Care Manag Res*. 2017. https://doi.org/10.2147/cwcmr.s101323.
112. Sen CK. Human wounds and its burden: an updated compendium of estimates. *Adv Wound Care*. 2019. https://doi.org/10.1089/wound.2019.0946.
113. Kim PJ, Evans KK, Steinberg JS, Pollard ME, Attinger CE. Critical elements to building an effective wound care center. *J Vasc Surg*. 2013;57(6):1703−1709. https://doi.org/10.1016/j.jvs.2012.11.112.
114. Hicks CW, Canner JK, Karagozlu H, et al. Quantifying the costs and profitability of care for diabetic foot ulcers treated in a multidisciplinary setting. *J Vasc Surg*. 2019;70(1):233−240. https://doi.org/10.1016/j.jvs.2018.10.097.
115. Sholar AD, Wong LK, Culpepper JW, Sargent LA. The specialized wound care center: a 7-year experience at a tertiary care hospital. *Ann Plast Surg*. 2007;58(3):279−284. https://doi.org/10.1097/01.sap.0000248116.28131.94.
116. Steed DL, Edington H, Moosa HH, Webster MW. Organization and development of a university multidisciplinary wound care clinic. *Surg*. 1993;114(4):775−778. https://doi.org/10.5555/uri:pii:003960609390267H.
117. Levinson AW, Lavery HJ, Santos AP, Parekh N, Ciminello FS, Marriott RJ. Effect of weekly specialized surgeon-led bedside wound care teams on pressure ulcer time-to-heal outcomes: results from a national dataset of long-term care facilities. *Wounds*. 2019;31(10):257−261.
118. Flores AM, Mell MW, Dalman RL, Chandra V. Benefit of multidisciplinary wound care center on the volume and outcomes of a vascular surgery practice. *J Vasc Surg*. 2019;70(5):1612−1619. https://doi.org/10.1016/j.jvs.2019.01.087.
119. Attinger CE, Hoang H, Steinberg J, et al. How to make a hospital-based wound center financially viable: the Georgetown University Hospital model. *Gynecol Oncol*. 2008;111(2):92−97. https://doi.org/10.1016/j.ygyno.2008.07.044.
120. Nieuwlaat R, Wilczynski N, Navarro T, et al. Interventions for enhancing medication adherence. *Cochrane Database Syst Rev*. 2014. https://doi.org/10.1002/14651858.CD000011.pub4.
121. Moffatt C, Kommala D, Dourdin N, Choe Y. Venous leg ulcers: patient concordance with compression therapy and its impact on healing and prevention of recurrence. *Int Wound J*. 2009. https://doi.org/10.1111/j.1742-481X.2009.00634.x.
122. Gast A, Mathes T. Medication adherence influencing factors - an (updated) overview of systematic reviews. *Syst Rev*. 2019;112(8). https://doi.org/10.1186/s13643-019-1014-8.
123. Finlayson K, Edwards H, Courtney M. The impact of psychoso factors cial on adherence to compression therapy to prevent recurrence of venous leg ulcers. *J Clin Nurs*. 2010. https://doi.org/10.1111/j.1365-2702.2009.03151.x.
124. Atreja A, Bellam N, Levy SR. Strategies to enhance patient adherence: making it simple. *MedGenMed Medscape Gen Med*. 2005;7(1):4−15.

CHAPTER 14

Negative pressure wound therapy for venous leg ulcers

Fedor Lurie[1,2] and Richard Simman[1,3,4]

[1]Jobst Vascular Institute of Promedica, Toledo, OH, United States; [2]Division of Vascular Surgery, University of Michigan, Ann Arbor, MI, United States; [3]Division of Plastic Surgery, Promedica Physician Group, Toledo, OH, United States; [4]University of Toledo, Department of Surgery, Toledo, OH, United States

The general principles of managing patients with venous leg ulcers (VLU) have remained fundamentally the same over the last several decades. They include compression therapy, wound care, and addressing treatable components of underlying venous pathology. Technological advances have significantly broadened the spectrum of treatable pathology. Superficial and perforator vein reflux can be treated with a variety of endovascular modalities that allow ablation of any refluxing vein.[1] Venoplasty and stenting greatly extend the ability to resolve deep vein obstruction,[2] and open surgical and evolving endovascular options can correct deep vein reflux.[3,4] However, as these interventional options have been implemented in general practice, it has become clear that correcting venous hemodynamic abnormalities can only partially address the issue of VLU burden. In community practice, healing rates were reported as low as 22% at 12 weeks and 69% in one year.[5,6] Even in randomized clinical trials (RCT), when the best clinical care was provided to enrolled patients, the healing rates of VLUs were far from desirable. The EVRA trial reported a healing rate of 63.5% at 12 weeks in the early intervention group and only 51% in the deferred-intervention group.[7] Correcting venous hemodynamic abnormalities is also insufficient for the prevention of VLU recurrence. More than 30% of VLUs recur during a year after healing.[5] In the ESCHAR trial, patients who were randomized to surgery and compression therapy had a 12% recurrence rate at 12 months, in addition to the 10.3% of patients who did not heal their ulcers.[8]

Broadly defined wound care was a component of all mentioned RCTs and of the community practices. However, the uniqueness of VLUs compared to other chronic wounds was rarely addressed. Patients with chronic venous disease (CVD) develop VLUs as the end stage of their condition. In the case of primary CVD, it takes 2–3 decades from initial clinical manifestations of CVD to the occurrence of VLUs. Even in the much more rapidly occurring secondary CVD after acute deep vein thrombosis (DVT), the majority of patients develop VLUs several years from the time of the DVT. It has been shown that CVD affects the basic structure and function of the skin long before ulceration occurs and may affect areas other than the lower extremities, such as the neck and thorax.[9,10] VLUs, therefore, occur in the areas of chronically diseased skin. A significant part of the skin pathology in CVD is the disintegration of the intercellular matrix, which leads to high protease activity of the VLU exudate.[11] These skin-related factors contribute not only to VLU chronicity and treatment resistance but also to the high frequency of recurrence. Both the large area of diseased skin and theik large amount of chemically active exudate are especially challenging to manage in cases of large VLUs and in the presence of significant leg edema.

Contemporary dressings can manage large amounts of exudate but require frequent changes, which limits their practicality. The issue of poorly controlled protease-rich exudate is critical after skin grafting, which continues to be one of the best options for healing a large VLU.[12] In such cases, negative pressure wound therapy (NPWT) becomes a reasonable option.

NPWT description and mechanism of action

The mechanism of action in NPWT is the application of negative pressure to the wound surface. It is achieved by using a specially designed wound dressing connected to the pump and a reservoir for collecting wound and tissue fluid. When introduced in the 1970s, NPWT used unsealed dressings.[13,14] The technique was modified by sealing the dressing in the late 1980s.[15,16] The topical negative pressure dressing in its most common form consists of an open-cell polyurethane sponge with 400–600 μm pores, a transparent adhesive covering, noncollapsible tubing, and a vacuum-generating device with a collection reservoir. It should be noted that there is also a polyvinyl alcohol sponge with a 60–270 μm open-cell design that can be utilized in certain circumstances. The device has a set of controls that will maintain a range of negative pressure settings and is capable of maintaining continuous or intermittent suction. A properly applied NPWT dressing should not have air leak. The introduction of commercially available NPWT systems led to their rapid adoption in wound care practice. From 2001 to 2007, Medicare payments for NPWT increased almost eightfold, from 24 to 164 million US dollars (USD).[17] Collecting and removing large volumes of exudate reduces the frequency of dressing changes as well as the time of wound exposure to exudate and the environment. It has been demonstrated that matrix metalloproteinases and the products of their breakdown that accumulate in the chronic wound fluid impair wound healing.[11,18,19] The removal of these products advances the wound healing process. Wound fluid from chronic venous ulcers may also inhibit cell growth by having fibroblasts enter a quiescent phase of the cell cycle; additionally, the chronic wound fluid also suppresses epithelial cells and endothelial cells, further inhibiting the wound healing process.[20] Another benefit of the topical negative pressure treatment is a potential decrease in wound bacterial burden. Animal studies showed that the wounds treated with NPWT dressing at −125 mmHg pressure had a significant decrease in the bacterial burden four days after treatment.[21] However other studies demonstrated that the bacterial burden may increase for some bacteria (in particular *S. aureus*).[22] The study also showed there was a decrease in the concentration of other bacteria, mainly aerobic gram-negative rods. Notably despite the elevation in the number of *S. aureus* colonies in the wound, there was no impairment in the wound-healing process. Several additional mechanisms of action associated with NPWT efficacy in promoting wound healing have been suggested, such as increasing perfusion of the ulcer bed and approximating wound edges.[23] However, the quality of evidence supporting these mechanisms is very low.[24,25] Further well-designed studies are needed to elucidate the mechanism(s) by which the potential benefits of NPWT are gained.

Of particular interest is the use of NPWT in conjunction with skin grafting. In this scenario, NPWT can change the passive inosculation into an acute process, decrease exudate and blood collection, and improve graft adherence.[18,26] However, NPWT is not free from complications and adverse events. An inadequate rate of fluid drainage results in skin maceration and overdraining may cause dressing retention and wound injury. Despite some experimental studies that have shown a decrease in bacterial counts in quantitative cultures after NPWT,[21] in real-world clinical practice NPWT has been reported to cause wound infection in previously contaminated wounds.[27,28] To reduce the risk of adverse events, some practitioners recommend placement of a nonadherent barrier between the open cell sponge and the split-thickness skin graft to avoid ingrowth of healthy tissue into the sponge and inadvertent damage to the graft with dressing changes. When used to bolster a split-thickness skin graft, the negative pressure is usually applied in a continuous, rather than in an intermittent fashion.

Use of NPWT in patients with VLUs

Several benefits of using NPWT in patients with VLUs have been reported. These benefits include an improved healing rate, an increase in achieving complete ulcer healing, a decrease in time to prepare for skin grafting, improved skin grafting outcomes, cost savings, and improved patient quality of life (QOL).[29–32]

Compression therapy is a mandatory component of the management of VLU patients.[33] Discontinuing compression therapy, even for a short time, can delay ulcer healing. The use of NPWT does not require the discontinuation of compression therapy; in fact, it can be successfully combined with it.[34–36] A combination of NPWT and intermittent pneumatic compression was reported in a series that included 11 patients with VLUs. It was noted that this combination significantly reduced edema and accelerated wound healing.[37] Acceleration of the healing rate has been reported as the benefit of NPWT, especially in patients with large ulcers resistant to conventional therapy.[38] In a case series of 15 patients with VLUs with an ulcer surface area from 50.8 to 76.2 cm^2 (mean 60.71 cm^2) that failed initial treatment ranging from 60 to 112 weeks (mean 76.3 weeks), the application of NPWT was able to reduce their ulcer surface area by 26% during the first three weeks of therapy.[39]

The success in complete ulcer healing was prospectively studied in an RCT of 60 patients (26 with VLU). Patients were randomly assigned to either treatment by NPWT or by conventional wound care. The median time to complete healing was 29 days (95% confidence interval [CI], 25.5–32.5) in the NPWT group compared with 45 days (95% CI, 36.2–53.8) in the control group ($P = 0.0001$).[29] Several studies have used QOL measures as a secondary outcome, consistently showing the benefits of NPWT.[29–31] The use of portable NPWT devices potentially carries more benefits, allowing patients to freely ambulate. A series of 10 patients with VLUs using a portable NPWT device for 14 days after skin grafting showed significant improvement in patient QOL.[32]

The mechanisms specific to the action of NPWT on VLU have not been extensively studied. An investigation of 30 patients with hard-to-heal VLUs compared compression therapy alone versus compression therapy plus NPWT. A series of biopsies with histochemistry demonstrated increased biomarkers for angiogenesis, lymphatic vessel proliferation, and macrophage activity at one week of NPWT.[40] Animal studies demonstrated that the reduction of tissue edema by NPWT accelerated the disappearance of the plane of separation after skin grafting.[41] These mechanisms are similar to those that are implied in poor graft take rates for skin grafts in VLUs.[42]

Several case reports have been published showing the benefit of NPWT for wound preparation before and after skin grafting of VLUs.[12,43–45] In the above-mentioned RCT, wound bed preparation during NPWT was significantly shorter at seven days (95% CI 5.7–8.3) when compared to conventional wound care at 17 days (95% CI, 10 to 24, $P = 0.005$). The use of NPWT was also associated with a significant increase in QOL at the end of therapy and a significant decrease in pain scores at the end of follow-up.[32] An industry-sponsored 2011 consensus statement recommended NPWT or wound preparation for surgical closure of VLUs.[46] However, despite listing the evidence support as Level 1 for this recommendation, the references in this consensus statement are almost exclusively case reports and case series. In fact, although the use of NPWT in preparation for and during skin grafting has been shown to be very effective in patients with chronic wounds, data specific to VLUs are less consistent. Some case series report a 100% graft take,[26] while other reports are less enthusiastic. In a consecutive case series of 54 patients with chronic leg ulcers, postoperative NPWT resulted in the complete healing of 92.9% of grafts, compared to 67.4% in the control group. While diabetic wounds had a 100% graft take, this outcome for VLUs was only 71.4%.[47] In a series of seven patients with VLUs that failed to heal for 12 weeks, who were treated with a combination of NPWT and multicomponent compression bandages for four weeks, the one-week reduction in ulcer area was greater than in the control group. However, there was no difference between the groups by the end of the four-week course of NPWT.[34] Other studies indirectly support the finding of a greater short-term benefit of NPWT.[38]

Several studies demonstrated a cost-benefit of NPWT.[29,48] A prospective observational study of seven patients with 10 large VLUs (112–325 cm^2) used a standardized NPWT protocol for both wound preparation and postoperative wound care during four days postskin grafting.[48] Patients that were enrolled in this study had ulcers for 38 months on average, which is not surprising as other studies showed that large VLUs had poor outcomes from standard therapy.[49] They also had a poor graft take.[42] Yet, in this study eight of the ten VLUs healed in six months, and the remaining two VLUs had a 70% and 80% graft take. The cost of the hospital stay for NPWT patients was 20,966 USD, and the total cost was 27,152 USD. The cost for the standard management of VLU patients in the same institution was 27,792 USD. Although the cost was comparable, the real-world cost for standard therapy of VLUs of that size and chronicity is much higher.[50]

The following case illustrates the use of NPWT for skin grafting of VLUs.

Case study

A 71-year-old female with a long history of primary CVD had a chronic VLU located in her left medial ankle and distal calf. Past radiofrequency thermal ablation of her left great saphenous vein had eliminated her varicose veins, but she had severe lipodermatosclerosis and transitory edema. A duplex ultrasound scan showed reflux in her left femoral and popliteal veins with no venous obstruction (C3,4b, 6$_S$,Ep, Ad, Pr$_{(FV,PopV)}$ class of the Clinical, Etiological, Anatomical, and Pathophysiological classification,[51] Fig. 14.1). This ulcer had remained active for 10 years and had failed the standard of care including

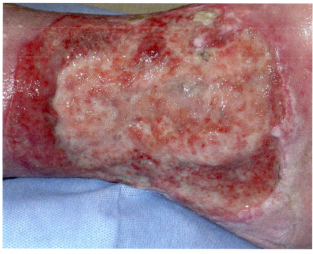

Figure 14.1 Left medial ankle chronic venous ulcer.

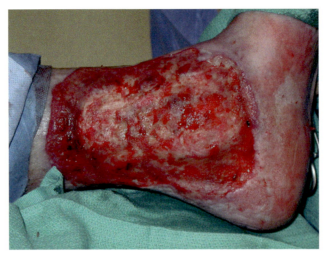

Figure 14.2 Intraoperative view of ulcer debridement in preparation for skin grafting.

compression therapy, debridement with cell and tissue product applications, and aggressive wound care. The patient was taken to the operating room (OR) and her ulcer was debrided aggressively (Fig. 14.2). A split-thickness skin grafting that was meshed to a 1.5 to 1 dimension was applied (Fig. 14.3). NPWT was used as a bolster dressing (Fig. 14.4). After six days, NPWT was discontinued and the graft showed a 100% take. The patient was discharged home with a wound vacuum-assisted cover (VAC) over the graft for a week with outpatient follow-up (Fig. 14.5). A multicomponent compression system

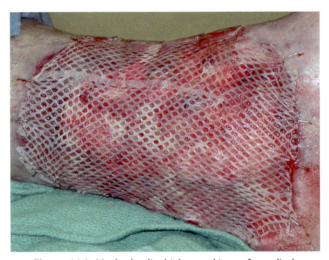

Figure 14.3 Meshed split-thickness skin graft applied.

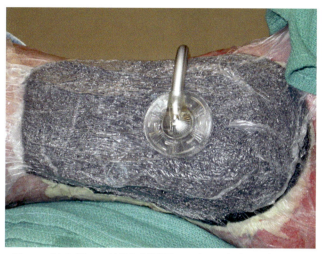

Figure 14.4 Wound VAC (NPWT) used as a bolster dressing.

was adjusted twice a week by home healthcare. On the graft, she received Xeroform gauze and calcium alginate. The donor site healed two weeks after surgery with one application of Xero form gauze in the OR. Figs. 14.6 and 14.7 show the results at five weeks and three months' follow-up. At four weeks after surgery, complete healing was achieved, and the compression therapy was continued using an adjustable nonelastic compression system (CircAid).

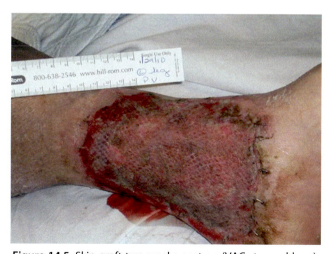

Figure 14.5 Skin graft two weeks post op (VAC stopped here).

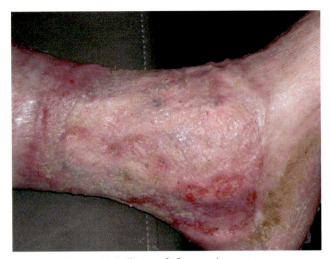

Figure 14.6 Skin graft five weeks post op.

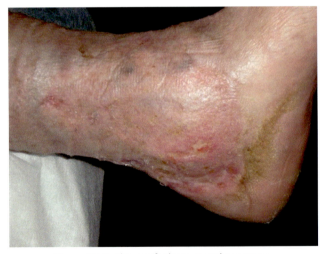

Figure 14.7 Skin graft three months post op.

Summary

In conclusion, although numerous theories exist pertinent to the utility of NPWT in the management of VLU the exact mechanism of action remains to be elucidated. Studies suggest based on available data that the efficacy of an NPWT may be a result of the removal of chronic wound fluid, increase in local blood flow, decrease in wound bacterial burden, effects of mechanical remodeling of the wound bed, decrease in interstitial edema, approximation of wound edges and the maintenance of a moist wound healing environment. Although the aforementioned processes, individually or combined, may account for the NPWT beneficial effects, further investigation with long-term follow-up studies and RCT is needed to establish a better understanding of the mode of action, clinical efficacy, and safety associated with NPWT treatment of VLU.

References

1. Pappas P, Gunnarsson C, David G. Evaluating patient preferences for thermal ablation versus nonthermal, nontumescent varicose vein treatments. *J Vasc Surg Venous Lymphat Disord*. 2021;9(2): 383–392.
2. Jayaraj A, Chandler N, Kuykendall R, Raju S, et al. Long-term outcomes following use of a composite Wallstent-Z stent approach to iliofemoral venous stenting. *J Vasc Surg Venous Lymphat Disord*. 2021; 9(2):393–400 e2.
3. Maleti O, Perrin M. Reconstructive surgery for deep vein reflux in the lower limbs: techniques, results and indications. *Eur J Vasc Endovasc Surg*. 2011;41(6):837–848.
4. Vasudevan T, Robinson DA, Hill AA, et al. Safety and feasibility report on nonimplantable endovenous valve formation for the treatment of deep vein reflux. *J Vasc Surg Venous Lymphat Disord*. 2021;9: 1200–1208.
5. Moffatt CJ, Franks PJ, Oldroyd M, et al. Community clinics for leg ulcers and impact on healing. *BMJ*. 1992;305(6866):1389–1392.
6. Monk BE, Sarkany I. Outcome of treatment of venous stasis ulcers. *Clin Exp Dermatol*. 1982;7(4): 397–400.
7. Gohel MS, Heatley F, Liu X, et al. A randomized trial of early endovenous ablation in venous ulceration. *N Engl J Med*. 2018;378(22):2105–2114.
8. Barwell JR, Davies CE, Deacon J, et al. Comparison of surgery and compression with compression alone in chronic venous ulceration (ESCHAR study): randomised controlled trial. *Lancet*. 2004; 363(9424):1854–1859.
9. Crawford JM, Lal BK, Duran WN, Pappas PJ, et al. Pathophysiology of venous ulceration. *J Vasc Surg Venous Lymphat Disord*. 2017;5(4):596–605.
10. Sansilvestri-Morel P, Rupin A, Jaisson S, et al. Synthesis of collagen is dysregulated in cultured fibroblasts derived from skin of subjects with varicose veins as it is in venous smooth muscle cells. *Circulation*. 2002;106(4):479–483.
11. Wysocki AB, Staiano-Coico L, Grinnell F. Wound fluid from chronic leg ulcers contains elevated levels of metalloproteinases MMP-2 and MMP-9. *J Invest Dermatol*. 1993;101(1):64–68.
12. Simman R, Phavixay L. Split-thickness skin grafts remain the gold standard for the closure of large acute and chronic wounds. *J Am Col Certif Wound Spec*. 2011;3(3):55–59.
13. Davydov IA, Larichev AB, Abramov AL, Men'kov KG, et al. Concept of clinico-biological control of the wound process in the treatment of suppurative wounds using vacuum therapy. *Vestn Khir Im I I Grekova*. 1991;146(2):132–136.
14. Davydov IA, Larichev AB, Smirnov AP, Flegontov VB, et al. Vacuum therapy of acute suppurative diseases of soft tissues and suppurative wounds. *Vestn Khir Im I I Grekova*. 1988;141(9):43–46.

15. Fleischmann W, Strekter W, Bombelli M, Kinzl L, et al. Vacuum sealing as treatment of soft tissue damage in open fractures. *Unfallchirurg*. 1993;96(9):488–492.
16. Fleischmann W, Becker U, Bischoff M, Hoekstra H. Vacuum sealing: indication, technique, and results. *Eur J Orthop Surg Traumatol*. 1995;5(1):37–40.
17. Department of Health and Human Services, O.O.I.G. *Comparison of Prices for Negative Pressure Wound Therapy Pumps*; 2009 [cited 2021 06/04/2021]; Available from: htps//:oig.hhs.gov/oei/reports/oei-02-07-00660.pdf.
18. Argenta LC, Morykwas MJ. Vacuum-assisted closure: a new method for wound control and treatment: clinical experience. *Ann Plast Surg*. 1997;38(6):563–576. discussion 577.
19. Shirakawa M, Isseroff RR. Topical negative pressure devices: use for enhancement of healing chronic wounds. *Arch Dermatol*. 2005;141(11):1449–1453.
20. Falanga V. Growth factors and chronic wounds: the need to understand the microenvironment. *J Dermatol*. 1992;19(11):667–672.
21. Morykwas MJ, Argenta LC, Shelton-Brown EI, McGuirt W. Vacuum-assisted closure: a new method for wound control and treatment: animal studies and basic foundation. *Ann Plast Surg*. 1997;38(6):553–562.
22. Moues CM, Vos MC, van der Bemd GJCM, Stijnen T, Hovius SER. Bacterial load in relation to vacuum-assisted closure wound therapy: a prospective randomized trial. *Wound Repair Regen*. 2004;12(1):11–17.
23. Huang C, Leavitt T, Bayer LR, Orgill DP. Effect of negative pressure wound therapy on wound healing. *Curr Probl Surg*. 2014;51(7):301–331.
24. Dumville JC, Land L, Evans D, Peinemann F. Negative pressure wound therapy for treating leg ulcers. *Cochrane Database Syst Rev*. 2015;7:CD011354.
25. Norman G, Shi G, Goh EL, et al. Negative pressure wound therapy for surgical wounds healing by primary closure. *Cochrane Database Syst Rev*. 2022;4(4):CD009261.
26. Egemen O, Ozkaya O, Ozturk MB, et al. Effective use of negative pressure wound therapy provides quick wound-bed preparation and complete graft take in the management of chronic venous ulcers. *Int Wound J*. 2012;9(2):199–205.
27. Hurd T, Kirsner RS, Sancho-Insenser JJ, et al. International consensus panel recommendations for the optimization of traditional and single-use negative pressure wound therapy in the treatment of acute and chronic wounds. *Wounds*. 2021;33(suppl 2):S1–S11.
28. US Food, Drug Administration. *FDA Safety Communication: Update on Serious Complications Associated with Negative Pressure Wound Therapysystems*; 2011. Available from: http://www.fda.gov/MedicalDevices/Safety/AlertsandNotices/ucm244211.htm.
29. Vuerstaek JD, Vainas T, Wuite J, et al. State-of-the-art treatment of chronic leg ulcers: a randomized controlled trial comparing vacuum-assisted closure (V.A.C.) with modern wound dressings. *J Vasc Surg*. 2006;44(5):1029–1037. discussion 1038.
30. Janssen AH, et al. Negative pressure wound therapy versus standard wound care on quality of life: a systematic review. *J Wound Care*. 2016;25(3):154, 156–159.
31. Cetinkaya OA, Celik SU, Boztag CY, Uncu H. Treatment of hard-to-heal leg ulcers with hyaluronic acid, sodium alginate and negative pressure wound therapy. *J Wound Care*. 2020;29(7):419–423.
32. Cuomo R, Nisi G, Gimaldi L, Brandi C, D'Aniello C. Use of ultraportable vacuum therapy systems in the treatment of venous leg ulcer. *Acta Biomed*. 2017;88(3):297–301.
33. O'Donnell Jr TF, Passman M, Marston W, et al. Management of venous leg ulcers: clinical practice guidelines of the Society for Vascular Surgery (R) and the American Venous Forum. *J Vasc Surg*. 2014;60(2 Suppl):3S–59S.
34. Kieser DC, Roake JA, Hammond C, Lewis DR. Negative pressure wound therapy as an adjunct to compression for healing chronic venous ulcers. *J Wound Care*. 2011;20(1):35–37.
35. Khashram M, Huggan P, Ikram R, et al. Effect of TNP on the microbiology of venous leg ulcers: a pilot study. *J Wound Care*. 2009;18(4):164–167.
36. Dowsett C, Grothier L, Henderson V, et al. Venous leg ulcer management: single use negative pressure wound therapy. *Br J Community Nurs*. 2013:S12–S15. Suppl: p. S6, S8–10.

37. Arvesen K, Nielsen CB, Fogh K. Accelerated wound healing with combined NPWT and IPC: a case series. *Br J Community Nurs*. 2017;22(Suppl 3(Sup3)):S41−S45.
38. Loree S, Dompmartin A, Panven K, Harel D, Leroy D. Is Vacuum Assisted Closure a valid technique for debriding chronic leg ulcers? *J Wound Care*. 2004;13(6):249−252.
39. Kucharzewski M, Mieczczanski P, Wilemska-Kucharzewska K, et al. The application of negative pressure wound therapy in the treatment of chronic venous leg ulceration: authors experience. *BioMed Res Int*. 2014;2014:297230.
40. Dini V, Miteva M, Romanelli P, et al. Immunohistochemical evaluation of venous leg ulcers before and after negative pressure wound therapy. *Wounds*. 2011;23(9):257−266.
41. Simman R. A comparative histological study of skin graft take with tie-over bolster dressing versus negative pressure wound therapy in a pig model. *Wounds*. 2004;16(2):76−80.
42. Jones JE, Nelson EA, Al-Hity A. Skin grafting for venous leg ulcers. *Cochrane Database Syst Rev*. 2013;(1):CD001737.
43. Morimoto N, Kuro A, Yamauchi T, et al. Combined use of fenestrated-type artificial dermis and topical negative pressure wound therapy for the venous leg ulcer of a rheumatoid arthritis patient. *Int Wound J*. 2016;13(1):137−140.
44. Scarpa C, Antoni E, Vindigni V, Bassetto F, et al. Efficacy of negative pressure wound therapy with instillation and dwell time for the treatment of a complex chronic venous leg ulcer. *Wounds*. 2020;32(12):372−374.
45. McElroy E, Lemay S, Reider K, Behnam A, et al. A case review of wound bed preparation in an infected venous leg ulcer utilizing novel reticulated open cell foam dressing with through holes during negative pressure wound therapy with instillation. *Cureus*. 2018;10(10):e3504.
46. Vig S, Dowsett C, Breg L, et al. Evidence-based recommendations for the use of negative pressure wound therapy in chronic wounds: steps towards an international consensus. *J Tissue Viability*. 2011;20(Suppl 1):S1−S18.
47. Korber A, Franckson T, Grabbe S, Dissemond J. Vacuum assisted closure device improves the take of mesh grafts in chronic leg ulcer patients. *Dermatology*. 2008;216(3):250−256.
48. Yang CK, Alcantara S, Gross S, Lantis JC, et al. Cost analysis of negative-pressure wound therapy with instillation for wound bed preparation preceding split-thickness skin grafts for massive (>100 cm(2)) chronic venous leg ulcers. *J Vasc Surg*. 2015;61(4):995−999.
49. Marston WA, Carlin RE, Passman M, Farber MA, Keagy BA, et al. Healing rates and cost efficacy of outpatient compression treatment for leg ulcers associated with venous insufficiency. *J Vasc Surg*. 1999;30(3):491−498.
50. de Araujo T, Valencia I, Federman DG, Kirsner RS. Managing the patient with venous ulcers. *Ann Intern Med*. 2003;138(4):326−334.
51. Lurie F, Passman M, Meissner M, et al. The 2020 update of the CEAP classification system and reporting standards. *J Vasc Surg Venous Lymphat Disord*. 2020;8(3):342−352.

CHAPTER 15

Medical therapies for chronic venous insufficiency

Mark D. Iafrati
Department of Vascular Surgery, Vanderbilt University Medical Center, Nashville, TN, United States

Venous reflux is the most common vascular disorder in the western world.[1,2] In recent years increasing attention has been focused on venous disease with entire journals and textbooks (such as this one) devoted to venous disease and its treatment. During the last 2 decades, there have been tremendous advances in minimally invasive surgical treatments for both venous reflux and venous obstructive diseases. In particular, endovenous ablation and endovenous stenting that address the underlying pathophysiology on a macroscopic hemodynamic level have proven safe and effective. While these procedures clearly improve venous symptoms, reduce infections, and promote ulcer healing and recurrence rates, they have their limitations. These procedures are expensive and insurance plans generally restrict their use to advanced diseases. Venous leg ulcers (VLUs) are costly to the individuals affected as well as to society as a whole. In our academic vascular surgery unit, the average cost to treat a VLU during the year following presentation exceeded $15,000, with some individual patient costs exceeding $50,000.[3]

While sclerotherapy and phlebectomy extend the benefits of invasive approaches beyond the axial system, there remain significant limitations to these approaches. Therefore in this chapter, we will explore the utility of medical (noninvasive) interventions for venous disease. Recommendations for lifestyle modifications, compression garment use, and medications will be explored. In particular, we will explore the mechanisms of actions of pharmacologic therapies and the clinical data on their effectiveness.

As the management of venous thrombotic events is relatively well defined, we will limit our discussion to the medical management of nonthrombotic venous disease. It is our hope that this analysis will provide the venous provider, patient, and insurer with the data necessary to determine if and under what conditions venotropic agents should be considered.

Pathophysiology

To understand the potential benefits of venous interventions, it is necessary to elucidate the underlying pathophysiology of varicose veins, chronic venous disease (CVD), and ultimately ulceration. Given the devastating consequences of VLU, much attention is paid to the treatment of this end-stage manifestation. While this attention is well deserved, understanding and mitigating earlier states of the venous disease have long-term potential benefits.[4] Venous reflux or obstruction that impairs drainage may cause venous hypertension. In some patients, this sets off a cascade of events that can further damage vein valves and induce further inflammation.[5] These inflammatory responses include leukocyte adhesion and endothelial activation. Leukocytes release leukotrienes, prostaglandins, bradykinins, oxygen-free radicals, and cytokines. TNF-a and VEGF are upregulated, which are correlated to capillary permeability, impaired lymphatic flow, and vascular proliferation.[6,7]

Increased capillary permeability and lymphatic damage results in chronic edema. Hyperpigmentation, induration, and skin ulceration are a direct result of diminished capillary blood flow and increased leakage. Thirty percent of varicose vein patients eventually progress to lipodermatosclerosis, with a high risk of ulceration, highlighting the potential advantage of treating the venous disease at an early stage.[8] Venous hypertension can be either the cause or result of this complex cycle involving inflammatory molecules and the endothelium of veins and lymphatics. Perturbation of the systems to regulate fluid and cellular flux across these vessels may result in edema, skin damage, and ulceration.[6,9] Unchecked venous collagen production and cellular proliferation result in varicose veins while the typical symptoms of pain, achiness, and heaviness result from inflammation and edema. The advanced skin changes of lipodermatosclerosis are associated with extracellular protein accumulation. Development of refractory edema with induration impairs skin healing and contributes to the development of VLUs and their slow healing once they do develop.[5] A primary goal of medical management of the venous disease is therefore to effectively utilize compression and medications to reduce leukocyte activation and infiltration and thereby improve the signs and symptoms of the venous disease at all clinical stages.[6,7]

Lifestyle considerations

Orthostatism (standing), obesity, estrogen therapy, and multiparity are generally considered risk factors for developing venous disease. In a survey of 40,000 Polish adults, time spent standing was correlated with CVD severity. Obesity and sedentary lifestyle were also more common in people with CVD.[10] First-line therapy generally recommends leg elevation, weight loss, and compression garments. Most insurance policies in the US mandate a trial of these interventions before initiation of any invasive therapies. Leg elevation aids venous drainage, reduces leg swelling, and improves microcirculation.[11] Doppler fluximetry has demonstrated improved perfusion and oxygenation with compression when applied in patients with C4–C6 disease. Increased use of leg elevation and compression in C5 patients, has been shown to reduce ulcer recurrence rates.[12] Recommending leg elevation for C3-6 patients results in reduced edema, symptomatic relief, and quicker ulcer healing.[13] Elevation was given a IA recommendation from the European Society of Vascular Surgery in 2015[14]

Kostas and associates investigated whether trying to change these factors might impact the rate of development of venous disease in 73 asymptomatic limbs contralateral to a limb undergoing venous surgery. While they had little success in counseling patients to improve adverse predisposing factors during 5 years of follow-up ($P=$NS), they were able to show that people who initiated orthostatism or became obese developed progressive venous symptoms more so than those who stabilized time standing and body weight ($P < .001$). Similarly, patients who did not use compression had more disease progression than people who initiated or continued compression ($P < .001$). They concluded that maintenance of a normal body weight, limitation of prolonged standing, and systematic compression use may be recommended in patients with CVD to limit future disease progression.[15]

Compression therapy

Normal lower extremity standing venous pressure is 60–80 mm Hg.[16] Significant hemodynamic impact on lower extremity pathologic venous return is seen with interface compression pressure >35 mmHg.[16] There is however a point at which compression becomes excessive and impairs venous return. The extreme example is demonstrated when a limb is exsanguinated and a pneumatic tourniquet is applied, as in some surgical procedures. Partsch over decades of study has convincingly demonstrated the safety of compression garments up to 60 mmHg. Even patients with moderate peripheral arterial disease (i.e., ABI $= 0.5$) can be treated with compression garments.[16] Contrary to the concern that compression would cause tissue injury, where moderate PAD is present, multiple studies have actually shown improved cutaneous microcirculation, edema control, and ulcer healing.[17]

Compression socks

Compression stockings are commercially sold in four strengths: 10–15 mm Hg (Class 1); 20–30 mm Hg (class 2); 30–40 mm Hg (class 3); and 40–50 mm Hg (class 4). While class 1 garments are sold over the counter, classes 2–4 require a prescription in the United States. Commonly classes 1 or 2 are recommended for CEAP class II–III disease, whereas Class 3 compression is recommended for advanced venous disease. Brown demonstrated that Class 3 compression stockings are optimal for advanced disease (C4–C6).[18] The effectiveness of compression stockings in the management of VLUs has been demonstrated in multiple prospective randomized trials and affirmed in well-done meta-analyses and systemic reviews.[19]

Inelastic compression garments

Inelastic compression garments have a primary piece that wraps behind the leg and is made of noncomplaint synthetic materials. These devices use pliable straps that cross in the front of the leg and are secured with Velco. The compression pressure is adjustable according to the length the straps are stretched. Some devices are marked to allow the patient to achieve compression strength from 20 to 50 mmHg, by stretching the straps to line up the desired marks. Many patients find these easier to don than compression stockings. This is especially true of the elderly who may be limited by orthopedic and rheumatologic frailties that diminish grip strength and ability to reach their toes. Because these devices have minimal elastic properties, they are better suited for ambulatory patients but they have been shown in multiple publications to effectively reduce edema, promote venous blood return, and improve VLU healing rates.[20,21]

Paste boots

Paste boots are effectively multilayered compression dressings that utilize a moist inner layer believed favorable to compromised skin. Depending on the manufacturer, the gauze inner layer may be treated with zinc oxide, glycerin, calamine, magnesium, aluminum, gelatin, and/or sorbitol. This inner wrap is effectively inelastic thus resisting expansion with ambulation. A nonmedicated outer wrap is then applied, which both provides graded elastic compression and prevents the paste from staining clothes and furniture. In a long-term retrospective study of nearly 1000 patients, VLUs healed in 73% of all patients treated with paste boots and in 91% of patients being treated for their first VLU.[22] Paste boots provided better healing rates than noncompressive polyurethane foam in a randomized study.[23] As paste boots are less elastic than compression socks or multilayer compression systems, they may need to be changed more frequently in the initiation phase of compression treatment if a large volume reduction is achieved. However paste boots are well tolerated by patients who often describe a soothing effect from the paste and frequently achieve improvement in the dry, flakey, skin associated with CVD. Additionally, paste boots will often remain in place better than socks or multilayer nonpaste systems in patients with extreme conical/champagne glass calf shapes.

Multilayer compression bandages

Traditional multilayer compression wraps include a padded inner layer (i.e., Webril), followed by 1 or 2 elastic layers, and a final relatively inelastic layer (i.e., Coban). Although multilayer compression wraps are widely considered superior to short stretch bandaging, De Carvalho recently performed a metanalysis to compare these modalities, They identified 7 reports including 1437 patients which fulfilled their meta-analysis inclusion criteria. The average age was 70 (range 23–97) years. Complete ulcer healing at 24 weeks was achieved in 268 ulcers (69%) in the 4-layer group and 257 (62%) in the short-stretch bandage groups ($P = 0.16$). Thus, the purported advantage to multilayer compression did not reach statistical significance.[24]

Today's medical marketplace contains many options to provide compression, with a wide variety of material properties. Products should be chosen based on comfort, ease of use, durability, cost, and other considerations. However, one must consider at all times that the fundamental purpose of these products is to provide optimal graduated and sustained subbandage compression pressure. Milic randomized a group of patients with chronic VLUs to one of 3 compression systems; group A: class III compression stockings, 36.2–43.9 mm Hg; group B: multilayer compression wrap with one elastic layer, 53.9–68.2 mm Hg; group C; multilayer compression wrap with two elastic layers, 74.0–87.4 mm Hg. At 26 weeks, complete VLU healing was achieved in 25% (13/42) in group A, 67.4% (31/46) in group B, and 74.4% (32/43) in group C, clearly demonstrating the value of sustained high subbandage compression pressure.[25] Not surprisingly they also demonstrated that small ulcer size (<5 cm^2) and small calf circumference (<33 cm) correlated with faster healing.[25]

The bottom line on compression

The impact of compression on VLU patients was comprehensively addressed in a 2021 Cochrane review of "compression bandages or stocking versus no compression for treating venous leg ulcers." This review analyzed prospective randomized trials which reported on any sort of compression versus no compression in patients with VLUs. They identified 14 studies suitable for analysis, including 1391 participants. The studies' mean ages ranged from 58 to 76 years, mean 70 years. Duration of VLU at presentation was 9 weeks–36 months, median = 22 mo; Ulcer size 5–20 cm^2; median follow-up was 12 weeks. Compression was with socks, paste boots, or multilayer compression wraps. The No-Compression group used standard-of-care wound therapies, which included a variety of dressing and pharmacologic therapies. The authors concluded that there is moderate-certainty evidence that: (1) use of compression shortens the time to complete healing of VLUs (pooled hazard ratio 2.17, 95% confidence interval (CI) 1.52–3.10; $I^2 = 59\%$; 5 studies, 733 patients); (2) use of compression reduces pain (four studies with 859 patients): pooled mean difference −1.39, 95% CI −1.79 to −0.98; $I^2 = 65\%$; (3) Compression improves the disease-specific quality of life (QoL), during follow-up from 12 weeks to 12 months (four studies with 859 patients). However, they were not able to establish the cost-effectiveness of compression in VLU patients in the 3 studies with 486 participants, which were suitable for analysis.[26]

Pneumatic compression

Use of intermittent pneumatic compression (IPC) has robust support in the treatment of lymphedema and severe lower extremity venous edema. In the case of VLU management, when used as an adjuvant therapy along with standard wound care and compression, IPC has been shown to improve healing rates.[27] However, data are lacking to support the use of IPC in lieu of other methods of sustained compression such as stockings or wraps. Current data support the use of IPC in VLU patients who cannot tolerate compression stockings or bandages, and those who have failed to heal despite standard/sustained compression therapy and wound care.[28]

Dietary supplements

Although many patients with venous leg ulcers are noted to be overweight or obese, many authors have reported nutritional deficiencies of various vitamins and minerals, as well as fatty acid imbalances.[29,30] The possible utility of nutrient supplementation in C5–C6 patients was explored in a metanalysis which included 20 studies. Supplementation of vitamin D, folic acid, and flavonoid administration was associated with modest benefits in VLU healing and recurrence. This report also confirmed obesity to be a risk factor for poor VLU healing. However, the authors conclude that the data to support vitamin and mineral supplantation does not support this as a therapeutic approach. They also note that despite obesity being significantly associated with refractory venous leg ulcers, attempts at weight control methods to support healing have not proven beneficial.[30] In a large systematic review of nutritional supplementation in chronic ulcers, Saeg and colleagues found that current data support supplementation of vitamins A, B_1, B_6, B_{12}, D, and E and the minerals calcium, copper, magnesium, selenium, and zinc. However, in the 7 papers reviewed, which included VLUs, only zinc supplementation demonstrated benefit in wound healing.[31] Overall, these reports suggest that obesity and unhealthy dietary intake of vitamins, minerals, and fatty acids likely contribute to poor venous wound healing. Pursuing a healthy diet and body weight is strongly endorsed, for the many well-demonstrated health benefits. However, attempting to address these chronic dietary imbalances, so common in VLU patients, in the acute treatment of a venous leg ulcer patient has not been found to have measurable benefits. Of the many proposed dietary interventions considered, zinc supplementation appears to have the most convincing support though it generally does not reach the level of being recommended in clinical practice guidelines.

Physical activity

The association between venous leg ulcers and diminished mobility is well documented. Walking speed, endurance, and ankle motion are all diminished in VLU patients[32] The mechanics of venous return from the lower extremity, against the force of gravity, relies largely on the calf venous pump system, which relies on active calf muscle contraction and ankle movement. Unfortunately, many of the therapies used in treating VLU patients have the unintended consequence of exacerbating this problem of diminished mobility. In particular, compression bandaging and leg elevation, which are the mainstays of VLU management, clearly impair mobility. Additionally, the pain associated with VLUs is commonly described as being most severe when the ankle flexes, further contributing to decreased ankle movement over time. A systematic review including 16 studies on the role of physical activity in VLU healing found that while improvement in healing rates could not be demonstrated, the use of exercise programs and therapy to improve mobility and range of motion did result in reduced venous ulcer recurrence rates.[18] These data demonstrate the importance of continued engagement with physical therapy and exercise counseling after VLU patients heal to promote long-term success.

Medications

Venoactive drugs have been widely available around the world for decades but have only been legally available in the United States in recent years. These vein-specific medications work by decreasing capillary permeability and release of inflammatory mediators while increasing venous tone. Other medications that do not have a direct action on the vein wall, like pentoxifylline and aspirin, have hematologic effects reducing, inflammation, white cell activation, and platelet activation. As signs and symptoms vary widely across the defined clinical classes of CVD (C0—C6), the goals of therapeutic venoactive drugs (VADs) likewise vary depending on the target group for treatment. Treatment with VADs in conjunction with vein ablation and compression is increasingly recognized as being highly effective in relieving pain and swelling in early-stage disease as well as in advanced C4—C6 disease.[28,33] Adjunctive use of VADs has been shown to improve quality of life (QoL). The following section will delineate the proposed mechanisms of action and data from human and animal trials defining the benefits of various VADs.

Animal models

Mechanistic exploration as well as subsequent refinements of drug combinations are accelerated when an effective animal model is available. The hamster cheek model has proven to be an extremely valuable model. Bouskela and coworkers have refined this model for over 3 decades, and have gained tremendous insights into the microvascular actions of VADs. This model allows placebo-controlled studies and in vivo visualization of the impact of oral VADs on vessels and cells, notably leukocytes.[34,35] Investigators apply agents topically to the inside of the animal's cheek and using intravital microscopy create video recordings of arterioles and venules that can later be measured. These studies allow for the assessment of macromolecular permeability, reflecting edema; the diameter of the microvessels, reflecting the distensibility of venules; and the state of leukocytes involved in inflammation. The most commonly utilized VADs in clinical use are Ruscus extract and micronized purified flavonoid fraction (MPFF). When tested in this model neither drug resulted in a change in arteriolar diameter, while Ruscus but not diosmine elicited significant venular constriction.[36] This hamster model can also be used to measure macromolecular permeability induced by either ischemia reperfusion or histamine. Both drugs demonstrated dose-dependent inhibition of permeability to ischemia reperfusion but only Ruscus diminished the leakiness associated with histamine application.

During the early phase of inflammation, leukocytes are attracted by the endothelium.[37] Rolling and sticking are two important initial steps taken by leukocytes interacting with endothelium and lead to their extravasation from the circulation and migration to the site of inflammation. Leukocytes are also involved in oxidation by releasing free superoxide radicals responsible for oxidative stress.[38] Leukocyte rolling and sticking, as is seen in CVD, were induced in the cheek model. At the doses tested, *Ruscus* extract application created a more profound dose-dependent reduction in leukocyte rolling and sticking compared to micronized diosmine.

The residual effects of VADs after cessation of treatment were also measured. *Ruscus* had a greater residual effect than micronized diosmine on leukocyte rolling at day 15 posttreatment, on permeability induced by histamine at day 20, and on leukocyte sticking at day 30. While *Ruscus* and MPFF demonstrated antiinflammatory and antioxidant effects, only *Ruscus* had a significant effect on venular constriction, at the doses used (50–450 mg/kg/day). *Ruscus* extract also had a greater residual effect on the increase in macromolecular permeability and on leukocyte rolling and sticking. These results support the demonstrated clinical benefits of *Ruscus* in terms of reducing symptoms and edema, and suggest that the action of *Ruscus* extract on CVD symptoms may last several days after treatment discontinuation.[39]

An effective hamster model of chronic venous hypertension is produced by ligating the femoral vein and tributaries or by ligating the iliac vein. This model has allowed for the investigation of the effects of VADs in the face of pathologic venous hypertension in the hamster epigastric veins.[40] Using this model, investigators are able to observe and quantitate leukocyte adhesion, leukocyte rolling, capillary density, and venular diameter. Importantly, investigators are able to compare the impact of long-term administration of VADs versus control agents, providing a more clinically relevant investigation of these agents. Using this venous ligation model, Das Gracas reported that at 6 weeks postligation, there were significant increases in leukocyte adhesion and rolling; an increase in venular diameter; and a decrease in capillary density.[40] These changes are all in line with findings in CVD. By administering MPFF for 2 weeks before and 6 weeks after venous ligation, they were able to demonstrate decreases in each of these pathologic changes ($P < .01$) compared to the unmedicated but ligated controls. These investigators also individually examined the activity of the components of MPFF (diosmine and flavonoids). They found that the combination of the components was more effective than either as a stand-alone therapy. These animal data support the utilization of MPFF in early-stage CVD, as they show a clear benefit in ameliorating the early pathologic events that are believed to trigger the long-term deterioration seen in Human CVD.[41] This model is a powerful tool for future investigations, as it allows scientists to test novel agents and combinations of agents quickly with the ability to measure multiple endpoints. Of course, as with all animal models, it is imperative that investigators seek the 3 R's of animal research (Replace, Reduce, and Refine).

Micronized purified flavonoid fraction

Daflon is the most prescribed VAD, marketed around most of the world, but not yet FDA cleared for sale in the United States. Daflon 500, an MPFF, comprises 10% flavonoids and 90% micronized diosmin.[33] MPFFs impact several biologic systems that can explain their purported benefit in CVD. In particular, they are antiinflammatory, inhibiting granulocyte and macrophage infiltration through the venous endothelium.[33] The reduction in leukocyte adhesion and rolling appears to be mediated by ICAM-1 and P-selectin expression, and endothelial apoptosis.[42] MPFF has also been demonstrated to create measurable decreases in circulating growth factors as well as the endothelial cell adhesion molecules.[5,28] At a macroscopic level, the net result is protection from vein valve damage and resultant reflux, with decreased capillary leakage and edema.[5,28] These properties are believed to result in clinical benefits for CVD patients including reduction of edema, skin hyperpigmentation and induration, time to heal ulcers, and time to ulcer recurrence ultimately venous QoL.[28] Despite their influence on many biologic pathways, MPFF therapy has been shown to be remarkably safe with side effects being generally rare and mild, most commonly gastrointestinal upset.

As the goals of therapy vary dramatically across the clinical classes, we will review data pertaining to early- and late-stage CVD. Several early studies identified the clinical benefit of MPFF treatment. Considering less severe diseases, MPFF has been shown in multiple studies to improve venous tone and thus venous symptoms.[43] The largest of these clinical trials, the RELIEF study included 5052 patients with C0—C4 disease. This single-arm study provided MPFF to all participants for 6 months. The response to therapy was measured over time and the analysis compared the outcomes for patients with and without reflux. This study demonstrated clinical benefit in patients with and without reflux with reduced reporting of heaviness, pain, and cramps in lower limbs over the treatment period. Similarly, the objective measurement of ankle swelling was improved.[44] An earlier prospective randomized controlled study found that MPFF reduced the time to ulcer healing.[45] They also reported improved subjective findings of leg pain and heaviness. Other studies in the same period confirmed these findings.[46,47] In 2005, Smith reported a meta-analysis of the 5 randomized trials which evaluated MPFF in the treatment of VLUs. This report included 723 VLU patients randomized to compression with or without MPFF (500 mg twice daily). Complete ulcer healing at 6 months was significantly improved in the MPFF group. (RR = 32%; 95%

CI, 3%—70%).⁴⁸ Subgroup analysis found the greatest benefit in VLUs 5—10 cm² size, and in VLUs present 6—12 months at study enrollment.⁴⁸ However, the validity of these early reports was questioned by a Cochrane review from 2013, which identified methodologic shortcomings in terms of blinding and allocation.⁴⁹

Collectively, these "early" publications suggested clinical benefits of MPFF pharmacotherapy for CVD; however, the conclusions were not uniform. VADs during this period were widely adopted in Europe but were largely ignored in the United States because they were not FDA cleared. More recent publications have continued to show benefits to MPFF across the spectrum of CVD. MPFF has been examined in women with leg symptoms (heavy, crampy, achy) but without visible venous disease (C0s) and was found to reduce evening venous reflux and associated symptoms by 85% after 2 months of MPFF administration.⁵⁰ Symptomatic women with C1 disease (telangiectasia), treated with 3 months of MPFF improved or eliminated symptoms of cramps, fatigue, heaviness, and pain in the majority of these women.⁵¹

The European Venous Forum and Servier conducted a large prospective observational study of the nonoperative management of CVD. This project called the VEIN ACT Program was conducted in Austria, Central America, Colombia, Romania, Russia, Spain, and the West Indies.⁵²,⁵³ All clinical classes of venous disease were enrolled. CEAP classification and symptoms were recorded, as were the therapies prescribed, compliance with the treatment plan, and the response to treatment. Similar recommendations were made from these treatment sites around the world. In most cases, conservative therapy included a recommendation for lifestyle modification, compression, and medications usually including VADs and over-the-counter pain medications. Primary care providers in Colombia enrolled 1570 patients, the majority (55%) were C2 or C3 and symptoms included pain (85%), heaviness (79%), sensation of swelling (57%), and cramps (46%). After approximately 60 days of conservative treatment, which included a VAD (mostly MPFF) for 99% of patients, compression in 49%, and nonsteroidal antiinflammatory drugs in 16%, patients reported that symptom intensity had improved by at least 50%. Vascular surgeons managed slightly fewer patients (1460) and were found to use similar recommendations with similar outcomes. In Russia, 1607 patients were enrolled, of whom 68% were C2 or C3.⁵³ Intensity of leg cramps, heaviness, pain, and sensation of swelling all improved by approximately 50% ($P = .0001$). Patient satisfaction was high at 95%. While not carrying the weight of a randomized trial, the VEIN ACT Program provided strong support for the benefit of conservative management of CVD. Despite enrolling patients from countries with diverse climate, healthcare systems, and economic standing, this study demonstrated consistent benefit in patient-reported outcomes with treatment programs largely centered on VAD use (mainly MPFF), with or without more comprehensive conservative interventions.⁴¹

Even when the venous disease is amenable to venous surgery or sclerotherapy, the benefits of VADs may be additive. In particular, perioperative administration of MPFF has been shown to improve venous clinical severity scores in the VEIN ACT program.[54] Mansilha et al. conducted a systematic review of perioperative VAD treatment, after endovenous ablation or phlebectomy.[55] Though few studies met inclusion criteria, their review overall concluded that there is a benefit to perioperative MPFF administration in reducing post-op pain, hematoma formation, and venous symptoms.

Diosmin is a flavonoid isolated from hesperidin found in citrus fruit. Vasculera is a proprietary MPFF formulation (Diosmiplex 630 mg). The companies' literature suggests that one tablet has the equivalent amount of diosmin found in 140 oranges. This proprietary formulation contains alka-4 complex thus diosmiplex. This buffering agent reduces acidity and GI discomfort. Vasculera is currently the only VAD cleared for use by the Food and Drug Administration (FDA). The FDA categorizes Vasculera as a medical food, thus requiring a prescription. Unlike unregulated, nonprescription supplements, medical foods are required to be indicated for use in the treatment of a specified diagnosis. After decades of watching patients around the world receive the benefits of these apparently safe and effective oral medications, US patients and physicians now have the ability to legally utilize a VAD for the management of CVD. It should however be noted that Vasculera has not been tested in pregnant or breastfeeding women.

Recent reviews of clinical trials, expert opinion papers, and clinical practice guidelines uniformly endorse MPFF treatment as beneficial in all stages of venous disease. Decades of accumulated evidence support the benefit of MPFF in reducing signs and symptoms of venous disease in early- and late-stage CVD.[5,28,56]

Ruscus extract

The mechanism of action of *Ruscus* extracts was initially elucidated 4 decades ago. Ruscus extracts directly activate postjunctional α1-and α2-adrenergic receptors and adrenergic nerve endings by displacing stored norepinephrine.[57] Studies have demonstrated a dose-dependent venoconstriction, in animal veins and in human veins (normal or varicose),[58] as well as constriction of lymphatic vessels,[59] a decreased adhesion of leukocytes to endothelial cells, likely from hypoxia inhibiting endothelial cell activation,[60] a decreased protein and fluorescein-labeled dextran extravasation after induced microvascular permeability,[61] increased capillary resistance,[62] and increased venous output and decreased venous pressure.[63] When Ruscus is used in combination with hesperidin methyl chalcone (HMC), there is an additive effect on lowering the capillary filtration rate[64] and increasing venoconstriction.[62,65]

A single-arm clinical trial was performed on symptomatic women with C_2–C_3 disease. Patients were treated with Cyclo 3 Fort (comprising *Ruscus*, HMC, and vitamin C (Pierre Fabre Medicament laboratories, Boulogne-Billancourt, France)). Pain was graded on a Visual Analog Scale (VAS) and objective assessment utilized venous refilling time by plethysmography. The investigators reported venous refilling time increased compared to baseline by 26% ($P < .0001$). They also reported that symptomatic improvement of

>30% on the VAS was significantly correlated with improvement of venous refill time ≥10%, $P = .04$. These data suggest that Cyclo 3 Fort provides symptomatic relief in uncomplicated varicose vein patients and that venous refill time is a useful surrogate measure in studies of VAD effectiveness.[66]

The effectiveness of Ruscus extract in venous disease was analyzed in a large number of reports of varying design and quality. Kakkos and Allaert published a thorough meta-analysis of human randomized control trials (RCTs) involving Ruscus extract, vitamin C, and HMC; all constituents of Cyclo 3 Fort.[67] These authors identified 10 studies with double-blind randomized design (before November 2016), reporting on 719 patients.[61,68–76] All treatment arms included Ruscus; however, the specific formulations included Cyclo 3 (150 mg *Ruscus* extract + 150 mg trimethyl hesperidin chalcone + 150 mg ascorbic acid); extracts of Ruscus aculeatus (60 mg) + extracts of Ribes nigrum (550 mg) + Ascorbic acid (200 mg); Phlebodril,Ruscus (75 mg) + trimethyl hesperidin chalcone (75 mg) + ascorbic acid (50 mg), and Fagorutin (dry extract 36–37.5 mg) from butcher's broom rhizome. They reported data on the effectiveness of Ruscus extracts in the management of leg symptoms and swelling. Compared to placebo, administration of Ruscus extract improved 7 specified symptoms, including leg cramps, fatigue, heaviness, itching, pain, paresthesia, and sensation of swelling. When they assessed relief of the various symptoms as categorical variables, the effect size for the benefit varied between approximately 0.25 and 0.50 indicating a robust treatment effect.[77] Even tingling which is thought to reflect venous hypertension was significantly improved by Ruscus, $RR = 0.26$ supporting the proposed hemodynamic mechanism of action.[78] Objective findings were also improved with Ruscus showing favorable impact in leg/foot volume (SMD = −0.61, 95% CI −0.91 to −0.31) and ankle circumference (SMD = −0.74, 95% CI −1.01 to −0.47).[67]

Calcium dobesilate

Calcium dobesilate, a synthetic VAD, has been studied in several randomized clinical trials with mixed results. Gohel published a 509-patient RCT, which failed to show clinical benefit for Calcium dobesilate in the management of CVD.[79] Later a placebo-controlled RCT randomized 256 patients with C3 CVD to treatment with dobesilate or placebo. They reported a significant reduction in calf volume in the active treatment group compared to placebo, -64.72 ± 111.93 cm^3 (mean ± SD). This benefit in calf volume reduction was not dependent on stocking use. Symptoms of cramps, discomfort, fatigue, heaviness, pain, pruritus, and tingling, as well as the global assessments by patients and investigators also improved more in the dobesilate group ($P < .05$).[80] The same group later published a 351-patient RCT evaluating 500 mg of calcium dobesilate three times daily versus placebo for 12 weeks in C3/C4 patients. In this trial, at the end of treatment, the relative volume change in the most pathological leg was $-0.6 \pm 4.8\%$ with calcium dobesilate, which was not significantly different than $-0.3 \pm 3.3\%$ with placebo ($P = .09$). At the end of the follow-up, this was $-1.01 \pm 5.4\%$ for calcium dobesilate versus $-0.08 \pm 3.5\%$ for placebo ($P = .002$). While statistically significant, the difference in leg volume was not a clinically robust finding.[81] Overall Calcium dobesilate though demonstrating a favorable safety profile has not consistently demonstrated clinically significant benefits in CVD management.

Pentoxifylline and sulodexide

Several drugs that lack vasoactive properties have long been employed in the management of venous disease. The most commonly used non-VAD pharmaceutical agents are sulodexide and pentoxifylline. These agents are commonly prescribed in the management of VLUs and as adjuncts to compression garments and wound care. Pentoxifylline, a methylated xanthine derivative, is a nonselective phosphodiesterase inhibitor.[82] Pentoxifylline has rheolytic activity that increases red blood cell deformability, decreases blood viscosity, and decreases platelet aggregation.[82] Additionally, pentoxifylline is antiinflammatory, selectively depressing tumor necrosis factor (TNF) expression and depression of granulocyte phagocytic activity and neutrophil-endothelium adhesion.[82]

Like the VADs, interest in pentoxifylline spans decades. Initially approved as a drug for the management of arterial claudication, it never delivered robust benefits in terms of walking distance for patients with peripheral arterial disease. Today, Cilostazol has largely replaced pentoxifylline in the management of claudication. However, a clinical benefit for pentoxifylline in VLU healing was suggested more than 20 years ago.[83,84] A well-designed and reasonably powered study published in 2007 randomized 245 patients with C6 disease to pentoxifylline (1200 mg daily) or placebo. All patients received compression therapy and wound care. The modest benefit in complete wound healing did not reach statistical significance (62% vs. 53%; $P = 0.21$).[85] However, when the authors used regression analysis, they did report significant improvement in ulcer healing in the pentoxifylline group (RR for healing = 1.4; 95% CI, 1.0–2.0).[85] As data accumulated, a Cochrane Review of prospective randomized trials comparing the effectiveness of pentoxifylline as an adjunct to compression in the healing of VLUs was undertaken. This study published in 2012 found 12 studies suitable for inclusion, which included 864 patients. They reported that pentoxifylline added to standard compression therapy significantly improved VLU healing. (RR = 1.56; 95% CI, 1.14–2.13).[86]

Sulodexide is a combination drug that comprises low molecular weight heparin (80%) and dermatan sulfate (20%). Sulodexide has antithrombotic and profibrinolytic properties as well as antiinflammatory effects. A 2016 Cochrane Review examined the role of sulodexide in the management of VLUs. They searched for studies that randomized patients with VLU to receive Sulodexide versus placebo or any other drug (such as aspirin, flavonoids, pentoxifylline), with or without compression. They selected four RCTs, including

463 patients. A meta-analysis found patients treated with Sulodexide healed their VLU in 49.4%, whereas conventional treatment without Sulodexide resulted in VLU healing in only 29.8%, RR 1.66 (95% CI 1.30–2.12). Despite the statistically significant benefit of Sulodexide demonstrated in their review, the authors cautioned that the evidence is low quality with a high risk of bias; therefore, the conclusions could change with future studies.[87] Similar results were obtained from another analysis that included two additional studies.[88] As of this writing Sulodexide is not approved by the FDA.

Horse chestnut

Horse chestnut is a traditional herbal remedy for CVD. Horse chestnut seed extract (HCSE) is derived from *Aesculus hippocastanum* L seeds. Escin, a saponin, is the bioactive component.[89] Escin inhibits hyaluronidase, an enzyme involved in proteoglycan degradation.[90] Modulation of hyaluronidase activity may play a role in the skin changes and refractory ulcers associated with advanced CVD. Given the widespread use of this product as a home remedy, the Cochrane Library has vigorously attempted to bring clarity to the utility of this product. Cochrane Reviews have addressed this topic five times, most recently in 2012. They evaluated 17 RCTs evaluating the effect of HCSE in CVD.[91] While the studies varied in reported endpoints, they generally supported the effectiveness of HCSE in CVD when compared to placebo. Seven RCTs reported on pain reduction. One study reported pain reduction compared to baseline while six studies reported less pain in HCSE-treated patients versus placebo. Change in leg volume was reported in 6 placebo-controlled trials ($n = 502$). HCSE administration resulted in significantly better limb volume reduction compared to placebo, WMD of 32.1 mL (95% CI 13.49–50.72). Studies comparing HCSE to rutosides, pycnogenol, or compression found no significant differences in pain relief.[91]

Acetylsalicylic acid

By virtue of its antiplatelet activity, some studies suggested a beneficial effect of acetylsalicylic acid on ulcer healing, but these were small studies with significant methodological flaws.[47] These purported benefits have not been identified in larger studies and Acetylsalicylic acid is NOT recommended to promote healing in VLU as a routine treatment but may be considered in resistant ulcers.[14]

Antibiotics

The benefit of systemic antibiotics for infected VLUs is uncontested, despite the lack of PRTs. However, the benefit of systemic or topical antibiotics in healing noninfected or colonized VLUs has been widely tested, yet remains highly contested. Routine use of systemic antibiotics for VLUs was not supported by a Cochrane review of 45 RCTs.[92] This lack of compelling efficacy data, in the face of increasing bacterial resistance, has led clinical practice guidelines to recommend that antimicrobial agents only be used for clinical infection and not for bacterial colonization.[14] The exception to this may be wound dressings: in particular, silver- and iodine-containing products. O'Meara's 2014 Cochrane review included 11 studies that compared cadexomer iodine to standard care and found an advantage to wound healing with antimicrobial dressings. In this meta-analysis, four RCTs reporting complete VLU healing at 4–12 weeks favored the antimicrobial dressing, RR 2.17 (95% CI 1.30–3.60). However, when iodine products were compared to dextranomer, hydrocolloids, paraffin gauze, or silver, no significant difference was found.[92] A later Cochrane review focused specifically on silver-containing wound care products. This 2014 Cochrane review found no advantage to silver dressings compared to nonantimicrobial dressings. Additionally, no silver product was shown to be superior to any other silver product. The most recent Cochrane review on this topic, a 2018 report of VLU dressing care found that silver dressings may improve venous leg ulcer healing, compared to nonadherent dressings: RR 2.43, 95% CI 1.58–3.74.[93] While these data suggest there is likely benefit to iodine or silver-containing dressings in VLU management, they lack the granularity to define the optimal dose or duration of these antimicrobial agents.

Biologic wound care products

Even with full compliance to the recommended lifestyle modifications, compression, medications, debridement, routine wound care, and venous surgery some VLUs will prove refractory. There are many factors that contribute to the cellular senescence seen in these chronic refractory ulcers.[94] In these cases, one may consider autologous skin grafting or the application of cellularized tissue products. Autologous skin grafting offers the opportunity to harvest healthy skin from a remote location and to bring those newly activated cells (by virtue of the harvest) to the surface of a VLU. When successful this surgical procedure may result in complete wound healing in as little as 2 weeks. For decades, skin grafting has been considered the gold standard for the treatment of these refractory wounds. In 2013, a Cochrane review identified 17 RCTs, including 1034 patients, suitable for analysis, which enrolled patients in one treatment arm who underwent autologous skin grafting.[95] Despite the relatively large number of studies and the large total number of patients, there was significant heterogeneity and multiple techniques/products tested. Although the lack of support for autologous skin grafting does not negate clinical experience, the only approach found to provide a statistically significant advantage in VLU healing was bilayered skin substitute (BSS) (Apligraf), which was found to be superior to autografts, compression therapy, freeze-dried keratinocyte allografts, and porcine dermis.[95] In 2021, a systematic review of RCTs studied VLU healing by comparing Advanced Wound Care Matrices (AWCMs) to Standard compression. The authors reported clinical trials that included financial analysis as well as healing benefits. Eight studies analyzing BSS, dehydrated human amnion/chorion membrane (dHACM) (Epifix), human fibroblast-derived dermal substitute (Dermagraft), extracellular wound matrix (ECM) (Oasis), advanced matrix (AM) (Talymed), and matrix wound dressing (Promogran) met inclusion criteria. Improved wound healing was only demonstrated in 4 trials comparing these AWCMs to standard care (BSS, dHACM, ECM, and AM). For the products that demonstrated clinical benefit, the incremental cost per additional VLU healed ranged from $2593 to $210,800.[96]

Guidelines

There is near universal support among vascular specialty societies around the world that compression, wound care, and VADs provide clinical benefits to patients with CVD. Among those societies that have included this endorsement in published guidelines are the American College of Chest Physicians, American Venous Forum, European Society for Vascular Surgery, European Venous Forum, International Union of Angiology, and the Union Internatinoale de Phlebologie. Not surprisingly there remain differences between guidelines based on the date that the guidelines were written and the weight of evidence assigned to various publications. Although there are subtle differences in the endorsements, the theme is that patients benefit from MPFF being added to conventional wound care and compression with reduction in time to heal, improved swelling, pain, and QoL.[14,97–100] The most contemporary, European guidelines (2018), site grade A evidence, in recommending MPFF, pentoxifylline and sulodexide as an adjunct to compression therapy in the management of VLUs.[28] Interestingly, the 2016 Dutch guidelines for the management of recurrent varicose veins, do not take a position on VADs.[101] Thus, there remains a recognition that despite decades of widespread utilization and accumulating scientific evidence, the published studies are frequently underpowered and subject to bias, resulting in overall low-quality evidence and weak recommendations. The need for high-quality clinical trial data remains in nearly all aspects of venous disease management.

Conclusions

Lifestyle modifications, compression wraps/garments, and skin/wound care products, as well as surgical procedures for CVD, are fundamental components of a comprehensive approach to the care of CVD patients. Increasingly VADs have been shown to be beneficial for patients presenting with all clinical classes of CVD. Medical management with VADs augments the benefits of compression therapy and reduces adverse perioperative consequences of venous surgery. These laudatory effects of VADs are realized even absent compression garment use. Given the large number of patients for whom compression is contraindicated (i.e., Severe peripheral arterial disease) or poorly tolerated because of a sensation of tightness or itching, hot and humid climate, or physical challenges with donning compression garments, VADs may be the most important intervention for this group of desperate patients with limited options.[14,98]

The world marketplace has for some time had access to several effective VADs, most notably Daflon 500 (90% micronized diosmin and 10% flavonoids) and Cyclo 3 Fort (Ruscus, HMC, and vitamin C combination). Current data supporting the efficacy of these VADs have resulted in a Grade 1A recommendation in CVD.[36] US providers were largely left as bystanders because of regulatory issues for many years, the recent FDA approval of Vasculera (Diosmiplex 630 mg) has changed the landscape for US providers and patients, opening for the first time legal access to a prescription MPFF. Moving forward, the goal should be to personalize venous care, optimizing outcomes based on the particular characteristics of the whole patient and their presenting clinical class of venous disease. At early stages, attention should be directed toward halting disease progression through lifestyle changes, compression, and VADs. If CVD does progress, patients may benefit from interventional procedures (i.e., endovenous ablation, phlebectomy, or sclerotherapy). Even then VADs have a role, as perioperative MPFF reduces periprocedural pain and bleeding (hematoma) and improves CVD symptoms. Venous leg ulcers remain the greatest clinical challenge for CVD patients and providers. Given the tremendous impact on patient quality of life, the cost to the health care system, and the high risk of recurrent infections and ulcers, it is appropriate to thoroughly evaluate and treat these patients in accordance with the evidence-based interventions supported in this chapter. Compression and wound care remain the foundation upon which VLU management is built. Early correction of superficial venous reflux supports enhanced wound healing. In addition, early prescription of a VAD (Diosmiplex now FDA approved) and non-VAD drugs (pentoxifylline and sulodexide) further speed up VLU healing and reduce symptoms.[102]

References

1. Beebe-Dimmer JL, Pfeifer JR, Engle JS, Schottenfeld D. The epidemiology of chronic venous insufficiency and varicose veins. *Ann Epidemiol*. 2005;15(3):175–184. https://doi.org/10.1016/j.annepidem.2004.05.015.
2. Criqui MH, Jamosmos M, Fronek A, et al. Chronic venous disease in an ethnically diverse population: the San Diego population study. *Am J Epidemiol*. 2003;158(5):448–456. https://doi.org/10.1093/aje/kwg166.
3. Ma H, O'Donnell Jr TF, Rosen NA, Iafrati MD. The real cost of treating venous ulcers in a contemporary vascular practice. *J Vasc Surg Venous Lymphat Disord*. 2014;2(4):355–361. https://doi.org/10.1016/j.jvsv.2014.04.006.
4. Nicolaides AN. The most severe stage of chronic venous disease: an update on the management of patients with venous leg ulcers. *Adv Ther*. 2020;37(Suppl 1):19–24. https://doi.org/10.1007/s12325-020-01219-y.

5. Mansilha A, Sousa J. Pathophysiological mechanisms of chronic venous disease and implications for venoactive drug therapy. *Int J Mol Sci*. 2018;19(6). https://doi.org/10.3390/ijms19061669.
6. Bush R, Comerota A, Meissner M, Raffetto JD, Hahn SR, Freeman K. Recommendations for the medical management of chronic venous disease: the role of Micronized Purified Flavanoid Fraction (MPFF). *Phlebology*. 2017;32(1_suppl):3–19. https://doi.org/10.1177/0268355517692221.
7. Bergan J, Shortell CK. *Venous Ulcers*. Burlington, MA: Academic Press, Elsevier; 2006.
8. Robertson LA, Evans CJ, Lee AJ, Allan PL, Ruckley CV, Fowkes FG. Incidence and risk factors for venous reflux in the general population: Edinburgh vein study. *Eur J Vasc Endovasc Surg*. 2014;48(2): 208–214. https://doi.org/10.1016/j.ejvs.2014.05.017.
9. Saharay M, Shields DA, Georgiannos SN, Porter JB, Scurr JH, Coleridge Smith PD. Endothelial activation in patients with chronic venous disease. *Eur J Vasc Endovasc Surg*. 1998;15(4):342–349. https://doi.org/10.1016/s1078-5884(98)80039-7.
10. Jawien A. The influence of environmental factors in chronic venous insufficiency. *Angiology*. 2003; 54(Suppl 1):S19–S31. https://doi.org/10.1177/0003319703054001S04.
11. Abu-Own A, Scurr JH, Coleridge Smith PD. Effect of leg elevation on the skin microcirculation in chronic venous insufficiency. *J Vasc Surg*. 1994;20(5):705–710. https://doi.org/10.1016/s0741-5214(94)70157-1.
12. Sindrup JH, Avnstorp C, Steenfos HH, Kristensen JK. Transcutaneous PO2 and laser doppler blood flow measurements in 40 patients with venous leg ulcers. *Acta Derm Venereol*. 1987;67(2):160–163. https://www.ncbi.nlm.nih.gov/pubmed/2438882.
13. Dix FP, Reilly B, David MC, simon D, Dowding E, Ivers L. Effect of leg elevation on healing, venous velocity,and ambulatory venous pressure in venous ulceration. *Phlebology*. 2005;20:87–94.
14. Wittens C, Davies AH, Baekgaard N, et al. Editor's choice—management of chronic venous disease: clinical practice guidelines of the European society for vascular surgery (ESVS). *Eur J Vasc Endovasc Surg*. 2015;49(6):678–737. https://doi.org/10.1016/j.ejvs.2015.02.007.
15. Kostas TI, Ioannou CV, Drygiannakis I, et al. Chronic venous disease progression and modification of predisposing factors. *J Vasc Surg*. 2010;51(4):900–907. https://doi.org/10.1016/j.jvs.2009.10.119.
16. Partsch B, Partsch H. Calf compression pressure required to achieve venous closure from supine to standing positions. *J Vasc Surg*. 2005;42(4):734–738. https://doi.org/10.1016/j.jvs.2005.06.030.
17. Klyscz T, Galler S, Steins A, Zuder D, Rassner G, Junger M. [The effect of compression therapy on the microcirculation of the skin in patients with chronic venous insufficiency (CVI)]. *Hautarzt*. 1997; 48(11):806–811. https://doi.org/10.1007/s001050050664.
18. Brown A. Life-style advice and self-care strategies for venous leg ulcer patients: what is the evidence? *J Wound Care*. 2012;21(7):342–344. https://doi.org/10.12968/jowc.2012.21.7.342, 346, 348–344.
19. Mauck KF, Asi N, Elraiyah TA, et al. Comparative systematic review and meta-analysis of compression modalities for the promotion of venous ulcer healing and reducing ulcer recurrence. *J Vasc Surg*. 2014;60(2 Suppl):71S–90S. https://doi.org/10.1016/j.jvs.2014.04.060. e1-2.
20. Lund E. Exploring the use of CircAid legging in the management of lymphoedema. *Int J Palliat Nurs*. 2000;6(8):383–391. https://doi.org/10.12968/ijpn.2000.6.8.9063.
21. Blecken SR, Villavicencio JL, Kao TC. Comparison of elastic versus nonelastic compression in bilateral venous ulcers: a randomized trial. *J Vasc Surg*. 2005;42(6):1150–1155. https://doi.org/10.1016/j.jvs.2005.08.015.
22. Lippmann HI, Fishman LM, Farrar RH, Bernstein RK, Zybert PA. Edema control in the management of disabling chronic venous insufficiency. *Arch Phys Med Rehabil*. 1994;75(4):436–441. https://doi.org/10.1016/0003-9993(94)90168-6.
23. Rubin JR, Alexander J, Plecha EJ, Marman C. Unna's boot vs polyurethane foam dressings for the treatment of venous ulceration. A randomized prospective study. *Arch Surg*. 1990;125(4):489–490. https://doi.org/10.1001/archsurg.1990.01410160075016.
24. De Carvalho MR, Peixoto BU, Silveira IA, Oliveria B. A meta-analysis to compare four-layer to short-stretch compression bandaging for venous leg ulcer healing. *Ostomy/Wound Manag*. 2018; 64(5):30–37. https://www.ncbi.nlm.nih.gov/pubmed/29847309.
25. Milic DJ, Zivic SS, Bogdanovic DC, et al. The influence of different sub-bandage pressure values on venous leg ulcers healing when treated with compression therapy. *J Vasc Surg*. 2010;51(3):655–661. https://doi.org/10.1016/j.jvs.2009.10.042.

26. Shi C, Dumville JC, Cullum N, Connaughton E, Norman G. Compression bandages or stockings versus no compression for treating venous leg ulcers. *Cochrane Database Syst Rev*. 2021;7: CD013397. https://doi.org/10.1002/14651858.CD013397.pub2.
27. Nelson EA, Hillman A, Thomas K. Intermittent pneumatic compression for treating venous leg ulcers. *Cochrane Database Syst Rev*. 2014;(5):CD001899. https://doi.org/10.1002/14651858.CD001899.pub4.
28. Nicolaides A, Kakkos S, Baekgaard N, et al. Management of chronic venous disorders of the lower limbs. Guidelines according to scientific evidence. Part I. *Int Angiol*. 2018;37(3):181−254. https://doi.org/10.23736/S0392-9590.18.03999-8.
29. Barber GA, Weller CD, Gibson SJ. Effects and associations of nutrition in patients with venous leg ulcers: a systematic review. *J Adv Nurs*. 2018;74(4):774−787. https://doi.org/10.1111/jan.13474.
30. Tobon J, Whitney JD, Jarrett M. Nutritional status and wound severity of overweight and obese patients with venous leg ulcers: a pilot study. *J Vasc Nurs*. 2008;26(2):43−52. https://doi.org/10.1016/j.jvn.2007.12.002.
31. Saeg F, Orazi R, Bowers GM, Janis JE. Evidence-based nutritional interventions in wound care. *Plast Reconstr Surg*. 2021;148(1):226−238. https://doi.org/10.1097/PRS.0000000000008061.
32. Roaldsen KS, Rollman O, Torebjork E, Olsson E, Stanghelle JK. Functional ability in female leg ulcer patients–a challenge for physiotherapy. *Physiother Res Int*. 2006;11(4):191−203. https://doi.org/10.1002/pri.337.
33. Coleridge-Smith P, Lok C, Ramelet AA. Venous leg ulcer: a meta-analysis of adjunctive therapy with micronized purified flavanoid fraction. *Eur J Vasc Endovasc Surg*. 2005;30:198−208.
34. Bouskela E, Cyrino F, Marcelon G. Possible mechanisms for the inhibitory effect of Ruscus extract on increased microvascular permeability induced by histamine in the hamster cheek pouch. *J Cardiovasc Pharmacol*. 1994;24:281−285.
35. Rauly-Lestienne I, Heusler P, Cussac D, Lantoine-Adam F, de Al-meida Cyrino FZG, Bouskela E. Contribution of muscarinic receptors to in vitro and in vivo effects of Ruscus extract. *Microvasc Res*. 2017;114:1−11.
36. Kakkos SK, Bouskela E, Jawien A, Nicolaides AN. New data on chronic venous disease: a new place for Cyclo 3(R) Fort. *Int Angiol*. 2018;37(1):85−92. https://doi.org/10.23736/S0392-9590.17.03935-9.
37. Langer HF, Chavakis T. Leukocyte-endothelial interactions in inflammation. *J Cell Mol Med*. 2009; 13(13):1211−1220.
38. Smith P. The causes of skin damage and leg ulceration in chronicvenous disease. *Int J Low Extrem Wounds*. 2006;5:160−168.
39. de Almeida Cyrino FZG, Balthazar DS, Sicuro FL, Bouskela E. Effects of venotonic drugs on the microcirculation: comparison between Ruscus extract and micronized diosmine1. *Clin Hemorheol Microcirc*. 2018;68(4):371−382. https://doi.org/10.3233/CH-170281.
40. das Gracas CSM, Cyrino FZ, de Carvalho JJ, Blanc-Guillemaud V, Bouskela E. Protective effects of micronized purified flavonoid fraction (MPFF) on a novel Experimental model of chronic venous hypertension. *Eur J Vasc Endovasc Surg*. 2018;55(5):694−702. https://doi.org/10.1016/j.ejvs.2018.02.009.
41. Ulloa JH. Micronized purified flavonoid fraction (MPFF) for patients suffering from chronic venous disease: a review of new evidence. *Adv Ther*. 2019;36(Suppl 1):20−25. https://doi.org/10.1007/s12325-019-0884-4.
42. Takase S, Pascarella L, Lerond L, Bergan JJ, Schmid-Schonbein GW. Venous hypertension, inflammation and valve remodeling. *Eur J Vasc Endovasc Surg*. 2004;28(5):484−493. https://doi.org/10.1016/j.ejvs.2004.05.012.
43. Allaert FA. Meta-analysis of the impact of the principal venoactive drugs agents on malleolar venous edema. *Int Angiol*. 2012;31(4):310−315. https://www.ncbi.nlm.nih.gov/pubmed/22801396.
44. Jantet G. Chronic venous insufficiency: worldwide results of the RELIEF study. Reflux assEssment and quaLity of lIfe improvEment with micronized Flavonoids. *Angiology*. 2002;53(3):245−256. https://doi.org/10.1177/000331970205300301.
45. Guihou JJ, Dereure O, Marzin L, O'uvry, Zuccarelli P, Dbure CMeara S. Efficacy of Dalfon 500 mg in venous leg ulcer healing: a double-blind, randomized, controlled versus placebo trial in 107 patients. *Angiology*. 1997;48:77−85.

46. Glinski W, Chodynicka B, Roszkiewcz J, Bogdanowski T, Kaszuba. The beneficial augmentative effect of micronized purified flavanoid fractoin (MPFF) on the healing of leg ulcers: an open, multi-centre, controlled randomized study. *Phelebology*. 1999;14:151–157.
47. Roztocil K, Stvrtinova V, Strejcek J. Efficacy of a 6-month treatment with Daflon 500 mg in patients with venous leg ulcers associated with chronic venous insufficiency. *Int Angiol*. 2003;22(1):24–31. https://www.ncbi.nlm.nih.gov/pubmed/12771852.
48. Smith PC. Daflon 500 mg and venous leg ulcer: new results from a meta-analysis. *Angiology*. 2005; 56(Suppl 1):S33–S39. https://doi.org/10.1177/00033197050560i106.
49. Scallon C, Bell-Syer SE, Aziz Z. Flavonoids for treating venous leg ulcers. *Cochrane Database Syst Rev*. 2013;(5):CD006477. https://doi.org/10.1002/14651858.CD006477.pub2.
50. Tsoukanov YT, Tsoukanov AY. Great saphenous vein transitory reflux in patients with symptoms related to chronic venous disorders, but without visible signs (C0), and its correction with MPFF treatment. *Phlebolymphology*. 2017;22(1):3–11.
51. Tsoukanov YT, Nikolaichuk AI. Orthostatic loading induced transient venous refluxes (day orthostatic loading test), and remedial effect of micronized purified flavonvoid fractoin in patients with telanjectasias and reticular veins. *Int Angiol*. 2017;36(2):189–196.
52. Ulloa JH, Guerra D, Cadavid LG, Fajardo D, Villarreal R, Bayon G. Nonoperative approach for symptomatic patients with chronic venous disease:results from the VEIN Act program. *Phlebolymphology*. 2018;25(2):123.
53. Lishkov DE, Kirienko AI, Larionov AA, Chernookov AI, Nikolaichuk. Patients seeking treatment for chronic venous disorders: Russian results from the VEIN Act program. *Phlebolymphology*. 2016;23(1):44.
54. Bogachev VY, Boldin BV, Turkin PY. Administration of micronized purified flavonoid fraction during sclerotherapy of reticular veins and telangiectasias: results of the national, multicenter, observational program VEIN ACT PROLONGED-C1. *Adv Ther*. 2018;35(7):1001–1008. https://doi.org/10.1007/s12325-018-0731-z.
55. Mansilha A, Sousa J. Benefits of venoactive drug therapy in surgical or endovenous treatment for varicose veins: a systematic review. *Int Angiol*. 2019;38(4):291–298. https://doi.org/10.23736/S0392-9590.19.04216-0.
56. Kakkos SK, Nicolaides AN. Efficacy of micronized purified flavonoid fraction (Daflon(R)) on improving individual symptoms, signs and quality of life in patients with chronic venous disease: a systematic review and meta-analysis of randomized double-blind placebo-controlled trials. *Int Angiol*. 2018;37(2):143–154. https://doi.org/10.23736/S0392-9590.18.03975-5.
57. Marcelon G, Vanhoutte PM. Mechanism of action of Ruscus extract. *Int Angiol*. 1984;3:74–76.
58. Bouskela E, Cyrino FZ, Marcelon G. Inhibitory effect of the Ruscus extract and of the flavonoid hesperidine methylchalcone on increased microvascular permeability induced by various agents in the hamster cheek pouch. *J Cardiovasc Pharmacol*. 1993;22(2):225–230. https://doi.org/10.1097/00005344-199308000-00009.
59. Marcelon G, Pouget G, Tisne-Versailles. Effects of Ruscus on the adrenoceptors of the canine lymphatic thoracis duct. *Phelbology*. 1988;3:109–112.
60. Bouaziz N, Michiels C, Janssens D, et al. Effect of Ruscus extract and hesperidin methylchalcone on hypoxia-induced activation of endothelial cells. *Int Angiol*. 1999;18(4):306–312. https://www.ncbi.nlm.nih.gov/pubmed/10811519.
61. Parrado F, Buzzi A. A study of the efficacy and tolerability of a preparation containing ruscus aculeatus in the treatment of chronic venous insufficiency of the lower limbs. *Clin Drug Invest*. 1999;(18): 255–261.
62. Thebault JJ. [Effects of a phlebotonic agent]. *Fortschr Med*. 1983;101(25):1206–1212. https://www.ncbi.nlm.nih.gov/pubmed/6350133.
63. Capelli R, Nicora M, Di Peri T. Use of extract of Ruscus aculeatus in venous disease in the lower limbs. *Drugs Exp Clin Res*. 1988;14:277–283.
64. Rudofsky G. Venentonisierung und kapillarabdichtung. Die wirkung der kombination aus ruscus-extrakt und trimethylhesperidinchalkon bei gesunden probanden unter warmebelastung. *Fortschr Med*. 1989;107(52):5–8.
65. Bouskela E, Cyrino FZ, Marcelon G. Effects of Ruscus extract on the internal diameter of arterioles and venules of the hamster cheek pouch microcirculation. *J Cardiovasc Pharmacol*. 1993;22(2): 221–224. https://doi.org/10.1097/00005344-199308000-00008.

66. Allaert FA, Hugue C, Cazaubon M, Renaudin JM, Clavel T, Escourrou P. Correlation between improvement in functional signs and plethysmographic parameters during venoactive treatment (Cyclo 3 Fort). *Int Angiol.* 2011;30(3):272–277. https://www.ncbi.nlm.nih.gov/pubmed/21617611.
67. Kakkos SK, Allaert FA. Efficacy of Ruscus extract, HMC and vitamin C, constituents of Cyclo 3 Fort®, on improving individual venous symptoms and edema: a systematic review and meta-analysis of randomized double-blind placebo-controlled trials. *Int Angiol.* 2017;36:93–106.
68. Altenkamp H. Efficacy of antivaricotic drugs can be measured objectively. *Phlebologie in der praxis.* 1987;(2):9–20.
69. Braun R, Hirche H, van Laak H. Die therapie der venösen insuffizienz: eine doppelblind-studie mit Phlebodril. *ZFA Zeitschrift für Allgemeinmedizin.* 1985;61:309–319.
70. Elbaz C, Nebot F, Reinharz D. Insuffisance veineuse des membres inférieurs étude controleé de l'action du Cyclo 3. *Phlebologie.* 1976;(29):77–84.
71. le Devehat C, Lemoine A, Roux E, Cirette B, Vimeux M, Martinaggi P. Aspects clinique et hémodynamique de Cyclo 3 dans l'insuffisance veineuse. *Angéiologie.* 1984;3:119–122.
72. Questel R, Walrant P, Questel R. Bilan de l'essai randomisé Veinobiase contre placebo dans l'unsuffisance veineuse: observation de la microcircula- tion per capillarographie conjonctivale. *Gaz Med Fr.* 1983;90:508–514. Xgazette Medicale de France 1983(90):508-514.
73. Rieger H. Efficacy of a combination drug in patients with chronic venous insufficiency under orthostatic conditions. *Phlebology.* 1988;3:127–130.
74. Rudofsky G, Diehm C, Grub JD, Hartmann M, Schultz-Ehrenburg HK, Bisler H. Chronic venous insufficiency. Treatment with Ruscus extract and trimethylhesperidin chalcone. *MMW Munch Med Wochenschr.* 1990;132:205–210.
75. Sentou Y, Bernard-Fernier MF, Demarez JP, Laurent D, Cauquil J. Symptomatologie et pléthysmographie: parallélisme des résultants obtenus lors d'un traitement par Cyclo 3 de patientes porteuses d'une insuffisance neineuse chronique (étude en double insu contre place- bo). *Gazette Medicale.* 1985;92: 73–77.
76. Vanscheidt W, Jost V, Wolna P, Lucker PW, Muller A, Theurer C. Efficacy and safety of a Butcher's broom preparation (Ruscus aculeatus l. extract) compared to placebo in patients suffering from chronic venous insufficiency. *Arzneimittelforschung.* 2002;52.
77. Perrin M, Eklof B, VANR A, et al. Venous symptoms: the SYM vein consensus statement developed under the auspices of the European venous Forum. *Int Angiol.* 2016;35(4):374–398. https://www.ncbi.nlm.nih.gov/pubmed/27081866.
78. Padberg Jr FT, Maniker AH, Carmel G, Pappas PJ, Silva Jr MB. Hobson RW, 2nd. Sensory impairment: a feature of chronic venous insufficiency. *J Vasc Surg.* 1999;30(5):836–842. https://doi.org/10.1016/s0741-5214(99)70008-x.
79. Gohel MS, Davies AH. Pharmacological agents in the treatment of venous disease: an update of the available evidence. *Curr Vasc Pharmacol.* 2009;7(3):303–308. https://doi.org/10.2174/157016109788340758.
80. Rabe E, Jaeger KA, Bulitta M, Pannier F. Calcium dobesilate in patients suffering from chronic venous insufficiency: a double-blind, placebo-controlled, clinical trial. *Phlebology.* 2011;26(4):162–168. https://doi.org/10.1258/phleb.2010.010051.
81. Rabe E, Ballarini S, Lehr L, Doxium E. A randomized, double-blind, placebo-controlled, clinical study on the efficacy and safety of calcium dobesilate in the treatment of chronic venous insufficiency. *Phlebology.* 2016;31(4):264–274.
82. Graninger W, Wenisch C. Pentoxifylline in severe inflammatory response syndrome. *J Cardiovasc Pharmacol.* 1995;25(Suppl 2):S134–S138. https://doi.org/10.1097/00005344-199500252-00028.
83. Dale JJ, Ruckley CV, Harper DR, Gibson B, Nelson EA, Prescott RJ. Randomised, double blind placebo controlled trial of pentoxifylline in the treatment of venous leg ulcers. *BMJ.* 1999; 319(7214):875–878. https://doi.org/10.1136/bmj.319.7214.875.
84. Falanga V, Fujitani RM, Diaz C, et al. Systemic treatment of venous leg ulcers with high doses of pentoxifylline: efficacy in a randomized, placebo-controlled trial. *Wound Repair Regen.* 1999;7(4): 208–213. https://doi.org/10.1046/j.1524-475x.1999.00208.x.

85. Nelson EA, Prescott RJ, Harper DR, Gibson B, Brown D, Ruckley CV. A factorial, randomized trial of pentoxifylline or placebo, four-layer or single-layer compression, and knitted viscose or hydrocolloid dressings for venous ulcers. *J Vasc Surg*. 2007;45(1):134–141. https://doi.org/10.1016/j.jvs.2006.09.043.
86. Jull AB, Arroll B, Parag V, Waters J. Pentoxifylline for treating venous leg ulcers. *Cochrane Database Syst Rev*. 2012;12:CD001733. https://doi.org/10.1002/14651858.CD001733.pub3.
87. Wu B, Lu J, Yang M, Xu T. Sulodexide for treating venous leg ulcers. *Cochrane Database Syst Rev*. 2016;6:CD010694. https://doi.org/10.1002/14651858.CD010694.pub2.
88. Coccheri S, Bignamini AA. Pharmacological adjuncts for chronic venous ulcer healing. *Phlebology*. 2016;31(5):366–367. https://doi.org/10.1177/0268355515619562.
89. Schrader E, Schwankl W, Sieder C, Christoffel V. [Comparison of the bioavailability of beta-aescin after single oral administration of two different drug formulations containing an extract of horsechestnut seeds]. *Pharmazie*. 1995;50(9):623–627. https://www.ncbi.nlm.nih.gov/pubmed/7480102.
90. Facino RM, Carini M, Stefani R, Aldini G, Saibene L. Anti-elastase and anti-hyaluronidase activities of saponins and sapogenins from Hedera helix, Aesculus hippocastanum, and Ruscus aculeatus: factors contributing to their efficacy in the treatment of venous insufficiency. *Arch Pharm (Weinheim)*. 1995;328(10):720–724. https://doi.org/10.1002/ardp.19953281006.
91. Pittler MH, Ernst E. Horse chestnut seed extract for chronic venous insufficiency. *Cochrane Database Syst Rev*. 2012;11:CD003230. https://doi.org/10.1002/14651858.CD003230.pub4.
92. O'Meara S, Al-Kurdi D, Ologun Y, Ovington LG, Martyn-St James M, Richardson R. Antibiotics and antiseptics for venous leg ulcers. *Cochrane Database Syst Rev*. 2014;1:CD003557. https://doi.org/10.1002/14651858.CD003557.pub5.
93. Norman G, Westby MJ, Rithalia AD, Stubbs N, Soares MO, Dumville JC. Dressings and topical agents for treating venous leg ulcers. *Cochrane Database Syst Rev*. 2018;6:CD012583. https://doi.org/10.1002/14651858.CD012583.pub2.
94. Raffetto JD, Ligi D, Maniscalco R, Khalil RA, Mannello F. Why venous leg ulcers have difficulty healing: overview on pathophysiology, clinical consequences, and treatment. *J Clin Med*. 2020;10(1). https://doi.org/10.3390/jcm10010029.
95. Jones JE, Nelson EA, Al-Hity A. Skin grafting for venous leg ulcers. *Cochrane Database Syst Rev*. 2013;(1):CD001737. https://doi.org/10.1002/14651858.CD001737.pub4.
96. Massand S, Lewcun JA, LaRosa CA. Clinical and cost efficacy of advanced wound care matrices in the treatment of venous leg ulcers: a systematic review. *J Wound Care*. 2021;30(7):553–561. https://doi.org/10.12968/jowc.2021.30.7.553.
97. O'Donnell Jr TF, Passman MA. Clinical practice guidelines of the society for vascular surgery (SVS) and the American venous Forum (AVF)–Management of venous leg ulcers. Introduction. *J Vasc Surg*. 2014;60(2 Suppl):1S–2S. https://doi.org/10.1016/j.jvs.2014.04.058.
98. Nicolaides A, Kakkos S, Eklof B, et al. Management of chronic venous disorders of the lower limbs - guidelines according to scientific evidence. *Int Angiol*. 2014;33(2):87–208. https://www.ncbi.nlm.nih.gov/pubmed/24780922.
99. Kearon C, Kahn SR, Agnelli G, Goldhaber S, Raskob GE, Comerota AJ. Antithrombotic therapy for venous thromboembolic disease: American College of chest physicians evidence-based clinical practice guidelines (8th Edition). *Chest*. 2008;133(6 Suppl):454S–545S. https://doi.org/10.1378/chest.08-0658.
100. Nicolaides AN, Allegra C, Bergan J, et al. Management of chronic venous disorders of the lower limbs: guidelines according to scientific evidence. *Int Angiol*. 2008;27(1):1–59. https://www.ncbi.nlm.nih.gov/pubmed/18277340.
101. Lawson JA, Toonder IM. A review of a new Dutch guideline for management of recurrent varicose veins. *Phlebology*. 2016;31(1 Suppl):114–124. https://doi.org/10.1177/0268355516631683.
102. Nicolaides AN. The benefits of micronized purified flavonoid fraction (MPFF) throughout the progression of chronic venous disease. *Adv Ther*. 2020;37(Suppl 1):1–5. https://doi.org/10.1007/s12325-019-01218-8.

CHAPTER 16

Treatment modalities for the management of nonhealing wounds in patients with chronic venous insufficiency

Gregory G. Westin, John G. Maijub and Michael C. Dalsing
Division of Vascular Surgery, Indiana University, Indianapolis, IN, United States

Introduction

Surgically eliminating major venous obstruction and axial reflux to the venous ulcer bed and providing adequate compression therapy are the two major treatment modalities used for venous ulcer healing.[1-4] Other chapters in this text deal with the diagnosis and treatment of venous outflow obstruction, superficial venous reflux, deep venous valve reflux, and concomitant arterial insufficiency. In addition to surgically addressing these hemodynamic considerations, other chapters in this text address important aspects relevant to ulcer healing including compression, potential adjuvant therapy, treatment compliance, negative pressure dressings, venoactive drug treatment, and advanced dressings and biologics for treating venous ulcers. The pathophysiology of the many complicated potential factors that can contribute to nonhealing venous ulceration was well outlined in a recent review.[5] These factors should all be considered in trying to develop and implement treatments for patients with refractory ulcers.

This chapter focuses on local treatment of venous ulcers and additional treatment modalities for nonhealing venous ulcers after first-line therapy has failed.

The Society for Vascular Surgery (SVS) and the American Venous Forum published guidelines in 2014 that outline major components of treatment for venous leg ulcers, along with further detail regarding prior research in each area.[6] The GRADE system used in those guidelines to evaluate strong (grade 1) versus weak (grade 2) recommendations and levels of evidence (A: high quality, B: moderate quality, C: low or very low quality) are noted with regard to each of the modalities discussed later.

Lifestyle modification

Healing of any wound regardless of underlying etiology can be inhibited by inadequate nutrition. As in a patient with any other nonhealing wound, if the patient has evidence of malnutrition a complete nutritional assessment should be conducted, and nutritional supplementation should be provided if malnutrition is identified.

The most clearly beneficial lifestyle change for venous ulcers is supervised exercise therapy to improve calf muscle pump function and reduce edema.[7,8] Six randomized controlled studies, including five combined and analyzed by meta-analysis, demonstrated improved venous ulcer healing at 12 weeks after an exercise program versus no exercise.[9] A combination of progressive resistance exercise plus prescribed physical activity appears to be most effective (additional 27 cases healed per 100 patients; $P = .004$). Supervised exercise improves muscle pump function and reduces pain and edema in patients with venous ulcers and thus is a 2B guideline recommendation from the SVS.[6] This simple and inexpensive therapy may be often neglected in the early stages of ulcer treatment, so special attention should be paid to readdressing it in the patient with a nonhealing wound.

Medical therapy

The use of pentoxifylline can improve venous ulcer healing in cases of long-standing venous ulcers, although this is an off-label use in the United States. A Cochrane review included 12 trials (864 participants) in a meta-analysis and concluded that pentoxifylline demonstrated its beneficial use in patients with difficult-to-heal venous ulcers when added to standard compression therapy. It is more effective than placebo in achieving complete ulcer healing or significant improvement (RR 1.70, 95% CI 1.30–2.24) and even appears beneficial in patients not receiving compression.[10] Pentoxifylline is associated with gastrointestinal side effects, has antiplatelet effects, and may cause a decrease in blood pressure. Overall, it appears to be a reasonable medical adjunct to add for patients with recalcitrant venous ulceration.

Untreated arterial or deep venous disease

The presence of the arterial occlusive disease may prevent an ulcer that is primarily due to venous causes from healing, even if the arterial disease was not the underlying cause of the ulcer. While one study found that patients with the moderate arterial occlusive disease (ankle-brachial index (ABI) of 0.5–0.8) can heal venous ulcers with gentle care including reduced compression therapy, the ulcers took longer for complete healing (52 vs. 48 weeks, $P = .009$) and had a delayed median healing time (25.5 vs. 23 weeks, $P = .03$), even after accounting for other factors delaying wound healing including deep venous insufficiency, ulcer surface area, and ulcer duration.[11] This is also supported by evidence from other studies showing a lower rate of healing in patients with uncorrected moderate arterial insufficiency. Even in patients with moderate arterial disease, but certainly in patients with critical limb ischemia, treating the arterial occlusive disease appears to improve the chances of rapid venous ulcer healing.[12,13] In patients with untreated arterial disease and an ABI of ≤ 0.5 or an absolute ankle pressure of <60 mmHg, the SVS guidelines suggest against compression therapy.[6] If this arterial disease is corrected, however, compression therapy could be pursued, which may help explain some of the benefits of arterial treatment in combined arterial and venous ulcers. In the authors' experience, large venous ulcers may interfere with the measurement of ankle pressure due to pain from the blood pressure cuff; in these cases, we suggest an arterial evaluation by measuring a toe-brachial index (TBI) and consideration of revascularization in patients with TBI of less than 0.5 or an absolute toe pressure of below 60 mmHg.

In the authors' experience, superficial and perforator vein insufficiency are commonly identified as underlying pathophysiology for venous leg ulcers and should be corrected. These should certainly be evaluated in the patient with a nonhealing venous ulcer, if they have not been already (2C recommendations), to expedite ulcer healing.[6] Similarly, endovascular therapy of iliac vein or inferior vena cava (IVC) obstruction is relatively common and recommended (1C).[6] However, deep venous obstruction or reflux (requiring open repair) is less commonly treated, and failure to treat this has been demonstrated to be a predictor of treatment failure in multiple studies.[11,14] We, therefore, recommend special attention to identifying uncorrected deep venous obstruction or reflux in patients with nonhealing venous ulcers. SVS recommendations include autogenous venous bypass or endophlebectomy for infrainguinal venous obstruction (2C); valve repair, transposition, transplantation, or autogenous substitution for deep venous reflux (2C); and open venous bypass for caval or iliac occlusive disease not treatable endovascularly (2C).[6]

Compression

In the patient without significant arterial insufficiency or after treatment as mentioned earlier, compression therapy has strong evidence to support its use in facilitating venous wound closure. With at least nine randomized clinical trials and a meta-analysis, the SVS gave a 1A recommendation for compression therapy.[6,15] There is lower-quality evidence regarding the many different systems for bandaging to provide compression, but the SVS recommends multicomponent compression bandages over single-component bandages (2B).[6]

Identification and treatment of infection

The prophylactic use of systemic antibiotics is not recommended (2C), as it does not improve ulcer healing and may cause the proliferation of resistant bacteria that are difficult to treat.[16,17] Multiple trials evaluating many different antibiotic strategies, including both oral and topical preparations, support this conclusion. There is conflicting evidence pertinent to the management of bacterial colonization and biofilm formation in wounds, as high bacterial counts correlate with delayed healing, and healing with reduced bacterial load, but the details of the bacterial bioburden and treatment with topical antibiotics have not been shown to make a difference in healing.[18–20]

In the face of healing resistance, however, cellulitis or a deeper infection must be considered and, if present, treated. Systemic antibiotics are recommended for established infections including cellulitis or deeper infections (1B).[6] In general, antibiotic therapy should be directed at the gram-positive organisms common in skin flora,[21] but a wound culture may help to direct therapy. Conventional wound cultures may not capture well the complexity of the bacterial burden in chronic venous wounds, however, and advanced methods of analysis or disruption of biofilm may be necessary for a culture-guided approach to be superior to empiric therapy for cellulitis associated with venous ulcers.[22] SVS recommendations include treating wounds with clinical evidence of infection and $>1 \times 10^6$ colony-forming unit (CFU)/g of bacteria on quantitative culture (2C) and virulent bacteria even with lower bacterial loads (2C), using cultures to guide antibiotic therapy (1C), using oral antibiotics limited to 2 weeks initially (1C), and performing concomitant mechanical disruption (2C).[6] In addition, in wounds that do not improve with standard wound and compression therapy after 4–6 weeks of treatment and in all ulcers with atypical features, a wound biopsy should be performed to exclude malignancy (1C). The rate of skin cancer in chronic leg ulcers has been estimated to be 4%–10%.[23,24]

Debridement

Debridement to remove slough, nonviable tissue, and eschar at initial wound evaluation (1B) and as needed to maintain a clean wound bed appropriate for healing (2B) are important components of venous ulcer care, and the appearance of the wound bed should be carefully reevaluated in the nonhealing venous ulcer[6] (Fig. 16.1A and B). In general, many types of debridement, including surgical debridement, can be performed under local anesthesia (1B), though in select cases regional or even general anesthesia may be necessary. Wounds treated with debridement appear to reduce in size more quickly when treated, but the frequency of debridement has not been clearly linked to differences in outcomes.[25,26]

The ideal type of debridement for venous ulcers has not been clearly established. The SVS gives its strongest recommendation (1B) to surgical debridement, but suggests that multiple types may be appropriate (2B). Hydrosurgical debridement (2B) may reduce the time required for debridement and has some evidence of comparable efficacy to surgical debridement, but may be associated with increased cost and postprocedural pain.[27,28] Larval debridement anecdotally leads to very clean wounds, in that wounds with associated larvae seen in the emergency room tend to be very clean once the larvae are removed. It also has a 2B recommendation from the SVS and is associated with decreased time to debridement, but was not better than hydrogel therapy in terms of cost or efficacy in a randomized trial.[29,30] Enzymatic debridement was recommended by the SVS when surgical debridement is not available (2C), but a subsequent systematic review and meta-analysis suggested that the data supporting it are at high risk of bias and patients receiving it may be at higher risk of adverse events.[31]

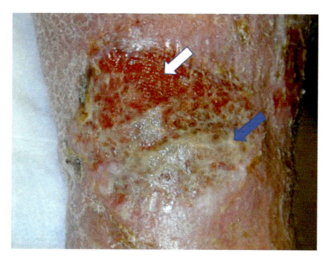
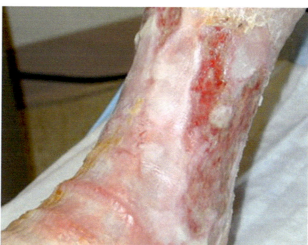

Figure 16.1 (A) Venous ulcer bed near medial malleolus demonstrating skin sloughing and debris (*blue arrow*) and need for mild/moderate debridement. Area of nicely granulating bed as well (*white arrow*). (B) Cleaned and granulating venous ulcer with areas of skin bridging where healing.

Ultrasonic debridement appeared to show some benefit in two studies in the 1990s,[32,33] but subsequent data are limited. One randomized trial reported improvement in wound "appearance" without significant differences in wound size compared to a control group at 12 weeks, with a dramatically higher rate of control group patients missing treatment visits and follow-up suggestive of trial limitations.[34] A comparison to historic controls suggested some benefit in healing time with ultrasonic debridement.[35] A meta-analysis written by a medical device consultant, relying on unpublished data from conference presentations, suggested that ultrasonic debridement was superior to surgical debridement at 3 and 5 months but not 6 months.[36] Overall, the SVS concluded that evidence for ultrasonic debridement is limited and suggested against it (2C).[6]

Overall, the data appears limited with regard to decision-making on the type of debridement, but there is a general consensus that debridement offers some benefit particularly for wounds with a large amount of nonviable material in them.[37]

Dressings and local agents

The availability of a wide variety of dressings to treat venous ulcers is impressive (alginate, foam, hydrocolloid, hydrofiber, hydrogel, honey, etc.) and is likely explained by the fact that none have been proven to be superior to others for venous ulcers. A systematic review of multiple dressings suggested that hydrogels offered some benefit over basic contact dressings (RR 1.80 [95% CI, 1.27−2.56]), but that no other dressings demonstrated significant differences.[38] A meta-analysis of honey dressings suggested some benefits in other types of wounds but no clear benefit in venous ulcers.[39] A meta-analysis of silver-based dressings found some improvement in the surrogate outcomes of wound healing rate and reduction of wound size, without a benefit in terms of complete wound healing.[40]

An evidence review from the Agency for Healthcare Research and Quality and guidelines from the National Institute for Health and Care Excellence do not recommend any particular dressing or topical agent to treat venous ulcers based on the available data.[41,42]

While no particular dressing is clearly superior in general, there are potential indications for each in particular instances. For example, in situations of very exudative wounds, a dressing that is very absorbent, such as alginate or foam, could be beneficial. The SVS also recommends using dressings that will protect the periulcer skin and applying skin lubricants to reduce periulcer dermatitis.

Many physicians have advocated the use of the Unna's boot as a dressing for venous wounds. A zinc-based bandage that hardens to become a semirigid mold against the leg skin after application, the Unna boot provides inelastic compression rather than elastic compression. This prevents the leg from swelling with dependency and may be more comfortable for patients at night. They are routinely left in place for a week at a time and can reduce the frequency of dressing changes. While some small studies have suggested a benefit of Unna boot therapy over other dressings and forms of compression,[43] most have suggested no benefit.[44] A recent meta-analysis suggested no statistically significant benefit of Unna boots in complete ulcer healing.[45] However, given their comfort for patients, several studies have reported positive impacts on patient-reported outcomes such as well-being, the hope of healing, self-esteem, and functional status.[46−48]

Topical antibiotics either alone or within dressings to treat noninfected ulcers are not generally recommended (2A recommendation against). Similarly, the SVS recommends against the routine use of topical antiinflammatory agents (2C against), unless there is severe dermatitis present, in which case topical steroids might reduce the symptoms and prevent secondary ulceration (2C in favor).[6]

Another type of topical treatment is that of "protease-modulating matrix" treatments. The theory behind these products is that nonhealing ulcers may have chronically elevated protease levels, which could cause tissue damage and impair healing. A meta-analysis of 12 studies of protease-modulating matrix treatments, however, demonstrated no clear benefit compared to other dressings.[49]

Isolated growth factors and cytokines do not appear to add benefit to the healing of venous ulcers and can add significant cost to care. Although more invasive than topical agents, there might be potential use for tissues that can alter the general environment of the local bed, such as the injection of centrifuged adipose tissue containing progenitor cells. In one trial, there were no adverse effects, the healing rate was significantly faster when locally injected (17.5 ± 7.0 weeks) than when not injected (24.5 ± 4.9 weeks, $P < .036$), and the patients reported reduced pain.[50]

External stimulants

Therapeutic ultrasound, both high and low frequency, has been evaluated as an adjuvant treatment for venous ulcer healing. Neither frequency has been shown statistically to improve healing time or proportion of ulcers healed. Moreover, there is at least one study that suggested an increased number of nonserious adverse events with the use of high-frequency ultrasound.[51,52] The SVS recommends against routine therapeutic ultrasound in the treatment of venous ulcers (2B).[6]

Extracorporeal shock therapy, as might be used to treat ureteral stones or improve bony healing, has been observed to potentially affect wound healing by stimulating angiogenesis and increasing collagen production in animal models.[53,54] A single-arm study demonstrated some tolerability, however, 15% of patient did not continue therapy after their first treatment with the technology.[55] One small nonrandomized study reported a significant benefit of extracorporeal shock wave therapy compared to control in the reduction of wound size in a mixture of various lower extremity wound types.[56] To date, there are no randomized controlled studies to support the use of this modality in treating recalcitrant venous ulcers.[57]

The use of electrical stimulation was not supported by data regarding the efficacy of this treatment modality in guideline recommendations to date, and the SVS recommended against it (2C).[14] Two subsequent meta-analyses suggest that electrical stimulation does appear to improve the speed of ulcer healing for leg ulcers in general compared to controls, but that there appears to be greater benefit in pressure ulcers than in venous ulcers.[58,59] There is also some early evidence from an animal study that dressings that deliver low-level electrical stimulation to wounds (electroceutical dressings) may disrupt biofilms to improve wound healing in wounds with difficult-to-treat polymicrobial infections.[60] One function of electrical stimulation may be a reduction in pain.[61]

Photobiomodulation, a technique using low-energy light to induce healing in recalcitrant wounds, has been investigated in depth in other areas such as maxillofacial wounds and has some research in lower extremity wound healing. A prospective uncontrolled trial demonstrated good tolerability in venous ulcers.[62] A trial of patients with multiple wounds, using each patient as their own control, demonstrated small residual wound area in wounds of multiple etiologies at 10 weeks (42% vs. 63%, $P = .03$), with more pronounced improvement in venous ulcers (33% vs. 60%, $P < .01$), and a significant reduction in pain compared to baseline ($P < .01$; could not be compared to controls due to design).[63] A small randomized pilot study also demonstrated decreased wound areas in pressure injuries compared to controls.[64] There have been many protocols for the wavelength of light and other parameters, and the potential mechanism of action is unclear. Ultraviolet light therapy, the only light therapy addressed in the SVS guidelines, was not recommended due to a paucity of evidence (2C).[6]

Intermittent pneumatic compression (IPC) has had variable results for venous ulcer healing. A Cochrane review of multiple studies found most to be at risk of bias, and based on the available evidence suggested that IPC appears to provide benefit compared to dressings alone (without compression) but showed mixed results when comparing IPC plus compression versus compression alone.[65] One trial deemed at low risk of bias compared a rapid inflation protocol to slow inflation and demonstrated a higher rate of healing at 6 months (86% vs. 61%, $P < .01$) and a faster rate of healing (0.09 cm2/day vs. 0.04 cm^2/day, $P < .01$) with rapid inflation.[66] The Centers for Medicare & Medicaid Services (CMS) will reimburse for IPC in patients with venous ulcers that have not healed after 6 months of standard compression. The SVS guidelines recommend IPC when other compression options are not available, cannot be used, or an ulcer has not healed despite prolonged therapy with standard compression.[6]

Negative pressure wound therapy (NPWT) has been studied in a variety of wounds and found to have some benefits, for example, in wounds healing by secondary intention in general.[67] The SVS guidelines suggested against NPWT based on insufficient evidence and noted that it may be beneficial for preparing a wound bed for skin grafting.[6] A randomized trial in hospitalized patients with chronic leg ulcers that compared NPWT along with punch skin grafting demonstrated reduced time for wound bed preparation (7 vs. 17 days, $P < .01$), reduced time to complete healing (29 vs. 45 days, $P < .01$), more successful skin grafts (83% vs. 70%, $P = .01$), less nursing time (232 vs. 386 min, $P < .01$), and lower total cost ($3881 vs. $5,452, $P < .01$).[68] However, concerns were raised about the methodology of this industry-sponsored trial, and a subsequent Cochrane review concluded that the evidence for NPWT in the context of venous ulcers was limited.[69,70] In a subsequent multicenter randomized trial comparing mechanically powered NPWT to electrically powered NPWT for venous ulcers, the mechanically powered device was found to be superior.[71]

Skin grafting, skin substitutes, and soft tissue substitutes

In trying to achieve closure of venous ulcers, one area of significant interest has been the use of skin or skin substitutes to achieve epithelialization. While autogenous split-thickness skin grafting does not have evidence to support its use as primary therapy for venous ulcers, it may be of benefit for chronic nonhealing ulcers, particularly large ones (2B).[6] Jankunas et al. demonstrated improved healing with skin grafting compared to conservative therapy for large venous wounds that were present for over 6 months, but the overall take of the split-thickness skin graft was only 67.5%.[72]

Of the many nonautogenous skin substitute products, the SVS recommends porcine small intestinal submucosa (OASIS, 2B) and a cultured allogeneic bilayer skin replacement product (Apligraf, 2A).[6] Small intestinal submucosa with compression versus compression only healed 55% versus 43% of venous ulcers at 12 weeks, respectively ($P = .02$).[73] Cultured allogenic skin replacement with compression versus Unna boot compression demonstrated 63% versus 48% wound healing at 6 months ($P = .02$), with an average time to healing of 61 versus 181 days ($P < .01$).[74] A Cochrane review in 2013 suggested only the bilayer artificial skin.[75] However, the number of skin substitute types continues to expand rapidly. For example, a recent multicenter randomized trial demonstrated a greater decrease in ulcer area at 14 weeks with autologous skin cell suspension plus compression compared to compression alone (8.94 vs. 1.23 cm^2, $P = .01$).[76]

Large, recurrent, or recalcitrant venous ulcers may require the replacement of the entire surrounding soft tissue bed to allow healing. One experience involving 25 free flap procedures in 22 such patients (nonhealing for a mean of 5.24 years) allowed healing without recurrence in the area of the flap for a mean of 58 months, but three patients did experience a new ulcer in the same leg after 6–77 months.[77]

Conclusion

Nonhealing venous ulcer care continues to be a challenging area with multiple competing possible treatment modalities, and thus well-designed randomized controlled trials are needed to resolve the clinical dilemma of "best care" faced daily by practitioners caring for patients with venous ulcers. While the quality of much of the evidence is low and comparisons between treatment options even more limited, we offer some possible structure for deciding between treatment choices in Table 16.1. Local experience and expertise will likely guide decision-making until better quality evidence emerges. Treatments to consider foremost, in addition to correcting underlying venous pathology, that are best supported by the literature are compression, lifestyle changes, treating associated significant arterial occlusive disease, surgical debridement to clean tissue, treatment of clinical infection, and skin substitutes.

Table 16.1 Findings in chronic venous ulcer patients and suggested treatment options.

Finding	Recommended further evaluation	Treatment option
Absent pedal pulses, no available recent ankle-brachial index (ABI), or other signs of ischemia	ABI or, in patients with diabetes or inability to measure ankle pressure due to pain, toe-brachial index (TBI)	Consider arterial revascularization if ABI ≤ 0.8 or TBI ≤ 0.5.
Known untreated venous insufficiency, or no recent evaluation of venous insufficiency	Venous duplex ultrasound to evaluate for obstruction and reflux. Consider further imaging if this suggests proximal venous obstruction.	Ablation of refluxing superficial or perforating veins, endovascular treatment of iliocaval obstruction, or surgical reconstruction of venous obstruction or reflux
Residual limb edema	N/A	Increase strength of compression, consider multilayered compression, and consider pneumatic compression
Wound bed with slough, nonviable tissue, or eschar	N/A	Debridement
Clinical evidence of infection such as surrounding erythema or purulent drainage	Wound culture	Oral antibiotic therapy, guided by culture results or focused on gram-positive organisms in the absence of culture data, with concomitant mechanical disruption, initially limited to 2 weeks of therapy.[6]
Large ulcer that has not epithelialized after 6 weeks of comprehensive therapy including compression	Clinical evaluation to ensure clean wound bed	Split-thickness skin graft, bilayer allogeneic skin replacement, or porcine small intestinal submucosa skin substitute application

References

1. Mayberry JC, Moneta GL, Taylor LM, Porter JM. Fifteen-year results of ambulatory compression therapy for chronic venous ulcers. *Surgery*. May 1991;109(5):575–581.
2. Erickson CA, Lanza DJ, Karp DL, et al. Healing of venous ulcers in an ambulatory care program: the roles of chronic venous insufficiency and patient compliance. *J Vasc Surg*. November 1995;22(5): 629–636.
3. Montminy ML, Jayaraj A, Raju S. A systematic review of the efficacy and limitations of venous intervention in stasis ulceration. *J Vasc Surg Venous Lymphat Disord*. May 2018;6(3):376–398. e1.

4. Raju S, Owen S, Neglen P. The clinical impact of iliac venous stents in the management of chronic venous insufficiency. *J Vasc Surg*. January 2002;35(1):8–15.
5. Raffetto JD, Ligi D, Maniscalco R, Khalil RA, Mannello F. Why venous leg ulcers have difficulty healing: overview on pathophysiology, clinical consequences, and treatment. *J Clin Med*. December 24, 2020;10(1):E29.
6. O'Donnell TF, Passman MA, Marston WA, et al. Management of venous leg ulcers: clinical practice guidelines of the society for vascular surgery and the American venous forum. *J Vasc Surg*. August 1, 2014;60(2):3S–59S.
7. Padberg FT, Johnston MV, Sisto SA. Structured exercise improves calf muscle pump function in chronic venous insufficiency: a randomized trial. *J Vasc Surg*. January 2004;39(1):79–87.
8. Kan YM, Delis KT. Hemodynamic effects of supervised calf muscle exercise in patients with venous leg ulceration: a prospective controlled study. *Arch Surg*. December 2001;136(12):1364–1369.
9. Jull A, Slark J, Parsons J. Prescribed exercise with compression vs compression alone in treating patients with venous leg ulcers: a systematic review and meta-analysis. *JAMA Dermatol*. November 1, 2018; 154(11):1304–1311.
10. Jull AB, Arroll B, Parag V, Waters J. Pentoxifylline for treating venous leg ulcers. *Cochrane Database Syst Rev*. December 12, 2012;12:CD001733.
11. Mosti G, Cavezzi A, Massimetti G, Partsch H. Recalcitrant venous leg ulcers may heal by outpatient treatment of venous disease even in the presence of concomitant arterial occlusive disease. *Eur J Vasc Endovasc Surg*. September 2016;52(3):385–391.
12. Lantis JC, Boone D, Lee L, Mendes D, Benvenisty A, Todd G. The effect of percutaneous intervention on wound healing in patients with mixed arterial venous disease. *Ann Vasc Surg*. January 2011;25(1): 79–86.
13. Treiman GS, Copland S, McNamara RM, Yellin AE, Schneider PA, Treiman RL. Factors influencing ulcer healing in patients with combined arterial and venous insufficiency. *J Vasc Surg*. June 2001;33(6): 1158–1164.
14. Melikian R, O'Donnell TF, Suarez L, Iafrati MD. Risk factors associated with the venous leg ulcer that fails to heal after 1 year of treatment. *J Vasc Surg Venous Lymphat Disord*. January 2019;7(1):98–105.
15. Mauck KF, Asi N, Elraiyah TA, et al. Comparative systematic review and meta-analysis of compression modalities for the promotion of venous ulcer healing and reducing ulcer recurrence. *J Vasc Surg*. August 1, 2014;60(2, Suppl.):71S–90S.e2.
16. O'Meara S, Al-Kurdi D, Ologun Y, Ovington LG, Martyn-St James M, Richardson R. Antibiotics and antiseptics for venous leg ulcers. *Cochrane Database Syst Rev*. January 10, 2014;1:CD003557.
17. Alinovi A, Bassissi P, Pini M. Systemic administration of antibiotics in the management of venous ulcers. A randomized clinical trial. *J Am Acad Dermatol*. August 1986;15(2 Pt 1):186–191.
18. Sibbald RG, Contreras-Ruiz J, Coutts P, Fierheller M, Rothman A, Woo K. Bacteriology, inflammation, and healing: a study of nanocrystalline silver dressings in chronic venous leg ulcers. *Adv Skin Wound Care*. 2007;20(10):549–558.
19. Lantis Jc, Gendics C. In vivo effect of sustained-release silver sulphadiazine foam on bioburden and wound closure in infected venous leg ulcers. *J Wound Care*. February 1, 2011;20(2):90–96.
20. Moore K, Hall V, Paull A, et al. Surface bacteriology of venous leg ulcers and healing outcome. *J Clin Pathol*. September 1, 2010;63(9):830–834.
21. Kilburn SA, Featherstone P, Higgins B, Brindle R. Interventions for cellulitis and erysipelas. *Cochrane Database Syst Rev*. 2010;2020(12).
22. Tuttle MS. Association between microbial bioburden and healing outcomes in venous leg ulcers: a review of the evidence. *Adv Wound Care*. January 1, 2015;4(1):1–11.
23. Senet P, Combemale P, Debure C, et al. Malignancy and chronic leg ulcers: the value of systematic wound biopsies: a prospective, multicenter, cross-sectional study. *Arch Dermatol*. 2012;148(6):704–708.
24. Yang D, Morrison BD, Vandongen YK, Singh A, Stacey MC. Malignancy in chronic leg ulcers. *Med J Aust*. 1996;164(12):718–720.
25. Williams D, Enoch S, Miller D, Harris K, Price P, Harding KG. Effect of sharp debridement using curette on recalcitrant nonhealing venous leg ulcers: a concurrently controlled, prospective cohort study. *Wound Repair Regen*. 2005;13(2):131–137.

26. Cardinal M, Eisenbud DE, Armstrong DG, et al. Serial surgical debridement: a retrospective study on clinical outcomes in chronic lower extremity wounds: original research—clinical science. *Wound Repair Regen*. 2009;17(3):306–311.
27. Caputo WJ, Beggs DJ, DeFede JL, Simm L, Dharma H. A prospective randomised controlled clinical trial comparing hydrosurgery debridement with conventional surgical debridement in lower extremity ulcers. *Int Wound J*. 2008;5(2):288–294.
28. Mosti G, Iabichella ML, Picerni P, Magliaro A, Mattaliano V. The debridement of hard to heal leg ulcers by means of a new device based on Fluidjet technology. *Int Wound J*. 2005;2(4):307–314.
29. Dumville JC, Worthy G, Bland JM, et al. Larval therapy for leg ulcers (VenUS II): randomised controlled trial. *BMJ*. March 20, 2009;338:b773.
30. Soares MO, Iglesias CP, Bland JM, et al. Cost effectiveness analysis of larval therapy for leg ulcers. *BMJ*. March 20, 2009;338:b825.
31. Patry J, Blanchette V. Enzymatic debridement with collagenase in wounds and ulcers: a systematic review and meta-analysis. *Int Wound J*. December 2017;14(6):1055–1065.
32. Weichenthal M, Mohr P, Stegmann W, Breitbart EW. Low-frequency ultrasound treatment of chronic venous ulcers. *Wound Repair Regen*. March 1997;5(1):18–22.
33. Peschen M, Weichenthal M, Schöpf E, Vanscheidt W. Low-frequency ultrasound treatment of chronic venous leg ulcers in an outpatient therapy. *Acta Derm Venereol*. July 1997;77(4):311–314.
34. Murphy CA, Houghton P, Brandys T, Rose G, Bryant D. The effect of 22.5 kHz low-frequency contact ultrasound debridement (LFCUD) on lower extremity wound healing for a vascular surgery population: a randomised controlled trial. *Int Wound J*. June 2018;15(3):460.
35. Ennis WJ, Valdes W, Gainer M, Meneses P. Evaluation of clinical effectiveness of MIST ultrasound therapy for the healing of chronic wounds. *Adv Skin Wound Care*. October 2006;19(8):437–446.
36. Voigt J, Wendelken M, Driver V, Alvarez OM. Low-frequency ultrasound (20–40 kHz) as an adjunctive therapy for chronic wound healing: a systematic review of the literature and meta-analysis of eight randomized controlled trials. *Int J Low Extrem Wounds*. December 1, 2011;10(4):190–199.
37. Gethin G, Cowman S, Kolbach DN. Debridement for venous leg ulcers. *Cochrane Database Syst Rev*. September 14, 2015;9:CD008599.
38. Saco M, Howe N, Nathoo R, Cherpelis B. Comparing the efficacies of alginate, foam, hydrocolloid, hydrofiber, and hydrogel dressings in the management of diabetic foot ulcers and venous leg ulcers: a systematic review and meta-analysis examining how to dress for success. *Dermatol Online J*. August 15, 2016;22(8), 13030/qt7ph5v17z.
39. Jull AB, Cullum N, Dumville JC, Westby MJ, Deshpande S, Walker N. Honey as a topical treatment for wounds. *Cochrane Database Syst Rev*. March 6, 2015;3:CD005083.
40. Zhao M, Zhang D, Tan L, Huang H. Silver dressings for the healing of venous leg ulcer: a meta-analysis and systematic review. *Medicine (Baltim)*. September 11, 2020;99(37):e22164.
41. Full evidence summary: medicines and prescribing briefing | Chronic wounds: advanced wound dressings and antimicrobial dressings | Advice | NICE [Internet]. NICE; [cited 2021 Sep 20]. Available from: https://www.nice.org.uk/advice/esmpb2/chapter/Full-evidence-summary-medicines-and-prescribing-briefing.
42. Zenilman J, Valle MF, Malas MB. Chronic Venous Ulcers: A Comparative Effectiveness Review of Treatment Modalities. :306.
43. de Abreu AM, de Oliveira BGRB. A study of the Unna Boot compared with the elastic bandage in venous ulcers: a randomized clinical trial. *Rev Lat Am Enfermagem*. August 2015;23(4):571–577.
44. Polignano R, Bonadeo P, Gasbarro S, Allegra C. A randomised controlled study of four-layer compression versus Unna's Boot for venous ulcers. *J Wound Care*. January 2004;13(1):21–24.
45. Paranhos T, Paiva CSB, Cardoso FCI, et al. Systematic review and meta-analysis of the efficacy of Unna boot in the treatment of venous leg ulcers. *Wound Repair Regen*. May 2021;29(3):443–451.
46. Faria EC, Loiola T, Salomé GM, Ferreira LM. Unna boot therapy impact on wellbeing, hope and spirituality in venous leg ulcer patients: a prospective clinical trial. *J Wound Care*. April 2, 2020;29(4):214–220.
47. Salome GM, de Brito MJA, Ferreira LM. Impact of compression therapy using Unna's boot on the self-esteem of patients with venous leg ulcers. *J Wound Care*. September 2014;23(9):442–444, 446.

48. L de Lima E, Salomé GM, de Brito Rocha MJA, Ferreira LM. The impact of compression therapy with Unna's boot on the functional status of VLU patients. *J Wound Care*. October 2013;22(10):558–561.
49. Westby MJ, Norman G, Dumville JC, Stubbs N, Cullum N. Protease-modulating matrix treatments for healing venous leg ulcers. *Cochrane Database Syst Rev*. December 15, 2016;12:CD011918.
50. Zollino I, Campioni D, Sibilla MG, Tessari M, Malagoni AM, Zamboni P. A phase II randomized clinical trial for the treatment of recalcitrant chronic leg ulcers using centrifuged adipose tissue containing progenitor cells. *Cytotherapy*. February 2019;21(2):200–211.
51. Cullum N, Liu Z. Therapeutic ultrasound for venous leg ulcers. *Cochrane Database Syst Rev*. May 15, 2017;5:CD001180.
52. Watson JM, Kang'ombe AR, Soares MO, et al. VenUS III: a randomised controlled trial of therapeutic ultrasound in the management of venous leg ulcers. *Health Technol Assess*. March 2011;15(13):1–192.
53. Mittermayr R, Hartinger J, Antonic V, et al. Extracorporeal shock wave therapy (ESWT) minimizes ischemic tissue necrosis irrespective of application time and promotes tissue revascularization by stimulating angiogenesis. *Ann Surg*. May 2011;253(5):1024–1032.
54. Yang G, Luo C, Yan X, Cheng L, Chai Y. Extracorporeal shock wave treatment improves incisional wound healing in diabetic rats. *Tohoku J Exp Med*. December 2011;225(4):285–292.
55. Schaden W, Thiele R, Kölpl C, et al. Shock wave therapy for acute and chronic soft tissue wounds: a feasibility study. *J Surg Res*. November 2007;143(1):1–12.
56. Saggini R, Figus A, Troccola A, Cocco V, Saggini A, Scuderi N. Extracorporeal shock wave therapy for management of chronic ulcers in the lower extremities. *Ultrasound Med Biol*. August 2008;34(8):1261–1271.
57. Cooper B, Bachoo P. Extracorporeal shock wave therapy for the healing and management of venous leg ulcers. *Cochrane Database Syst Rev*. June 11, 2018;6:CD011842.
58. Barnes R, Shahin Y, Gohil R, Chetter I. Electrical stimulation vs. standard care for chronic ulcer healing: a systematic review and meta-analysis of randomised controlled trials. *Eur J Clin Invest*. April 2014;44(4):429–440.
59. Khouri C, Kotzki S, Roustit M, Blaise S, Gueyffier F, Cracowski J-L. Hierarchical evaluation of electrical stimulation protocols for chronic wound healing: an effect size meta-analysis. *Wound Repair Regen*. September 2017;25(5):883–891.
60. Barki KG, Das A, Dixith S, et al. Electric field based dressing disrupts mixed-species bacterial biofilm infection and restores functional wound healing. *Ann Surg*. 2019;269(4):756–766.
61. Guest JF, Singh H, Rana K, Vowden P. Cost-effectiveness of an electroceutical device in treating non-healing venous leg ulcers: results of an RCT. *J Wound Care*. April 2, 2018;27(4):230–243.
62. Romanelli M, Piaggesi A, Scapagnini G, et al. Evaluation of fluorescence biomodulation in the real-life management of chronic wounds: the EUREKA trial. *J Wound Care*. November 2, 2018;27(11):744–753.
63. Fraccalvieri M, Amadeo G, Bortolotti P, et al. Effectiveness of blue light photobiomodulation therapy in the treatment of chronic wounds. Results of the blue light for ulcer reduction (B.L.U.R.) study. *Ital J Dermatol Venereol*. April 2022;157(2):187–194. https://doi.org/10.23736/S2784-8671.21.07067-5.
64. Baracho V da S, Chaves MEDA, Huebner R, Oliveira MX, Ferreira PH da C, Lucas TC. Phototherapy (cluster multi-diode 630 nm and 940 nm) on the healing of pressure injury: a pilot study. *J Vasc Nurs*. September 2021;39(3):67–75.
65. Nelson EA, Hillman A, Thomas K. Intermittent pneumatic compression for treating venous leg ulcers. *Cochrane Database Syst Rev*. May 12, 2014;5:CD001899.
66. Nikolovska S, Arsovski A, Damevska K, Gocev G, Pavlova L. Evaluation of two different intermittent pneumatic compression cycle settings in the healing of venous ulcers: a randomized trial. *Med Sci Mon Int Med J Exp Clin Res*. July 2005;11(7):CR337–343.
67. Zens Y, Barth M, Bucher HC, et al. Negative pressure wound therapy in patients with wounds healing by secondary intention: a systematic review and meta-analysis of randomised controlled trials. *Syst Rev*. October 10, 2020;9(1):238.
68. Vuerstaek JDD, Vainas T, Wuite J, Nelemans P, Neumann MHA, Veraart JCJM. State-of-the-art treatment of chronic leg ulcers: a randomized controlled trial comparing vacuum-assisted closure (V.A.C.) with modern wound dressings. *J Vasc Surg*. November 2006;44(5):1029–1037. discussion 1038.

69. Ahmed M, Soskova T, Williams DT. Regarding "state-of-the-art treatment of chronic leg ulcers: a randomized controlled trial comparing vacuum-assisted closure (V.A.C.) with modern wound dressings". *J Vasc Surg*. September 1, 2007;46(3):614–615.
70. Dumville JC, Land L, Evans D, Peinemann F. Negative pressure wound therapy for treating leg ulcers. *Cochrane Database Syst Rev*. July 14, 2015;7:CD011354.
71. Marston WA, Armstrong DG, Reyzelman AM, Kirsner RS. A multicenter randomized controlled trial comparing treatment of venous leg ulcers using mechanically versus electrically powered negative pressure wound therapy. *Adv Wound Care*. February 1, 2015;4(2):75–82.
72. Jankunas V, Bagdonas R, Samsanavicius D, Rimdeika R. An analysis of the effectiveness of skin grafting to treat chronic venous leg ulcers. *Wounds*. May 2007;19(5):128–137.
73. Mostow EN, Haraway GD, Dalsing M, Hodde JP, King D. Effectiveness of an extracellular matrix graft (OASIS Wound Matrix) in the treatment of chronic leg ulcers: a randomized clinical trial. *J Vasc Surg*. May 1, 2005;41(5):837–843.
74. Falanga V, Margolis D, Alvarez O, et al. Rapid healing of venous ulcers and lack of clinical rejection with an allogeneic cultured human skin equivalent. *Arch Dermatol*. March 1, 1998;134(3):293–300.
75. Jones JE, Nelson EA, Al-Hity A. Skin grafting for venous leg ulcers. *Cochrane Database Syst Rev*. January 31, 2013;1:CD001737.
76. Hayes PD, Harding KG, Johnson SM, et al. A pilot multi-centre prospective randomised controlled trial of RECELL for the treatment of venous leg ulcers. *Int Wound J*. June 2020;17(3):742–752.
77. Kumins NH, Weinzweig N, Schuler JJ. Free tissue transfer provides durable treatment for large non-healing venous ulcers. *J Vasc Surg*. November 2000;32(5):848–854.

CHAPTER 17

Deep vein thrombosis and prevention of postthrombotic syndrome

Matthew Sussman[2] and Jose Almeida[1]

[1]Miami Vein, Division of Vascular and Endovascular Surgery, University of Miami Miller School of Medicine, Miami, FL, United States; [2]Division of Vascular and Endovascular Surgery, University of Miami Miller School of Medicine, Miami, FL, United States

The primary purpose of this chapter is to identify appropriate preventive and treatment modalities in patients who have endured lower extremity thromboembolic sequelae—specifically those patients at increased risk of developing postthrombotic syndrome (PTS). The secondary goal is to provide details on the diagnosis and management of patients with developed PTS and resulting chronic venous ulceration.

Traditionally, the focus of PTS management was directed to the afflicted lower extremity, and evaluation of venous blood flow that required a biomechanical understanding of the entire venous system (complex network consisting of a large number of vessels), and the interactions with the surrounding environment referred to as "boundary conditions" (abdominal, thoracic, and skeletal muscle pump).

One of the key functions of the venous system is to ensure adequate preload to the heart. Accordingly, the size of the venous conduit largely governs flow velocity, and hence, the propensity to thrombosis, as suggested by Virchow's triad. Obstruction of flow may also be secondary to external compression, as seen in May–Thurner syndrome. The local features of the flow, particularly wall shear stress, venous pressures, valve function, and endothelial function can influence thrombosis. These parameters determine different hemodynamic properties that do not function in isolation.

The venous obstruction and valvular incompetence that follows the resolution of acute venous thrombosis produces chronic venous hypertension that can lead to the development of PTS. Although early manifestations of this condition may be confined to mild edema and dull aching pain, eventually dermatitis, induration, ulceration, and severe pain commonly develop. Although this chapter will discuss strategies to mitigate the risks of severe PTS, the authors opine that ultimately the outcome depends on the thrombus load, and the systemic ability (reserves) of the venous network to compensate for the disruption of flow. Rapidly progressive thrombosis of common femoral vein, deep femoral vein, and femoral veins may lead to gangrene whereas when the thrombotic process is more gradual the occlusion can be relatively well compensated by concomitant collateral development. An occlusion that has been present for a long time will usually produce less obstruction than a more acute lesion of a similar extent, principally because of the additional time allowed to develop a more extensive collateral network.

Similarly, disability due to valvular incompetence is related to the location of the damaged valves, and possibly to the distensibility of the venous wall. Unlike obstruction, valvular incompetence may become worse with time. The healing process causes fibrosis of valvular cusps and subsequent retrograde flow; increased hydrostatic pressure or inherent wall defects lead to failure of normal cusps to coapt; and collateral vessels dilate beyond the capacity of their valves to function properly.

Two-thirds of patients with deep venous thrombosis (DVT) will not develop signs and symptoms of PTS, and the rest will develop PTS with variable degrees of severity.[1] Among the predisposing factors for PTS, ipsilateral recurrent DVT appeared to be the strongest predictor in several prospective studies.[2,3] Iliofemoral DVT, higher Villalta scores at 1-month post DVT, non-optimal anticoagulation, and higher BMI are the other strong predisposing risk factors found in the literature.[4]

Anatomic considerations in PTS

The calf possesses a venous system consisting of six axial tibial veins, multiple intramuscular gastrocnemius and soleal veins, and the great and small saphenous vein. Multiple connections of these veins exist via the extensive network within muscles, subcutaneous fat, and skin. This small-caliber calf network responds to increased venous pressure signaling and goes on to form robust collateral connections.[5,6] This calf network forms the femoropopliteal trunk as the venous flow progresses in the cephalad direction. As the venous network enters the thigh, the femoral, profunda, and great saphenous veins conduct most of the flow. The iliac veins on the other hand do not form collateral veins as readily as the veins of the calf and thigh. Thousands of collateral veins are needed to replicate the physiology of one 16 mm common iliac vein.[6,7]

A distinct compensatory mechanism of embryologic origin present in the thigh is the axial transformation of the profunda femoris vein.[7] The profunda femoris is believed to rapidly restore adequate flow volume after most acute femoral vein occlusions. The profunda will enlarge and increase its flow in response to an increase in venous pressure within 1 h of onset of a femoral vein DVT.[7] Femoral DVT-related symptoms will not improve in most cases of concomitant profunda femoris vein involvement. The great saphenous vein is an inadequate conduit in most cases of femoropopliteal DVT.[8] Sixteen 4 mm diameter saphenous veins are required to equal the conductance of a single 8 mm diameter femoral vein.[6,7] Many patients with iliac vein obstruction have extensive collaterals visible with contrast imaging yet continue to be symptomatic because the collateral flow will not correct venous hypertension despite conducting adequate flow.[9] Collaterals disappear only when a large caliber, low resistance conduit is provided, and relatively recently venous stenting with designated venous stents became increasingly utilized treatment modality to efficiently restore venous outflow. Investigators have found that venous pressure drops after stent correction of iliac vein stenosis, but iliac venous flow volume does not change.[10,11]

Hemodynamics of PTS

There are two essential hemodynamic alterations associated with PTS. The first is *venous obstruction*. Although many of the veins involved in the thrombotic process will become recanalized to some degree,[12] there is usually some residual venous obstruction.[13] When the obstructed segments are not in a critical location (i.e., iliac vein) and are limited in extent, the available venous collaterals are usually adequate to keep the venous resistance low. The second, and perhaps more severe hemodynamic alteration, is venous valvular insufficiency (reflux). Often, the recanalization process involved in thrombus resolution damages the affected valves and renders them incompetent.[14] The valve cusps may be fragmented or fibrotic, thickened and contracted, and may adhere to the venous wall. When these veins become collaterals, their lumens become dilated two or more times, thus preventing the physiologic coaptation of the valvular cusps.[14] Thus, the valves of veins that were never involved in the original thrombotic process may become functionally incompetent.

Sakaguchi et al.[15] examined venous outflow and venous reflux in a series of postthrombotic patients using a plethysmographic technique. In 30% of these patients, they could demonstrate only venous obstruction; in 24%, only valvular incompetence; and in 27%, both obstruction and valvular incompetence. On average, the resistance offered by the venous system in postthrombotic extremities is somewhat increased over the normal value but distinctly less than that observed in patients with acute thrombophlebitis. Barnes et al.[16] compared the maximum venous outflow (MVO) in postthrombotic extremities and extremities without evidence of current or previous thrombosis. The study showed MVO of 34 ± 15 cc/100 cc/min and 41 ± 11 cc/100 cc/min, in postthrombotic and nonaffected groups, respectively. Venographic studies have confirmed that the lowest values for MVO are obtained in those extremities that have persistent deep venous occlusion and that normal values occur in those limbs in which the deep system is patent even though there may be obvious postthrombotic changes.[17]

It has been previously shown that the ability of postthrombotic limbs to expand in response to venous congestion is usually somewhat less than that of normal limbs but more than that of limbs with acute phlebitis. This illustrates the negative impact that vessel wall fibrosis has on venous capacitance. Dahn and Eiriksson equated this ability with venous capacity.[18] Most investigators who attempted to quantitate the volume or rate of venous reflux have found it greater in patients with the PTS than in controls. Sakaguchi et al.[15,19] measured the rate at which the calf blood vessels refill after being emptied by manual squeezing with the patient sitting. By subtracting the rate of arterial inflow, they were able to estimate the rapidity with which blood flowed backward to fill the void left in the veins. The rate of reflux in postthrombotic extremities was about 11 times that in control limbs and more than double that found in acute phlebitis.

Barnes et al.[16] used a proximally placed thigh cuff inflated to 250–300 mmHg to prevent venous outflow and arterial inflow in the leg. A distal thigh cuff was rapidly inflated to 50 mmHg. Blood was displaced from the thigh to the calf at a rate of 4 ± 2 cc/100 cc/min in unaffected limbs. However, the rate of retrograde flow increased to 14 ± 7 cc/100 cc/min in patients with venous reflux. Using a tilt method, Bydeman et al.[20] found a volume increase of >2% in postthrombotic limbs within 10 s compared to 0.4%–1.5% in unaffected limbs. In 1971, Rutherford et al.[21] evaluated the volume of blood in the calf using Indium-133 isotope. When unaffected subjects were placed from the supine to a 45-degree foot-down position, the isotope counts increased 1.1 ± 0.4% per second. Although counts in postthrombotic extremities increased more rapidly (1.6 ± 0.8% per second), the difference was not significant.

In summary, studies confirm the following: venous outflow obstruction varies in patients with PTS which, on average, is less severe than that seen in acute DVT; the rate of venous reflux increases in PTS patients. These changes affect pressure-flow relationships and microcirculation.

Collateralization and flow patterns in acute DVT

Venous collaterals are more robust in venous occlusions when compared to arterial occlusions.[22] Venous hypertension is worse when large-caliber veins occlude. In 1963, DeWeese and Rogoff published a study that showed venous pressures in acute thrombosis of popliteal vein, femoral vein, and iliac vein of 18 mmHg, 51 mmHg vein, and 83 mmHg, respectively.[23] The pressure elevation caused by thrombosis stimulates the enlargement of the secondary venous (back-up) network that was previously quiescent. This backup system, otherwise known as collateral veins, will reduce peripheral venous pressure.[22] However, in the femoropopliteal segment, smaller-caliber collateral formation alone is usually unable to eradicate symptoms of venous hypertension.

Phlegmasia cerulea dolens

Phlegmasia cerulea dolens may develop from massive venous thrombosis of the lower extremities when the ilio-femoral veins are completely thrombosed, as well as the vena cava and the cross pelvic collaterals.[24] In addition, thrombosis is extensive in the femoral, popliteal, and below-knee veins. Thus, phlegmasia cerulea dolens may be considered the terminal stage of progressive ilio-femoral thrombosis. Massive edema usually forms rapidly and cyanosis begins distally and then progresses cephalad to involve the entire limb. On palpation, the limb feels cold, tense, and indurated and the skin becomes shiny and tight. The foot and toes may become necrotic and the lower leg develops a violaceous appearance. Cutaneous blebs, which occasionally contain hemorrhagic fluid, frequently develop. Intense agonizing pain, often described as a "bursting" sensation, is invariably present. The disease is characterized by a stagnant circulatory insufficiency and can be associated with extremity compartment syndrome. Distal arterial pulses are palpable in only 17% of the afflicted extremities. Gangrene, which occurs in 55% of the patients, requires an above-knee amputation in about one-half of the patients, the others recover to varying degrees. An occasional patient may need a hip disarticulation. Shock develops in 30% of the patients, pulmonary embolism in 22%, and death in 32%.[25] Prompt recognition of the disorder and intervention are important to emphasize because the early stage of phlegmasia cerulea dolens is reversible.

Microcirculation and capillary exchange

The exchange of biomaterials between the interstitial fluid and the bloodstream happens at the capillary level. This exchange is controlled by many factors such as the gap junctions of the endothelium, Starling forces, and glycocalyx to name a few. The edema that occurs with venous obstruction is explained by the loss of equilibrium of the Starling forces. Increased hydrostatic pressures in the venules favor the diffusion of fluid from the plasma into the interstitium. The quantity of fluid movement that occurs is regulated by the blood pressure (P_c) and the osmotic pressure of the plasma (πc) within the capillary plus the hydrostatic pressure (*PIF*) and the osmotic pressure (πIF) of the surrounding interstitial fluid.[26–29] The effective hydrostatic pressure forcing fluid out of the capillary is: *Pc−PIF*; and the osmotic pressure that facilitates the reabsorption of fluid back into the capillary is given by $\pi c - \pi IF$. Therefore, the net pressure moving fluid out of the capillary (P) is P = (*Pc−PIF*)−($\pi c - \pi IF$).[30] Under normal conditions, a net pressure of about 0.3 mmHg moves fluid out of the capillary. The excess fluid is removed by lymphatics.[31] Any increase in the arterial blood pressure (P_c) would increase the net filtration across the capillary membrane and thereby increase the quantity of interstitial fluid. Assuming that the osmotic pressure gradient remained constant, filtration would continue until the interstitial fluid pressure (*PIF*) increases to the point that equilibrium is reestablished. The role of glycocalyx and especially its role in stabilizing the endothelium is currently under investigation and likely will affect our understanding of the Starling model in the future.

Anticoagulation to prevent PTS

As stated earlier strongest predictor of PTS remains ipsilateral recurrent DVT.[2,3] For this reason, it is critical to direct treatment of these patients at mitigating the risk of recurrent DVT. To date, only anticoagulation at therapeutic values is an effective treatment for recurrent DVT and subsequent PTS development.[32] Subtherapeutic anticoagulation with warfarin in patients who had an international normalized ratio (INR) <2.0 is associated with a threefold risk increase for PTS.[33,34] Kahn et al.[35] reported that the severity of symptoms 1 month post-DVT was a strong predictor of PTS development. These findings reinforce the importance of therapeutic anticoagulation to reduce the severity of thrombotic damage to the microcirculation.

Venoactive drugs

The high-quality evidence in the literature on venoactive drugs in PTS prevention is lacking.[35] Sulodexide is a new agent that protects the vascular glycocalyx. An analysis based on an observational registry in Italy showed the benefits of sulodexide in protecting against PTS when combined with standard medical management.[36]

Direct oral anticoagulants (dabigatran, rivaroxaban, apixaban, and edoxaban) will likely be used in the future as anticoagulants to prevent PTS, however, their efficacy and safety have yet to be shown in the trials.[36] Optimization of calf muscle function through supervised exercise programs can improve venous return and reduce edema formation. This benefit has been demonstrated in a small, randomized control trial (RCT) of 95 patients with postthrombotic syndrome, 69 were eligible, 43 consented and were randomized, and 39 completed the study. Exercise training was associated with improvement in VEINES-QOL scores (exercise training mean change 6.0, standard deviation [SD] 5.1 v. control mean change 1.4, SD 7.2; difference 4.6, 95% CI 0.54 to 8.7; $P = .027$) and improvement in scores on the Villalta scale (exercise training mean change -3.6, SD 3.7 v. control mean change -1.6, SD 4.3; difference -2.0, 95% CI -4.6 to 0.6; $P = .14$).[37]

Elastic compression therapy

Compression therapy is the cornerstone of treatment for acute and chronic venous disease. In fact, two RCTs showed a decreased risk of PTS 2 years after an acute DVT in patients using compression stockings versus a control group without compression.[38,39] However, a widely cited RCT—Compression Stockings to Prevent the Post-Thrombotic Syndrome after Symptomatic Proximal Deep Vein Thrombosis—the SOX trial published in 2014, found little benefit in using compression stockings after DVT.[40] The cumulative incidence of PTS in elastic compression group was 14.2% versus 12.7% in placebo group, and there was statistical significance among groups ($P = .58$). This trial used the Villalta scale as the main measurement tool for PTS which has its limitations. One major limitation of using the Villalta scale for "measurement" of PTS was highlighted by Ning et al.,[41] who showed that without a preintervention Villalta score known before a DVT episode, classification of PTS postintervention may be inaccurate any as many as 40% of patients. In addition, it was argued that the study had limitations pertinent to selection bias, study endpoints, primarily related to the timing of application of elastic compression stockings, the application itself, and definition and evaluation of compliance, as well as inadequate treatment adherence. Nevertheless, the SOX trial has brought some reservations that compression stockings will reduce PTS.[40]

From a basic science standpoint, compression reduces cytokine levels in limbs with venous ulcers.[42] The latest guidelines from the American College of Chest Physicians support compression stocking use for mitigating symptoms of DVT.[43] Most recently, a single-center RCT evaluated patients with iliofemoral and femoro-popliteal DVT and concluded that elastic compression stockings mitigate the development of PTS.[44]

As briefly stated earlier, the boundary conditions of the limb are important. Transmural pressure is defined by the difference in pressures between the inside and outside of the vessel wall. Transmural pressure alterations cause edema as described earlier. DeWeese and Rogoff[23] measured foot pressures in the supine position of patients with thrombosis. They found that pressures ranged from 8.5 to 18.4 mmHg. Because these pressures are low, edema was minimal or absent altogether. However, when venous pressures were higher (50 mmHg range), edema was invariably present. This emphasized the concept of interface pressure which is the pressure exerted over the surface (boundary) of the skin. Compression stockings aid in interface pressure control in patients presenting with edema. Therefore, graduated elastic compression should be used in PTS patients until high-quality data from the literature rejects its benefits.[45]

Thrombolysis/endovascular therapies to prevent PTS

The "open vein hypothesis" is a widely accepted concept among venous disease practitioners. It suggests that intervention to quickly restore flow after venous thrombosis will preserve valvular function and thus reduce venous hypertension.[46,47] The Catheter-Directed Venous Thrombolysis in Acute Iliofemoral Vein Thrombosis (CaVenT) trial[48a] randomized 209 patients with acute DVT into two groups: (1) catheter-directed thrombolysis (CDT) plus standard anticoagulation, with (2) anticoagulation alone. The authors reported a 26% relative reduction in risk of PTS at 2 years in the CDT group with a number needed to treat of seven, and noted that 41% of CDT patients still developed PTS.[48a] The efficacy was confirmed in a 5-year follow-up study that demonstrated 28% absolute risk reduction (95% CI 14–42) and a number needed to treat of four (95%; CI 2–7).[48b]

The Acute Venous Thrombosis: Thrombus Removal with Adjunctive Catheter-Directed Thrombolysis study (the ATTRACT Trial) randomized 692 patients with acute DVT into two groups: (1) anticoagulation alone or (2) anticoagulation plus pharmaco-mechanical thrombolysis (PMT). The hypothesis of the study was that prompt treatment of the occluded vein would reduce the incidence of PTS. The investigators concluded that combined therapy with PMT and anticoagulation resulted in a higher risk of bleeding within 10 days without a risk reduction of PTS.[49]

Surgical thrombectomy

Surgical thrombectomy is another option for DVT patients who are not candidates for lytics because of bleeding risk. Casey et al. found a 33% RR reduction (95% CI, 13–48) in the incidence of PTS in 611 patients treated with surgical thrombectomy.[50] There are no trials that compare surgical thrombectomy versus anticoagulation or versus percutaneous thrombectomy with or without lytics.

Our experience

At Miami Vein Center, we have witnessed firsthand the devastating sequela of venous thrombo-embolic events in our patients. For the most part, we agree with and believe that the majority of recommendations in the literature should be part of the standard of care for the treatment of DVT and the prevention of progression to PTS. Based on our continuous, 25-years of experience we emphasize that each case must be managed on an individualized basis. It has been our impression that outcomes are strongly correlated with the extent, severity, and location of the initial DVT; but ultimately, early and effective treatment can mitigate the harmful effects of the thrombus burden inside of the venous system. It is worth emphasizing that the underlying condition of the patient and cardiovascular reserve play important roles for the outcomes as well.

Currently, our ability to effectively treat patients with DVT and predict their outcomes is poor. Duplex imaging has been a pivotal addition to the venous practice in terms of identifying the location and extent of the thrombus. However, duplex ultrasound does not offer much in terms of quantifying the thrombus duration, and the tissue damage, caused by thrombosis. Because much of the tissue damage occurs in the skin, in response to venous hypertension, it is the damage of the microcirculation where the "end target organ" effects seem to occur. Our inability to image the microcirculation, in a clinical setting, is a disadvantage when planning treatment strategies for patients with DVT.

The vascular system is a circuit consisting of the heart pump, the arteries, the arterioles, the capillaries, the venules, and the veins, which increase in size as they return centrally to the right atrium. There is a large pressure differential between the left ventricle and right atrium. As blood leaves the left ventricle, the pressure reduces as the arterial tree diverges, and it is greatly affected at the arteriole level before entering the capillary network. Dynamic pressure from the arterial side "pushes" through the capillaries in the supine position. In the upright position, pressure is added to the arterioles by the hydrostatic column. If it were not for the Bayliss effect,[51] which mitigates the arteriolar pressure (dynamic plus hydrostatic) entering the capillary bed, the Starling forces at the capillary level would be catastrophic. The blood leaves the capillary bed and enters the venules that converge into veins of increasing size as they enter the right atrium. The lymphatic system carries away the "spillover" from the capillary bed. As the thrombus burden occupies increasing amounts of the system of venules and small veins, the ability of the venous system to compensate for occlusive thrombus deteriorates. Herein, lies a critical missing link in our understanding of PTS.

A case study

A 70-year-old male patient with PTS presents with a right lower extremity ulcer and severe left lower extremity lipodermatosclerosis. He developed acute thrombo-embolic complications from the Covid-19 virus and had evidence of thrombosis in both iliac veins and the inferior vena cava. He had a more than 20 years history of thrombophilia with protein S deficiency and prior bilateral iliofemoral DVT. Fig. 17.1 shows a photograph of the right lower extremity ulcer (panel A) and venogram demonstrating bilateral iliocaval occlusion (panel B).

In Fig. 17.2, panel A shows cone-beam CT demonstrating a recanalized right iliac vein and vena cava with guidewire present. It is worth noticing that the occluded left lower extremity outflow is conducted via two collateral systems: (1) the hemiazygos posteriorly and (2) the ascending lumbar anteriorly. Fig. 17.2, panel B, shows the right iliocaval endovascular stent stack reconstruction.

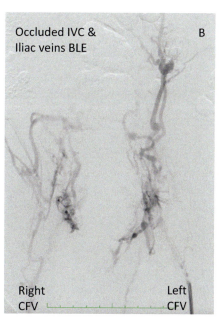

Figure 17.1 (A) 70 yo WM who presents with history of recurrent bilateral DVT who presents with nonhealing ulcer of RLE. (B) Venography shows bilateral iliocaval occlusion.

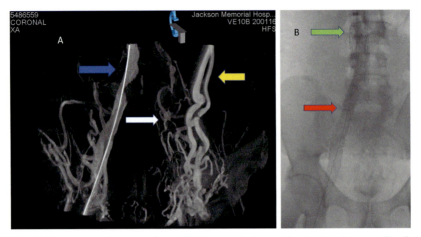

Figure 17.2 (A) Venogram using cone beam CT demonstrates guidewire inside recanalized right iliocaval system (*blue arrow*), left lower extremity collateral outflow via hemi azygous system (*white arrow*) and ascending lumbar system (*yellow arrow*); (B) Completion venogram shows rapid flow of contrast through right iliac vein stent (*red arrow*) and vena cava z-stent (*green arrow*).

Lastly, Fig. 17.3 demonstrates postreconstruction IVUS images demonstrating reconstitution of the right lower extremity outflow of adequate luminal area. Fig. 17.3, panel B, shows a healed right lower extremity venous ulcer 2 months status postendovascular venous reconstruction.

Open surgical reconstructions

Patients with common femoral vein and iliocaval outflow obstruction present a unique problem when the common femoral vein fibrosis is extensive. Common femoral vein endovenectomy was introduced as a technique to improve inflow from the infrainguinal

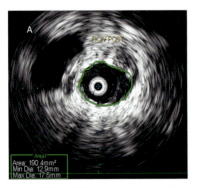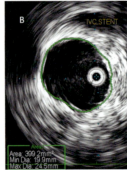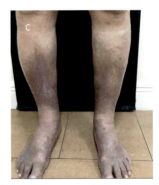

Figure 17.3 IVUS postprocedure shows satisfactory luminal area of (A) Right common iliac vein and (B) Inferior vena cava poststenting. (C) Healed right lower extremity venous ulcer.

venous system through a competent common femoral vein and thus improve patency of concomitant iliac vein recanalization and stenting. Adjunctive arteriovenous fistulas may or may not be added to maintain patency; however, this will unlikely add to the reduction of peripheral venous pressure as it mainly is used to perfuse the iliac vein conduit. Masuda[52] looked at 51 patients with severe reflux and PTS, she reported 43% success with open valvuloplasty versus 73% success in patients with non-thrombotic reflux. Comerota[53] reported open surgery results in a small group of venous ulcer patients with iliofemoral PTS. Patients who underwent endovenectomy in addition to iliofemoral stenting reported improved quality of life and clinical severity scores improved postprocedure. The complications included bleeding, rethrombosis, and acute lymphedema.

Ultrasound-accelerated thrombolysis

Much of the mitigation strategy of PTS is directed to the management of acute venous occlusion. For patients that have already developed PTS, options are limited. Unlike in an acute thrombus where pharmacomechanical interventions have shown great success, the treatment of a chronic thrombus is more resistant to such treatments. This is thought to be a result of increased collagen content in the chronic thrombus. One strategy that has been used to address this issue is the use of high-frequency ultrasound, which is thought to disrupt the intricate collagen and fibrin structure of a chronic thrombus. The Accelerated Thrombolysis for Post-Thrombotic Syndrome Using the Acoustic Pulse Thrombolysis Ekosonic Endovascular System (ACCESS PTS) study[54] assessed the ability of ultrasound-accelerated thrombolysis in PTS patients. Patency of the limb with the confirmed flow by ultrasound was observed in 90% of limbs treated with ultrasound-accelerated thrombolysis (USAT) at 1 year. Furthermore, there was a significant improvement in symptoms in 64.6% of patients at 30 days and 77.3% of patients at 1 year with a reduction of Villalta score >4. Patients that underwent USAT also experienced, on average, an absolute increase of 20.7 points in their VEINES-QOL score. While randomized studies on USAT need to be conducted, the ACCESS PTS study showed promising results with regard to improving symptoms of PTS.[54]

Venous ulcer care

Ten % of DVT patients develop PTS; and 5% will ulcerate at 10 years.[55] Compression therapy remains the cornerstone of care for patients with venous ulcers. Seven RCTs reported that multicomponent compression is better than single-component compression.[56]

Pharmacological management with pentoxifylline also showed a benefit in ulcer healing (RR, 1.70; 95% CI, 1.30—2.24) in a meta-analysis of 11 trials.[57] Additional wound care such as maintaining a moist environment with a protective covering, and aggressive debridement to control infection are often also required to achieve ulcer healing.[58,59]

In patients with chronic venous obstruction, stenting can also be of significant therapeutic benefit. Neglen et al.[60] prospectively assessed the outcomes of stenting and found that ulcer healing occurred in 55% of patients with PTS. Of note, patency rates in PTS limbs 72 months after stenting were significantly lower than in patients with nonthrombotic reflux suggesting the benefit of stenting is greater in primary versus secondary chronic venous disease.[60]

Finally, open surgical venous valvular repair may be considered for refractory venous ulcers in PTS patients with deep venous incompetence. Lugli et al. reported at a 2-year follow-up that ulcer healing occurred in 90% of patients who underwent monocusp neo-valve reconstructions.[61]

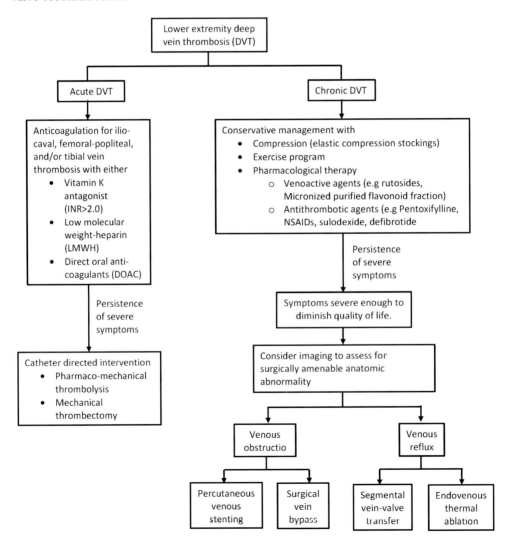

Conclusion

The severity of PTS is directly correlated with the nature and extent of the thrombosis. That is, thrombus burden, the number of segments involved, and whether or not the iliocaval outflow is affected are the main determinants of prognosis. Ipsilateral recurrent DVT has been shown to worsen PTS. Damage to microcirculation is an important aspect that has not been emphasized in the past. This is probably due to our inability to image the microcirculation in the clinical setting with practical devices such as duplex ultrasound. Early and effective anticoagulation is critical. Likely direct oral anticoagulants will take over this role in the future. The role of venotonics is unclear. Compression stockings are important for the control of edema and may be beneficial in the course of PTS development or progression. Exercise therapy is an important adjunct mainly to stimulate the growth of a new venous collateral network. Axial transformation of the profunda femoris vein is probably the most important early event to mitigate chronic damage from femoral-popliteal vein thrombosis. When considering endovascular reconstruction of a PTS patient, the goal is to restore the continuity of axial venous blood flow from the ankle to the right atrium.

Trial	Type	Number of patients	Intervention	Findings
Six-month exercise training program to treat postthrombotic syndrome[37]	Randomized control trial	69	6-month trainer-supervised exercise training program versus control (education with monthly phone follow-ups)	Improvement in VEINES-QOL score in exercise training program (exercise training mean change in VEINES-QOL score of 6.0, v. control mean change 1.4, difference 4.6, 95% CI 0.54−8.7; $P = 0.027$)
Randomized trial of effect of compression stockings in patients with symptomatic proximal-vein thrombosis[38]	Randomized control trial	194	Compression stocking or no stockings	Significant decrease in risk for developing postthrombotic syndrome in patients treated with compression stockings 2 years after acute DVT (20% in stocking group vs. 47% in control, $P < 0.001$)
Below-knee elastic compression stockings to prevent the	Randomized control trial	180	30−40 mmHg compression	25% absolute risk reduction in the development of

Continued

Trial	Type	Number of patients	Intervention	Findings
postthrombotic syndrome[39]			stockings or no stockings	postthrombotic syndrome after 2 years ($P = 0.011$).
Compression stockings to prevent postthrombotic syndrome—SOX trial[40]	Randomized control trial	806	Active versus placebo elastic compression stockings (ECS)	No significant difference in the development of postthrombotic syndrome at 2 years in active versus placebo ECS (hazard ratio 1.13, 95% 0.73–1.76, $P = 0.58$).
Catheter-directed venous thrombolysis trial (CaVenT)[48a]	Randomized control trial	209	Standard anticoagulation (control) or catheter-directed thrombolysis w/alteplase	Significant decrease in the risk of PTS in the thrombolysis group versus control—absolute risk reduction of 14.4% (95% CI, 0.2–27.9) at 2yr and 28% (95% CI 14–42) at 5yr follow-up.
Acute venous thrombosis: thrombus removal with adjunctive catheter-Directed thrombolysis (ATTRACT)[49]	Randomized control trial	692	Anticoagulation (control) or pharmaco mechanical thrombolysis (treatment)	No significant difference in the development of the postthrombotic syndrome at 6 and 24 months (risk ratio, 0.96; 95% confidence interval [CI], 0.82–1.11; $P = 0.56$).
Long-term results of venous valve reconstruction[52]	Case series	51	Venous valve reconstruction surgery	Clinical success at 10-year follow-up in 43% of patients with postthrombotic syndrome patients versus 73% of patients with primary venous insufficiency. Success defined by the ability to resume full activity.
Accelerated thrombolysis for postthrombotic Syndrome using the acoustic Pulse thrombolysis ekosonic endovascular system (access PTS)[54]	Prospective study	78	Ultrasound-accelerated thrombolysis (USAT) and percutaneous transluminal venoplasty	At 1 year, 90% of limbs had patency with ultrasound flow. In addition, 64.6% of patients at 30 days and 77.3% of patients at 1 year had significant reductions in Villalta score (defined as a reduction of >4). There was also a statistically significant improvement in VEINES-QOL score
		864		

Trial	Type	Number of patients	Intervention	Findings
Pentoxifylline for treating venous leg ulcers[57]	Meta-analysis of 11 RCTs		Pentoxifylline versus placebo	Pentoxifylline is more effective than placebo in improving symptoms of venous insufficiency and venous ulcer healing (RR 1.70, 95% CI 1.30–2.24). Pentoxifylline + compression was more effective than placebo + compression (RR 1.56, 95% CI 1.14–2.13) and pentoxifylline in the absence of compression is more effective than placebo without compression (RR 2.25, 95% CI 1.49–3.39).
Stenting of the venous outflow in chronic venous disease: ong-term stent-related outcome, clinical, and hemodynamic result[60]	Prospective study	464	Intravascular ultrasound-guided venous stenting	Ulcer healing occurred in 55% of patients.

References

1. Tran NT, Meissner MH. The epidemiology, pathophysiology, and natural history of chronic venous disease. *Semin Vasc Surg.* 2002;15(1):5–12, 11840420.
2. Labropoulos N, Gasparis AP, Tassiopoulos AK. Prospective evaluation of the clinical deterioration in post-thrombotic limbs. *J Vasc Surg.* 2009;50(4):826–830, 19628354.
3. Kahn SR, Kearon C, Julian JA, et al. Predictors of the post-thrombotic syndrome during long-term treatment of proximal deep vein thrombosis. *J Thromb Haemostasis.* 2005;3(4):718–723, 15733061.
4. Malgor RD, Labropoulos N. Post thrombotic syndrome. In: Almeida JI, ed. *Atlas of Endovascular Venous Surgery.* 2nd ed. Philadelphia, PA: Elsevier; 2019:409–430.
5. Sarin S, Scurr JH, Smith PD. Medial calf perforators in venous disease: the significance of outward flow. *J Vasc Surg.* 1992;16:40–46.
6. Ludbrook J. Functional aspects of the veins of the leg. *Am Heart J.* 1962;64:706–713.
7. Raju S, Fountain T, Neglen P, Devidas M. Axial transformation of the profunda femoris vein. *J Vasc Surg.* 1998;27:651–659.
8. Labropoulos N, Volteas N, Leon M, et al. The role of venous outflow obstruction in patients with chronic venous dysfunction. *Arch Surg.* 1997;132:46–51.
9. Neglen P, Raju S. Intravascular ultrasound scan evaluation of the obstructed vein. *J Vasc Surg.* 2002;35:694–700.
10. Raju S, Kirk O, Davis M, Olivier J. Hemodynamics of "critical" venous stenosis and stent treatment. *J Vasc Surg Venous Lymphat Disord.* 2014;2:52–59.

11. Raju S, Buck W, Jayaraj A, Crim W, Murphy EH. Peripheral venous pressure before and after iliac vein stenting. *J Vasc Surg Venous Lymphat Disord*. 2017;5:148.
12. Ludbrook J. *Aspects of venous function in the lower limbs*. Springfield, I11; 1966. Charles C. Thomas.
13. Bauer G. A roentgenological and clinical study of the sequels of thrombosis. *Acta Chir Scand*. 1942; 86(Suppl 74):1–116.
14. Edwards EA, Edwards JE. The effect of thrombophlebitis on the venous valve. *Surg Gynccol Obstet*. 1937;65:310.
15. Sakaguchi S, lshitobi K, Kameda T. Functional segmental plethysmography with mercury strain gauge. *Angiology*. 1972;23(127).
16. Barnes RW, Collicott PE, Mozcrsky DJ, Sumner DS, Strandness Jr DE. Noninvasive quantitation of venous hemodynamics in the postphlebitic syndrome. *Arch Surg*. 1973;107:807.
17. Boijsen E, Eiriksson E. Plethysmographic and pblebographic findings in venous thrombosis of the leg. *Acta Chir Scand*. 1968;398:43.
18. Dahn I, Eiriksson E. Plethysmographic diagnosis of deep venous thrombosis of the leg. *Acta Chir Scand*. 1968;398:33.
19. Sakaguchi S, Tomita T, Endo I, Ishitobi K. Functional segmental plethysmography: a new venous function test. *J Cardiovasc Surg*. 1968;9:87.
20. Bygdeman S, Aschberg S, Hindmarsh T. Venous plethysmography in the diagnosis of chronic venous insufficiency. *Acta Chir Scand*. 1971;137:423.
21. Rutherford RB, Reddy CMK, Walker FG, Wagner Jr HN. A new quantitative method of assessing the functional status of the leg veins. *Am J Surg*. 1971;122:594.
22. Strandness DE, Sumner DS. *Hemodynamics for Surgeons*. New York: Grune & Stratton; 1975.
23. Deweese JA, Rogoff SM. Phlebographic patterns of acute deep venous thrombosis of the leg. *Surgery*. 1963;53:99–108.
24. Negus D, Cockett FB. Femoral vein pressures in postphlebitic iliac vein obstruction. *Br J Surg*. 1967;54: 522.
25. Walker AJ, Longland CJ. Venous pressure measurement in the foot in exercise as an aid to investigation of venous disease in the leg. *Clin Sci*. 1950;9:101.
26. Starling EH. On the absorption of fluids from the connective tissue spaces. *J Physiol*. 1896;19:312.
27. Landis EM. Capillary permeability and the factors affecting the composition of capillary filtrate. *Ann N Y Acad Sci*. 1946;46:713.
28. Pappenheimer JR, Soto-Rivera A. Effective osmotic pressure of the plasma proteins and other quantities associated with the capillary circulation in the hind limbs of cats and dogs. *Am J Physiol*. 1948;152: 471.
29. Landis EM, Pappenheimer JR. Exchange of sub-stances through capillary walls. In: Hamilton WF, Dow P, eds. *Handbook of Physiology. Section 2. Circulation*. Vol. II. Washington D.C: American Physiological Soc; 1963:961–1034. chap. 29.
30. Raju S, Buck W, Crim w, Jayaraj A. Optimal sizing of iliac vein stents. *Phlebology*. 2017;0(0):1–7.
31. Guyton AC, Granger HJ, Taylor AE. Interstitial fluid pressure. *Physiol Rev*. 1971;51:527.
32. Kearon C, Akl EA, Comerota AJ, et al. Antithrombotic therapy for VTE disease: antithrombotic therapy and prevention of thrombosis, 9th ed: American College of chest Physicians evidence-based clinical practice guidelines. *Chest*. 2012;141(suppl):e419S–e494S.
33. van Dongen CJ, Prandoni P, Frulla M, Marchiori A, Prins MH, Hutten BA. Relation between quality of anticoagulant treatment and the development of the postthrombotic syndrome. *J Thromb Haemostasis*. 2005;3:939–942.
34. Chitsike RS, Rodger MA, Kovacs MJ, et al. Risk of post-thrombotic syndrome after subtherapeutic warfarin anticoagulation for a first unprovoked deep vein thrombosis: results from the REVERSE study. *J Thromb Haemostasis*. 2012;10:2039–2044.
35. Kahn SR, Shrier I, Julian JA, et al. Determinants and time course of the post-thrombotic syndrome after acute deep venous thrombosis. *Ann Intern Med*. 2008;149:698–707.
36. Baglin T. Prevention of post-thrombotic syndrome: a case for new oral anticoagulant drugs or for heparins? *J Thromb Haemostasis*. 2012;10:1702–1703.

37. Kahn SR, Shrier I, Shapiro S, et al. Six-month exercise training program to treat post-thrombotic syndrome: a randomized controlled two-centre trial. *CMAJ (Can Med Assoc J)*. 2011;183:37−44.
38. Brandjes DPM, Buller HR, Heijboer H, et al. Randomised trial of effect of compression stockings in patients with symptomatic proximal-vein thrombosis. *Lancet*. 1997;349:759−762.
39. Prandoni P, Lensing AWA, Prins MH, et al. Below-knee elastic compression stockings to prevent the post-thrombotic syndrome: a randomized, controlled trial. *Ann Intern Med*. 2004;141:249−256.
40. Kahn SR, Shapiro S, Wells PS, et al. Compression stockings to prevent post-thrombotic syndrome: a randomised placebo-controlled trial. *Lancet*. 2014;383(9920):880−888.
41. Ning J, Ma W, Fish J, Trihn F, Lurie F. Biases of Villalta scale in classifying post-thrombotic syndrome in patients with pre-existing chronic venous disease. *J Vasc Surg: Venous and Lym Dis*. 2020;8:1025−1030.
42. Beidler SK, Douillet CD, Berndt DF, et al. Inflammatory cytokine levels in chronic venous insufficiency ulcer tissue before and after compression therapy. *J Vasc Surg*. 2009;49(4):1013−1020.
43. Kearon C, Akl EA, Ornelas J, et al. Antithrombotic therapy for VTE disease: CHEST guideline and expert panel report. *Chest*. 2016;149(2):315−352.
44. Yang X, Zhang X, Yin M, Wang R, Lu X, Ye K. Elastic compression stockings to prevent post-thrombotic syndrome in proximal deep venous thrombosis patients without thrombus removal DOI: https://doi.org/10.1016/j.jvsv.2021.06.023.
45. Subbiah R, Aggarwal V, Zhao H, et al. Effect of compression stockings on post thrombotic syndrome in patients with deep vein thrombosis: a meta-analysis of randomised controlled trials. *Lancet Haematol*. 2016;3(6):e293−e300.
46. Comerota AJ, Grewal N, Martinez JT, et al. Postthrombotic morbidity correlates with residual thrombus following catheter-directed thrombolysis for iliofemoral deep vein thrombosis. *J Vasc Surg*. 2012;55:768−773.
47. Vedantham S. Valvular dysfunction and venous obstruction in the postthrombotic syndrome. *Thromb Res*. 2009;123:S62−S65.
48. a Enden T, Haig Y, Kløw N-E, et al, CaVenT Study Group. Long-term outcome after additional catheter directed thrombolysis versus standard treatment for acute iliofemoral deep vein thrombosis (the CaVenT study): a randomised controlled trial. *Lancet*. 2012;379:31−38.
    b Haig Y, Enden T, Grotta O, et al. Post-thrombotic syndrome after catheter-directed thrombolysis for deep vein thrombosis (CaVenT): 5-year follow-up results of an open-label, randomised controlled trial. Lancet Haematol.
49. Vedantham S, Goldhaber SZ, Julian JA, et al. ATTRACT trial investigators. Pharmacomechanical catheter-directed thrombolysis for deep-vein thrombosis. *N Engl J Med*. December 7, 2017;377(23):2240−2252.
50. Casey ET, Murad MH, Zumaeta-Garcia M, et al. Treatment of acute iliofemoral deep vein thrombosis. *J Vasc Surg*. 2012;55:1463−1473.
51. Raju S, Knight A, Lamanilao L, Pace N, Jones T. Peripheral venous hypertension in chronic venous disease. *J Vasc Surg: Venous and Lym Dis*. 2019;7:706−714.
52. Masuda EM, Kistner RL. Long-term results of venous valve reconstruction: a four- to twenty-one-year follow-up. *J Vasc Surg*. 1994;19:391−403.
53. Comerota AJ. Contemporary concepts in the management of acute iliofemoral DVT and chronic postthrombotic iliofemoral venous obstruction. In: Charles J, Tegtmeyer MD, eds. *Annual Lecture Presented at: 25th Annual International Symposium on Endovascular Therapy (ISET)*. January 23, 2013 (Miami Beach, FL).
54. Garcia MJ, Sterling KM, Kahn SR, et al, ACCESS PTS Investigators. Ultrasound-accelerated thrombolysis and venoplasty for the treatment of the postthrombotic syndrome: results of the ACCESS PTS study. *J Am Heart Assoc*. February 4, 2020;9(3):e013398. https://doi.org/10.1161/JAHA.119.013398.
55. Schulman S, Lindmarker P, Holmstrom M, et al. Post-thrombotic syndrome, recurrence, and death 10 years after the first episode of venous thromboembolism treated with warfarin for 6 weeks or 6 months. *J Thromb Haemostasis*. 2006;4:734−742.
56. O'Meara S, Cullum N, Nelson EA, Dumville JC. Compression for venous leg ulcers. *Cochrane Database Syst Rev*. 2012;11:CD000265.

57. Jull AB, Arroll B, Parag V, Waters J. Pentoxifylline for treating venous leg ulcers. *Cochrane Database Syst Rev.* 2012;12:CD001733.
58. Rippon M, Davies P, White R. Taking the trauma out of wound care: the importance of undisturbed healing. *J Wound Care.* 2012;21:359–360, 362, 364–368.
59. Douglas WS, Simpson NB. Guidelines for the management of chronic venous leg ulceration: report of a multidisciplinary workshop: British Association of Dermatologists and the Research Unit of the Royal College of Physicians. *Br J Dermatol.* 1995;132:446–452.
60. Neglen P, Hollis KC, Olivier J, Raju S. Stenting of the venous outflow in chronic venous disease: long-term stent-related outcome, clinical, and hemodynamic result. *J Vasc Surg.* 2007;46:979–990.
61. Lugli M, Guerzoni S, Garofalo M, Smedile G, Maleti O. Neovalve construction in deep venous incompetence. *J Vasc Surg.* 2009;49(156–162):162e1–162e2.

CHAPTER 18

Improving treatment outcomes—management of coexisting comorbidities in patients with venous ulcers

Giovanni Mosti[1] and Alberto Caggiati[2]
[1]Head Angiology Department, MD Barbantini Hospital, Lucca, Italy; [2]Department of Anatomy, Sapienza University of Rome, Rome, Italy

Introduction

Venous leg ulcers (VLUs) can affect patients who have other coexisting medical conditions due to the age-related prevalence of VLUs. These lesions are significantly more prevalent in advanced-age patients when other medical conditions are more frequent. In an epidemiological study from the United Kingdom, published in 2004[1] among 74,346 patients who were older than 65, a clinical diagnosis of VLU was performed in 2371 individuals (3.19%). The risk of developing a VLU increased with age and was more frequent in women than in men; age and sex were two predicting factors. In this study, several medical conditions commonly occurred in patients with VLU, including anemia, angina, asthma, lower extremity cellulitis, depression, diabetes, lower limb edema, hypertension, osteoarthritis, pneumonia, and urinary tract infection. After a statistical analysis to exclude fortuitous associations, many medical conditions were statistically significantly associated with VLU development including asthma, lower extremity cellulitis, congestive heart failure, diabetes, deep venous thrombosis, lower limb edema, osteoarthritis, lower extremity peripheral vascular arterial disease, rheumatoid arthritis, hip surgery history, and venous surgery/ligation history. Unexpectedly, some disease occurrences (including angina, cerebral vascular accident, depression, malignancy, myocardial infarction, pneumonia, and urinary tract infection) were inversely associated with a recent VLU onset. This inverse correlation probably occurred because many of these diseases were associated with decreased patient longevity. Furthermore, individuals with these medical conditions interact more frequently with physicians and therefore may receive better medical care, which can prevent leg ulcer development.

Many of the above-mentioned medical conditions (such as diabetes, advancing age, and lower limb arterial disease) are associated with a reduced wound healing ability. Other conditions may impair VLU healing due to a lower limb muscle pump dysfunction. Muscle pump function, one of the main mechanisms of venous return, can be impaired in patients with clinical conditions such as osteoarthritis; rheumatoid arthritis; a history of spine, hip, knee, or ankle surgery; reduced ankle mobility; foot static disorders; reduced calf muscle strength; a sedentary lifestyle. Such an impairment in muscle pump function results in venous stasis that can cause leg ulcers independently from a venous disease.[2,3] Alternatively, it can worsen venous hemodynamic impairment due to venous disease and contribute to leg ulcer occurrence and maintenance.

Hence, when managing patients with VLUs, a detailed and systematic investigation for coexisting medical conditions must be performed and an appropriate treatment plan must be established to address identified comorbidities that can affect VLU healing process. Herein, we attempted to provide an overview of the main medical conditions that are associated with VLU. Some comorbidities related to ulcer healing delays, such as arterial disease and deep venous thrombosis, will not be considered here, as they are described in other chapters (see Chapters 10 and 17).

Diabetes mellitus

Diabetes mellitus can certainly impair VLU healing and is responsible for arterial wall degeneration, especially when poorly controlled or uncontrolled.[4] Chronic nonhealing wounds in these patients are common.[5] Impaired wound healing in patients with diabetes is the result of complex pathophysiologic factors involving vascular, neuropathic, immunologic, and biochemical components.[6] Fundamentally the healing process in patients with diabetes is mainly characterized by the chronicization of inflammatory conditions, disrupted angiogenic process, a reduction in the number of endothelial progenitor cells, and an imbalance in extracellular matrix regulation.[7]

Although delayed wound healing is known to be common in patients with diabetes, sound data regarding the impact of diabetes on VLU healing are missing, as diabetes is considered either as an exclusion criterion or is not controlled for in studies focusing on VLU healing.[8] As a consequence, a difference in terms of wound healing between patients with and without diabetes cannot be assessed. One study reported that the median VLU healing times in patients with and without diabetes were 25 and 28 weeks, respectively, which represented a slight but statistically significant difference ($P = .09$).[9] Another study suggested that diabetes seems to be correlated with ulcer recurrence.[10] Due to a lack of data, diabetes is usually not considered a risk factor for delayed healing.[11–13] However, diabetes should be effectively treated in all patients independently from coexisting leg ulcers.

Arterial hypertension

Arterial hypertension is widespread in the general population. Approximately 30% of the US population is affected by arterial hypertension, although this prevalence increases to

65.2% in persons aged 60 years or older,[14] which is also the usual age range of patients with VLU. Consequently, arterial hypertension is so common in patients with VLU that the potential VLU healing delay in patients with and without arterial hypertension has never been assessed and compared. Moreover, arterial hypertension is not considered as a risk factor for delayed VLU healing,[11–13] but must be considered and treated effectively. In patients with arterial hypertension, a specific leg ulcer can occur: the so-called hypertensive ulcer. In hypertensive ulcers, arterial hypertension is the first pathophysiological factor, producing small artery wall thickening and, finally, occlusion.[15] Hypertensive ulcers are completely different from venous ulcers, and their characterization is beyond the scope of this chapter.

Coronary artery disease and congestive heart failure

As mentioned above, angina and myocardial infarction seem to be negatively associated with VLU occurrence. Advanced stages of coronary artery disease can lead to congestive heart failure, which is frequently associated with VLU occurrence. Sedentary lifestyle and leg edema are conditions that favor venous ulcer occurrence. Leg edema can result from heart failure due to upstream overload resulting from a poor-functioning left heart, and from sitting position maintained in the long term, resulting in venous stasis. Systemic treatment with different combinations of diuretics, angiotensin convertin enzyme (ACE) inhibitors, angiotensin-2 receptor blockers, β-blockers, and digoxin is mandatory for heart failure. Additional measures involve lifestyle changes, including eating a well-balanced diet, low sodium intake, physical exercise based on heart failure severity, and smoking cessation. Compression therapy has long been considered as a contraindication in patients with heart failure.[16] Nowadays, it is indicated in these patients provided that special precautions are taken to avoid a massive shift of fluids from the lower extremities into the systemic and cardio-pulmonary circulation, which can cause pulmonary edema.[17] For instance it is recommended to start compression therapy when systemic treatment has been established and to begin with light compression in one leg for patients who are in the first three stages of heart failure according to the New York Heart Association (NYHA) classification. This allows for a gradual shift of fluids into the systemic circulation. When edema is slowly reduced in one leg, the other leg can also undergo light compression.[17] It is strongly recommended to avoid compression therapy in patients with severe heart failure (stage 4 according to the NYHA classification).[17] No better specified cardiac diseases seem to be correlated with ulcer recurrence.[10]

Reduced muscle pumping function

Abnormal venous pump function is a well-known risk factor for ulcer occurrence, delayed ulcer healing, and ulcer recurrence.[18] Venous hypertension may be caused by venous disease (occlusion, incompetence, or both) and reduced leg muscle pumping efficiency. Several clinical conditions, that may impair the muscle pumping function, are here reported.

Abnormal plantar load

The plantar pump is the most peripheral pump of the lower extremity[18,19] and plays an important role in venous return.[20] It was demonstrated that the hemodynamic effects of the plantar pump are quantitatively similar to those of the calf.[21] Although no study has shown a correlation between abnormal plantar load incidence and VLU onset or recurrence, it has been demonstrated that static foot disorders, such as hollow foot and flat foot, represent risk factors for lower limb venous disease onset and worsening.[21] Static foot disorders reduce the effectiveness of the plantar pump and modify the gait pattern, thus affecting the synchronization of leg pumps.[21] It was demonstrated that the correction of plantar loading obtained using properly placed insoles improves venous return and quality of life (QoL), with an efficacy that is almost equal to that provided by compression stockings.[21–23]

Gait disorders

Venous return from the lower extremity is ensured by a coordinated sequence of muscular pumps (plantar, anterior, and posterior leg, anterior and posterior thigh, and gluteal muscles).[20] The synchronized activation of these overlapping pumps mainly depends on a regular gait (rhythmic cadence, step length, and speed) and a correct sequence of weight bearing on the foot: on the heel first; then, on the mid-foot; and, finally, on the anterior foot.[24]

The gait pattern is strongly influenced by obesity and osteoarticular, muscular, or neurologic disorders.[25] The prevalence of gait disorders markedly increases with age (exceeding 60% after the age of 80 years); however, their prevalence is higher in patients with severe venous disorders than in similarly aged individuals.[26] This is probably due to the impairment of ankle mobility and proprioceptive and muscular functions typical of long-lasting severe venous insufficiency.[27–29]

The goal of gait reeducation is to regain the correct sequence of weight bearing ("heel-to-toe" pattern), gait rhythmic cadence, step length, and speed, and to discourage the shuffling gait. Although gait reeducation has been demonstrated to significantly increase the efficacy of plantar and leg pumps,[26,27] no study has evaluated the effects of gait reeducation on venous ulcer healing.

Physical activity

Physical inactivity is a well-known risk factor of venous ulcers in patients with venous disorders.[30] It was demonstrated that patients with severe venous disorders are less mobile than their age-matched counterparts.[31] Forty-five percent of patients with leg ulcers show reduced mobility,[32] which may result from musculoskeletal or neurologic comorbidities, work activity, lifestyle, psychiatric impairment, and social isolation. Prolonged inactivity and bed rest can lead to muscle atrophy, contractures, and degenerative joint diseases.

Adapted physical activities (APAs) consist of protocols of physical activity tailored according to the patient's abilities and comorbidities. APAs are indicated in sedentary patients with venous disease, and in individuals who do not perform physical exercises on a regular basis,[31] with the aim of increasing muscle strength, articular flexibility, and breathing dynamics.[21] Finally, adherence to APA protocols should also improve the psychosocial conditions of these patients.[21]

The efficacy of APA in patients with venous disorders was proven using short- and long-term protocols that showed symptom and muscle pumping function improvements. Furthermore, adherence to APA protocols is correlated with a lower incidence of ulcer recurrence.[10]

Ankle motility and leg muscle strength

Normal ankle mobility and ankle range of motion (aROM) is required for the effectiveness of the leg pumps.[33] A reduced aROM prevents leg muscle excursion and modifies the gait pattern. Even if the aROM spontaneously decreases with increasing age,[34] it is further reduced in patients with venous disorders more than in similarly aged individuals without venous disorders.[35] McRorie et al. reported that 32% of venous ulcer legs had severe ankle movement limitation, albeit only 9% had arthritis,[36] probably due to lipodermatosclerosis, fascial fibrosis, and venous stasis-related periostitis.[37] Abnormalities in nerve functions due to venous insufficiency have also been considered as a possible cause of reduced aROM.[38,39] A reduction in the aROM also occurs in legs with less severe venous insufficiency,[35] although the aROM reduction correlates with venous disease severity,[21] ulcer healing reduction, and ulcer recurrence.[33] People with impaired calf muscle pumps show significantly lower ejection volumes and fractions as assessed by air plethysmography.

Increasing aROM results in an improvement of the calf pump effectiveness,[40] and it is mandatory to obtain a physiologic gait to attain this effectiveness.[21] In addition to foot and ankle skeletal anomalies, calf muscle pumping function reduction may be related to abnormal muscle strength.[40] Muscle dysfunction is physiologically related to age and muscle inactivity with resulting atrophy. However, it was demonstrated that venous incompetence is a possible cause of muscle atrophy and peripheral neuropathy.[38,41]

Physical therapy (PT) is the main treatment used to increase aROM and calf muscle strength. Individually tailored protocols of PT include a mix of stretching exercises followed by repeated contractions and relaxations.[21] Passive movements of foot dorsiflexion are followed by ankle active nonweight-bearing movements.[42] Further, resistive exercises are followed by weight-bearing activities.[21] Moreover, it has been demonstrated that ankle and calf exercises significantly improve muscle strength, thereby increasing ejection fraction and reducing residual volume.[21] Soft tissue manipulations may be necessary when subcutaneous fibrosis coexists with a reduced aROM.[43]

Treatment of reduced aROM and calf muscle deficiency

Exercise has been shown to decrease the detrimental effects associated with impaired calf muscle function. A structured program of calf muscle exercise may improve hemodynamic performance and prevent ulcer recurrence[44]; an implementation of walking programs is suggested for ambulatory patients. Higher levels of muscle activity and greater muscle mass may enhance the emptying of veins in the calf.[45]

A relatively recent report highlighted the positive effect of different exercises in combination with compression therapy in patients with VLUs.[46] A simple progressive resistance and aerobic activity is recommended for suitable patients with VLUs.

Finally, the use of an exercise program in combination with compression therapy in patients with VLUs results in better outcomes and cost reduction for the English National Health Service.[47]

In conclusion, we suggest the routine evaluation of aROM and leg muscle strength in legs with venous ulcers, and the referral to ankle and calf muscle rehabilitation, when necessary.

Other clinical conditions concurring with venous leg ulcers, venous ulcer occurrence, healing failure, and recurrence

The risk of developing venous ulcers may be increased in patients with obesity, malnutrition, and impaired psychosocial conditions, as well as in smokers.[30,48–65] All these conditions must be considered and managed.

Cigarette smoking

Few studies have examined the association between smoking and the risk of ulceration; therefore, evidence supporting this association is poor. However, cigarette smoking seems to be associated with an increased risk of venous ulceration. A case-control study estimated that patients who smoked 10–19 cigarettes daily were 1.8 times more likely (95% CI, 1.4–2.2) to develop chronic venous insufficiency and leg ulcers than nonsmokers.[50]

The mechanisms explaining the correlation between smoking and venous ulcer occurrence were not explained. However, another study suggested the possibility of skin hypoxia due to nicotine, carbon monoxide, and hydrogen cyanide.[51] Therefore, wound care clinicians should incorporate personalized, evidence-based smoking cessation strategies in their care plans to increase the likelihood of improving healing outcomes in patients with leg ulcers.

Obesity

High body mass index (BMI) is considered to be a risk factor for the development of severe chronic venous disease (CVD)[52] and venous ulcers.[53] According to a previous report, *"the CEAP clinical stage of venous disease is more advanced in obese patients than in non-obese patients with comparable anatomical patterns of venous incompetence,"*[54] which is currently attributed to two causes:

1. Elevated intraabdominal pressure due to exaggerated fat deposition results in increased iliofemoral vein pressure, which increases vein diameter and valve dysfunction, and causes chronic venous hypertension.[55]
2. Limitations of physical activity and exercise: Obese patients have more difficulties with walking and performing exercises than those with normal weight, which further worsens the calf muscle pump function and causes venous hypertension.

Notably, in patients with morbid obesity (BMI >40 kg/m^2), two-thirds of the limbs showing severe symptoms typical of chronic venosu insufficiency (CVI) have no anatomic evidence of venous disease.[56] The negative effects of obesity on venous ulceration were recently demonstrated,[57] ulcer healing rates, the occurrence of other CVD symptoms such as venous claudication, QoL were assessed in a series of patients who underwent bariatric surgery (BS) (group A), and compared with those of patients who did not undergo BS and did not lose weight (group B). Patients in group A showed a significant improvement in QoL, an increased ulcer healing rate, and a decreased incidence of venous claudication.[57] The ulcer healing rate in group A was striking (82%), especially when compared with the healing rate in group B (18%). The mean ulcer sizes in groups A and B decreased from 33.8 ± 2.3 mm to 16.3 ± 0.2 mm ($P = 0.001$) and 41.3 ± 5.6 mm to 38.4 ± 6.1 mm ($P = 0.126$), respectively.

Malnutrition

Nutritional deficiency seems to be relatively common in patients with leg ulcers.[58] Several studies have shown that many patients with leg ulcers have low levels of vitamins A and C, zinc, and carotenes. Although this is well demonstrated in patients with pressure ulcers,[59] current literature provides no solid evidence concerning a direct relationship between nutritional status or the use of nutritional supplements and the VLU outcome.[60] Only zinc supplementation shows slight evidence of effectiveness in increasing VLU healing rate in patients with low serum zinc levels.[61]

Guidelines on leg ulcers often report the nutritional studies for chronic wounds, but not VLUs. Larger-scale trials are required before recommending the use of zinc or other nutritional supplements in patients with VLUs. Furthermore, future studies should assess patient nutritional status to identify and treat possible nutritional defects.

Psychosocial condition

Despite the common belief that leg ulceration was associated with social factors, supporting evidence was anecdotal until 2006, when the social class, ethnicity, marital status, living standard, and social support were assessed in a population of patients with chronic leg ulceration. It was concluded that chronic leg ulceration was associated with poor socioeconomic status and factors related to social isolation.[62] Subsequently, the same authors examined the mental health of patients with chronic leg ulceration and found a need for social support. Many patients had a very poor health-related QoL. Compared with the general population, patients with chronic leg ulceration had more severe depression combined with lower levels of social support.[63] For both studies, it was not possible to determine whether these associations were causes or consequences of leg ulceration. Recently, it was confirmed that approximately one-third of patients with C5—C6 disease are at risk of depression.[64] In addition, VLUs are significantly associated with being single, low self-management ability, and low levels of social support.[65]

The treatment of psychosocial impairment is significantly associated with reduced ulcer recurrence rates.[64,65] In addition to a direct effect on the immune system and normal healing processes, the above-mentioned association can also result from the impact of self-care, physical activities, and a better adherence to compression therapy.[66,67] The strong and statistically significant relationship between social support and ulcer recurrence highlights the need for an increased use of social interaction measures in patients with severe CVI.[64]

In conclusion, compared to the general population, patients with leg ulceration have poor mental health with a greater risk of depression, less perceived social support, and greater social isolation. Care systems should offer an environment that reduces social isolation and increases levels of social support for these patients. Psychological support and social interaction measures influence the level of physical activity and self-management[21]; thus, allowing a greater adherence to treatment, especially the use of compression stockings.[66]

References

1. Margolis DJ, Knauss J, Bilker W. Epidemiology and health services research medical conditions associated with venous leg ulcers. *Br J Dermatol*. 2004;150:267—273.
2. Gaylarde PM, Dodd HJ, Sarkany I. Venous leg ulcers and arthropathy. *Br J Rheumatol*. 1990;29:142—144.
3. Seitz CS, Berens N, Bröcker EB, Trautmann A. Leg ulceration in rheumatoid arthritis–an underreported multicausal complication with considerable morbidity: analysis of thirty-six patients and review of the literature. *Dermatology*. 2010;220:268—273.
4. Katakami N. Mechanism of development of atherosclerosis and cardiovascular disease in diabetes mellitus. *J Atherosclerosis Thromb*. 2018;25:27—39.
5. Dinh T, Elder S, Veves A. Delayed wound healing in diabetes: considering future treatments. *Diabetes Manag*. 2011;1:509—519.
6. Greenhalgh DG. Wound healing and diabetes mellitus. *Clin Plast Surg*. 2003;30:37—45.

7. Spampinato FS, Caruso GI, De Pasquale R, Sortino AM, Merlo S. The treatment of impaired wound healing in diabetes: looking among old drugs. *Pharmaceuticals*. 2020;13:60.
8. O'Meara S, Cullum N, Nelson EA, Dumville JC. Compression for venous leg ulcers. *Cochrane Database Syst Rev*. November 14, 2012;11(11):CD000265.
9. Mosti G, Cavezzi A, Bastiani L, Partsch H. Compression therapy is not contraindicated in diabetic patients with venous or mixed leg ulcer. *J Clin Med*. 2020;9:3709.
10. Finlayson K, Edwards H, Courtney M. Factors associated with recurrence of venous leg ulcers: a survey and retrospective chart review. *Int J Nurs Stud*. 2009;46:1071—1078.
11. Margolis DJ, Berlin JA, Strom BL. Risk factors associated with the failure of a venous leg ulcer to heal. *Arch Dermatol*. 1999;135:920—926.
12. Milic DJ, Zivic SS, Bogdanovic DC, Karanovic ND, Golubovic ZV. Risk factors related to the failure of venous leg ulcers to heal with compression treatment. *J Vasc Surg*. 2009;49:1242—1247.
13. Parker CN, Finlayson KJ, Shuter P, Edwards HE. Risk factors for delayed healing in venous leg ulcers: a review of the literature. *Int J Clin Pract*. 2015;69:967—977.
14. Hajjar I, Kotchen JM, Kotchen TA. Hypertension: trends in prevalence, incidence and control. *Annu Rev Publ Health*. 2006;27:465—490.
15. Martorell F. Hypertensive ulcer of the leg. *Angiology*. 1950;1:133—140.
16. Andriessen A, Apelqvist J, Mosti G, Partsch H, Gonska C, Abel M. Compression therapy for venous leg ulcers: risk factors for adverse events and complications, contraindications—a review of present guidelines. *J Eur Acad Dermatol Venereol*. 2017;31:1562—1568.
17. Rabe E, Partsch H, Morrison N, et al. Risks and contraindications of medical compression treatment - a critical reappraisal. An international consensus statement. *Phlebology*. 2020;35:447—460.
18. Uhl JF, Gillot C. Anatomy of the veno-muscular pumps of the lower limb. *Phlebology*. 2015;30:180—193.
19. Ludbrook J. The musculovenous pumps of the human lower limb. *Am Heart J*. 1966;71:635—641.
20. Gardner AMN, Fox RH. The venous pump of the human foot—preliminary report. *Bristol Medico-Chirurgical J*. July 1983;98:109—112.
21. Caggiati A, De Maeseneer M, Cavezzi A, Mosti G, Morrison N. Rehabilitation of patients with venous diseases of the lower limbs: state of the art. *Phlebology*. 2018;33:663—671.
22. Uhl JF, Chahim M, Allaert FA. Static foot disorders: a major risk factor for chronic venous disease? *Phlebology*. 2012;27:13—18.
23. Uhl JF, Chahim M, Allaert FA. Compression versus inner sole for venous patients with foot static disorders: a prospective trial comparing symptoms and quality of life. *Phlebology*. 2013;27:19—30.
24. Pirker W, Katzenschlager R. Gait disorders in adults and the elderly : a clinical guide. *Wien Klin Wochenschr*. 2017;129:81—95.
25. Mahlknecht P, Kiechl S, Bloem BR, et al. Prevalence and burden of gait disorders in elderly men and women aged 60-97 years: a population-based study. PLoS One. *PLoS One*. 2013;24;8(7):e69627.
26. van Uden CJ, van der Vleuten CJ, Kooloos JG, Haenen JH, Wollersheim H. Gait and calf muscle endurance in patients with chronic venous insufficiency. *Clin Rehabil*. 2005;19:339—344.
27. Yang D, Vandongen YK, Stacey MC. Changes in calf muscle function in chronic venous disease. *Cardiovasc Surg*. 1999;7:451—456.
28. Yim E, Vivas A, Maderal A, Kirsner RS. Neuropathy and ankle mobility abnormalities in patients with chronic venous disease. *JAMA Dermatol*. 2014;150:385—389.
29. de Moura RM, GomesHde A, da Silva SL, Britto RR, Dias RC. Analysis of the physical and functional parameters of older adults with chronic venous disease. *Arch Gerontol Geriatr*. 2012;55:696—701.
30. Abelyan G, Abrahamyan L, Yenokyan G. A case-control study of risk factors of chronic venous ulceration in patients with varicose veins. *Phlebology*. 2018;33:60—67.
31. Roaldsen KS, Biguet G, Elfving B. Physical activity in patients with venous leg ulcer–between engagement and avoidance. A patient perspective. *Clin Rehabil*. 2011;25:275—286.
32. Callam M, Harper D, Dale J, &Ruckley C. Arterial disease in chronic leg ulceration: an underestimated hazard? Lothian and forth valley leg ulcer study. *Br Med J*. 1987;294(6577):929—931.
33. Back TL, Padberg FT, Araki CT, Thompson PN, Bobson RW. Limited range of motion of the ankle joint is a significant factor in venous ulceration. *J Vasc Surg*. 1995;22:519—523.

34. Grimston SK, Nigg BM, Hanley DA, Engsberg JR. Differences in ankle joint complex range of motion as a function of age. *Foot Ankle*. May 1993;14:215−222.
35. Dix FP, Brooke R, McCollum CN. Venous disease is associated with an impaired range of ankle movement. *Eur J Vasc Endovasc Surg*. 2003;25:556−561.
36. McRorie ER, Ruckley CV, Nuki G. The relevance of large-vessel vascular disease and restricted ankle movement to the aetiology of leg ulceration in rheumatoid arthritis. *Br J Rheumatol*. 1998;37:1295−1298.
37. Gylarde PM Dodd HJ, Sarkani J. Venous leg ulcers and arthropathy. *Br J Rheumatol*. 1990;29:142−144.
38. Yim E, Vivas A, Maderal A, et al. Neuropathy and ankle abnormalities in patients with chronic venous disease. *JAMA Dermatol*. 2014;150:385−389.
39. Newland MR, Patel AR, Prieto L, Boulton AJ. , Pacheco M, Kirsner RS. Neuropathy and gait disturbances in patients with venous disease: a pilot study. *Arch Dermatol*. 2009;145:485−486.
40. Williams KJL, Avekolove O, Moore HM, Davies AH. The calf muscle revisited. *J Vasc Surg Ven Lym Dis*. 2014;2:329−334.
41. Taheri SA, Heffner R, Pendergast D, Pisano SM. Myopathy in venous insufficiency. *Phlebology*. 1987;2:7−12.
42. Klyscz T, Ritter-Schempp C, Junger M, et al. Biomechanical stimulation therapy as physical treatment of arthrogenic venous insufficiency. *Hautarzt*. 1997;48:318−322.
43. Pereira de Godoy JM, Braile DM, de Fatima Guerreiro Godoy M. Lymph drainage in patients with joint immobility due to chronic ulcerated lesions. *Phlebology*. 2008;23:32−34.
44. Padberg Jr FT, Johnston MV, Sisto SA. Structured exercise improves calf muscle pump function in chronic venous insufficiency: a randomized trial. *J Vasc Surg*. 2004;39:79−87.
45. Kugler C, Strunk M, Rudofsky G. Venous pressure dynamics of the healthy human leg. Role of muscle activity, joint mobility and anthropometric factors. *J Vasc Res*. 2001;38:20−29.
46. Jull A, Slark J, Parsons J. Prescribed exercise with compression vs compression alone in treating patients with venous leg ulcers: a systematic review and meta-analysis. *JAMA Dermatol*. 2018;154:1304−1311.
47. Klonizakis M, Tew GA, Gumber A, et al. Supervised exercise training as an adjunct therapy for venous leg ulcers: a randomized controlled feasibility trial. *Br J Dermatol*. 2018;178:1072−1082.
48. Robertson L, Lee AJ, Gallagher K, et al. Risk factors for chronic ulceration in patients with varicose veins: a case control study. *J Vasc Surg*. 2009;49:1490−1498.
49. Meulendijks AM, Welbie M, Tjin EPM, Schoonhoven L, Neumann HAM. A qualitative study on the patient's narrative in the progression of chronic venous disease into a first venous leg ulcer: a series of events. *Br J Dermatol*. 2020;183:332−339.
50. Gourgou S, Dedieu F, Sancho-Garnier H. Lower limb venous insufficiency and tobacco smoking: a case-control study. *Am J Epidemiol*. 2002;155:1007−1015.
51. McDaniel JC, Browning KK. Smoking, chronic wound healing, and implications for evidence-based practice. *J Wound, Ostomy Cont Nurs*. 2014;41:415−E2.
52. Davies HO, Popplewell M, Singhal R, Smith N, Bradbury AW. Obesity and lower limb venous disease - the epidemic of phlebesity. *Phlebology*. 2017;32:227−233.
53. Schneider C, Stratman S, Kirsner RS. Lower extremity ulcers. *Med Clin*. 2021;105:663−679.
54. van Rij AM, De Alwis CS, Jiang P, et al. Obesity and impaired venous function. *Eur J Vasc Endovasc Surg*. 2008;35:739−744.
55. Danielsson G, Eklof B, Grandinetti A, Kistner RL. The influence of obesity on chronic venous disease. *Vasc Endovasc Surg*. 2002;36:271−276.
56. Padberg Jr F, Cerveira JJ, Lal BK, et al. Does severe venous insufficiency have a different etiology in the morbidly obese? Is it venous? *J Vasc Surg*. 2003;37:79−85.
57. Shaalan W, El Emam A, Lotfy H, Naga A. Clinical and hemodynamic outcome of morbidly obese patients with severe chronic venous insufficiency with and without bariatric surgery: a comparative study. *J Vasc Surg Venous Lymphat Disord*. February 2, 2021;S2213333X(21). https://doi.org/10.1016/j.jvsv.2021.01.005, 00064-0.
58. Haughey L, Barbul A. Nutrition and lower extremity ulcers: causality and/or treatment. *Int J Low Extrem Wounds*. 2017;16:238−243.

59. Breslow RA, Bergstrom N. Nutritional prediction of pressure ulcers. *J Am Diet Assoc.* 1994;94:1301−1304.
60. SzewczykMT, Jawien A, Kedziora-Kornatowska K, et al. The nutritional status of older adults with and without venous ulcers: a comparative, descriptive study. *Ostomy/Wound Manag.* 2008;54:34−36.
61. Tobon J, Whitney JD, Jarrett M. Nutritional status and wound severity of overweight and obese patients with venous leg ulcers: a pilot study. *J Vasc Nurs.* 2008;26:43−52.
62. Moffatt CJ, Franks PJ, Doherty DC, Smithdale R, Martin R. Sociodemographic factors in chronic leg ulceration. *Br J Dermatol.* 2006;155:307−312.
63. Moffatt CJ, Franks PJ, Doherty DC, Smithdale R, Steptoe A. Psychological factors in leg ulceration: a case-control study. *Br J Dermatol.* 2009;161:750−756.
64. Finlayson K, Edwards H, Courtney M. Relationships between preventive activities, psychosocial factors and recurrence of venous leg ulcers: a prospective study. *J Adv Nurs.* 2011;67:2180−2190.
65. Souza Nogueira G, Rodrigues Zanin C, Miyazaki MC, et al. Quality of life of patients with chronic venous ulcers and socio-demographic factors. *Wounds.* 2012;24:289−292.
66. Heinen MM, van der Vleuten C, de Rooij MJ, et al. Physical activity and adherence to compression therapy in patients with venous leg ulcers. *Arch Dermatol.* 2007;143:1283−1288.
67. Weller CD, Buchbinder R, Johnston RV. Interventions for helping people adhere to compression treatments for venous leg ulceration. *Cochrane Database Syst Rev.* 2013;9:CD008378.

CHAPTER 19

Emerging modalities in local treatment of venous ulcers: advanced dressings, bioengineering, and biologics

Mabel Chan, Jani Lee and John C. Lantis, II
Department of Surgery, Mount Sinai West Hospital, Icahn School of Medicine, New York, NY, United States

Introduction

The initial step when addressing advanced dressings and cellular and tissue-based therapies for venous leg ulcers (VLUs) is to determine indications and good candidates for their use. It has long been reported that approximately 50% of VLUs will heal even after 24 weeks of best practice care.[1] David Margolis' work supports that a VLU that does not close by 40% after 4 weeks of compression is much less likely to heal at 24 weeks.[2] Guest et al. also reported that only 50%–70% of VLUs will heal within a 24-week period.[3] The question then becomes which wounds will heal or not heal? And what are the indicators to help predict wound healing or nonhealing? There are numerous publications with different opinions around these questions. However, the general consensus on the most important indicators for VLU healing are as follows: size ($>12-20$ cm^2), duration ($>50-52$ weeks), previous history of ulceration, deep vein insufficiency, history of deep vein thrombosis, failure to use adequate compression, obesity, sleep apnea, decreased walking distance, poor ankle range of motion, and bacterial burden.[4,5]

None of the cellular and tissue-based products (CTPs) that we will discuss in this chapter are used alone and should be applied in conjunction with the best clinical therapies (i.e., compression therapy, correction of underlying venous pathology) for VLUs. These therapies will be addressed in the remainder of this book. Furthermore, it must be emphasized that appropriate diagnostic studies should be performed to help direct appropriate individualized venous care for the patient in terms of compression, and wound bed preparation to allow advanced dressings to be effective. Most of the studies described in this chapter have had diagnostic imaging before compression therapy. Venous interventions and bacterial burden evaluation may also be included as part of comprehensive venous care.

For the advanced vein practitioner, the inclusion criteria for patients to undergo treatment with CTPs is based primarily on clinical diagnosis in almost all cases. Although many of the clinical trials require some form of venous imaging as well as assessment of arterial inflow to meet inclusion criteria, many of the patients are enrolled from wound care centers that often do not have good access to high-quality venous ultrasounds. Some trials have unsuccessfully attempted to obtain Clinical, Etiological, Anatomical, and Pathophysiological classification for patients. Furthermore, patients are usually excluded for at least 30—90 days if they have recently undergone venous or arterial intervention. While this may not seem ideal for comparison studies, it does reflect the vast majority of patients being treated for VLUs in the United States.[5]

The notion of using a biologic agent to improve the healing rates of VLUs was nurtured and developed fully in 1998. The first and only agent to obtain a United States Food and Drug Administration (FDA) approval for the treatment of the VLUs is a bioengineered bilayered skin equivalent (BSE). It is composed of an epidermal layer (from human foreskin-derived neonatal keratinocytes allowed to stratify) and a dermal layer (cultured from active human foreskin-derived fibroblasts in a bovine type I collagen matrix) (Organogenesis, Canton MA). The makers of this product completed a prospective multicenter study of 293 VLU patients in an outpatient setting, comparing compression therapy alone and serial (up to five) applications of BSE with compression therapy. This study found that treatment with bioengineered skin was more effective than compression therapy alone, closing 63% versus 49% of VLUs at 6 months. In addition, the median time to complete wound closure was 61 versus 181 days, respectively. More interestingly, BSE was found to be effective to a greater extent in larger wounds (>10 cm^2) and wounds that have been present for more than 6 months.[6] It should be noted that therapy made very little difference in patients with ulcers less than a year in duration.

With evidence that a cellular-based product could accelerate the healing rate of VLUs, the Center for Medicare Services and other payers began to recognize the possible benefit of these products. From 2000 to now, there has been a slow and steady increase in recognition that such therapy could be valuable, and many novel products have been placed on the market. It must be noted that not all commercially available products discussed in this chapter are covered by any health insurance. Furthermore, it must be noted that there are generally three health care settings (types of location) in which these products are applied—the hospital outpatient department—which is how most "Wound Care Centers" are billed, the doctor's private office and the inpatient hospital setting. All three of these locations have their own nuances to billing and reimbursement. We do not intend to address these in this book chapter as they tend to change over time. We will note that payment for advanced therapies is usually withheld until a trial of a minimum of 4 weeks of best clinical practice has been applied and has shown failure to reduce wound area by 40%. And as noted previously, the practice should include diagnostic imaging, debridement, topical microbial control, and appropriate compression according to pathophysiology. However, we believe that large wound size (>12 cm^2) and wounds that have been present for longer than a year should be considered for advanced therapies earlier, as these wounds are demonstrated to have prolonged closure times.

In this chapter, we will first discuss the use and reasoning behind biologic therapies for wound bed preparation of the VLU. We will then discuss the currently available studies for CTPs and explore single and multi-stage VLU closure strategies. Upon review of the current literature in this field, we know that there is not one absolute best approach to VLUs. Overall, CTPs in most cases have been shown to improve the rate or speed of wound closure. On the other hand, we will discuss some therapies which have also not shown significant improvement.

Mechanism of action

It must be noted that venous ulcers are full-thickness lesions with the skin breaking down from the inside out. Therefore, when applying a topical agent, one must realize the core of the problem is in the dermis. Venous ulcers are in a state of pathological inflammation with disorganized and unbalanced healing. The edema from venous insufficiency in the skin increases the distance between the capillaries and the tissue leading to decreased oxygen diffusion.[7] As the hypoxic skin breaks down in a wound, bacterial colonization stimulates the inflammatory response with host leukocytes, particularly neutrophils and macrophages, creating a hostile environment with the release of reactive oxygen species and proteases.[7] The proinflammatory state in chronic wounds with disruption of normal signaling pathways leads to damage of cellular components, extracellular matrix (ECM), and protective growth factors.[7] Histologic examination of VLUs show hyperproliferative wound edges with nonmigratory keratinocytes, decreased angiogenesis, and an increase in protease activity degrading wound repair mediators, which may all contribute to the delay in epithelialization in these chronic wounds.[8]

Skin substitutes have been engineered to help accelerate wound healing by providing a tissue graft to the chronic wound bed to stimulate cell migration, angiogenesis, and epithelialization.[7] Stone et al. conducted a translational study evaluating gene expression in venous ulcers after the application of bioengineered bilayer living cellular construct (BLCC/BSE) compared to VLUs treated with compression therapy alone. BLCC/BSE skin substitute application compared to the control group was shown to induce more gene expression identified in the Ingenuity Pathway Analysis related to many aspects of the innate and adaptive immune response. It was hypothesized that BLCC/BSE may secrete growth factors and cytokines, which in turn activates prohealing signaling pathways within VLUs, thus, stimulating acute inflammatory healing pathways to theoretically reverse the chronic nonhealing environment.[8]

The practitioner must question why there are so many products potentially being used on VLUs when BLCC/BSE is the only FDA-indicated product for the closure of VLUs. It should be noted that there are several large prospective trials that failed to achieve improved closure in VLUs.[9,10] In addition, many companies have a certain hesitation of moving into this area as the chronic VLU wound has a more complex pathophysiology than the diabetic foot ulcer patient. While standard diabetic foot care and CTPs have been shown to demonstrate improved healing in diabetic foot ulcers (DFUs), the results for VLUs vary and have been most significant especially if the ulcer has been present for more than 6 months or extends to subcutaneous tissue.[7] Many VLU patients most likely have rheumatologic and hematologic disorders or ulcers that occur secondary to an underlying condition that is inadvertently included in many large randomized trials. The closure of these ulcers may be worsened by the underlying venous pathology, but the actual etiology of the ulcer is not venous. While this reflects real-world medicine, it can cause healing rates to appear quite low. Some recent trials that did not show improvement in VLUs have in part failed because the standard of care healing rates has been much higher than predicted.[9,10] One could infer from this that the ideal inclusion and exclusion criteria have not yet been identified for VLU trials. Even though BSE has FDA approval demonstrating improved efficacy, it is far from perfect. Therefore, it is necessary to have a continuous evaluation of many products on a commercial and individual level.

Practitioners will often make choices based on personal experience, reimbursement, and ease of use. On the other hand, some choices may be based upon a lack of knowledge or influence of marketing by industry representatives. A lot of the claims that are available for each product are from expert opinion, underpowered studies, or nonconclusive systematic reviews. To compare the various types of skin substitutes currently available commercially, this review will organize skin substitutes into three major categories: Human Skin Therapy, ECM, and Placenta Derived Therapy (PDT). However, we will first examine the concept of wound bed preparation or modulation using advanced and biologic dressings.

Wound bed preparation

One of the key issues in evaluating this section is that wound bed preparation is not a very objective finding. The quality of wound bed preparation is difficult to measure, and most studies indicate wound closure as a primary endpoint. Therefore, it is very difficult to design studies looking at therapies that enhance the wound bed or hasten wound closure. Newer techniques which include matrix metalloprotease (MMP) levels, bacterial imaging, and local tissue oxygenation may make this evaluation more objective.

Negative pressure wound therapy

Negative pressure wound therapy (NPWT) is a well-known therapy used throughout the world for improving granulation tissue, managing exudate, and promoting a healthy wound bed. However, it is not widely used for the treatment of VLUs. In 2004, the original makers of NPWT (KCI, San Antonio TX) presented, but never published two prospective studies regarding treating VLUs with NPWT.[11] In 2015, a Cochrane review noted that there was a significant lack of rigorous randomized controlled trials (RCTs) to support NPWT in the treatment of VLUs. However, they noted that there is some evidence that the treatment may reduce healing time when used with compression therapy.[12] Yang et al. published a small study showing the use of NPWT to prepare the wound bed for split-thickness skin grafting (STSG).[13] Another study looked at noninferiority of mechanical (nonpowered spring device) versus electrical NWPT therapy, where approximately 75% of the 115 patients had a venous disease. Overall, the study showed that 50% of wounds were healed in 8–12 weeks and 65% in 12–16 weeks.[14] Similar results of wound area reduction (32%) were seen in a small study with nine patients with VLUs treated with single-use disposable NPWT (Smith andand Nephew, Hull, UK) over 4 weeks.[15] There are some nuances to combining NPWT with compression, which can be done, especially by using long foam bridge dressings.

In general, we use NPWT only for wound bed preparation in the inpatient setting for VLUs, with an inclusion criterion of having a wound greater than 100 cm^2. Otherwise, we use NPWT as an adjunct dressing for wounds greater than 40 cm^2 that are either treated with CTPs in the inpatient setting or to affix STSG postoperatively.

Matrix metalloprotease modulation

In the clinic setting, we often use a dressing composed of 55% bovine collagen and 45% oxidized regenerated cellulose (ORC) product, which is created as a sterile, freeze-dried composite (3M—St Paul, MN). When the collagen-ORC matrix is placed in the wound, the material is readily broken down and reabsorbed to form a soft gel conforming to the shape of the wound.[16] The gel then binds and inactivates potential harmful effects of proteases, including metalloproteinases and elastases that cause the destruction of ECM components at elevated levels.[16] Studies have shown that chronic wounds like VLUs have high levels of proteases compared with healthy tissue, which have been shown to decrease after compression treatment in healing ulcers.[17] These absorbent dressings targeting proteases have been categorized as protease-modulating matrix (PMM) treatments. A large Cochrane systematic review by Westby et al. included 12 RCTs involving 784 participants, in which nine studies compared PMM dressing regimens with other non-PMM dressings, and one study directly compared two different PMM dressings.[18] This systematic review assessed a range of PMM treatments that included interventions varying in the presence of silver, collagen, or ORC, as a single treatment class. Although some studies reported increased healing rate and median time to wound closure (estimated 1.5 months faster for PMM dressing regimens), the overall evidence was found to be of low or very low certainty. The authors concluded that there was limited evidence with a high risk of bias for outcomes in most studies and it is unclear whether PMM treatments are superior in the rate of healing of VLUs over non-PMM treatments in the short, medium, or long term.

In the first RCT for only collagen-ORC dressing,[19] 37 patients with VLUs were allocated to collagen-ORC and 36 to the control group using nonadherent silicone dressing. Eleven VLUs healed in the control group compared to 15 VLUs in the collagen-ORC group ($P = .373$), and a greater size reduction of the VLU was noted in the collagen-ORC cohort ($P < .001$). Another prospective study by Wollina et al. also showed a statistically significant reduction of VLU area and improvement in wound score (score assessed by granulation, color, consistency, discharge/exudates).[20] The authors used remission spectroscopy to measure oxygen saturation of the wound as a reflection of the microcirculation and showed a marked decrease in remission in VLUs after collagen-ORC treatment, demonstrating improvement in microcirculation. They hypothesized that the collagen-ORC matrix has a beneficial effect on microcirculation, thus improving the level of collagen and ECM components, in addition to reducing protease activity in VLUs.[20]

We use PMM dressings early in therapy as part of moisture management or as a bolster on top of CTPs to help reduce the negative effect of MMP. In general, we will initiate compression and debridement with a topical antimicrobial dressing for 2 weeks. This will then be followed by 2 weeks of topical PMM dressing. At the end of 4 weeks, the wound is measured, and the wound area reduction is calculated to determine the next therapeutic steps. Over the years, we have used many methods to measure wounds including digital imaging; although length × width × depth is still used. However, we strive to use a three-dimensional wound scanner (eKare Inc, Fairfax, VA) as a more reliable tool for measuring and monitoring wound areas for the majority of our patients.

Debridement in venous leg ulcers

Although there is not a large amount of prospective randomized data to show that debridement is mandatory for the healing of VLUs, there is significant retrospective evidence to support the use of debridement to facilitate ongoing closure in VLUs.[21] We tend to place topical 4% lidocaine on VLUs for 15 min before undergoing sharp debridement in the outpatient setting, as sharp debridement is not usually well tolerated by many patients. Because few prospective studies support sharp debridement, the practitioner should always evaluate the need for repeat debridement periodically and if the procedure is performed for remuneration instead of clinical efficacy.

Due to the significant pain involved in debriding VLU, the use of topical enzymatic debriding ointments is extensive in the United States. A 2015 Cochrane review could not find significant evidence to support active enzymatic debridement over autolytic debridement.[22] Most topical dressings, in short, did not appear to make a significant reduction in wound size. With this admonition, practitioners should continuously evaluate contact dressings on VLUs based on wound characteristics, including exudate, size, and duration. VLUs are known to exude a large amount of MMP, which has its own collagenase, and so it is possible that autolytic debridement is all that is required to facilitate wound closure.

In the near future, there may be more evidence to support the use of topical enzymes in lower extremity wounds. Currently, there is a large prospective randomized trial assessing the efficacy of Bromelain (proteolytic enzyme) for adequate debridement of VLUs. Another multicenter, prospective, randomized, and adaptive design study is underway to assess the safety and the efficacy of 5% EscharEx (EX-02) compared to gel vehicle (placebo) and to nonsurgical standard of care (enzymatic and autolytic debridement), in debridement of VLU (in a ratio of 1:1:1). The study includes 174 randomized adult patients with VLU that have failed to heal for 4 weeks to 2 years and with >50% of nonviable tissue (necrotic/slough/fibrin). The maximum number of patients to be enrolled is 225. As of July 2021, approximately half of the patients have been enrolled in this trial (clinicaltrials.gov/ct2/show/NCT03588130).

When evaluating VLUs for treatment, we usually delineate those that can undergo outpatient debridement versus those that require debridement in the operating room. Wounds larger than 40 cm^2, patients on significant anticoagulation, and with intractable pain are usually taken to the operating room. During the procedure, wounds larger than 40 cm^2 are debrided with tangential hydrosurgery (Smith and Nephew, Hull UK), while wounds smaller than 40 cm^2 are debrided with ultrasound-enhanced sharp debridement (Misonix, Farmingdale NY).

Topical antimicrobial therapy

The effect of the microbiome on VLUs healing process has long been a question. New imaging strategies and culturing techniques are starting to help elucidate how practitioners may need to use topical antimicrobials as part of VLU care. Wounds with clinical signs of infection or high bacterial colonization have been shown to have poor healing.[23] A study by Schwartz et al. demonstrated serial sharp debridement of chronic lower extremity wounds alone does not reduce planktonic bacterial wound burden significantly.[24] Thus, an antimicrobial therapy after debridement can be considered to prevent delay in healing. In 2014, a Cochrane review published an article that evaluated and compared 53 antibiotic and antiseptic preparations among a total of 4486 participants. The aforementioned review included many RCTs that had a small number of patients and had a high risk of bias. The studies also had varying definitions of what constituted an infected ulcer and wound closure as an indicator of reduction in bacterial contamination. While the review could not recommend discontinuation of a specific antimicrobial topical therapy, it did find substantial support for the use of cadexomer iodine (povidone-iodine-based topical) in promoting healing of VLUs.[23]

In the outpatient setting, we prefer to use cadexomer iodine after sharp debridement of VLU. It is applied twice a week with a multilayer compression for 2 weeks. We have been unable to demonstrate significant bacterial reduction with the application of honey (unpublished data), sustained release silver sulfadiazine,[25] bovine native collagen, and silver or sodium carboxymethylcellulose and ionic silver.[26]

For larger wounds (>40 cm^2) that undergo debridement in the operating room, we have extensive experience using NPWT with the instillation of dilute hypochlorite solution.[27] This has been effective in reducing the bacterial burden and essentially sterilizing wounds that are greater than 100 cm^2 and preparing them for STSG. However, NPWT with instillation (3M, St Paul, MN) is an inpatient therapy and the length of hospital stay for this therapy including STSG can take approximately 11 days with costs approaching $25,000.[13]

Combination therapy

An alternative strategy is to place a modulatory ECM on the chronic wound to theoretically reduce the bacterial burden and provide an ECM to enhance closure rates. One example is a porcine-derived small intestinal submucosa (SIS) coated with polyhexamethylenbiguanide hydrochloride (PHMB) (Organogenesis, Canton MA). As a topical antimicrobial, PHMB strongly binds to bacterial cell walls and membranes, and it disrupts the transport, biosynthesis, and catabolic functions of the bacteria.[28,29] PHMB works best when in intimate contact with the wound bed for more than 20 min. Another such product that is theoretically designed to provide similar benefits is the acellular fetal bovine dermis (FBD) with a silver (Integra Lifesciences, Princeton, NJ). These products do not have antimicrobials placed in the dressing to eliminate bacteria from the wound but serve to prevent bacterial growth.

In a registry study of SIS with PHMB (porcine collagen matrix with PHMB (PCMP)) applied weekly, 22%[66] of the 307 patients treated had VLUs. Bain et al. reported a 64% closure rate at 26 weeks.[31] In this paper, 43 out 63 wounds studied achieved complete wound closure after PCMP treatment, but 2 of the 43 closed only after bridging to other modalities and surgical closure. Overall, wounds that did not achieve complete closure after 4–6 weeks (12 of 63) were switched to a cellular graft after SIS with PHMB treatment. In theory, a reasonable application of this algorithm of care would be a period of SIS with PHMB to prepare the wound bed, which would modulate bacteria followed by the application of the FDA-indicated bilayered skin equivalent. However, we have no experience with this algorithm.

An FBD with the silver trial was initiated in June 2012 but was terminated due to poor enrollment. It was designed as a 3-pronged multicenter trial looking at the percentage of VLUs healed by week 12, after treatment with FBD with and without silver and moist wound care. There are 31 patients reported as enrolled, but no peer-reviewed reporting is available (clinicaltrials.gov/ct2/show/NCT01612806). We have only anecdotal experience with this therapy.

Cellular and tissue-based therapies

Here, we review the data supporting particular CTPs that have been studied in VLUs. As a group, we accept the categorization of these products set out in the 2020 Agency for the Healthcare Research and Quality, which was classified by Davison-Kotler et al.[33,32] as follows: acellular dermal replacement from synthetic materials, acellular dermal replacement from synthetic and animal sources, dermal replacement from donated human dermis, dermal replacement from donated placental membrane, dermal replacement from human placental membrane for epidermis and dermis, dermal replacement from the animal source, cellular, dermal replacement from the placental membrane, non—autogenous cultured cells with or without animal components.[33] However, we find this categorization cumbersome and difficult to follow. We prefer to group the products as living skin cellular products, ECM (xenograft vs. allograft), placental-based products (cellular vs. acellular), biosynthetic and synthetic. Within the ECM category, we consider there to be a continuum that starts with products that serve primarily as a scaffold to those that serve primarily in the modulation of inflammation. The modulatory products often require more frequent applications and are more likely to dissolve and disintegrate into the wound, while scaffold products maintain their structure for a longer period of time.

Living skin cellular products

Living skin cell products donate cells to the wound bed with varying amounts of ECM. However, due to the host immune response, the donated or engrafted cells do not stay in the wound bed for very long. Studies have shown that they are present in the wound bed for less than 4 weeks for BSE and 7–10 days for cryopreserved human skin split-thickness allograft (CSSA).[35,34] They function to provide several potential progrowth ingredients that serve multiple benefits including acting as a biologic dressing with healthy responsive cells that provide growth factors to stimulate healing and with keratinocytes providing bacteria-suppressing peptides in the wound.[36]

We have already discussed BSE (Organogenesis, Canton, MA) previously, which was the first BLCC skin substitute and only FDA-indicated CTP for venous leg ulcers. All the other products to be discussed after, have gained FDA clearance through the less rigorous 510-K pathway or minimally manipulated human tissue "361" pathway. As noted, BSE was found in a large randomized clinical trial to increase the rate of complete healing in venous ulcer disease.[6]

Due to BSE being first in the US market and the only CTP with a specific VLU indication, BSE has been exposed to more comparative effectiveness trials than any other CTPs. Treadwell et al. further evaluated the effectiveness of BSE with a comparative study with CSSA. They found that VLUs with greater than 28-day history treated with either BSE or CSSA had no significant difference in wound characteristics, but showed BSE significantly improved the median time to wound closure within 15.1 weeks compared to 31.3 weeks by CSSA. BSE treatment doubled the probability of healing compared to CSSA, and 65% of VLUs healed at 24 weeks with the BSE treatment.[36] Another comparative study by Marston et al. compared the effectiveness of BSE to porcine SIS (Smith and Nephew, Hull, UK), which is an acellular collagen scaffold derived from the porcine small intestinal mucosa. This product is the predicate product in the FDA 510-K regulatory pathway used for almost all Xenografts. This retrospective review included 1457 venous ulcers treated with BSE and 350 treated with SIS and found BSE improved wound closure rates with 61% closed at 36 weeks compared to 46% in the SIS group.[37] It should be noted that there is a very significant product cost between these two products (BSE costing $29/cm^2 vs. SIS costing $8/cm^2).

Another living cellular product is a human fibroblast-derived dermal substitute (HFDS), which consists of cultured dermal fibroblasts derived from neonatal human foreskin tissue and cultured onto a bioresorbable polyglactin (Vicryl) mesh, requiring cryopreservation (Organogenesis, Canton MA).[38] HFDS resorbs in three to 4 weeks and is replaced with ECM components produced by the fibroblasts secreting growth factors, which promotes wound remodeling.[39] The allogeneic fibroblasts have underdeveloped human leukocyte antigens and do not show clinical signs of rejection.[40] In initial studies a regimen of four pieces of HFDS applied to VLUs resulted in a 38% closure rate versus 15% for the standard of care.[41] This led to a pivotal 316-patient prospective randomized open-label trial of HFDS versus compression alone. Both arms exhibited 34% and 31% of wounds closed at 12 weeks and showed no significant difference.[42] However, the subanalysis did find that the duration of VLUs less than 12 months showed improved healing at week 12 compared to the control group. At 24 weeks, 15% of the healed ulcers recurred in the HFDS group, whereas 23% recurred in the control group. These results were not significant as well and Baber et al. concluded that they could not find support for using HFDS over the standard of care.[43] Based on all these facts, HFDS has not gained significant coverage for the treatment of VLUs.

We have briefly mentioned CSSA (Misonix, Framingham NY) in comparative studies with other CTPs. This is a CSSA harvested from donors within 24 h of death which therefore retains human fibroblasts and keratinocytes. The allograft is processed by hair removal, meshing, and cleaning with antibiotics and reagents. Landsman et al. published a study characterizing CSSA and determined that the amount of collagen and the ratio of collagen types were similar to the composition of collagen in unprocessed human split thickness skin. Along with collagen, CSSA and fresh human skin have equivalent amounts of vascular endothelial growth factor, insulin-like growth factor 1, fibroblast growth factor 2, and transforming growth factor β1. Whereas the average percentage of apoptotic cells in CSSA was 34.3%, fresh skin had only 3.1% apoptotic cells.[44] As a result, there were significantly more dead cells in CSSA than STSG. The same lead investigator also conducted a retrospective study of 134 VLUs that was not controlled nor randomized and noted closure rates of 60% of VLUs at 12 weeks and 75% at 20 weeks.[45]

As mentioned previously, there is a highly endorsed study that shows the superiority of BSE to CSSA for VLU closure. One disadvantage of using BSE is that due to its complex manufacturing process, it is only commercially available in one size (44 cm^2), therefore creating significant wastage if the wound is not large. Towler et al. performed a prospective single-site randomized trial for the treatment of VLUs and showed no difference in clinical outcomes between BSE and CSAA, with approximately 50% cost savings by using CSSA.[46] Clinicaltrials.gov database (NCT03935386) reports the initiation of a prospective 100 patient trial studying CSSA versus compression alone. Per the principal investigator (Philip Garrett, DPM), this study has been completed and may be published in 2021.

There is growing interest in autologous cultured or expanded skin grafts, pinch grafts, and traditional STSG for the closure of VLUs. As surgeons, many may argue that we have a strong bias toward operating with surgical debridement and traditional STSG. There are very few reviews that address the efficacy of STSG for VLUs. The Cochrane review in 2013 noted that after reviewing 17 trials with 1034 participants in which studies included autografts, frozen allografts, and BSE, they could only recommend the use of BSE. Alternative skin grafting techniques were compared in six trials such as autografts compared with frozen allograft, pinch autograft with the porcine dermis, growth-arrested human keratinocytes and fibroblasts with placebo, autograft on porcine pads with an autograft on porcine gelatin microbeads, meshed graft with a cultured keratinocyte autograft, and a frozen keratinocyte allograft with lyophilized keratinocyte allografts. Despite the many alternative comparisons, the review noted that only BSE had significant support for its use in increasing the closure rate of VLUs.[47] However, as it is a procedure and not a commercial product there are almost no studies reviewing the efficacy of STSG for the treatment of VLU.

If the patient has a VLU under 40 cm^2 and the wound bed is well prepared and free of bacterial contamination, the current prospective randomized data do support the use of BSE. However, the practitioner must be prepared to apply the graft weekly or every other week, up to 5 or even 8 weeks consecutively. This is of course in conjunction with appropriate compression therapy. However, if the wound is greater than 40 cm^2, we tend to proceed with wound bed preparation and placement of STSG. We will further elaborate on different treatment options for patients with wounds between 40 to 100 cm^2 in the next section.

Extracellular matrix therapy (ECM)

The human-derived dermal substrates of ECM are all noted to act like a template. Most of these products are fundamentally acellular human dermis. While the ECM provides necessary proteoglycans and glycosaminoglycans to promote growth, there appears to be a continuum of the speed of engraftment in these products. For example, some dissolve into the wound quickly, while others are present for up to a month later. There are three strategies for replacing the missing ECM in a chronic wound; these are to protect the breakdown, to grow, or most notably to try to replace (by direct application) the ECM of the wound. The human-derived materials have passed through the regulatory pathway by being minimally manipulated human tissue, while the xenografts have come to market by following the 510-K pathway, using porcine submucosa as the predicate product. Neither of these regulatory pathways requires stringent prospective RCTs. In general, all these products are FDA cleared (but not indicated) for use in full-thickness wounds, which includes venous stasis ulcers.

Acellular human tissue

There exists a freeze-dried, decellularized dermis (DCD) human cadaver skin substitute (Wright Medical Technology, Inc., Arlington, TN), which is an immunologically inert human dermal matrix with a natural porosity and a basement membrane that provides human ECM to promote healing.[48] There is limited research evaluating the use of DCD in VLUs. Greaves et al. conducted a clinical study evaluating the use of DCD from human cadaveric tissue donors on 22 patients; 16 of which had pure VLUs.[49] The patients included in the study had chronic ulcers for at least 3 months with a mean duration of 4.76 years, and the study was concluded at 6 months. In the first week of therapy, these patients were treated with the addition of NPWT. Among the patients treated with DCD, 60% of patients completely healed their ulcers during the 6-month study period, and those with partially healed ulcers showed an 87% reduction in wound size. They also evaluated the immunohistochemistry of the wounds, and there was strong evidence of angiogenesis with wound biopsies showing host cell migration, proliferation, and conversion of chronic ulcers to wounds with acute wound characteristics.[49] After 6 weeks of DCD treatment, wounds had higher concentrations of fibronectin, collagen I, and collagen III indicating increased deposition by migrating fibroblasts.

Another human decellularized-acellular dermis matrix (D-ADM) (LifeNet Health (Virginia Beach, VA) was investigated in a 2:1 prospective randomized trial of 28 patients.[50] Eighteen patients were included in the D-ADM arm and 10 patients in the control arm. For the ADM-treated patients, the wound area reduction was 59.6% at 24 weeks versus 8.1% at 24 weeks for control patients. It was also noted that the healed ulcers in the D-ADM arm remained closed at a substantially higher rate after termination than healed ulcers in the control group.[50]

In our extensive experience using DCD, we have primarily used it as a bridge therapy, with a plan of eventual STSG. We have over 200 venous and diabetic ulcer cases experience using DCD in the operating room after debriding larger wounds. Our method of application of DCD is similar to the study by Cazzell et al. with the addition of NPWT for 1 week. We have previously presented but not published our experience about the need to debride to remove the product from the wound before skin grafting at 4 weeks.[51] This extra debridement step has been secondary to DCD not completely engrafting in the wound. We continue to experiment with utilizing and optimizing DCD use for chronic wounds.

Acellular xenograft tissue

Porcine small intestinal submucosa (first developed by Cook Medical as a possible vascular graft) was the first animal-sourced ECM membrane to be available for commercial use in the chronic wound market. In 2005, Mostow et al. published a 120-patient prospective study evaluating its use in VLUs.[52] Sixty-two of the patients were treated with SIS and compression, while 58 were treated with compression alone. At the end of 12 weeks, 55% of the patients treated with SIS had a closed VLU versus 34% for the standard of care. At follow-up, none of the patients treated with SIS had a recurrence of their VLU. This study also had a crossover component where 26% of the patients that failed to heal with compression alone healed after applications of SIS. Furthermore, patients who underwent debridement at the beginning of the study had better results with SIS application than those who were not well debrided before SIS.[52] The mean number of SIS applications over the 12-week period was eight 21 cm^2 sheets. This would cost approximately $2400 in products alone in 2021, compared to five applications of BSE that would cost $8250. However, the application cost for BSE would be $504 dollars less, based on fewer applications.

Dr. Romanelli in Italy looked at the cost effectiveness of SIS versus standard of care. He included patients with VLU and mixed arterial and venous ulcers.[53] From 50 patients included, 27 of them had a pure venous disease and 23 of them had mixed disease. Twenty-five were treated with SIS and the average wound size was roughly 24 cm^2. At the end of 8 weeks, 80% of the patients treated with SIS had closed wounds versus 65% treated with standard of care. Those treated with SIS were seen to have wound closure at an average of 5 weeks, which is 3 weeks earlier than those treated with standard of care. Also, there was a much lower recurrence rate at 32 weeks for those treated with SIS. Although the relative upfront cost of SIS was higher, due to earlier improvement to the wound with the application (reduction of 7 weeks that wound was open during a 32-week period) it is less costly in the long term.

Other xenografts have been studied for the treatment of VLUs, but almost all studies have been small and often without a control. One of these products is FBD (Integra Lifesciences, Princeton, NJ). It is an acellular dermal matrix with type I and III collagen fibers derived from the cadaveric skin of fetal bovine (FBDM; fetal bovine dermal matrix).[54] The product is sterilized and stored at room temperature with a 3-year shelf life. It is available premeshed or in thin strips for irregular wounds. This product acts like a template of dermal matrix and engrafts into the wound bed over the course of 4 weeks.[55] A retrospective, single-center clinical comparative study of 28 venous ulcers with either BSE or FBDM treatment showed that ulcers treated with FBDM had complete healing in 32 days, whereas the BSE group had complete healing in 63 days.[56] The wound characteristics were similar for both groups and each patient had sharp debridement before the application of the skin substitutes. Although it was not a large study, it is interesting to note that a single-application product such as FBDM can outperform a multiapplication product such as BSE.

Our group has extensive experience with FBDM, as well. We have reported treating 40 wounds in 33 patients with VLUs.[57] All these patients had various difficult-to-heal criteria including obesity, poor goniometry, and failed healing with multilayer compression for more than 4 weeks. With a single application of FBDM, we saw a 24% median reduction of wound area after 4 weeks, wherein 40% of VLUs had a greater than 40% area reduction. This was a retrospective study, and in most cases, we used FBDM as a staged application before STSG or using another adjunctive method to close the wound.

Acellular intact fish skin (IFS) is another newer modulatory xenograft (Kerecis, Arlington, VA), which we have been using frequently. It contains high levels of Omega 3 fatty acids, which may provide some benefits in terms of reducing inflammation of the wound. In addition, the product has a natural porosity that is conducive to cellular ingrowth. In our experience, weekly or biweekly application of this product is appropriate. In 2016, we reviewed 18 patients who had hard-to-heal criteria for VLUs who were treated for 5 weeks consecutively with IFS followed by 3 weeks of the standard of care. There was a 40% wound area reduction over 6 weeks and complete closure was seen in three of 18 patients. Of note, all these patients had at least one other advanced tissue skin therapy application prior to IFS.[58]

Xenografts are much less expensive to manufacture than harvesting material from humans or creating large skin banks. They are mostly sterilized and decellularized animal products that would otherwise have been discarded. While the cost-reimbursement structure in the United States leans heavily toward lower actual dollar reimbursements for ECM products, studies have shown that they are very cost effective in the long term. Hankin et al. evaluated the clinical and economic efficacy of different types of ECM products (BSE, HFDS, human skin equivalent, and a sterile wound matrix composed of shortened fibers of poly-N-acetyl glucosamine (pGlcNAc) isolated from microalgae) compared to standard of care (SOC) for VLUs. The cost in 2012 per each additional successfully treated patient was $1600 for pGlcNAc, $3150 for ECM, and nearly $29,952 for living human skin equivalent after 24 weeks.[59] Wide variations exist in the cost of ECM products with almost similar clinical effectiveness on VLUs, so it is important for practitioners to consider different therapies without creating a financial burden for the patient.

More often than not we find ECM products to be very helpful in patients who have wound beds that are not well prepared. The studies behind SIS and PHMB[31] provide a very strong argument for wound bed preparation with ECMs that have antimicrobial property. In the future, more high-powered VLU studies with other ECM products are needed to compare and guide treatment decisions. FBD is also available with a silver component that may have antimicrobial activity, but its benefit to healing wounds has not been studied sufficiently. Likewise, Omega 3 fatty acids in IFS provide intrinsic antimicrobial properties. However once again, this property has not been clinically evaluated for its efficacy that would be translatable to VLUs.

Biosynthetic

The aforementioned collagen-ORC and DRT products fall under the biosynthetic ECM category. Of note, most studies for DRT have been on its application in traumatic wounds, diabetic foot ulcers, and in burns and not VLUs. Many RCTs have shown higher graft take in poorly vascularized wounds, improved STSG take on avascular structures (bone or tendon-exposed wounds), and good cosmesis with reduced rates of wound contraction and scarring.[60,61]

Hyaluronic acid (HA) is a prominent component in ECM and is very abundant in the skin with the dermis containing 50% of the total HA of the body.[62] As part of ECM, they play a key role in skin integrity, particularly in the migration and proliferation of antiinflammatory cells and angiogenesis.[62,63] Hyaluronic acid is a linear nonsulfated glycosaminoglycan that has a short half-life and has been made into an esterified matrix (benzyl esters of hyaluronic acid, HYAFF) to make a stable product for wounds (Fidia Advanced Biopolymers, Abano Terme, Italy). The HYAFF layer is coupled with a silicone layer that serves to provide immediate protection against dehydration and bacteria. This bilayer matrix is sterile and acts as a flexible three-dimensional matrix conforming to the wound bed and releasing a high concentration of HA.[62] The largest study to date on chronic ulcers (of which 46% were of vascular origin) is a multicenter, prospective observational study involving 262 elderly patients.[64] Patients were observed from the start of treatment until greater than 10% of reepithelization was achieved at the wound edge. This endpoint was met in 83% of ulcers in a median time of 16 days, and a 75% reepithelization was seen in 26% with only HYAFF.[64] A smaller study of 16 patients with VLUs was treated with HYAFF if they had a failure of reepithelization after 4 weeks of SOC.[62] HYAFF was applied once with nonabsorbable stitches and covered with a noncompressive dressing. The silicone film was removed at days 18–21 and a 0.1 mm autograft was applied to 12 patients. Four patients had complete reepithelization and did not require an epidermal graft. Only one patient had a recurrence of VLU after a 6-month follow-up.[62] One other prospective study of 16 patients showed that for the HA group, wound closure at 12 weeks was 66.6% and 14% for the control.[65] At week 16, 87.5% were healed in the HA group compared with 42.8% in the control, the mean time to healing in the HA-treated group was 41 days compared with 104 days in the control.[65] Despite the strong evidence of this small trial, a larger follow-up study was not pursued further.

Placenta-derived tissue

Placental-derived group of products has gained access to the wound care market by being classified by the FDA as minimally manipulated human tissue regulated under 21 CFR 1271.3(d)[1] and Section 361 of the Public Health Service Act (PHSA), which is often referred to as the "361" pathway. The federal, and in some cases state, governments are imposing stricter regulatory hurdles on PDTs. In November 2017, the FDA released four guidance documents that attempt to clarify the distinctions between products that are subject to the agency's full drug approval requirements and those that are not, while streamlining the review process and reducing some regulatory requirements. In August 2019, the FDA stated that "all allograft stem cell like Wharton's Jelly products currently on the market in the US are being sold in violation of FDA regulations" and should be regulated as drugs, which led to states such as New York to ban the sale of all products containing Wharton's jelly. Following this event, we may be expecting more well-designed prospective trials using PDT for VLUs.

PDT should be at minimum broken down into products that attempt to retain native immune-protected stem cells and those which do not. In general, cryopreservation or flash freezing of the tissue theoretically retains these cells while dehydration does not.[66] However, frozen products are more cumbersome to ship, store and possibly use than dehydrated PDT. Two companies (Organogenesis, Canton MA, and Smith and Nephew, Hull UK) have developed novel technologies to reduce this difficulty, thereby preventing the issue of thawing the product.[67,68] There is also much debate surrounding the layers of the tissue matter with many products trying to differentiate between amnion, amnion and chorion or even containing umbilical cord. The importance of the composition or layers unfortunately appears to be in many cases more industry profit driven than science related.[69]

Dehydrated PDT

Another product is a dehydrated human amnion/chorion membrane (DHACM) (MiMedix Group Inc., Marietta, GA) composed of a single layer of epithelial cells, basement membrane, and an avascular connective tissue matrix. Studies of DHACM for VLUs have shown a reduction in wound area of 40% at 4 weeks.[70] Serena et al. enrolled 84 patients of which 53 patients received DHACM and 31 had a multilayer wrap without DHACM. At 4 weeks, there was a significant difference between the allograft-treated group with 62% showing greater than 40% wound closure compared to only 32% in the multilayer compression therapy alone control group with greater than 40% wound closure ($P = .005$). Allograft treatment wounds had reduced in size by a mean of 48.1% compared to 19.0% for controls after 4 weeks.[70] In 2015, a multicenter RCT evaluated two application regimens of a DHACM (MTF Biologics, Edison, NJ) and SOC versus SOC alone in the treatment of VLUs. This trial (clinicaltrials.gov/ct2/history/NCT02609594) was taken over by a new research organization in 2018 and the now 240-patient trial has been, by report, closed to enrollment and accepted for publication as of July 2021 (personal communication).

Cryopreserved and other "viable PDT"

Another small prospective trial using a cryopreserved human umbilical cord product derived from donated human placental tissue following healthy, live, caesarian section, full-term births, has been completed (Tissue tech, Miami FL). The product is developed by using a "proprietary" process, which devitalizes the living cells, but retains the natural structural and relevant biological characteristics. The company undertook a 31-patient trial comparing their cryopreserved amniotic tissue versus multilayer compression. The trial has closed to enrollment on January 28, 2020 (clinicaltrials.gov/ct2/show/NCT03818828). This study will never be published but may be used by the company to construct a larger trial (personal communication). Currently, a hypothermically preserved PDT is being evaluated in a large pivotal trial versus SOC alone and this trial is currently underway (Organogenesis, Canton MA).

At present, our opinion is that there is neither a clear need nor clinical indication for using amniotic tissue for VLUs. However, we recognize that there is an extensive amount of research that is ongoing in PDTs and future opinions may change.

Compared to VLUs, DFUs have underlying pathophysiology (prolonged inflammatory phase with poor skin integrity) that is appropriate to treat with the regenerative properties of amniotic tissue. The concepts of supplying growth factors and stem cell activation also is compatible for treating ischemic or arterial wounds. VLUs on average are much larger in size than DFUs, which at the current frequency of application that is studied in most amniotic tissue trials would make treatment too costly. We recognize that treatment algorithms and costs may change over time.

Our treatment algorithm

Effective management of VLUs remains difficult with a significant burden on the patient, decreased quality of life, and increasing economic burden on the US healthcare system and payers. The SOC for the treatment of VLUs is compression therapy, but it has limitations of its efficacy for improving the healing rate and only provides little benefit for large and exudative VLUs.

We have attempted to maximize outpatient management for wounds of a certain size. However, wounds in unfavorable locations or of large size are not economically viable to be managed in an outpatient setting or have high failure rates to close. Because there are so many significant factors to take into consideration for VLU management, we will detail our algorithm using surgical intervention, advanced topical treatments, CTPs in conjunction with other modalities discussed throughout the book.

Before starting any interventions, we first address several patient factors; anatomy, need for venous or arterial intervention, compliance with compression therapy, ankle range of motion, walking status, obesity, sleep apnea, and chronicity of the wound. We base our algorithm on existing data and our experience but continue to enroll and participate in numerous prospective RCTs to achieve better outcomes for our patients. For VLUs that do not show a 40% reduction in area at 4 weeks, or if present for more than 1 year or greater than 12 cm^2, we manage with the treatment algorithm below (Fig. 19.1). All treatment pathways adhere to 4 weeks of compression therapy after the intervention. If the patient has poor ambulatory status, obesity, or limited range of motion of the ankle, we will add pneumatic compression therapy. Ideally, all these patients are seen weekly in the outpatient setting and their wounds are wrapped with multilayer compression.

- Wounds <12 cm^2 with health vascular wound base (postdebridement): while five applications of BSE remain the therapy that is best supported by the literature, the amount of wastage and cost needs to be considered due to its set commercially available size. Therefore, the patient will have up to 5 weekly or biweekly applications of IFS. If multiple applications are prohibited due to lack of health insurance or pain, we consider FBD as it often only requires one application.
- Wounds <12 cm^2 with poor/avascular wound base (postdebridement): in our experience, BSE on a poorly vascularized wound bed is not efficacious. We have seen IFS to develop granulation tissue quickly in the wound bed after SIS for undetermined reasons. Therefore, the patient will have up to five weekly or biweekly applications of IFS. If multiple applications are prohibited due to lack of health insurance or pain, we consider FBD as it often only requires one application.
- Wounds 12–40 cm^2 with healthy wound base (postdebridement): this size is cost-effective for using BSE and is even more advantageous if the wound is greater than a year in duration. The patient will have up to five weekly or biweekly applications of BSE.
- Wounds 12–40 cm^2 with poor/avascular wound base (postdebridement): this is another area where it makes sense to explore up to four weekly applications of SIS with PHMB followed by BSE. However, this will require up to nine applications of therapy and a minimum of 10 weeks to close. Our current recommendation is a single application of FBD followed by STSG at 28 days, leading to complete closure at 8 weeks with two applications.

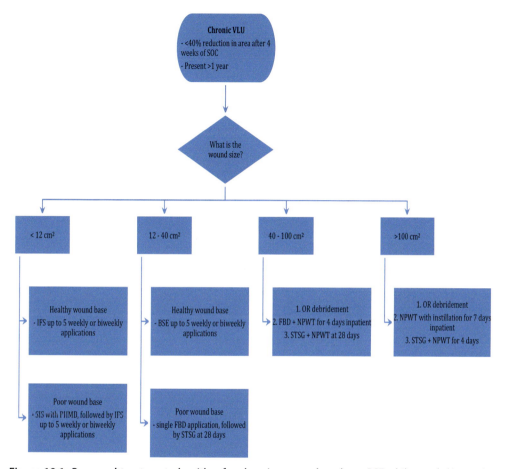

Figure 19.1 *Proposed treatment algorithm for chronic venous leg ulcers.* BSE—bilayered skin equivalent; FBD—fetal bovine dermis; IFS—acellular intact fish skin; PHMB—polyhexamethylenbiguanide; NPWT—negative pressure wound therapy; OR—operating room; SIS—porcine-derived intestinal submucosa; SOC—standard of care; STSG—split thickness skin graft; VLU—venous leg ulcer.

- Wounds 40–100 cm^2: we recommend surgical debridement in the operating room. FBD is then applied with NPWT for 4 days with bed rest as an inpatient. This is followed by STSG and NPWT at 28 days with 4 days of bed rest as well.
- Wounds >100 cm^2: we recommend surgical debridement in the operating room and 7 days of NPWT with instillation as an inpatient. This is followed by STSG and NPWT for 4 days at postoperative day 7.

Conclusion

There exist many effective treatments and adjunct therapies to provide closure to VLU wounds. However, the hostile wound environment and difficulty of treating especially large VLUs have led researchers and clinicians to shy away from discourse. Our recommendations are based on published data and closely monitored trial and error, which have led us to develop treatment algorithms over time. It must be emphasized that there is no one-size-fits-all approach for advanced therapy of VLUs, but the treatment can begin with an effective algorithm pathway that can be adjusted and tailored to individuals. With the advent of better diagnostic tools such as microbial imaging cameras, MMP testing, and theranostics, we will continue to improve our ability and detail to manage VLUs.[30]

References

1. Chaby G, Senet P, Ganry O, et al. Prognostic factors associated with healing of venous leg ulcers: a multicentre, prospective, cohort study. *Br J Dermatol*. 2013;169(5):1106−1113. https://doi.org/10.1111/bjd.12570.
2. Kantor J, Margolis DJ. A multicentre study of percentage change in venous leg ulcer area as a prognostic index of healing at 24 weeks. *Br J Dermatol*. 2000;142(5):960−964. https://doi.org/10.1046/j.1365-2133.2000.03478.x.
3. Guest M, Smith JJ, Sira MS, Madden P, Greenhalgh RM, Davies AH. Venous ulcer healing by four-layer compression bandaging is not influenced by the pattern of venous incompetence. *Br J Surg*. 1999;86(11):1437−1440. https://doi.org/10.1046/j.1365-2168.1999.01288.x.
4. Parker CN, Finlayson KJ, Shuter P, Edwards HE. Risk factors for delayed healing in venous leg ulcers: a review of the literature. *Int J Clin Pract*. 2015;69(9):967−977. https://doi.org/10.1111/ijcp.12635.
5. Marston WA, Ennis WJ, Lantis 2nd JC, et al. Baseline factors affecting closure of venous leg ulcers. *J Vasc Surg Venous Lymphat Disord*. 2017;5(6):829−835.e1. https://doi.org/10.1016/j.jvsv.2017.06.017.
6. Falanga V, Margolis D, Alvarez O, et al. Rapid healing of venous ulcers and lack of clinical rejection with an allogeneic cultured human skin equivalent. Human skin equivalent investigators group. *Arch Dermatol*. 1998;134(3):293−300. https://doi.org/10.1001/archderm.134.3.293.
7. Greaves NS, Iqbal SA, Bagueneid M, Bayat A. The role of skin substitutes in the management of chronic cutaneous wounds. *Wound Repair Regen*. 2013;21(2):194−210. https://doi.org/10.1111/wrr.12029.
8. Stone RC, Stojadinovic O, Rosa AM, et al. A bioengineered living cell construct activates an acute wound healing response in venous leg ulcers. *Sci Transl Med*. 2017;9(371):eaaf8611. https://doi.org/10.1126/scitranslmed.aaf8611.
9. Kirsner RS, Vanscheidt W, Keast DH, et al. Phase 3 evaluation of HP802-247 in the treatment of chronic venous leg ulcers. *Wound Repair Regen*. 2016;24(5):894−903. https://doi.org/10.1111/wrr.12467.
10. Harding K, Sumner M, Cardinal M. A prospective, multicentre, randomised controlled study of human fibroblast-derived dermal substitute (Dermagraft) in patients with venous leg ulcers. *Int Wound J*. 2013;10(2):132−137. https://doi.org/10.1111/iwj.12053.
11. Dumville JC, Land L, Evans D, Peinemann F. Negative pressure wound therapy for treating leg ulcers. *Cochrane Database Syst Rev*. 2015;2015(7):CD011354. https://doi.org/10.1002/14651858.CD011354.pub2. Published 2015 Jul 14.
12. John Lantis JC, Gendics C. *Vac Therapy Appears to Facilitate STSG Take when Applied to Venous Leg Ulcers*. Paris: 2nd World Union of Wound Healing Societies' Meeting; 2004:42−43.
13. Yang CK, Alcantara S, Goss S, Lantis 2nd JC. Cost analysis of negative-pressure wound therapy with instillation for wound bed preparation preceding split-thickness skin grafts for massive (>100 cm(2)) chronic venous leg ulcers. *J Vasc Surg*. 2015;61(4):995−999. https://doi.org/10.1016/j.jvs.2014.11.076.
14. Armstrong DG, Marston WA, Reyzelman AM, Kirsner RS. Comparative effectiveness of mechanically and electrically powered negative pressure wound therapy devices: a multicenter randomized

controlled trial. *Wound Repair Regen.* 2012;20(3):332−341. https://doi.org/10.1111/j.1524-475X.2012.00780.x.

15. Schwartz JA, Goss SG, Facchin F, Gendics C, Lantis JC. Single-use negative pressure wound therapy for the treatment of chronic lower leg wounds. *J Wound Care.* 2015;24(Suppl 2):S4−S9. https://doi.org/10.12968/jowc.2015.24.Sup2.S4.
16. Cullen B, Smith R, McCulloch E, Silcock D, Morrison L. Mechanism of action of PROMOGRAN, a protease modulating matrix, for the treatment of diabetic foot ulcers. *Wound Repair Regen.* 2002;10(1):16−25. https://doi.org/10.1046/j.1524-475x.2002.10703.x.
17. Beidler SK, Douillet CD, Berndt DF, Keagy BA, Rich PB, Marston WA. Multiplexed analysis of matrix metalloproteinases in leg ulcer tissue of patients with chronic venous insufficiency before and after compression therapy. *Wound Repair Regen.* 2008;16(5):642−648. https://doi.org/10.1111/j.1524-475X.2008.00415.x.
18. Westby MJ, Norman G, Dumville JC, Stubbs N, Cullum N. Protease-modulating matrix treatments for healing venous leg ulcers. *Cochrane Database Syst Rev.* 2016;12(12):CD011918. https://doi.org/10.1002/14651858.CD011918.pub2. Published 2016 Dec 15.
19. Vin F, Teot L, Meaume S. The healing properties of Promogran in venous leg ulcers. *J Wound Care.* 2002;11(9):335−341. https://doi.org/10.12968/jowc.2002.11.9.26438.
20. Wollina U, Schmidt WD, Krönert C, Nelskamp C, Scheibe A, Fassler D. Some effects of a topical collagen-based matrix on the microcirculation and wound healing in patients with chronic venous leg ulcers: preliminary observations. *Int J Low Extrem Wounds.* 2005;4(4):214−224. https://doi.org/10.1177/1534734605283001.
21. Doerler M, Reich-Schupke S, Altmeyer P, Stücker M. Impact on wound healing and efficacy of various leg ulcer debridement techniques. *J Dtsch Dermatol Ges.* 2012;10(9):624−632. https://doi.org/10.1111/j.1610-0387.2012.07952.x.
22. Gethin G, Cowman S, Kolbach DN. Debridement for venous leg ulcers. *Cochrane Database Syst Rev.* 2015;2015(9):CD008599. https://doi.org/10.1002/14651858.CD008599.pub2. Published 2015 Sep. 14.
23. O'Meara S, Al-Kurdi D, Ologun Y, Ovington LG, Martyn-St James M, Richardson R. Antibiotics and antiseptics for venous leg ulcers. *Cochrane Database Syst Rev.* 2014;1:CD003557. https://doi.org/10.1002/14651858.CD003557.pub5. Published 2014 Jan 10.
24. Schwartz JA, Goss SG, Facchin F, Avdagic E, Lantis JC. Surgical debridement alone does not adequately reduce planktonic bioburden in chronic lower extremity wounds. *J Wound Care.* 2014;23(9). https://doi.org/10.12968/jowc.2014.23.Sup9.S4.
25. Lantis 2nd JC, Gendics C. In vivo effect of sustained-release silver sulphadiazine foam on bioburden and wound closure in infected venous leg ulcers. *J Wound Care.* 2011;20(2):90−96. https://doi.org/10.12968/jowc.2011.20.2.90.
26. Manizate F, Fuller A, Gendics C, Lantis 2nd JC. A prospective, single-center, nonblinded, comparative, postmarket clinical evaluation of a bovine-derived collagen with ionic silver dressing versus a carboxymethylcellulose and ionic silver dressing for the reduction of bioburden in variable-etiology, bilateral lower-extremity wounds. *Adv Skin Wound Care.* 2012;25(5):220−225. https://doi.org/10.1097/01.ASW.0000414705.56138.65.
27. Goss SG, Schwartz JA, Facchin F, Avdagic E, Gendics C, Lantis 2nd JC. Negative pressure wound therapy with instillation (NPWTi) better reduces post-debridement bioburden in chronically infected lower extremity wounds than NPWT alone. *J Am Coll Clin Wound Spec.* 2014;4(4):74−80. https://doi.org/10.1016/j.jccw.2014.02.001. Published 2014 Feb 20.
28. Hübner NO, Kramer A. Review on the efficacy, safety and clinical applications of polihexanide, a modern wound antiseptic. *Skin Pharmacol Physiol.* 2010;23(suppl):17−27.
29. Kaehn K. Polihexanide: a safe and highly effective biocide. *Skin Pharmacol Physiol.* 2010;23(suppl):7−16.
30. Bain MA, Koullias GJ, Morse K, Wendling S, Sabolinski ML. Type I collagen matrix plus polyhexamethylene biguanide antimicrobial for the treatment of cutaneous wounds. *J Comp Eff Res.* 2020;9(10):691−703. https://doi.org/10.2217/cer-2020-0058.

31. Snyder D, Sullivan N, Margolis D, Schoelles K. *Skin Substitutes for Treating Chronic Wounds*. Rockville (MD): Agency for Healthcare Research and Quality (US); February 2, 2020.
32. Davison-Kotler E, Sharma V, Kang NV, García-Gareta E. A universal classification system of skin substitutes inspired by factorial design. *Tissue Eng B Rev*. 2018;24(4):279—288. https://doi.org/10.1089/ten.TEB.2017.0477.
33. Phillips TJ, Manzoor J, Rojas A, et al. The longevity of a bilayered skin substitute after application to venous ulcers. *Arch Dermatol*. 2002;138(8):1079—1081. https://doi.org/10.1001/archderm.138.8.1079.
34. Vig K, Chaudhari A, Tripathi S, et al. Advances in skin regeneration using tissue engineering. *Int J Mol Sci*. 2017;18(4):789. https://doi.org/10.3390/ijms18040789. Published 2017 Apr 7.
35. Nicholas MN, Yeung J. Current status and future of skin substitutes for chronic wound healing. *J Cutan Med Surg*. 2017;21(1):23—30. https://doi.org/10.1177/1203475416664037.
36. Treadwell T, Sabolinski ML, Skornicki M, Parsons NB. Comparative effectiveness of a bioengineered living cellular construct and cryopreserved cadaveric skin allograft for the treatment of venous leg ulcers in a real-world setting. *Adv Wound Care*. 2018;7(3):69—76. https://doi.org/10.1089/wound.2017.0738.
37. Marston WA, Sabolinski ML, Parsons NB, Kirsner RS. Comparative effectiveness of a bilayered living cellular construct and a porcine collagen wound dressing in the treatment of venous leg ulcers. *Wound Repair Regen*. 2014 May-Jun;22(3):334—340. https://doi.org/10.1111/wrr.12156. Epub 2014 Mar 13. PMID: 24628712; PMCID: PMC4257085.
38. Mansbridge J, Liu K, Patch R, Symons K, Pinney E. Three-dimensional fibroblast culture implant for the treatment of diabetic foot ulcers: metabolic activity and therapeutic range. *Tissue Eng*. 1998;4(4):403—414. https://doi.org/10.1089/ten.1998.4.403.
39. Mansbridge JN, Liu K, Pinney RE, Patch R, Ratcliffe A, Naughton GK. Growth factors secreted by fibroblasts: role in healing diabetic foot ulcers. *Diabetes Obes Metabol*. 1999;1(5):265—279. https://doi.org/10.1046/j.1463-1326.1999.00032.x.
40. Nicholas MN, Yeung J. Current status and future of skin substitutes for chronic wound healing. *J Cutan Med Surg*. 2017 Jan/Feb;21(1):23—30. https://doi.org/10.1177/1203475416664037. Epub 2016 Aug 20. PMID: 27530398.
41. Krishnamoorthy L, Harding K, Griffiths D, et al. The clinical and histological effects of Dermagraft® in the healing of chronic venous leg ulcers. *Phlebology*. 2003;18:12—22.
42. Harding K, Sumner M, Cardinal M. A prospective, multicentre, randomised controlled study of human fibroblast-derived dermal substitute (Dermagraft) in patients with venous leg ulcers. *Int Wound J*. April 2013;10(2):132—137. https://doi.org/10.1111/iwj.12053. PMID: 23506344; PMCID: PMC7950758.
43. Barber C, Watt A, Pham C, et al. Influence of bioengineered skin substitutes on diabetic foot ulcer and venous leg ulcer outcomes. *J Wound Care*. 2008;17(12):517—527. https://doi.org/10.12968/jowc.2008.17.12.31766.
44. Landsman, Adam DPM, PhD, Rosines, PhD E, Houck AMS, Murchison ABS, Jones APD, Qin XMD, PhD, Chen SPD, , MBA, Landsman AR. DPM characterization of a cryopreserved split-thickness human skin allograft—TheraSkin. *Adv Skin Wound Care*. September 2016;29(9):399—406. https://doi.org/10.1097/01.ASW.0000489991.32684.9.
45. Landsman A, Rosines E, Houck A, et al. Characterization of a cryopreserved split-thickness human skin allograft-TheraSkin. *Adv Skin Wound Care*. September 2016;29(9):399—406. https://doi.org/10.1097/01.ASW.0000489991.32684.9e. PMID: 27538107.
46. Towler MA, Rush EW, Richardson MK, Williams CL. Randomized, prospective, blinded-enrollment, head-to-head venous leg ulcer healing trial comparing living, bioengineered skin graft substitute (apligraf) with living, cryopreserved, human skin allograft (TheraSkin). *Clin Podiatr Med Surg*. July 2018;35(3):357—365. https://doi.org/10.1016/j.cpm.2018.02.006. Epub 2018 Apr 14. PMID: 29861018.
47. Jones JE, Nelson EA, Al-Hity A. Skin grafting for venous leg ulcers. *Cochrane Database Syst Rev*. January 31, 2013;2013(1):CD001737. https://doi.org/10.1002/14651858.CD001737.pub4. PMID: 23440784; PMCID: PMC7061325.

48. Stocum DL. *Regenerative Biology and Medicine*. Elsevier/Ap; 2012:451–461.
49. Greaves NS, Benatar B, Baguneid M, Bayat A. Single-stage application of a novel decellularized dermis for treatment-resistant lower limb ulcers: positive outcomes assessed by SIAscopy, laser perfusion, and 3D imaging, with sequential timed histological analysis. *Wound Repair Regen*. 2013;21(6):813–822. https://doi.org/10.1111/wrr.12113.
50. Cazzell S. A randomized controlled trial comparing a human acellular dermal matrix versus conventional care for the treatment of venous leg ulcers. *Wounds*. 2019;31(3):68–74.
51. Polanco T, Lantis J. Tissue generation with acellular dermal collagen matrices: clinical comparison of human and fetal bovine matrices. In: 2016;9th Symposium on Biologic Scaffolds for Regenerative Medicine. Napa, California.
52. Mostow EN, Haraway GD, Dalsing M, Hodde JP, King D, OASIS Venus Ulcer Study Group. Effectiveness of an extracellular matrix graft (OASIS Wound Matrix) in the treatment of chronic leg ulcers: a randomized clinical trial. *J Vasc Surg*. 2005;41(5):837–843. https://doi.org/10.1016/j.jvs.2005.01.042.
53. Romanelli M, Dini V, Bertone MS. Randomized comparison of OASIS wound matrix versus moist wound dressing in the treatment of difficult-to-heal wounds of mixed arterial/venous etiology. *Adv Skin Wound Care*. 2010;23(1):34–38. https://doi.org/10.1097/01.ASW.0000363485.17224.26.
54. Cornwall KG, Landsman A, Kames KS. Extracellular matrix biomaterials for soft tissue repair. *Clin Podiatr Med Surg*. 2009;26:507–523.
55. Lineaweaver W, Bush K, James K. Suppression of α smooth muscle actin accumulation by bovine fetal dermal collagen matrix in full thickness skin wounds. *Ann Plast Surg*. 2015;74(Suppl 4(Suppl 4)): S255–S258. https://doi.org/10.1097/SAP.0000000000000449.
56. Karr JC. Retrospective comparison of diabetic foot ulcer and venous stasis ulcer healing outcome between a dermal repair scaffold (PriMatrix) and a bilayered living cell therapy (Apligraf). *Adv Skin Wound Care*. 2011;24(3):119–125. https://doi.org/10.1097/01.ASW.0000395038.28398.88.
57. Paredes JA, Bhagwandin S, Polanco T, Lantis JC. Managing real world venous leg ulcers with fetal bovine acellular dermal matrix: a single centre retrospective study. *J Wound Care*. 2017;26(Sup10): S12–S19. https://doi.org/10.12968/jowc.2017.26.Sup10.S12.
58. Yang CK, Polanco TO, Lantis 2nd JC. A prospective, postmarket, compassionate clinical evaluation of a novel acellular fish-skin graft which contains omega-3 fatty acids for the closure of hard-to-heal lower extremity chronic ulcers. *Wounds*. 2016;28(4):112–118.
59. Hankin CS, Knispel J, Lopes M, Bronstone A, Maus E. Clinical and cost efficacy of advanced wound care matrices for venous ulcers. *J Manag Care Pharm*. 2012;18:375–384.
60. De Angelis B, Orlandi F, Fernandes Lopes Morais D'Autilio M, et al. Long-term follow-up comparison of two different bi-layer dermal substitutes in tissue regeneration: clinical outcomes and histological findings. *Int Wound J*. 2018;15(5):695–706. https://doi.org/10.1111/iwj.12912.
61. Giovannini UM, Teot L. Long-term follow-up comparison of two different bi-layer dermal substitutes in tissue regeneration: clinical outcomes and histological findings. *Int Wound J*. 2020;17(5):1545–1547. https://doi.org/10.1111/iwj.13381.
62. Motolese A, Vignati F, Brambilla R, Cerati M, Passi A. Interaction between a regenerative matrix and wound bed in nonhealing ulcers: results with 16 cases. *BioMed Res Int*. 2013;2013:849321. https://doi.org/10.1155/2013/849321.
63. Simman R. The role of an esterified hyaluronic acid matrix in wound healing, a case series. *J Am Coll Clin Wound Spec*. 2018;8(1–3):10–11. https://doi.org/10.1016/j.jccw.2018.01.006. Published 2018 Feb 2.
64. Alvarez OM, Makowitz L, Patel M. Venous ulcers treated with a hyaluronic acid extracellular matrix and compression therapy: interim analysis of a randomized controlled trial. *Wounds*. 2017;29(7): E51–E54.
65. Caravaggi C, Grigoletto F, Scuderi N. Wound bed preparation with a dermal substitute (Hyalomatrix® PA) facilitates Re-epithelialization and healing: results of a multicenter, prospective, observational study on complex chronic ulcers (the FAST study). *Wounds*. 2011;23(8):228–235.
66. Johnson A, Gyurdieva A, Dhall S, Danilkovitch A, Duan-Arnold Y. Understanding the impact of preservation methods on the integrity and functionality of placental allografts. *Ann Plast Surg*. 2017;79(2): 203–213. https://doi.org/10.1097/SAP.0000000000001101.

67. McQuilling JP, Vines JB, Mowry KC. In vitro assessment of a novel, hypothermically stored amniotic membrane for use in a chronic wound environment. *Int Wound J*. December 2017;14(6):993–1005. https://doi.org/10.1111/iwj.12748. Epub 2017 Mar 29. PMID: 28370981; PMCID: PMC7949938.
68. Mao Y, Hoffman T, Dhall S, et al. Endogenous viable cells in lyopreserved amnion retain differentiation potential and anti-fibrotic activity in vitro. *Acta Biomater*. 2019;94:330–339. https://doi.org/10.1016/j.actbio.2019.06.002.
69. Bullard JD, Lei J, Lim JJ, Massee M, Fallon AM, Koob TJ. Evaluation of dehydrated human umbilical cord biological properties for wound care and soft tissue healing. *J Biomed Mater Res B Appl Biomater*. 2019;107(4):1035–1046. https://doi.org/10.1002/jbm.b.34196.
70. Serena TE, Carter MJ, Le LT, Sabo MJ, DiMarco DT, EpiFix VLU Study Group. A multicenter, randomized, controlled clinical trial evaluating the use of dehydrated human amnion/chorion membrane allografts and multilayer compression therapy vs. multilayer compression therapy alone in the treatment of venous leg ulcers. *Wound Repair Regen*. 2014;22(6):688–693. https://doi.org/10.1111/wrr.12227.

Further reading

1. Ronfard V, Williams T. Developments in cell-based therapy for wounds. In: Sen C, ed. *Advances in Wound Care*. New Rochelle, NY: Mary Ann Liebert, Inc Publications; 2010:412–418; Vol. 1.

SECTION 4

Operative and endovascular procedures for chronic venous insufficiency

CHAPTER 20

Benefits of superficial venous intervention (surgery or endovenous ablation) in the treatment of venous leg ulceration

Manjit Gohel
Cambridge Vascular Unit, Cambridge University Hospitals, Cambridge, United Kingdom

Introduction

Epidemiology and extent of the problem

Venous leg ulceration (VLU) (Fig. 20.1) is increasingly common and represents a major cause of patient distress, clinical workload, and health-service expense throughout the western world. Inconsistencies in care are common, and the fact that patients with lower extremity ulceration may present in a wide range of healthcare settings is a major impediment to standardization of care. The overall prevalence of VLU is thought to be between 0.3% and 1.0% in the adult population, increasing in those aged >65 years.[1−3] As age and obesity are risk factors for the development of VLU, the prevalence of this condition is likely to increase further.[4]

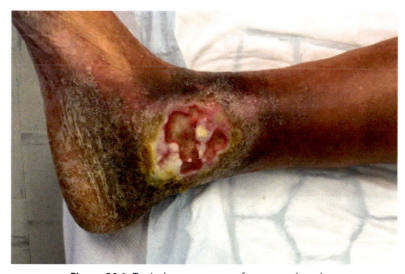

Figure 20.1 Typical appearances of a venous leg ulcer.

The natural history of venous ulceration involves protracted healing and frequent episodes of ulcer recurrence.[5] Consequently, the impact on the quality of life of patients and carers is significant. Patients often suffer significant embarrassment and social isolation due to the odor and stigma associated with their ulceration. The ability to work is often impaired and patients find it difficult to exercise or bathe due to the dressings and bandaging. Pain may be severe and is often underreported.

The underlying pathophysiological cause of venous skin changes and ulceration is chronic venous hypertension, which may be secondary to several potentially reversible (including superficial venous reflux, deep obstruction, and deep venous reflux) and some irreversible factors (including poor mobility, calf-muscle-pump failure). The mainstay of treatment for VLU is graduated compression therapy, ideally delivering >40 mmHg of pressure at the ankle.[6] Compression has been shown consistently to improve ulcer healing and a range of compression options are now available, including multilayer bandages, compression stockings, and adjustable compression garments.

Rationale for treating superficial venous reflux

Incompetence of superficial veins is an important and entirely treatable cause of chronic venous hypertension and VLU. Duplex studies have indicated that 60%—80% of patients with venous ulceration have superficial venous reflux, usually in isolation, but sometimes in combination with reflux in popliteal or femoral veins.[7,8] Although clinically effective, the application of compression therapy can be resource intensive and patient compliance with compression is variable. Leg elevation is also very effective at reducing venous hypertension but is not practical for most patients. There is a cogent case for treating superficial reflux to offer a permanent reduction in venous hypertension, not dependent on compliance with compression therapy.

Treatment for superficial venous reflux has traditionally been surgical stripping, performed under general anesthesia. In recent years, conventional surgical management has been superseded by local anesthetic, minimally invasive endovenous interventions for ablation of superficial venous reflux. Endovenous thermal ablation (such as laser or radiofrequency ablation), catheter-based nonthermal ablation (such as cyanoacrylate glue closure or mechanochemical ablation), and ultrasound-guided foam sclerotherapy are readily available in many units and are generally used in preference to traditional surgery.[9] As patients with VLU are often elderly, frail and may not be suitable for surgical stripping under general anesthesia, the development of endovenous modalities has made superficial venous interventions available to a greater number of patients with VLU.

Summary of ESCHAR trial

The Effect of Surgery and Compression on Healing And Recurrence (ESCHAR) trial recruited 500 participants with C5 and C6 disease in three centers between 1999 and

2002.[10,11] Randomization was between compression therapy alone ($n = 258$) and compression with superficial venous surgery (ligation or saphenous stripping, $n = 242$). Despite randomization to compression with surgery, 47/242 (19%) refused to undergo surgery, often because the ulcer was healing with compression therapy and the potential value of surgery was less apparent to the participants. Moreover, 48/242 (20%) of participants were treated with saphenous ligation procedures alone, performed under local anesthesia. Nevertheless, all analyses were performed on intention to treat.

The mortality rate at 3 years was 17%, highlighting the high level of comorbidity and frailty in the C6 population. The overall 3-year healing rates were similar for the two groups (89% for compression alone and 93% for compression plus surgery).[11] It should be noted that the median time to surgery was nearly 2 months, so patients did not receive the potential benefits of surgery immediately after randomization. However, the group randomized to compression and surgery did suffer significantly fewer ulcer recurrences at 3 years, compared to the group randomized to compression alone (21% vs. 51%, respectively). A significant improvement in ulcer recurrence was seen in participants with isolated superficial reflux and those with superficial with segmental deep reflux. In the ESCHAR study, the technical success of interventions was also evaluated, and a significant proportion of patients had residual saphenous reflux, despite surgery. This was partly due to the high proportion of procedures performed as junction ligations alone, but some technical failures were also seen.[12] Despite the suboptimal technical success of superficial venous surgery in the ESCHAR trial, the clinical benefits were highly significant.

The EVRA study
Background to EVRA study

The ESCHAR study clearly demonstrated the potential role of superficial venous surgery for patients with venous ulceration. However, in many centers, traditional surgical interventions have been superseded by minimally invasive thermal (laser or radiofrequency ablation) and nonthermal (ultrasound-guided foam sclerotherapy, cyanoacrylate glue, or mechanochemical ablation) procedures for ablation of superficial reflux. These procedures can usually be performed under local anesthesia and offer a potentially much more suitable intervention for the elderly venous ulcer patient population. Several nonrandomized studies have reported excellent clinical outcomes after endovenous interventions in patients with C6 disease, with healing rates higher than those reported in the ESCHAR trial. This was considered entirely plausible, as the technical success of endovenous ablation procedures, performed under ultrasound guidance would be expected to be higher than the technical success seen in the ESCHAR trial. The aim of the Early Venous Reflux Ablation (EVRA) study was to evaluate the clinical and health-economic effectiveness of early endovenous ablation for patients with VLU.

Study design and population

The EVRA study was a multicenter, randomized clinical trial conducted in 20 sites in the United Kingdom.[13] Consecutive patients with active VLU (C6 disease) between 6 weeks and 6 months duration, with clinically significant superficial venous reflux (as judged by the treating physician) were recruited. All received compression therapy (as per local policies). Participants were randomly allocated to receive either early endovenous ablation (within 2 weeks of randomization) or deferred endovenous ablation performed either when the ulcer had healed or if the ulcer remained unhealed at 6 months. The choice of endovenous ablation procedure was left to the discretion of the treating clinical team, all of whom had extensive experience in the management of venous disease. The primary outcome for the early phase of the study was time to ulcer healing, with healing adjudication performed by a blinded expert assessment of photographs. For the long-term follow-up, the primary outcome measure was time to ulcer recurrence. Throughout the study, a detailed patient-level health economic evaluation was performed from the perspective of the healthcare provider.[14]

Patient population

6555 patients were screened for potential inclusion and a total of 450 participants were randomized between October 2013 and September 2016. The most common reasons for noneligibility were ulcers that were present for >6 months (1772) or already healed (610) at the time of assessment.[13,14] The presence of venous outflow obstruction was an exclusion criterion, but imaging of the iliac veins was left to local policies. The median ages for participants were 67 and 68.9 years in early and deferred intervention groups respectively. There was a slight male preponderance (247/450, 55%) and 414/450 (92%) were Caucasian. Ulcers were medially located in 234 (52%), laterally in 185 (41%) and circumferential in 16 (4%). Ulcers were recurrent in 235/450 (52%) participants, and the location of superficial reflux was the great saphenous vein (GSV) in 248 (55%), small saphenous vein (SSV) in 55 (12%), and both GSV and SSV in 121 (27%). Around a third of patients had concomitant deep venous reflux. Overall, the EVRA study population was considered representative of the overall leg ulcer population, and there were no significant differences between groups.

In the early-intervention group, 203/224 (91%) of participants received endovenous ablation within 2 weeks of intervention as planned. In the deferred intervention group, 47/226 (21%) never received endovenous ablation, despite the intention to treat superficial reflux once the ulcer had healed or after 6 months if the ulcer remained unhealed. This observation demonstrated that for some patients with venous ulceration, the opportunity to treat superficial reflux may be lost if early intervention is not performed. For 111/224 participants randomized to early intervention, ultrasound-guided foam sclerotherapy was the sole modality used for treatment. The remaining treatments consisted of endovenous thermal ablation, either alone or with foam sclerotherapy. Formal follow-up was continued for 1-year postrandomization, with subsequent follow-up performed by telephone and using medical records. Median follow-up was 3.5 years.[15]

Clinical outcomes
Ulcer healing
Time to ulcer healing was the primary outcome measure for the first phase of the EVRA trial and the sample size was powered to detect a 15% difference in ulcer healing at 24 weeks. The group randomized to early intervention had significantly shorter time to ulcer healing compared to the deferred-intervention group (median 56 vs. 82 days; hazard ratio for healing 1.38, 95% confidence interval 1.13—1.68, $P = .001$). The healing rates at 24 weeks were 85.6% in the early-intervention group, compared to 76.3% in the deferred-intervention group (Fig. 20.2).[13,14] Despite excellent compression therapy (as evidenced by the exceptional healing rates in the deferred intervention group), endovenous ablation conferred an additional healing advantage. Prespecified analyses showed that the healing benefit in subgroups did not differ significantly to the overall healing benefit. Importantly, healing benefits were similar irrespective of the endovenous ablation modality used, indicating that the timing of intervention, rather than modality, is likely to be the most important factor.

Ulcer recurrence
For the extended follow-up phase, the study was adequately powered to detect a 15% difference in ulcer recurrence rates. A total of 426 participants had healed the primary ulcer. A total of 121/426 (28%) experienced at least one recurrent ulcer during study follow-up. There was no difference in time to first ulcer recurrence in the two groups (hazard ratio for recurrence from time of ulcer healing 0.86, 95% confidence interval 0.60—1.24, $P =$ ns) (Fig. 20.3). Ulcer recurrence rates at 4 years were 34.6% for the early-intervention group and 38.4% for the deferred-intervention group. However, in the early-intervention group, a total of 72 recurrent ulcers occurred in 675.5 years of follow-up, compared to 103 recurrent ulcers in 636 years of follow-up in the deferred-intervention group. The ulcer recurrence per year of follow-up after healing was significantly lower in the early intervention group (incidence ratio, 0.66; 95% CI, 0.48—0.89, $P = .003$).[15]

Patient-reported outcomes
Patient-reported health-related quality of life was evaluated using disease-specific (AVVQ) and generic (EQ-5D-5L and SF-36) tools at baseline, 6 weeks, 6 months, 12 months, and during long-term follow-up. Baseline scores in all quality-of-life tools were similar in the two groups. There was no clear difference in scores over the follow-up period, although the AVVQ scores were consistently lower in the early-intervention group, indicating better quality of life.

Health economic outcomes

The EVRA study protocol included a detailed, patient-level cost-effectiveness analysis. For the first 12 months postrandomization, participants were contacted monthly to gather information about all primary and secondary care resource use. At 1 year, the incremental cost-effectiveness ratio was £3976 per Quality Adjusted Life Year (QALY) gained with early intervention. At 3 years, the mean cost of early intervention was £155 lower per participant, and there was a mean QALY gain of 0.073. With a cost saving and quality of life gain, early intervention was demonstrated as being "dominant" from a health-economic perspective. The evaluation was conducted from the perspective of the healthcare provider and social or personal costs were not considered.[14–16]

Summary of conclusions from EVRA study

The EVRA study was the first adequately powered randomized clinical trial to evaluate the role of early endovenous ablation of superficial venous reflux in the VLU population. The study evaluated the impact of early intervention on ulcer healing, ulcer recurrence, and health-economic outcomes. In all three domains, a clear benefit for early endovenous ablation was demonstrated with follow-up to 5 years. Ulcer healing was accelerated, there were fewer ulcer recurrences and there was a cost saving, with better quality of life in the early intervention group. Several nonrandomized studies had suggested that endovenous interventions may be associated with good clinical outcomes, but level 1 evidence was lacking. Importantly, the EVRA study created a solid evidence platform on which to build appropriate changes in clinical care pathways internationally.

Limitations of the EVRA study

As with most large multicenter trials, only 7% of screened patients were randomized, raising questions about the generalisability of the study. The EVRA study was performed in a state-funded healthcare setting (United Kingdom National Health Service) and delays in referral or seeing patients were common. A large proportion of excluded patients had already healed when assessed, or ulcers had been present for >6 months. Chronic ulcers were excluded as EVRA investigators were concerned about withholding endovenous interventions for patients with chronic, intractable ulcers where aggressive interventions would normally be adopted.

The type of endovenous intervention was left to the discretion and preference of the treating clinical team, leading to a wide variation in procedures used in the study. Standardization of interventions was considered but deemed to be unworkable as many endovenous modalities are in use. However, prespecified subgroup analyses were planned (which did not show any difference in outcomes for different endovenous interventions).[13] By allowing any endovenous intervention in the trial and demonstrating the benefit of the early intervention treatment strategy, this may avoid unhelpful attention on the specific modality and allow greater focus on care pathways and rapid access to diagnostics for patients with C6 disease. Finally, a significant proportion of participants randomized to deferred intervention did not undergo endovenous ablation. While this was a deviation from the study protocol, the refusal of venous interventions in patients with venous ulceration has been observed previously and is likely to be representative of "real-life" practice.[10]

In recent years, endovascular stenting of postthrombotic and nonthrombotic venous outflow obstruction has become increasingly popular.[17] In the EVRA study, significant venous outflow obstruction was an exclusion criterion. However, the investigation of the iliac venous tract was left to the discretion of the treating clinical team. As the study was based in a UK setting, specific imaging of the venous outflow was performed rarely. It is possible that some participants were recruited with unrecognized venous occlusion, and this may have influenced some outcomes. In a randomized trial, variabilities in patient population should be equally distributed across the groups. However, the potential benefits of a more aggressive iliac outflow imaging and stenting policy were not assessed and remain unclear.

Implementation of ESCHAR and EVRA study results
Challenges to implementation

The positive results from ESCHAR and EVRA studies have shown that prompt duplex ultrasound assessment and ablation of superficial reflux should be an integral component in VLU care, in addition to effective compression therapy. The benefits were seen not only from a patient perspective, but also for the healthcare provider where early intervention resulted in a cost saving. With such a cogent case, it would be reasonable to expect rapid changes in leg ulcer care pathways. However, numerous challenges have been encountered in the implementation of the EVRA study results into routine clinical care. Since the EVRA study results were published in 2018, implementation efforts have been impacted enormously by the CoViD-19 pandemic, which has resulted in a reprioritization of healthcare resources away from chronic, nonlife-threatening conditions such as VLU in many countries.[18]

Education of healthcare professionals

Leg ulcer care is provided by a wide range of medical and nursing healthcare professionals in primary and secondary care settings. Internationally, there are enormous variations in care pathways, and it is common for the majority of leg ulcer care to be provided in a setting (such as primary care clinics, wound care centers, or tissue viability clinics), which may be separate and remote to locations where duplex ultrasound scanning or endovenous ablation procedures may be available. It is incumbent on enthusiasts and advocates of evidence-based leg ulcer care to reach out and proactively disseminate the results of the EVRA study and other trials to all professionals involved in leg ulcer care.

Updated policies and guidelines

As guideline documents and policies may not get reviewed or updated for several years, there is an inevitable lag between publication of research and changes to operating procedures and protocols.

Resource implications

Although early endovenous ablation has clear clinical benefits and is a cost-saving strategy, there are major resource implications to delivering venous interventions to all patients with VLU. As the prevalence of VLU is thought to be around 0.3%–1% of the adult population, there are likely to be millions of patients who would potentially benefit from endovenous ablation procedures. Providing diagnostics and interventions to this entire population would overwhelm even the most affluent healthcare systems. Therefore, the implementation of an aggressive endovenous ablation pathway should be performed in a staged manner to avoid excessive strain on staff and systems. Further work is needed to guide the prioritization of interventions. Teams may also need to embrace novel models of care to better create the capacity to meet this unmet need.

Potential levers for driving change

In recent years, there has been a growing recognition of the importance and clinical significance of leg ulceration. Awareness of the condition has increased among patients, carers, healthcare professionals, and policy makers. Societies, patient representatives,

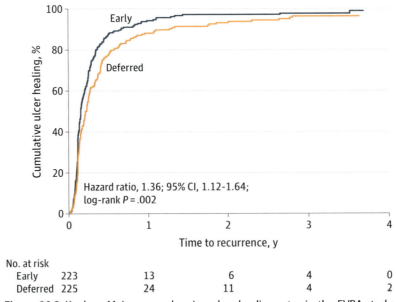

Figure 20.2 Kaplan–Meier curve showing ulcer healing rates in the EVRA study.

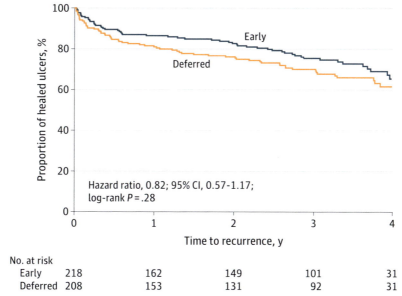

Figure 20.3 Kaplan–Meier curve showing ulcer recurrence rates in the EVRA study.

and researchers should ensure a clarity and unity of message. It is likely that a "bottom-up" approach, empowering patients and frontline health professionals, will need to be combined with a "top-down" strategy engaging policymakers and commissioners. The volume of potential workload may make leg ulcer care reform an intimidating and unappealing prospect, but pathways need to be redesigned to ensure that the correct clinical decisions are also the easiest to deliver.

Conclusions

Findings from the ESCHAR study clearly demonstrated that superficial venous surgery had significant clinical benefits for patients with VLU.[10,11] The EVRA study has provided further clarity that interventions should be endovenous and delivered as soon as possible to provide maximum benefit.[13,15] The timing of the intervention is likely to be more important than the choice of endovenous modality. These findings should drive a review of leg ulcer pathways to include prompt duplex ultrasound assessment and venous intervention. Implementation of these results will be challenging and likely to take many years. Further work is needed to better understand which patients with VLU benefit most from endovenous interventions.

References

1. Ruckley CV, Evans CJ, Allan PL, Lee AJ, Fowkes FG. Chronic venous insufficiency: clinical and duplex correlations. The Edinburgh vein study of venous disorders in the general population. *J Vasc Surg*. 2002;36(3):520−525.
2. Carpentier PH, Maricq HR, Biro C, Poncot-Makinen CO, Franco A. Prevalence, risk factors, and clinical patterns of chronic venous disorders of lower limbs: a population-based study in France. *J Vasc Surg*. 2004;40(4):650−659.
3. Margolis DJ, Bilker W, Santanna J, Baumgarten M. Venous leg ulcer: incidence and prevalence in the elderly. *J Am Acad Dermatol*. 2002;46(3):381−386.
4. Onida S, Davies AH. Predicted burden of venous disease. *Phlebology*. 2016;31(1 Suppl):74−79.
5. Franks PJ, Barker J, Collier M, et al. Management of patients with venous leg ulcers: challenges and current best practice. *J Wound Care*. 2016;25(Suppl 6):S1−S67.
6. O'Meara S, Cullum N, Nelson EA, Dumville JC. Compression for venous leg ulcers. *Cochrane Database Syst Rev*. 2012;11:CD000265.
7. Grabs AJ, Wakely MC, Nyamekye I, Ghauri AS, Poskitt KR. Colour duplex ultrasonography in the rational management of chronic venous leg ulcers. *Br J Surg*. 1996;83(10):1380−1382.
8. Magnusson MB, Nelzen O, Risberg B, Sivertsson R. A colour doppler ultrasound study of venous reflux in patients with chronic leg ulcers. *Eur J Vasc Endovasc Surg*. 2001;21(4):353−360.
9. De Maeseneer MG, Kakkos SK, Aherne T, et al. Editor's Choice - European Society for Vascular Surgery (ESVS) - Clinical practice guidelines on the management of chronic venous disease of the lower limbs. *Eur J Vasc Endovasc Surg*. 2022;63(2):184−267.
10. Barwell JR, Davies CE, Deacon J, et al. Comparison of surgery and compression with compression alone in chronic venous ulceration (ESCHAR study): randomised controlled trial. *Lancet*. 2004;363(9424):1854−1859.
11. Gohel MS, Barwell JR, Taylor M, et al. Long term results of compression therapy alone versus compression plus surgery in chronic venous ulceration (ESCHAR): randomised controlled trial. *BMJ*. 2007;335(7610):83.
12. Gohel MS, Barwell JR, Earnshaw JJ, et al. Randomized clinical trial of compression plus surgery versus compression alone in chronic venous ulceration (ESCHAR study)—haemodynamic and anatomical changes. *Br J Surg*. 2005;92(3):291−297.
13. Gohel MS, Heatley F, Liu X, et al. A randomized trial of early endovenous ablation in venous ulceration. *N Engl J Med*. 2018;378(22):2105−2114.
14. Gohel MS, Heatley F, Liu X, et al. Early versus deferred endovenous ablation of superficial venous reflux in patients with venous ulceration: the EVRA RCT. *Health Technol Assess*. 2019;23(24):1−96.
15. Gohel MS, Mora MJ, Szigeti M, et al. Long-term clinical and cost-effectiveness of early endovenous ablation in venous ulceration: a randomized clinical trial. *JAMA Surg*. 2020;155(12):1113−1121.
16. Epstein DM, Gohel MS, Heatley F, et al. Cost-effectiveness analysis of a randomized clinical trial of early versus deferred endovenous ablation of superficial venous reflux in patients with venous ulceration. *Br J Surg*. 2019;106(5):555−562.
17. Raju S, Darcey R, Neglen P. Unexpected major role for venous stenting in deep reflux disease. *J Vasc Surg*. 2010;51(2):401−408. discussion 8.
18. Tonna JE, Hanson HA, Cohan JN, et al. Balancing revenue generation with capacity generation: case distribution, financial impact and hospital capacity changes from cancelling or resuming elective surgeries in the US during COVID-19. *BMC Health Serv Res*. 2020;20(1):1119.

CHAPTER 21

Superficial surgery and perforator interruption in the treatment of venous leg ulcers

Peter F. Lawrence
Gonda Vascular Center and Division of Vascular and Endovascular Surgery, David Geffen School of Medicine, UCLA, Los Angeles, CA, United States

Although venous leg ulcers (Fig. 21.1) cause an enormous economic burden on healthcare, there are very few prospective studies that guide treatment decisions and compare the role of each treatment modality on venous leg ulcer healing. This has led to varied procedures performed in various sequences. As the pathogenesis of venous ulcers and the best methods of diagnosing ambulatory venous hypertension and the abnormal veins involved have been covered in other chapters, this chapter will focus on specific superficial and perforator procedures to heal venous leg ulcers and the reported effectiveness of each. There have been practice guidelines published for both varicose veins[1] and venous ulcers.[2] They provide evidence-based guidelines on the diagnosis and treatment, as well as outcomes. In a recent publication on appropriate use criteria for venous insufficiency,[3] experts in venous disease management used an iterative consensus technique to determine best practices with respect to varicose veins and perforator vein treatment in different clinical scenarios, including venous ulcers. There have also been large retrospective multiinstitutional trials[4] that provide higher levels of evidence regarding superficial and perforator vein treatment and their roles in healing venous leg ulcers.

Diagnosis

The goal of all procedures involving the superficial and perforator veins for the treatment of venous leg ulcers is to correct ambulatory venous hypertension, which occurs primarily with sitting and standing. Ulcer healing can be improved by procedures that eliminate superficial vein, tributary vein, or perforator vein reflux and reduce or eliminate the ambulatory venous hypertension. Each category of veins (superficial, perforator, and deep) can be evaluated for incompetence by a combination of physical exam and duplex ultrasound (DUS). Duplex ultrasound of the perforator veins is particularly challenging, even for experienced practitioners, as there are many locations for the origin of perforator veins, including paratibial, medial ankle, and lateral calf locations (Fig. 21.2). Based on level 2B evidence, the authors of the Society of Vascular Surgery (SVS) and the American

a.

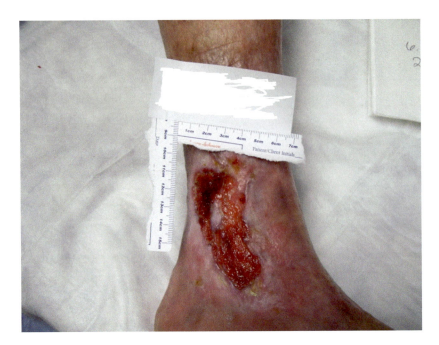

b.

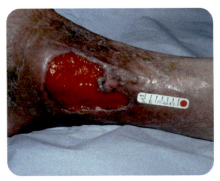

Figure. 21.1 (A) and (B) The skin around a longstanding venous ulcer is often fibrotic and susceptible to infection and ulceration. Longstanding venous ulcers often have extensive lipodermatosclerosis.

Venous Forum (AVF) Chronic Venous Disease Practice Guidelines recommend selective evaluation and treatment of the perforating veins. Perforator vein treatment should be reserved for patients with skin changes, healed or active venous ulcers and in patients with recurrent symptoms despite previous vein intervention(s).[1] The recommended

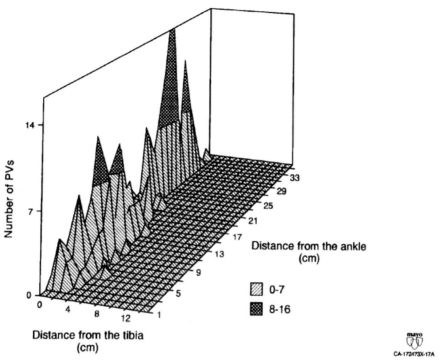

Figure. 21.2 This graph shows the multiple locations of the perforator veins in the medial leg. The traditional locations of "Cockett's perforators" were not confirmed by this study. Incompetent perforator veins can also occur in the paratibial location and in the lateral ankle.

criteria used to define "pathologic" perforating veins included reflux duration >0.5 s, perforating vein diameter >3.5 mm, and a location adjacent to or beneath a healed or open ulcer. Duplex ultrasound is used to both evaluate incompetent veins and to guide therapy during catheter-, glue-, and foam-based therapy.

Computed tomography or magnetic resonance imaging venograms and intravascular ultrasound are rarely used to assess the superficial or perforator veins, although they are often used to assess the suprainguinal deep system in patients who not only have superficial and/or perforator vein incompetence, and clinical signs and symptoms, but also ultrasonographic findings suggestive of iliofemoral venous outflow obstruction. The most important factor in treating venous leg ulcers is understanding that each abnormal superficial, perforator, and deep vein contributes to ambulatory venous hypertension, and therefore all that are incompetent or obstructed may need to be treated to heal a venous leg ulcer.

Once the anatomic site(s) of pathology has been identified in a patient with a non-healing venous ulcer, determining which abnormal veins should be treated and which should be treated first when multiple sites of reflux or occlusion are found is an important initial step in treatment planning.

Treatment options for incompetent superficial truncal veins

The mode of treatment, whether it be thermal ablation, nonthermal ablation, vein removal, or sclerotherapy has been studied retrospectively for healing and other outcomes in patients with venous leg ulcers. Any technique that closes or removes incompetent superficial veins has been shown to accelerate and improve ulcer healing, particularly when measured by healing rate,[5] rather than simply by absolute wound healing, as venous ulcers have different initial sizes and may have different degrees of scarring and have been present for different lengths of time.

Treatment options for incompetent perforator veins

For perforator veins, there are also several options for treatment, from the historic Linton procedure and subfascial endoscopic perforator surgery (SEPS) procedure, which are no longer in use, to treating incompetent perforator veins with thermal or nonthermal ablation. There are currently no accepted options for chemical ablation of perforator veins, due to the proximity of the perforator vein to the deep veins in the calf, as there is a risk of creating calf vein thrombosis if the sclerosant enters the deep system.

Sequence of treatment of abnormal superficial, perforator, and deep veins

There are different strategies regarding which abnormal veins should be treated first when multiple veins have reflux or obstruction, and whether more than one system should be treated simultaneously. Currently, many venous specialists use the most technically successful approach to treat refluxing truncal veins first, including the great, small, or anterior accessory saphenous veins (AASV). However, there are other venous specialists who treat iliofemoral deep venous reflux first and report comparable results. Rarely, do venous specialists treat incompetent perforator veins first, even in patients with a venous leg ulcer associated with a very large, incompetent perforator vein adjacent to the ulcer.

Facility requirements for treatment of superficial and perforator veins

All superficial and perforator vein procedures can be performed in a sterile outpatient treatment room that has adequate OR lighting, a mechanical tilt table, a portable ultrasound on wheels, room for an energy generator for either a radiofrequency catheter or laser, a storage cabinet area for supplies, a mobile Mayo stand or back table for placement of catheters and wires, and room for both the proceduralist and an assistant. A vascular lab tech is not required for procedural ultrasound—DUS can be performed, using a small mobile ultrasound designed for these procedures, by the interventionalist. The proceduralist needs to wear a sterile gown, gloves, and mask, and therefore the assistant must provide sterile supplies such as needles and syringes.

Technique: superficial saphenous vein treatments

Great saphenous vein closure. Although the rationale for the closure of the great saphenous vein is to reduce ambulatory venous hypertension, there are little data available on the extent of the saphenous vein closure required to heal venous leg ulcers. Thermal techniques to ablate the saphenous vein, using either radiofrequency energy or laser, are usually performed from the knee or upper calf to the proximal saphenous vein at the saphenous femoral junction, to avoid saphenous nerve injury in the lower leg (Fig. 21.3). The saphenous nerve is immediately adjacent to the saphenous vein from the ankle to the upper mid-calf. Nonthermal techniques, where the risk of saphenous nerve injury is very low, can initiate the ablation at the ankle and close the entire saphenous vein. Complete saphenous ablation, as opposed to above knee saphenous ablation, has theoretical advantages and has been reported in small series to improve venous ulcer wound healing.[6] If a thermal technique is used to close only the proximal saphenous vein and the wound does not heal, ablation of the below knee saphenous vein, by either a thermal technique, accepting the risk of thermal injury to the saphenous nerve, or using a nonthermal technique for the below knee saphenous vein, may be used to improve ulcer healing.

Anterior accessory saphenous vein treatment. The AASV has increasingly been implicated in contributing to the etiology of ambulatory vein hypertension and nonhealing venous ulcers. This vein is easily visualized by DUS and may be the primary source of venous reflux. The difference between this vein and the great saphenous vein is that it is laterally located above the fascia and is often more tortuous and difficult to treat with rigid thermal catheters. Therefore, the most proximal AASV may be treated by short thermal or nonthermal rigid ablation catheters, while the rest of the vein, which is tortuous and often extends laterally and into the calf, must be treated by microphlebectomy removal or foam sclerotherapy.

Small saphenous vein closure. Although the majority of venous leg ulcers are in the "gaiter zone" of the medial ankle and therefore are best treated by ablation of the great saphenous vein and posterior arch veins, some venous ulcers are located in the lateral ankle and therefore caused by small saphenous vein reflux and associated perforator reflux. Small saphenous vein closure can be successfully accomplished by either thermal or nonthermal ablation, although the thermal technique should be limited to the mid-calf and above, to avoid sural nerve injury. Nonthermal techniques can ablate the entire small saphenous vein and are used by some venous specialists due to the more complete ablation. For operative details regarding the treatment of superficial vein reflux disease, see Chapter 22.

a.

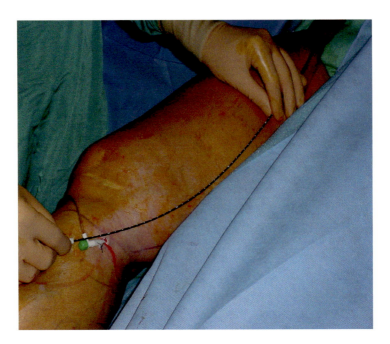

b.

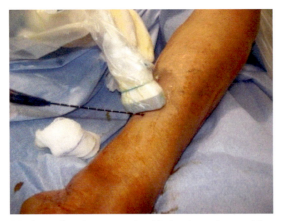

Figure. 21.3 (A) Thermal saphenous vein ablation in patients with a venous ulcer is usually initiated immediately below the knee, rather than at the ankle, to avoid saphenous nerve injury. Nonthermal techniques can be initiated at the ankle and therefore treat the entire saphenous vein, and (B) Access to a perforator vein often requires penetration through fibrotic tissue. A rigid catheter is useful to penetrate the tissue.

Perforator vein treatment

There is little evidence to support the simultaneous ablation of both the great saphenous vein and perforator veins, but when a venous ulcer is not healing after truncal vein ablation and there is an incompetent, refluxing perforator vein (or veins), closure of the perforator has been shown to improve venous ulcer healing.[7] The perforator vein can be closed with a radiofrequency catheter or a laser. However, closure with physician-modified or commercial foam is not recommended, due to the risk that foam will reflux into the deep venous system and cause calf vein thrombosis. Nevertheless, some practitioners still use foam sclerotherapy to treat perforating veins.

Success rates reported for closure of perforator veins range from 60% to 70%, so these veins frequently require a second treatment.[7] When the perforator vein is successfully closed, there have been increasing reports of using sclerotherapy or foam sclerotherapy to then close the subdermal plexus of incompetent veins which are connected to the perforator veins, and which are located under the ulcer.

Perforator vein closure technique

The technique of perforator vein closure is similar, whether using a radiofrequency catheter or a laser. The goal is to enter an incompetent perforator vein at the level of the fascia of the leg, or just above it. If the vein is accessed below the fascia, it is difficult to advance wires or sheaths through the fascia, and there are also artery, nerve, and veins below the fascia, so the risk of creating a deep venous thrombosis, arterial injury, or nerve injury is high, particularly in the lower leg. The initial identification of incompetent perforator veins is usually performed in an accredited vascular lab, as meeting the criteria for an incompetent perforator vein is critical to obtaining insurance authorization. Using a duplex scanner with high image quality and color flow helps to both determine the size of a perforator vein, which should be >3.5 mm and reflux of >0.5 s. As incompetent perforator veins reflux into the subcutaneous tissue and create ambulatory venous hypertension, their relationship to a venous ulcer is critical. Ideally, the perforators treated should be adjacent to and immediately above the ulcer. When incompetent perforators are located far away from the ulcer, closure is unlikely to heal the ulcer.

We ask our vascular lab to measure veins in two dimensions: from the plantar surface of the heel, and from the tibia for medial ulcers, or fibula for lateral ulcers. This assures the treating physician that the perforator vein being treated is the same as the perforator identified in the vascular lab.

Once the patient is in the procedure room, the correct equipment is needed for the successful closure of the incompetent perforator vein (Fig. 21.4A and B). A tilt table helps to institute reversed Trendelenburg and visualize the incompetent vein with a high-resolution duplex scanner.

a.

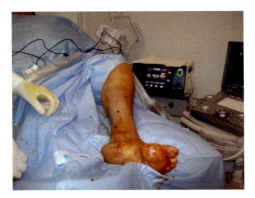

b.

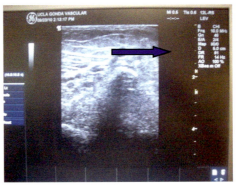

Figure. 21.4 (A)The patient is placed in reversed trendelenberg position on an electronic tilt table, and (B) A portable duplex scanner with color flow is used for imaging. The incompetent vein adjacent to the ulcer with both reflux and a diameter >3 mm is selected for treatment.

After the skin is anesthetized, a stylet or needle is placed into the incompetent perforator vein at a 45-degree angle (Fig. 21.5A and B), just above the fascia. Although it is ideal to withdraw venous blood, this is rarely successful. It is usually possible, though, to confirm that the catheter is in the vein in two dimensions. Once the catheter is in the vein, the tissue around the catheter should be anesthetized, down to the fascia. Large volumes of anesthetic can pull the catheter out of the vein and should be avoided, using constant imaging while delivering the local anesthesia. Once the catheter has been confirmed to be in the vein and the tissue around it is anesthetized, the patient should then be placed in Trendelenburg position to reduce the size of the vein. The delivery of thermal energy should then follow the instructions for use (IFU) for the laser or radio frequency device.

a.

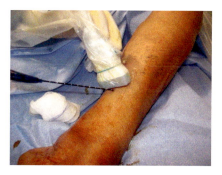

b.

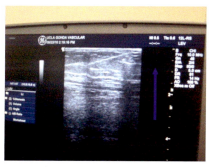

Figure. 21.5 (A) A radiofrequency catheter or a micropuncture needle for a laser fiber is placed at a 45-degree angle and inserted into the perforator vein just above the fascia, and (B) It must be viewed in both transverse and longitudinal directions to confirm presence within the vein. The vein is then treated with radiofrequency energy for 1 min in each of four quadrants. There should be an attempt to confirm successful ablation postprocedure by being unable to demonstrate color flow within the perforator vein.

Following thermal treatment of the incompetent perforator vein (Fig. 21.6A and B), it is important to leave markings on the leg, so that the noninvasive vascular lab sees the location of the closure when the patient returns for dressing changes. The locations of incompetence perforators (Fig. 21.7A and B) and their relationship to venous ulcers are demonstrated in these illustrations.

Follow-up recommendations

The following considerations are recommended as the basis for treatment strategies to improve outcomes following the completion of the aforementioned procedure(s):
1. Treatment of incompetent tributaries

a.

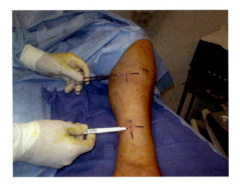

b.

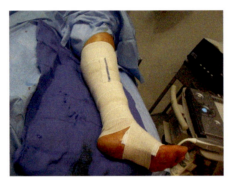

Figure. 21.6 (A) The sites of perforator closure should be marked in two dimensions to facilitate confirmation of closure at next clinic visit, ideally by an independent vascular lab technician, and (B) If an ulcer is present, then a compression dressing should be placed after marking the leg.

Once a truncal or perforator vein has been closed by one of the techniques discussed earlier, there may still be large, incompetent tributary veins that are immediately adjacent to the ulcer and that cause reflux and venous hypertension in the region of the ulcer. They often need to also be removed or closed to reduce reflux; based on our experience, treatment of these veins will often promote additional wound healing. There are several techniques to treat these veins, from microphlebectomy, accessed through the skin that is not fibrotic, to sclerotherapy using either commercial foam or liquid sclerotherapy. It is important to not make incisions into the lipodermatosclerotic tissue, as it may not heal and may develop into another venous ulcer. The results of these tributary procedures are generally excellent and low risk. They may be performed either concomitant with ablation techniques or after a period of observation to determine the rate of wound healing.

a.

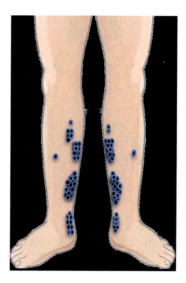

b.

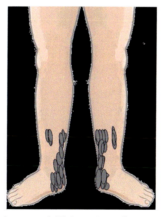

Figure. 21.7 (A) Location of venous ulcers, and (B) Location of incompetent perforator veins associated with venous ulcers.

2. Monitoring venous ulcer healing after superficial and perforator vein treatment
   a. To determine the impact of any of the above treatments on venous leg ulcers, measurement of the wound area, using planimetry (Fig. 21.8), is the best method to determine if there has been the elimination of all incompetent veins, which is often required for successful wound healing. If ablation treatments are successful and reduce ambulatory venous hypertension, it should be able to be documented by planimetry.

a.

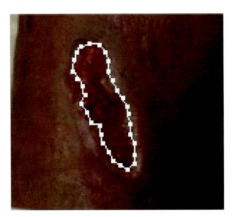

b.

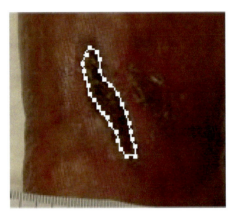

Figure. 21.8 (A) and (B) Planimetry objectively measures the area of the ulcer and is used, sequentially, to assess wound healing rates and determine if treatment of superficial or perforator veins has been effective in accelerating wound healing.

   b. Another important step in monitoring is to document the successful closure of the treated vein(s) postprocedure. For truncal veins, this requires repeat DUS of the treated vein. For incompetent perforator veins, it is critical that the treated vein be marked with indelible ink for the exact coordinates, as there are many perforator veins, and several are often incompetent. It is easy to confuse competent and incompetent veins and assume that an incompetent vein has been ablated, when in fact, the unmarked, imaged vein is actually another competent perforator vein.
3. *Failures of truncal vein and perforator closure.* When there is little evidence of venous ulcer healing, despite good wound care, then both the truncal and perforator veins that have been ablated must be reassessed by DUS to determine if recanalization or incomplete ablation has occurred.

Complications of superficial and perforator procedures

Complications related to saphenous vein, tributary, and perforator ablation procedures have been discussed in other chapters. As the procedures discussed in this chapter are being performed for venous ulcers, great care must be taken to avoid incisions in the lipodermatosclerotic tissue that often surrounds a venous ulcer. Although needle puncture incisions for catheter and wire access, and even for a small crochet hook to remove incompetent tributaries, are usually well tolerated in the scar tissue, larger incisions often do not heal and may either become infected or develop into additional ulcers.

The major complication related to superficial venous and perforator procedures is the extension of the thermal energy, sclerosant solution, or glue into the deep system. A recent classification system has been developed by the AVF and SVS to guide the treatment of endovenous heat-induced thrombosis (EHIT).[7] The risk of this complication can be reduced in virtually all procedures by adhering to the IFU for each device or procedure. When there is concern about the extension of energy, sclerosant solution, or glue into the deep system, a postprocedure DUS can easily identify EHIT extension into the deep venous system 2—4 days postprocedure, with treatment determined by the extent of the extension.

Literature review of superficial and perforator procedure outcomes for venous ulcers

The number of endovenous ablation procedures performed in the United States has dramatically increased over the past 10 years. The treatment options have also increased greatly, but there is a paucity of high-quality evidence regarding optimal devices and procedures.

In the 2004 ESCHAR trial,[8] 500 patients with venous leg ulcers were randomized to compression alone versus compression plus surgery on the superficial venous system. Surgical treatment of superficial venous reflux did not reduce ulcer healing time but did reduce the ulcer recurrence at 12 months. In another retrospective study published in 2011, 86 incompetent perforating veins were treated in patients with recalcitrant venous ulcers (Fig. 21.9).[7] Technically successful ablation occurred in 58% and increased to 90% with repeat ablation. In patients who had successful ablation, 90% of the ulcers healed. Harlander-Locke et al. measured ulcer healing rate using planimetry, and demonstrated improved healing rates following ablation of both refluxing truncal veins and perforator veins (Fig. 21.10).[5] 76% of ulcers healed in a mean of 142 days, and the recurrence rate was only 7% at 12 months.

O'Donnell et al., in the AVF/SVS Venous Ulcer Practice Guidelines, commented on the weakness of the evidence related to superficial and perforator vein ablation for venous ulcers, but did recommend treatment of superficial and perforator venous reflux when there are open (CEAP 6) venous ulcers (Grade 2, Level of Evidence-C).[2] They did

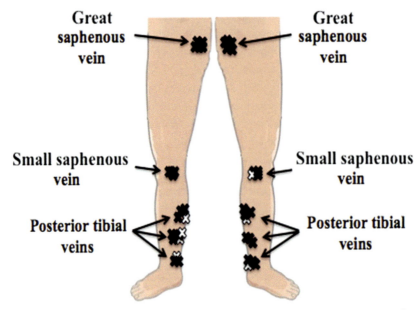

Figure. 21.9 In this study of venous leg ulcers treated with truncal vein ablation and perforator vein ablation, the location of the treated veins is shown.

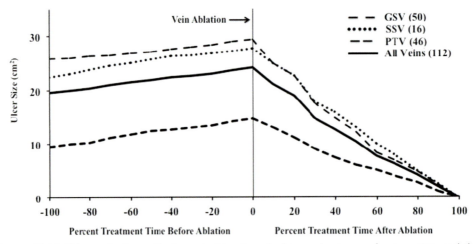

Figure. 21.10 This graph shows the impact of treatment of truncal veins, perforator veins, and the combination of both on ulcer healing, using planimetry as an objective measure of area reduction.

not recommend the sequence of treatment when both superficial and perforator veins are incompetent. In 2013, Samuel et al.[9] attempted to publish a Cochrane study, comparing endovenous thermal ablation with conservative treatment for venous ulcers, but they found no prospective randomized trials, demonstrating the lack of Level 1 evidence.

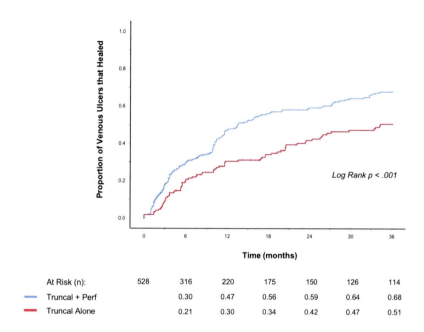

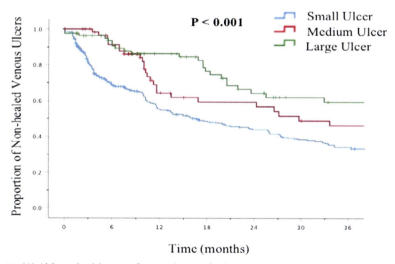

Figure. 21.11 (A) Although ablation of truncal veins facilitates venous ulcer healing at all time intervals, the addition of perforator ablation adds considerably to wound healing, and (B) The initial size of the ulcer is important; smaller ulcers heal more rapidly than larger ulcers, but all venous ulcers show evidence of healing at each time interval.

The EVRA trial,[10] published in 2018, was the first to show that early treatment of truncal reflux leads to improved healing. Four hundred and fifty patients with venous ulcers from 20 centers in the United Kingdom were randomized to early ablation of superficial venous reflux within 2 weeks of presentation versus compression therapy alone. More patients with early intervention healed and the time to heal was much shorter. In addition, ulcer-free time in the early intervention group was longer than in the conservatively treated group.

Recently, Lawrence et al.[4] published a multiinstitutional study of 832 patients with recalcitrant chronic venous leg ulcers who were initially treated with compression therapy for 2 months. The patients were then divided into saphenous vein ablation versus saphenous vein ablation plus treatment of incompetent perforator veins. Patients treated with compression and wound care alone had the lowest healing rates. Patients treated with ablation alone (Fig. 21.11) had excellent healing rates, but those who also had perforator ablation, in addition to truncal ablations, had the best outcomes.

The conclusions of the many retrospective studies addressing venous ulcers are that an aggressive treatment of truncal vein reflux, whether it be great saphenous, small saphenous, or AASV reflux, as well as the closure of refluxing perforator veins will lead to the greatest venous ulcer healing. The remaining questions in patients with venous ulcers are as follows: (1) Should superficial or deep venous disease be treated first? (2) Should incompetent perforator veins be treated concurrently with truncal disease? (3) How long should we wait, in patients with recalcitrant venous ulcers, before investigating for venous outflow obstruction?

References

1. Gloviczki P, Comerota AJ, Dalsing MC, et al. The care of patients with varicose veins and associated chronic venous diseases: clinical practice guidelines of the society for vascular surgery and the American venous forum. *J Vasc Surg*. May 2011;53(5 Suppl):2S–48S.
2. O'Donnell TF, Passman MA, Marston WA, et al. Management of venous leg ulcers: clinical practice guidelines of the society for vascular surgery ® and the American venous forum. *J Vasc Surg*. August 2014;60(2 Suppl):3S–59S.
3. Masuda E, Ozsvath K, Vossler J, et al. The 2020 appropriate use criteria for chronic lower extremity venous disease of the American venous forum, the society for vascular surgery. *The American Vein and Lymphatic Society, and the Society of Interventional Radiology*. July 2020;8(4):505–525.e4.
4. Lawrence PF, Hager ES, Harlander-Locke MP. Treatment of superficial and perforator reflux and deep venous stenosis improves healing of chronic venous leg ulcers. *J Vasc Surg Venous Lymphat Disord*. July 2020;8:601–609.
5. Harlander-Locke M, Lawrence PF, Alktaifi A, et al. The impact of ablation of incompetent superficial and perforator veins on ulcer healing rates. *J Vasc Surg*. February 2012;55(2):458–464.
6. Chan SSJ, Yap CJQ, Tan SG, Choke ETC, Chong TT, Tang TY. The utility of endovenous cyanoacrylate glue ablation for incompetent saphenous veins in the setting of venous leg ulcers. *J Vasc Surg Venous Lymphat Disord*. November 2020;8(6):1041–1048. https://doi.org/10.1016/j.jvsv.2020.01.013. Epub 2020 Mar 21. PMID: 32205130.
7. Lawrence PF, Alktaifi A, Rigberg DA, et al. Endovenous ablation of incompetent perforating veins is effective treatment for recalcitrant venous ulcers. *J Vasc Surg*. September 2011;54(3):737–742.

8. Barwell JR, Davies CE, Deacon J, et al. Comparison of surgery and compression with compression alone in chronic venous ulceration (ESCHAR study): randomised controlled trial. *Lancet*. June 5, 2004;363(9424):1854–1859.
9. Samuel N, Carradice D, Wallace T, et al. *Cochrane Database of Systematic Reviews—Endovenous Thermal Ablation for Healing Venous Ulcers and Preventing Recurrence*. October 04, 2013.
10. Gohel MS, Heatley F, Liu X, et al. A randomized trial of early endovenous ablation in venous ulceration. *N Engl J Med*. May 31, 2018;378:2105–2114.

Further Reading

1. Kabnick LS, Sadek M, Bjarnason H, et al. Classification and treatment of endothermal heat-induced thrombosis: recommendations from the American venous Forum and the society for vascular surgery. *J Vasc Surg Venous Lymphat Disord*. January 2021;9(1):6–22.

CHAPTER 22

Endovenous techniques for superficial vein ablation for treatment of venous ulcers

Monika L. Gloviczki[1] and Peter Gloviczki[2]

[1]Department of Internal Medicine and Gonda Vascular Center, Mayo Clinic, Rochester, MN, United States; [2]Mayo Clinic College of Medicine, Division of Vascular and Endovascular Surgery, Gonda Vascular Center, Mayo Clinic, Rochester, MN, United States

Introduction

Over the last decades, endovenous ablations have become preferred methods of treatment for patients with symptomatic great saphenous vein (GSV) reflux.[1] Endovenous thermal ablation techniques include radiofrequency ablation (RFA)[2] and endovenous laser ablation (EVLA)[3] while the most frequently used nonthermal techniques in the United States include endovenous cyanoacrylate embolization (CAE),[4] the mechanical-chemical ablation (MOCA) procedure,[5] and ultrasound-guided foam sclerotherapy (UGFS) using sodium tetradecyl sulfate (STS) or polidocanol.[6] The foam can be prepared at the bedside using the Tessari technique[7] or it can be a premade injectable polidocanol foam (Varithena).[8] These minimally invasive procedures reduce the risk of complications, discomfort, and hospitalization that were traditionally associated with open surgery and general anesthesia. They permit treatment of those patients who were previously considered high risk for perioperative complications or those who could not be treated because of prolonged recovery time. As approximately 56% of venous ulcer limbs have superficial vein incompetence and 29% do not have associated deep vein incompetence, ablation of the incompetent superficial system has become an effective strategy to treat many patients with venous ulcers.[9] The objective of this chapter is to discuss technical details, preoperative planning, operative technique, and postoperative care of endovenous ablations, used for the management of patients with venous ulcers and to provide evidence-based insights on the safety and efficacy of these endovascular treatment modalities.

Preoperative planning

Preoperative evaluation for endovenous ablation includes a thorough history and physical examination, complemented with venous duplex scanning, as discussed in detail in previous chapters of this book (see Chapters 5 and 6). Duplex evaluation is important to define the role of the superficial, deep, and perforating venous system in the pathophysiology of venous ulcers. Patients who are selected for ablation should have pathologic axial reflux (>500 ms), directed to the bed of the healed or active venous ulcer (Fig. 22.1A and B). Patency of the deep veins or sufficient drainage of venous blood toward the deep system must be confirmed in postthrombotic syndrome before large superficial veins are ablated. Patency and reflux in perforating veins should also be assessed. In patients with associated pelvic venous obstruction decision on the sequence of the venous procedures has to be made in advance.

Contraindications for endovenous ablation include restricted ambulation, infected ulcer or other acute infection, acute deep vein thrombosis (DVT), or extensive chronic deep vein obstruction. Relative contraindications are a large vein diameter or superficial venous aneurysm, extreme tortuosity of the axial vein for thermal ablations, and veins located immediately under the skin. Patients with venous or venolymphatic malformations should be treated only after detailed evaluation by an expert, multidisciplinary team (see Chapter 28).

Thermal ablation is not recommended for veins >3 cm in diameter because of the high risk of treatment failure, recanalization, perioperative thrombophlebitis, and thromboembolism. Anticoagulation treatment, hormone replacement therapy, and thrombophilia with increased risks for DVT need to be recognized before the procedure, but they are not absolute contraindications. In a recent prospective study, vein closure rates were the same in patients who were already anticoagulated or who were started on a standard enoxaparin protocol because of the high risk for DVT versus those who were

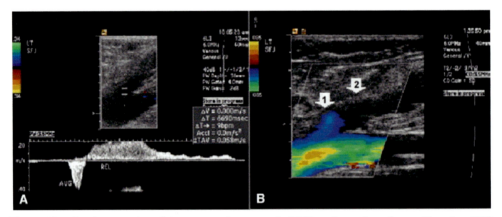

Figure. 22.1 Duplex scanning of the great saphenous vein (GSV) at the saphenofemoral junction (SFJ). (A) Pretreatment scan shows incompetent SFJ after augmentation. (B) Intraoperative duplex scan shows successful occlusion of the GSV with a patent, 3-mm proximal stump (*arrow 1*) and absence of flow within the treated segment (*arrow 2*). (From Puggioni A, Kalra M, Carmo M, Mozes G, Gloviczki P. *Endovenous laser therapy and radiofrequency ablation of the great saphenous vein: analysis of early efficacy and complications.* J Vasc Surg. 2005;42:488–493, with permission.)

not anticoagulated. There were no preprocedure DVTs nor pulmonary embolisms (PE), and there were no major bleeding complications. In another study, 36% of VLU patients, who were fully anticoagulated, underwent RFA and 41% underwent EVLA.[10] In our review, however, we found that being on anticoagulation affected long-term occlusion rates of the saphenous vein after endovenous ablations.[11]

Thrombophilia is suspected in individuals with a personal or family history of DVT, with recurrent superficial venous thrombosis and with multiple spontaneous abortions. The American College of Chest Physicians guidelines on risk assessment for deep vein thrombosis[12] provide recommendations for prophylactic anticoagulation. The SVS/AVF guidelines recommend perioperative low molecular weight heparin and compression stockings for patients who are at high risk for DVT.[1] An informed consent should be obtained before the procedure from each patient, after a clear explanation of the intervention's steps, its expected benefits, and the potential risks. It is advisable to take a photograph of the varicose veins before treatment and at follow-up exams to document any visible complications and the efficacy of treatment.

Endovenous thermal ablation techniques

Radiofrequency ablation

Approved in 1999 by the Food and Drug Administration (FDA), RFA has rapidly gained popularity as one of the endovenous thermal techniques of the incompetent GSV or small saphenous vein (SSV) (Fig. 22.2). It is an outpatient office procedure, a percutaneous technique for ablation, which can be performed using local tumescent anesthesia. The currently available RFA device used in the United States is the ClosureFast (Medtronic, Minneapolis, MN).

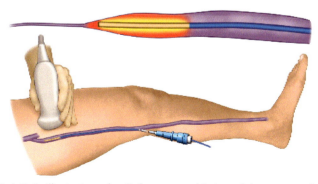

Figure. 22.2 Artist's illustration of radiofrequency ablation of the great saphenous vein.

General principles

During the RFA procedure, radiofrequency energy is converted into thermal energy at the level of the venous wall. In the first-generation RFA catheters, the radiofrequency energy was produced with bipolar electrodes, with an alternating electrical current of 200–1200 kHz. In the current ClosureFast device, the tip of the catheter has a 7 cm segment, which is heated to 120°C using the radiofrequency energy produced by the generator. The high temperature in the vein wall results in collagen denaturation, vein contraction, and collapse. To be efficient, the electrode needs to touch the endoluminal surface of the vein. It is maintained in the position for 20 s, heating the wall to 100–110°C. The catheter with a 7 cm heating element is then moved by 6.5 cm distally, and the process is repeated with 0.5 cm overlap over the initial location. This newer RF technology decreased the interoperator variability and increased the intervention efficacy, as well as procedural time (3–5 min per 45 cm vein length) when compared to laser.[13]

ClosureFast is available in two lengths of the heating element (3 and 7 cm), both available with a 60 cm catheter. The 7 cm element is also available with a longer, 100 cm catheter. The generator is equipped with a real-time feedback mechanism for intravascular temperature and automatic regulation of the power. The efficacy of the procedure depends on appropriate venous emptying and contact of the heated element with the vessel wall. ClosureFast monitors generated impedance and controls the vein wall temperature. This feedback loop protects against overheating and its secondary effects such as vaporization, boiling, coagulation, and carbonization.[13] When high power output is displayed, this suggests the possibility of poor contact with the vein wall. The correction of a possible cause (insufficient vein exsanguination or vein compression) may then be corrected by the operator. The use of tumescent anesthesia with subcutaneous injection of local anesthetic in the perivenous space provides optimal compression of the target vein around the catheter and conditions for successful ablation.

Operative technique with ClosureFast

The procedure is performed under local anesthesia for vein access and under tumescent anesthesia for RFA. Sedation is optional for patients who experience anxiety. The intervention starts with duplex ultrasound[14–16] to evaluate the location and size of the GSV and its tributaries and locate the site for access and percutaneous introduction of a guidewire into the target vein. To facilitate the identification of the optimal vein segment as a target for establishing venous access, the patient can be placed in the reverse Trendelenburg position. Following this, the GSV is cannulated via the Seldinger technique at the knee level or alternatively via the stab wound—Mueller hook approach.[17] After venous access is established, the patient is placed in Trendelenburg position and a 7-Fr introducer sheath is then inserted and, through it, the RFA catheter is advanced to the saphenofemoral junction (SFJ) under continuous ultrasound visualization. The catheter should be advanced with no resistance and without discomfort or pain. If GSV tortuosity renders navigation difficult, gentle compression applied over the proximal element of the catheter might slightly change its position and let it follow the vein trajectory. Occasionally a 0.025 or 0.018 inch guidewire can be used to advance the catheter. If necessary, new access and placement of a second sheath may be used. The tip of the catheter is positioned at a 2 cm distance from SFJ. For the SSV treatment, the tip of the catheter is placed at the turning point of the SSV toward the saphenopopliteal junction (SPJ), at least greater than 2 cm from SPJ. Treatment of the caudal segments of the saphenous veins is ideally avoided to prevent thermal injury of the saphenous or sural nerves.[18]

The patient is placed in the Trendelenburg position and the tip of the catheter is confirmed again with ultrasonography. The saphenous subcompartment along the GSV (the saphenous sheath) is then infiltrated with tumescent anesthetic solution (50 mL of 1% lidocaine and 1 mL of epinephrine [1:1000] diluted in 1L of normal saline)[19] under duplex ultrasound guidance around the catheter achieving depth of >1 cm and adequate "halo" effect. Additional tumescent above the catheter tip is recommended, especially for larger vein diameters. The injection causes hydrostatic pressure and vein spasm induced by the anesthetic solution and the epinephrine. In addition, tumescent anesthesia reduces patient discomfort and protects against injury of the vein's surrounding nerves, tissues, and skin, decreasing the risk of paresthesia and thermal injury to the skin.

Radiofrequency energy delivery is initiated using the start button situated on the catheter handle. Compression of the heating element using the duplex probe tightens the contact with the vein wall. The device automatically stops the energy delivery after 20 s. The first segment close to the junction requires a second cycle, and this second cycle delivery is also recommended for dilated vein segments and for zones with large tributaries. The catheter is withdrawn then by 6.5 cm increments, as delineated by shaft markers on the catheter guide. Once a cycle is completed, the catheter should not be reintroduced into the treated segment to avoid embolization. Treatment is frequently limited to the area above the medial condyle of the tibia to avoid injury to the saphenous nerve.[20,21] However, intraoperative ultrasound and generous use of tumescent anesthesia are helpful to avoid nerve injury and perform ablation in the distal third of the leg as well.[22] At the end of the intervention, after the catheter is removed, pressure is applied to the entry point for hemostasis. At the completion of the operation, a two-layer compressive dressing (Kerlix [Kendall Co, Mansfield, MA] and ACE bandages [BD, Franklin, NJ]) or elastic stockings (20−30 or 30−40 mmHg) are applied from toes to the groin to be maintained for the following 2 days. Patients are discharged the same day upon recovery from the anesthesia[13,23] with instructions to ambulate immediately and regularly to reduce the risk of venous thromboembolic events.

Duplex examination at the end of the procedure confirms successful closure of the saphenous vein and patency of the common femoral or popliteal vein. Stab avulsion phlebectomies, sclerotherapy of varicose vein tributaries, or perforator vein ablation can be performed at this time when indicated.

Endovenous laser ablation

Laser ablation is a thermal ablation procedure in which laser energy is employed to create heat delivery to the vein using a laser fiber and generator (Fig. 22.3). Laser power, pullback velocity, and vein diameter affect optimal and safe thermal ablation. Smaller veins were confirmed to have the most efficient treatment, by having the walls closer to the steam bubbles and the layer of the heated blood surrounding the hot tip.[24]

General principles

Laser generators are manufactured by several companies and characterized by different wavelengths: AngioDynamics offers 940, 980, and 1470 nm generators, CoolTouch 1320 nm, and Vari-Lase 810 nm. Lower wave-length lasers target hemoglobin as the primary chromophore (<1300 nm) while higher wavelength lasers target the water in the vein wall, causing direct thermal destruction.[25–27] The transmitting laser energy fibers are specific for each generator design. Fibers with a coated tip or "jacket-tipped" fiber decrease direct contact of the laser fiber with the wall and likely decrease the risk of perforation.[3] Radial tip fiber causes less trauma to the vein wall as well by dispersing the emitted energy. Kabnick and Sadek found that radial and jacket-tipped fibers decreased postoperative pain and bruising when compared to bare-tip fibers.[28]

Operative technique

As described earlier for radiofrequency endovenous procedures, laser venous ablation is most frequently done in outpatient office settings, using local and tumescent anesthesia, and sedation as deemed necessary by the operator.

As with RFA, duplex ultrasound mapping of the saphenous veins is performed before treatment. The initial steps of the procedure are identical to those discussed for RFA, including a selection of the access site, advancing the catheter in the GSV or SSV and administration of tumescent anesthesia.

Final localization of the laser fiber tip should be confirmed with ultrasound examination: for the GSV ablation optimal placement is 2 cm below the SFJ (below the entrance

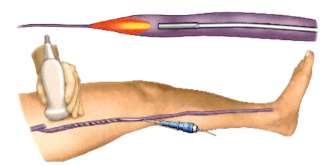

Figure. 22.3 Artist's illustration of endovenous laser therapy of the great saphenous vein.

of the superficial epigastric vein) and for the SSV below the level of the angulation. In contrast to RFA, which uses a segmental energy delivery, delivery of laser energy is done continuously, with the sheath and/or laser fiber withdrawn at a speed determined by the manufacturer, but on average with a speed of 6—8 s per cm. The energy required, per centimeter of the vein treated, is 60—100 J for low-wavelength lasers and 40—60 J for high-wavelength lasers.

Similar to the RFA procedure, a Duplex examination at the end of the EVLA procedure also confirms the successful closure of the saphenous vein and patency of the common femoral or popliteal vein. Stab avulsion phlebectomies, sclerotherapy of varicose vein tributaries or perforator vein ablation can be performed at this time when indicated.

At the completion of the procedure, a two-layer compressive dressing (Kerlix [Kendall Co, Mansfield, MA] and ACE bandages [BD, Franklin, NJ]) is applied from the toes to the groin to be maintained for the following 2 days.[19] Patients are also discharged the same day upon recovery from sedation with instruction to ambulate immediately. At the time of discharge, all patients are prescribed a standard dose of analgesics.

Postoperative care

Compression stockings or bandages are frequently advised to reduce postoperative pain, bruising, and edema.[29,30] The literature is somewhat inconsistent around the use of compression postoperatively, however. One meta-analysis[29] demonstrated no benefit from postoperative compression therapy for longer than one week postprocedure, as compression beyond that time did not significantly affect pain scores, leg volume, complication rate, and recovery time. A second meta-analysis[30] that included 775 patients from five studies considered postoperative pain intensity at 1—6 weeks as a primary outcome. The compression therapy applied was either short-duration (24—48 h of standard elastic bandages) or long-duration (1—2 weeks of compression stockings following 24 h of postoperative bandages). The long-duration treatment in four randomized controlled trials (RCTs) was associated with significantly reduced pain (mean difference [MD] 1.19 on the 10-point visual analog scale; 95% confidence interval [CI] 0.58—1.8) at 1 week. The recovery time off work was shorter in the long-duration group (MD = 1.01 day, 95% CI 0.06—1.96). Complications were similar in both groups. The authors of this study recommended 1—2 weeks of compression stockings as a routine therapy after endovenous ablation.

The routine performance of postoperative duplex within 1 week to assess for DVT or extension of the saphenous thrombus into the femoral vein, termed endovenous heat-induced thrombosis (EHIT) remains controversial. The Society for Vascular Surgery/American Venous Forum (SVS/AVF) guidelines suggest postprocedural duplex scanning within 24—72 h to exclude any thrombotic complication.[1] The strength of the recommendation, however, is weak and the evidence to support it is low (Grade 2C). The recently published EHIT guidelines by Kabnick et al. did not alter these recommendations but emphasized the controversial practice of routine postoperative duplex scanning.[31]

Nonthermal endovenous ablation techniques
Endovenous cyanoacrylate embolization
Endovenous cyanoacrylate embolization[4] for treatment of the incompetent GSV and other axial veins (VenaSeal; Sapheon, Inc, Morrisville, NC) is one of the nonthermal ablation techniques that avoid the use of tumescent anesthesia needed for RFA and EVLA treatments and thereby decreases procedure time, perioperative discomfort, pain, and bruising.

Mechanical-chemical ablation procedure
The MOCA procedure[5] (ClariVein, MeritMedical, South Jordan, UT) is another nonthermal ablation technique that is approved by the FDA. The device has a rotating wire tip at the end of a catheter that causes mechanical damage to the endothelium of the treated vein and the detergent sclerotic solution, STS or polidocanol, causes thrombosis and sclerosis of the vein.

Ultrasound guided foam sclerotherapy
Ultrasound-guided foam sclerotherapy (UGFS) is performed with either STS or polidocanol.[6] The foam can be prepared at the bedside using the Tessari technique[7] or it can be a premade injectable polydocanol foam (Varithena).[8] For detailed information on how to prepare the foam, for the techniques of saphenous and varicose vein treatments, and for results, the reader is referred to Chapter 23.

Outcomes
Early results of radiofrequency ablation
The procedural efficacy and safety of RFA were confirmed in early clinical trials and registries.[32–43] Earlier generation devices and initial lack of experience resulted in higher complication rates compared to subsequent trials, but major complications in recent trials are rare.

Bruising/ecchymosis scores, assessed by Visual Analog Scale or by specific digital image analysis software, are much better after RFA than after stripping and ligation ($P = .02$).[44] In the RECOVERY study, moderate-to-severe ecchymosis (>25% of the treated zone) was recorded in 2.2% of limbs from the RFA group and in the 51.3% of limbs from the 980-nm laser group.[43]

A systematic review and meta-analysis of *thrombotic events* following endovenous thermal ablation of the GSV[45] included 52 studies with 16,398 patients. EHIT was divided into four subtypes: Type 1, thrombus flash with the junction between superficial and deep vein; Type 2, thrombus extension into a deep vein, cross-sectional area <50%; Type 3, thrombus extension into a deep vein, cross-sectional area >50%; Type 4, complete occlusion of the deep vein. EHIT type 2–4 was reported in 1.4% of cases. DVT occurred in 0.3%, PE in 0.1%. The incidence of events was similar with RFA and EVLA techniques. Only eight out of 20 patients with DVT were symptomatic, and there was no propagation of thrombosis or PE. An update of the aforementioned systematic review,[46] published in 2021 by the same group, analyzed 75 studies with 23,265 patients confirming low incidence and benign clinical course of EHIT and associated thrombotic complications. The study emphasized the significance of the SVS/AFV guidelines pertinent to directing clinical decision-making in the management of EHIT.

Superficial thrombophlebitis may be the result of residual blood contained in the treated vein or branches of the treated vein, and it occurs in 10% of patients treated with the ClosureFast catheter.[42] Treatment is rarely required for symptom relief.

Saphenous nerve injury is rare (2.4%) with RFA.[41] *Paresthesia* (or hypoesthesia) and *dysesthesia* can occur within 1–3 weeks, but most resolve with time. Nerve injury is more frequent in the below knee location,[38] therefore, the nerve should be localized by ultrasound examination and set apart of the vein using tumescent anesthesia infiltration if it is determined that treatment below this level is necessary.

Bleeding/hematoma complications are extremely rare. In a prospective nonrandomized study, minor bleeding was more frequent in patients who were anticoagulated, especially in those receiving triple therapy (such as aspirin, clopidogrel, and warfarin combination). Leg elevation and manual compression stopped bleeding in each case.[47]

Early saphenous occlusion rates with recent generation RFA devices have been excellent, Morrison et al. reported a three month closure rate of 96%.[48]

Early results of endovenous laser ablation

Successful ablation was confirmed by duplex ultrasound in 100% of cases in multiple reports.[17,49] Minimal *ecchymosis/bruising* is generally of short duration.[50,51] In a recent review on EVLA, Teter et al. found that the use of longer wavelength may lead to less pain and bruising and the use of radial fibers and jacket-tipped fibers cause less perforation and pain.[3] *Paresthesia* (<1%[52]) is rare, and it likely correlates with lack of experience and insufficient tumescent anesthesia. It is often mild and transient.

Thrombotic complications have been rare and had an incidence similar to those observed after RFA (see above). Early case reports, however, included more severe complications like arterio-venous fistula,[53] retention of the sheath/fiber,[54] and stroke.[55]

Long-term results of endovenous thermal ablation
Anatomic success
A five year occlusion rate of RFA-treated veins of 91.9% was reported in the European multicenter cohort study.[2] No reflux was seen in an additional 3% raising the total percentage of anatomic success to 94.9%.[2] In the VeClose study, five year Kaplan–Meier estimates for freedom from recanalization was 85.2%.[48]

Vein occlusion rate up to five years after EVLA has varied between 80% and 96.8%.[56–61] A meta-analysis of RCTs evaluating long-term outcomes of endovenous thermal ablation[62] showed no significant difference in the recurrence of GSV incompetence between RFA, EVLA, and surgery, with similar efficacy in the treatment of saphenous venous insufficiency for all three methods.

Comparison of endovenous thermal versus nonthermal ablations

In a recent systematic review and meta-analysis, results of thermal versus nonthermal ablations were compared in 6549 patients reported in 16 RCTs and 11 comparative observational studies.[63] Using nonthermal ablations (CEA, UGFS) significantly decreased periprocedural pain when compared to treatment with EVLA. Patients who had EVLA reported worse early quality-of-life (QoL) scores than those who had UGFS or the MOCA procedure. Recurrent reflux was, however, frequently higher with UGFS than with any other modalities. Most importantly, anatomic closure rates were lower and reinterventions, usually with additional sclerotherapy, were more frequent with UGFS than after treatment with RFA or EVLA.

Clinical success in venous ulcer patients

The Effect of Surgery and Compression on Healing and Recurrence (ESCHAR) study[64,65] revealed that surgical treatment of superficial reflux added significant value to the strategy of venous leg ulcer management. This RCT included 500 patients with chronic venous ulcers, 242 in the surgery plus compression group and 258 in the compression alone group. Among the primary outcomes, the ulcer healing rate at the three year follow-up was not affected by surgical treatment (93% vs. 89%, $P = .73$), but rates of ulcer recurrences were significantly lower for the surgery plus compression group (31% vs. 56%, $P<.01$) at the three year follow-up. A systematic review found a similar ulcer healing rate after surgical treatment and compression treatment, and a significant reduction of recurrence rate, even in patients with combined superficial and deep venous insufficiency.[66]

The efficacy of endovenous ablations in the management of patients with venous ulcers was confirmed in the Early Venous Reflux Ablation (EVRA) trial.[67] UGFS, MOCA, and CEA methods were included in the treatment of superficial incompetence. This trial, analyzed in much detail in an earlier section of this book (see Chapter 20), found that in the early intervention group there was a shorter time to ulcer healing and more time free from ulcers than in the deferred endovenous ablation group. An online international survey after the EVRA study publication[68] estimated that 31% of clinicians changed their practice following EVRA.

A recent study of healing of chronic venous ulcers,[69] in 832 consecutive patients from 11 centers in the United States, found the recurrence rate at 6, 12, and 24 months of 3%, 5%, and 15%, respectively, when no surgery was performed. Truncal vein ablation resulted in ulcer healing in 51% of patients at 36 months, and the perforator's ablation added a 17% improvement with no difference however in the recurrence rates. The ulcer healing further increased to 87% at three years follow-up with ulcer recurrence of 26% at two years, in patients treated by deep venous stenting for stenosis and ablation of incompetent saphenous and perforating veins. Table 22.1 includes data of 20 publications that reported results on ulcer healing and recurrence in patients who underwent endovenous thermal ablations.

Improved Venous Clinical Severity Score (VCSS) and CEAP class as a result of RFA or EVLA treatment were recorded by Marston et al.[88] in patients with CEAP clinical class 3—6 and superficial venous reflux. VCSS scores improved significatly, from 11.5 ± 4.5 before to 4.4 ± 2.3 after ablation. A recent systematic review of four studies included 161 patients with nonhealing venous ulcers after elimination of superficial venous reflux.[89] Interventions to ablate superficial venous reflux were RFA or laser ablation (two studies),[71,72] RFA with or without microphlebectomies (one study),[90] and foam sclerotherapy (one study).[91] Sixty-two patients were treated for persistent or recurrent venous ulcers after treatment of superficial reflux, and the rates of persistent ulcers were between 21.1% at 1 year (for sclerotherapy) and 2.3% at two years. The authors concluded that additional ablative procedures to address persistent superficial reflux and perforator incompetence in combination with compression therapy are effective in healing persistent or recurrent venous ulcers after the elimination of superficial venous reflux.

Besides RFA and EVLA, UGFS was also successfully used for ulcer treatment, often in association with RFA[84,76] or EVLA[81,83,78] (see Table 22.1). UGFS showed excellent results in an observational study in VLU patients[92]: 79.4% of ulcers healed at 24 weeks and 95.6% at a median of 30 months. In another study of 130 patients with healed or active venous ulcers (C5-6)[93] observed an 82% healing rate at 1—2 months and a recurrence of 4.9% at 2 years after UGFS. A prospective study of RFA and UGFS treatment for GSV confirmed the safety of the treatment in VLU patients with large vein diameters (>13 mm). At 3 years, VCSS and QoL scores improved and 12 of 13 ulcers healed permanently.[94]

The effectiveness of the MOCA procedure was compared with RFA and EVLA in a retrospective study of 66 patients with open venous ulcers[82] (see Table 22.1). The healing rate was superior in the MOCA group (74% vs. 35%, $P = .01$) with a shorter time to heal (2.3 vs. 4.4 months, $P = .01$) and shorter mean length of the follow-up (7.9 vs. 12.8 months, $P = .02$). The MOCA procedure and younger age predicted ulcer healing in this study.

Cyanoacrylate embolization CAE in 39 patients with VLU treated for superficial venous reflux observed a mean time to healing of 73.6 days; closure rate was 100% at three months, with significant improvement in pain.[95] In a retrospective review of ulcer patients, shorter median healing time (43 vs. 104 days, $P = .001$) was observed in those who had CAE than those who underwent RFA.[86] The recurrence rates, however, were similar for both procedures (13.7% vs. 22.1%, $P = .25$) (Table 22.1).

Table 22.1 Results of endovenous thermal ablations for treatment of venous leg ulcers.

First author (year)	Type of the study; N. of patients	Endovenous techniques	Comparators	Outcome	Results
Viarengo et al.[70]	RCT, $N = 52$	EVLA of the GSV, SSV or both $N = 27$	Elastic compression or unna boot $N = 25$	• Healing rate at 12 months • Recurrence rate • Mean wound area	• 81.5% versus 24% • 44.4% with compression • From 22.3 to 2.7 cm^2 versus from 17.5 to 12.8 cm^2
Teo et al.[71]	Cohort study, $N = 44$	EVLA	None	• Healing rates at 1, 3, 6, and 12 months • Recurrence rate at 1 year	• 82.1%, 92.5%, 92.5%, and 97.4% • No recurrence at 1 year, 5 recurrences at 14 (x2), 23, 35, and 52 months
Sufian et al.[72]	Retrospective study, $N = 18$	RFA (ClosureFAST)	None	• Healing rate at 6–12 months • Recurrences	• 17/18 (94,4%) • 1 recurrence after 6 months
Harlander-locke et al.[73]	Cohort study, 140 consecutive procedures, $N = 110$ ulcers in 88 limbs	RFA (ClosureFAST and ClosureRFS)	None	• Ulcer area • Healing rate • Recurrences	• Improvement in healing to −4.4 cm^2/month • 76.3% ulcer healed in 142 + 14 days • 12 patients did not heal • 7.1% recurrences
Harlander-locke et al.[74]	Cohort study, $N = 20$ CEAP C5	RFA ClosureFAST 28 procedures: Superficial veins (19), perforators (9)	Prior compression studies with ulcer recurrence up to 67% at 12 months	• Ulcer recurrence	• 0% at 6 months, 4.8% at 12 and 18 months
Abdul-haqq et al.[75]	Retrospective cohort study (2005–10), $N = 95$	EVLA of GSV, $N = 78$; 35 without [Group 1], 43 with [Group 2] IPVs	EVLA of the GSV plus IPVs $N = 17$ [Group 3]	• Healing rate • Ulcer healing time	• 46%, 33% and 71% • Groups 1 to 3: 14.8, 11.2, and 13.2 weeks

Study	Design	Intervention	Comparison	Outcomes	Results
Alden et al.[76]	Cohort, $N = 86$	RFA = 33% alone + 29% with FS intervention group	Intervention group versus compression group	• Healing speed • Recurrences at 1-year	• 9.7% versus 4.2% per week ($P = 0.001$) • 27.1% versus 48.9% ($P<.015$)
Raju et al.[77]	Cohort study, 192 consecutive limbs with VLUs	EVLA ($N = 20$) or iliac vein stent ($n = 89$) or both ($N = 69$)	Nonthrombotic (NT) versus postthrombotic (PTS) limbs	• Healing rate	• 81% at 14 weeks for VLUs <1 inch • 75% at 5 years: 87% in NT limbs, 66% in PTS limbs
Wysong et al.[78]	Prospective cohort study (2007–14), $N = 31$	EVLA plus FS and phlebectomy if needed ($N = 30$)	None	• Healing rate • Healing time	• 93.1% • Median 55 days, median FU 448 days
Marston et al.[79]	Retrospective cohort study CEAP C5 and C6, $N = 173$ limbs (CEAP C5 in 101, C6 in 72)	EVLA of GSV, SSV, or both	None	• Healing rate at 3, 6, and 12 months • Recurrences at 1, 2, and 3 years	• 57%, 74% and 78% • 9%, 20% and 29%, +++ in patients with deep venous insufficiency
Sinabulya et al.[80]	Cohort study of consecutive patients (2006–13), $N = 170$ with healed or active ulcers	EVLA	None	• Recurrences in CEAP 6 • Recurrences in CEAP 5 • Complications	• 16% • 16% • Sensory loss (8%), DVT (1%)
Starodubtsev et al.[81]	Prospective cohort study of consecutive patients with primary CVI (CEAP C4 to C6), $N = 476$	EVLA and FS: Group 1 with GSV <15 mm, $N = 281$ cases (with 77 C5 and 55 C6)	EVLA with increased linear endovenous energy density (LEED): Group 2 with GSV >15 mm, $N = 210$ cases (with 64 C5 and 35 C6)	• Ulcer healing at 6 months • Reduction in ulcer area • Recurrences at 3 years	• 82% in G1/88% in G2 • 18%/15% • 7.3%/8.6%

Continued

Table 22.1 Results of endovenous thermal ablations for treatment of venous leg ulcers.—cont'd

First author (year)	Type of the study; N. of patients	Endovenous techniques	Comparators	Outcome	Results
Gohel et al.[67]	RCT, EVRA trial N = 450	RFA/EVLA; 31.7% of early interventions, 23.9% of deferred interventions, with FS in the additional 12.1% and 7.1% respectively	Early versus deferred interventions	• Rate of ulcer healing • Ulcer-free time during the first year	• 85.6% versus 76.3% • 306 versus 278 days
Kim et al.[82]	Review of the electronic medical records (2012–15), N = 66 (82 venous segments treated)	RFA and EVLA, N = 29/82	MOCA, N = 53/82	• Healing rate • Time to heal • Recurrences	• 35% versus 74% • 4.4 versus 2.3 months • 3 with MOCA • FU 12.8 and 7.9 months
Liu et al.[83]	Retrospective cohort study (2013–17) N = 350	EVLA-HL-FS plus compression N = 193	Compression alone N = 157	• Healing time • Recurrences at 12 months • Healing rate	• Hazard ratio (HR) 1.845 for EVLA plus; 1.1 versus 2.2 months • HR 0.42; 14% versus 29.4% • 93% versus 97.5%
Lawrence et al.[69]	Multicenter retrospective cohort study (2013–17) N = 832 consecutive patients from 11 centers	528 patients with ablation of superficial veins (344 + perforators ablation) 132 patients with deep veins stenosis—71% stented	187 patients with compression and wound care	• Healing rate • Recurrences • Stented patients healing and recurrence	• 51% for truncal ablation/68% for perforators ablation//75% at 36 months • 77% and 27% at 36 months; for extended ablation plus stent: 87% at 36 months and 26% at 24 months

Study	Design	Intervention	Comparator	Outcomes	Results
Semsathanasawadi et al.[84]	Retrospective cohort study (2011–17) N = 62	RFA + FS in 19% and + phlebectomy in 31%	None	• Healing rate • Recurrences • Time to healing • Time to recurrence	• 31%, 56%, and 66%, respectively, at 3, 6, and 12 months after RFA; better if with FS of perforators • 8%, 14%, and 23% respectively at 1, 2, and 3 years, worst with deep vein reflux
Kuserli et al.[85]	Cohort study (2009–14) N = 195	Group B: RFA of GSV + perforators Group C: RFA of GSV and SEPS	Group A: High saphenous ligation and stripping	• VCSS • Rate of occlusion at 6 and 12 months • Rate of ulcers at 2, 3, 4, and 5 years	Better results in Group C
O'Banion et al.[86]	Retrospective multicenter review (2015–20) N = 119	RFA (ClosureFAST) N = 68 More DVT—29% versus 10% and DV reflux—82% versus 51%	Cyanoacrylate glue ablation (VenaSeal) N = 51 Older patients—72 versus 65 years and more coronary artery disease—37% versus 16%	• Time to healing • Recurrences and infection rates	• 104 versus 43 days • 22.1% versus 13.7%; postprocedure infection in 2 RFA patients
Yang et al.[87]	Retrospective analysis (2017–19) N = 157	EVLA N = 64 Mean iliac vein stenosis 77%	EVLA plus stenting N = 93 Mean iliac vein stenosis 78%	• Ulcer healing at 1 year • VCSS at 22 months FU	• 65.6% versus 86.8% • 11.7 versus 8.3

CEAP, clinical class, etiology, anatomy, pathology; *EVLA*, endovenous laser ablation; *FS*, foam sclerotherapy; *GSV*, Great saphenous vein; *IPV*, incompetent perforating vein; *RCT*, randomized controlled trial; *RFA*, radiofrequency ablation; *SEPS*, subfascial endoscopic perforator surgery; *SSV*, small saphenous vein; *VCSS*, venous clinical severity score.

Endovenous ablation of the small saphenous vein

Around 15% of patients with varicose veins have significant small saphenous vein (SSV) incompetence.[96,79] SSV reflux correlates with increasing severity of CEAP class with deep venous reflux, and it can occur in patients who have venous ulcers.[97]

Knowing the anatomy of the SSV and its relation to the tibial nerve, the sural nerve, and the interlaying deep fascia is important to avoid any nerve injury during ablation.[98] To minimize the risk of tibial nerve injury, the tip of the catheter should stay close to the skin in the saphenopopliteal region and the last 3–4 cm of the SSV, where it turns deep toward the popliteal vein should not be treated. In the leg, the safest portion is the proximal one-third of the lower leg where the fascia is still between the sural nerve and the SSV.

A Cochrane review in 2016 compared the effectiveness of EVLA, RFA, and UGFS with conventional surgery for the treatment of SSV incompetence.[96] No comparative study was found for RFA but there were three trials comparing EVLA to surgery. At 1 year there was less reflux after EVLA but no difference between groups in QoL. Sural nerve injury at six weeks was reported in 6.8% after EVLA and 28.8% after surgery, but most improved with time. There were not enough data to compare UGFS with surgery.

In a more recent systematic review,[63] early results did not show a difference in closure rates or minor adverse events between CAE and RFA.

Another systematic review[99] concluded that endovenous thermal ablation (EVLA and RFA) is the method of choice because the closure rates were superior for EVLA (98.5%) and for RFA (97.1%) than those of surgery (58%) or UGFS (63.6%). Nerve injury was highest, 19.6% after surgery versus 4.8% for EVLA, 9.7% for RFA. In one study, MOCA resulted in a 94% early success rate. A successful case report with the MOCA procedure on ulcer healing was also reported.[100]

Endovenous ablation for anterior accessory vein

In patients with competent anterior accessory saphenous veins (AASVs), ablation concomitant to incompetent saphenous ablation is rarely appropriate.[101] In a retrospective review of ulcer patients treated with RFA,[84] the AASV was treated in six out of 62 limbs with an active or healed venous ulcer. VLU healing rate was 66% at 1 year and recurrence rate 8% at 1 year and 23% at three years. In a recent study of CEA and RFA for the treatment of varicose veins,[102] 5% of the procedures were ablations of the AASV, with an overall improvement of the technical success.

The Guideline Committee of the American College of Phlebology (American Vein and Lymphatic Society) recommends that symptomatic incompetence of the accessory great saphenous veins (anterior and posterior) should "be treated with endovenous thermal ablation (laser or radiofrequency) or ultrasound-guided foam sclerotherapy (Grade 1C)".[11,103]

Other endovenous procedures for venous ulcers treatment

Several endovenous minimally invasive techniques are presently available for endovenous ablation of incompetent veins contributing to CVI (chronic vennous insufficiency) and the development of venous ulcers.[104] They are reasonable alternatives to surgical stripping and high ligation. All these methods are the subject of separate chapters.

Recommendations of the SVS/AVF guidelines

The 2014 Venous Ulcer Guidelines of the SVS and AVF[105] were based on a systematic review of seven RCTs and four observational studies for ulcer healing outcome.[106] Only two RCTs compared compression alone versus compression plus surgery for prevention of ulcer recurrence.[107,65,64] The weak suggestion was that in the presence of superficial axial venous reflux directed to the bed of active ulcer, correction of the reflux of the incompetent veins is performed in addition to standard compressive therapy to improve ulcer healing (Grade 2C) and to prevent recurrence (Grade 1B). In the upcoming new SVS/AVF/American Vein and Lymphatic Society guidelines, based on the results of the EVRA trial,[108,67] early endovenous ablation with compression therapy will be recommended for accelerated healing of venous ulcers and both the level of evidence and the strength of the recommendation will be upgraded.

Conclusion

More than half of the patients with venous ulcers have superficial vein incompetence. Elimination of superficial incompetence using endovenous thermal or nonthermal ablation is a low-risk, effective, minimally invasive outpatient procedure with excellent long-term results. Evidence supports early intervention for superficial vein incompetence combined with compression therapy to accelerate ulcer healing and prevent recurrence. Repeat superficial ablation procedures are also effective in patients with recurrent and persistent venous ulcers.

References

1. Gloviczki P, Comerota AJ, Dalsing MC, et al. The care of patients with varicose veins and associated chronic venous diseases: clinical practice guidelines of the Society for Vascular Surgery and the American Venous Forum. *J Vasc Surg*. 2011;53:2S—48S.
2. Proebstle TM, Alm BJ, Gockeritz O, et al. Five-year results from the prospective European multicentre cohort study on radiofrequency segmental thermal ablation for incompetent great saphenous veins. *Br J Surg*. 2015;102:212—218.
3. Teter KA, Kabnick LS, Sadek M. Endovenous laser ablation: a comprehensive review. *Phlebology*. 2020;35:656—662.
4. Morrison N, Gibson K, Vasquez M, Weiss R, Jones A. Five-year extension study of patients from a randomized clinical trial (VeClose) comparing cyanoacrylate closure versus radiofrequency ablation for the treatment of incompetent great saphenous veins. *J Vasc Surg Venous Lymphat Disord*. 2020;8: 978—989.

5. Elias S, Lam YL, Wittens CH. Mechanochemical ablation: status and results. *Phlebology*. 2013;28(Suppl 1): 10−14.
6. Cartee TV, Wirth P, Greene A, et al. Ultrasound-guided foam sclerotherapy is safe and effective in the management of superficial venous insufficiency of the lower extremity. *J Vasc Surg Venous Lymphat Disord*. 2021;9:1031−1040.
7. Cavezzi A, Tessari L. Foam sclerotherapy techniques: different gases and methods of preparation, catheter versus direct injection. *Phlebology*. 2009;24:247−251.
8. Gibson K, Kabnick L. Varithena 013 Investigator G. A multicenter, randomized, placebo-controlled study to evaluate the efficacy and safety of Varithena(R) (polidocanol endovenous microfoam 1%) for symptomatic, visible varicose veins with saphenofemoral junction incompetence. *Phlebology*. 2017;32: 185−193.
9. Marston WA, Carlin RE, Passman MA, Farber MA, Keagy BA. Healing rates and cost efficacy of outpatient compression treatment for leg ulcers associated with venous insufficiency. *J Vasc Surg*. 1999;30:491−498.
10. Watson J, Mansour D, Shell W, et al. Outcomes of patients at risk for deep venous thrombosis undergoing endovenous ablation on systemic anticoagulation. *J Vasc Surg Venous Lymphat Disord*. 2020;8:319.
11. Erben Y, Vasquez I, Li Y, et al. A multi-institutional review of endovenous thermal ablation of the saphenous vein finds male sex and use of anticoagulation are predictors of long-term failure. *Phlebology*. 2021;36:283−289.
12. Kearon C, Akl EA, Comerota AJ, et al. Antithrombotic therapy for VTE disease: antithrombotic therapy and prevention of thrombosis, 9th ed: American College of chest Physicians evidence-based clinical practice guidelines. *Chest*. 2012;141:e419S−e496S.
13. Dietzek AM. Endovenous radiofrequency ablation for the treatment of varicose veins. *Vascular*. 2007; 15:255−261.
14. Khilnani NM, Min RJ. Duplex ultrasound for superficial venous insufficiency. *Tech Vasc Interv Radiol*. 2003;6:111−115.
15. Khilnani NM, Grassi CJ, Kundu S, et al. Multi-society consensus quality improvement guidelines for the treatment of lower-extremity superficial venous insufficiency with endovenous thermal ablation from the society of interventional radiology, cardiovascular interventional radiological society of europe, American College of Phlebology and Canadian interventional radiology association. *J Vasc Interv Radiol*. 2010;21:14−31.
16. Zygmunt Jr J. What is new in duplex scanning of the venous system? *Perspect Vasc Surg Endovasc Ther*. 2009;21:94−104.
17. Navarro L, Min RJ, Bone C. Endovenous laser: a new minimally invasive method of treatment for varicose veins–preliminary observations using an 810 nm diode laser. *Dermatol Surg*. 2001;27: 117−122.
18. Theivacumar NS, Dellagrammaticas D, Mavor AI, Gough MJ. Endovenous laser ablation: does standard above-knee great saphenous vein ablation provide optimum results in patients with both above- and below-knee reflux? A randomized controlled trial. *J Vasc Surg*. 2008;48:173−178.
19. Puggioni A, Kalra M, Carmo M, Mozes G, Gloviczki P. Endovenous laser therapy and radiofrequency ablation of the great saphenous vein: analysis of early efficacy and complications. *J Vasc Surg*. 2005;42: 488−493.
20. Min RJ, Khilnani NM. Re: cutaneous thermal injury after endovenous laser ablation of the great saphenous vein. *J Vasc Interv Radiol*. 2005;16:564. author reply -5.
21. Navarro TP, Nunes TA, Ribeiro AL, Castro-Silva M. Is total abolishment of great saphenous reflux in the invasive treatment of superficial chronic venous insufficiency always necessary? *Int Angiol*. 2009;28: 4−11.
22. Yamamoto K, Miwa S, Yamada T, et al. Strategy to prevent nerve injury and deep vein thrombosis in radiofrequency segmental thermal ablation of the saphenous veins using a new objective pain scale. *Phlebology*. 2021;36:659−664.

23. Dietzek AM, Blackwood S. Radiofrequency treatment of the incompetent saphenous vein. In: Gloviczki P, ed. *Handbook of Venous and Lymphatic Disorders*. 4th ed. Boca Raton, FL: CRC Press Taylor & Francis Group; 2017.
24. Malskat WS, Poluektova AA, van der Geld CW, et al. Endovenous laser ablation (EVLA): a review of mechanisms, modeling outcomes, and issues for debate. *Laser Med Sci*. 2014;29:393—403.
25. Fan CM, Rox-Anderson R. Endovenous laser ablation: mechanism of action. *Phlebology*. 2008;23: 206—213.
26. Proebstle TM, Sandhofer M, Kargl A, et al. Thermal damage of the inner vein wall during endovenous laser treatment: key role of energy absorption by intravascular blood. *Dermatol Surg*. 2002;28: 596—600.
27. Proebstle TM, Lehr HA, Kargl A, et al. Endovenous treatment of the greater saphenous vein with a 940-nm diode laser: thrombotic occlusion after endoluminal thermal damage by laser-generated steam bubbles. *J Vasc Surg*. 2002;35:729—736.
28. Kabnick LS, Sadek M. Fiber type as compared to wavelength may contribute more to improving postoperative recovery following endovenous laser ablation. *J Vasc Surg Venous Lymphat Disord*. 2016;4: 286—292.
29. Huang TW, Chen SL, Bai CH, Wu CH, Tam KW. The optimal duration of compression therapy following varicose vein surgery: a meta-analysis of randomized controlled trials. *Eur J Vasc Endovasc Surg*. 2013;45:397—402.
30. Chou JH, Chen SY, Chen YT, Hsieh CH, Huang TW, Tam KW. Optimal duration of compression stocking therapy following endovenous thermal ablation for great saphenous vein insufficiency: a meta-analysis. *Int J Surg*. 2019;65:113—119.
31. Kabnick LS, Sadek M, Bjarnason H, et al. Classification and treatment of endothermal heat-induced thrombosis: recommendations from the American venous Forum and the society for vascular surgery. *J Vasc Surg Venous Lymphat Disord*. 2021;9:6—22.
32. Rautio T, Ohinmaa A, Perala J, et al. Endovenous obliteration versus conventional stripping operation in the treatment of primary varicose veins: a randomized controlled trial with comparison of the costs. *J Vasc Surg*. 2002;35:958—965.
33. Lurie F, Creton D, Eklof B, et al. Prospective randomized study of endovenous radiofrequency obliteration (closure procedure) versus ligation and stripping in a selected patient population (EVOLVeS Study). *J Vasc Surg*. 2003;38:207—214.
34. Weiss RA, Weiss MA. Controlled radiofrequency endovenous occlusion using a unique radiofrequency catheter under duplex guidance to eliminate saphenous varicose vein reflux: a 2-year follow-up. *Dermatol Surg*. 2002;28:38—42.
35. Manfrini S, Gasbarro V, Danielsson G, et al. Endovenous management of saphenous vein reflux. Endovenous reflux management study group. *J Vasc Surg*. 2000;32:330—342.
36. Lurie F, Creton D, Eklof B, et al. Prospective randomised study of endovenous radiofrequency obliteration (closure) versus ligation and vein stripping (EVOLVeS): two-year follow-up. *Eur J Vasc Endovasc Surg*. 2005;29:67—73.
37. Goldman MP, Amiry S. Closure of the greater saphenous vein with endoluminal radiofrequency thermal heating of the vein wall in combination with ambulatory phlebectomy: 50 patients with more than 6-month follow-up. *Dermatol Surg*. 2002;28:29—31.
38. Merchant RF, DePalma RG, Kabnick LS. Endovascular obliteration of saphenous reflux: a multicenter study. *J Vasc Surg*. 2002;35:1190—1196.
39. Rautio TT, Perala JM, Wiik HT, Juvonen TS, Haukipuro KA. Endovenous obliteration with radiofrequency-resistive heating for greater saphenous vein insufficiency: a feasibility study. *J Vasc Interv Radiol*. 2002;13:569—575.
40. Calcagno D, Rossi JA, Ha C. Effect of saphenous vein diameter on closure rate with ClosureFAST radiofrequency catheter. *Vasc Endovasc Surg*. 2009;43:567—570.
41. Medical Advisory S. Endovascular radiofrequency ablation for varicose veins: an evidence-based analysis. *Ont Health Technol Assess Ser*. 2011;11:1—93.

42. Zuniga JM, Hingorani A, Ascher E, et al. Short-term outcome analysis of radiofrequency ablation using ClosurePlus vs ClosureFast catheters in the treatment of incompetent great saphenous vein. *J Vasc Surg*. 2012;55:1048–1051.
43. Almeida JI, Kaufman J, Gockeritz O, et al. Radiofrequency endovenous ClosureFAST versus laser ablation for the treatment of great saphenous reflux: a multicenter, single-blinded, randomized study (RECOVERY study). *J Vasc Interv Radiol*. 2009;20:752–759.
44. Hinchliffe RJ, Ubhi J, Beech A, Ellison J, Braithwaite BD. A prospective randomised controlled trial of VNUS closure versus surgery for the treatment of recurrent long saphenous varicose veins. *Eur J Vasc Endovasc Surg*. 2006;31:212–218.
45. Healy DA, Kimura S, Power D, et al. A systematic review and meta-analysis of thrombotic events following endovenous thermal ablation of the great saphenous vein. *Eur J Vasc Endovasc Surg*. 2018;56:410–424.
46. Healy DA, Twyford M, Moloney T, Kavanagh EG. Systematic review on the incidence and management of endovenous heat-induced thrombosis following endovenous thermal ablation of the great saphenous vein. *J Vasc Surg Venous Lymphat Disord*. 2021;9:1312–1320 e10.
47. Sharifi M, Mehdipour M, Bay C, Emrani F, Sharifi J. Effect of anticoagulation on endothermal ablation of the great saphenous vein. *J Vasc Surg*. 2011;53:147–149.
48. Morrison N, Gibson K, McEnroe S, et al. Randomized trial comparing cyanoacrylate embolization and radiofrequency ablation for incompetent great saphenous veins (VeClose). *J Vasc Surg*. 2015;61:985–994.
49. Proebstle TM, Gul D, Kargl A, Knop J. Endovenous laser treatment of the lesser saphenous vein with a 940-nm diode laser: early results. *Dermatol Surg*. 2003;29:357–361.
50. Doganci S, Demirkilic U. Comparison of 980 nm laser and bare-tip fibre with 1470 nm laser and radial fibre in the treatment of great saphenous vein varicosities: a prospective randomised clinical trial. *Eur J Vasc Endovasc Surg*. 2010;40:254–259.
51. Vuylsteke ME, Thomis S, Mahieu P, Mordon S, Fourneau I. Endovenous laser ablation of the great saphenous vein using a bare fibre versus a tulip fibre: a randomised clinical trial. *Eur J Vasc Endovasc Surg*. 2012;44:587–592.
52. Pannier F, Rabe E. Endovenous laser therapy and radiofrequency ablation of saphenous varicose veins. *J Cardiovasc Surg*. 2006;47:3–8.
53. Ziporin SJ, Ifune CK, MacConmara MP, Geraghty PJ, Choi ET. A case of external iliac arteriovenous fistula and high-output cardiac failure after endovenous laser treatment of great saphenous vein. *J Vasc Surg*. 2010;51:715–719.
54. Lekich C, Hannah P. Retained laser fibre: insights and management. *Phlebology*. 2014;29:318–324.
55. Caggiati A, Franceschini M. Stroke following endovenous laser treatment of varicose veins. *J Vasc Surg*. 2010;51:218–220.
56. Disselhoff BC, der Kinderen DJ, Moll FL. Is there recanalization of the great saphenous vein 2 years after endovenous laser treatment? *J Endovasc Ther*. 2005;12:731–738.
57. Nandhra S, El-sheikha J, Carradice D, et al. A randomized clinical trial of endovenous laser ablation versus conventional surgery for small saphenous varicose veins. *J Vasc Surg*. 2015;61:741–746.
58. Myers KA, Jolley D. Outcome of endovenous laser therapy for saphenous reflux and varicose veins: medium-term results assessed by ultrasound surveillance. *Eur J Vasc Endovasc Surg*. 2009;37:239–245.
59. Chang CJ, Chua JJ. Endovenous laser photocoagulation (EVLP) for varicose veins. *Laser Surg Med*. 2002;31:257–262.
60. Rasmussen L, Lawaetz M, Bjoern L, Blemings A, Eklof B. Randomized clinical trial comparing endovenous laser ablation and stripping of the great saphenous vein with clinical and duplex outcome after 5 years. *J Vasc Surg*. 2013;58:421–426.
61. Samuel N, Wallace T, Carradice D, Mazari FA, Chetter IC. Comparison of 12-w versus 14-w endovenous laser ablation in the treatment of great saphenous varicose veins: 5-year outcomes from a randomized controlled trial. *Vasc Endovasc Surg*. 2013;47:346–352.
62. Kheirelseid EAH, Crowe G, Sehgal R, et al. Systematic review and meta-analysis of randomized controlled trials evaluating long-term outcomes of endovenous management of lower extremity varicose veins. *J Vasc Surg Venous Lymphat Disord*. 2018;6:256–270.

63. Farah MH, Nayfeh T, Urtecho M, et al. A systematic review supporting the society for vascular surgery, the American venous Forum and the American vein and lymphatic society guidelines on the management of varicose veins. *J Vasc Surg Venous Lymphat Disord*. 2021;10:1155−1171.
64. Gohel MS, Barwell JR, Taylor M, et al. Long term results of compression therapy alone versus compression plus surgery in chronic venous ulceration (ESCHAR): randomised controlled trial. *BMJ*. 2007;335:83.
65. Barwell JR, Davies CE, Deacon J, et al. Comparison of surgery and compression with compression alone in chronic venous ulceration (ESCHAR study): randomised controlled trial. *Lancet*. 2004; 363:1854−1859.
66. Howard DP, Howard A, Kothari A, Wales L, Guest M, Davies AH. The role of superficial venous surgery in the management of venous ulcers: a systematic review. *Eur J Vasc Endovasc Surg*. 2008; 36:458−465.
67. Gohel MS, Heatley F, Liu X, et al. A randomized trial of early endovenous ablation in venous ulceration. *N Engl J Med*. 2018;378:2105−2114.
68. Salim S, Heatley F, Bolton L, Khatri A, Onida S, Davies AH. The management of venous leg ulceration post the EVRA (early venous reflux ablation) ulcer trial: management of venous ulceration post EVRA. *Phlebology*. 2021;36:203−208.
69. Lawrence PF, Hager ES, Harlander-Locke MP, et al. Treatment of superficial and perforator reflux and deep venous stenosis improves healing of chronic venous leg ulcers. *J Vasc Surg Venous Lymphat Disord*. 2020;8:601−609.
70. Viarengo LM, Poterio-Filho J, Poterio GM, Menezes FH, Meirelles GV. Endovenous laser treatment for varicose veins in patients with active ulcers: measurement of intravenous and perivenous temperatures during the procedure. *Dermatol Surg*. 2007;33:1234−1242. ; discussion 41-2.
71. Teo TK, Tay KH, Lin SE, et al. Endovenous laser therapy in the treatment of lower-limb venous ulcers. *J Vasc Interv Radiol*. 2010;21:657−662.
72. Sufian S, Lakhanpal S, Marquez J. Superficial vein ablation for the treatment of primary chronic venous ulcers. *Phlebology*. 2011;26:301−306.
73. Harlander-Locke M, Lawrence PF, Alktaifi A, Jimenez JC, Rigberg D, DeRubertis B. The impact of ablation of incompetent superficial and perforator veins on ulcer healing rates. *J Vasc Surg*. 2012;55: 458−464.
74. Harlander-Locke M, Lawrence P, Jimenez JC, Rigberg D, DeRubertis B, Gelabert H. Combined treatment with compression therapy and ablation of incompetent superficial and perforating veins reduces ulcer recurrence in patients with CEAP 5 venous disease. *J Vasc Surg*. 2012;55:446−450.
75. Abdul-Haqq R, Almaroof B, Chen BL, Panneton JM, Parent FN. Endovenous laser ablation of great saphenous vein and perforator veins improves venous stasis ulcer healing. *Ann Vasc Surg*. 2013;27: 932−939.
76. Alden PB, Lips EM, Zimmerman KP, et al. Chronic venous ulcer: minimally invasive treatment of superficial axial and perforator vein reflux speeds healing and reduces recurrence. *Ann Vasc Surg*. 2013;27:75−83.
77. Raju S, Kirk OK, Jones TL. Endovenous management of venous leg ulcers. *J Vasc Surg Venous Lymphat Disord*. 2013;1:165−172.
78. Wysong A, Taylor BR, Graves M, et al. Successful treatment of chronic venous ulcers with a 1,320-nm endovenous laser combined with other minimally invasive venous procedures. *Dermatol Surg*. 2016;42:961−966.
79. Marston WA, Crowner J, Kouri A, Kalbaugh CA. Incidence of venous leg ulcer healing and recurrence after treatment with endovenous laser ablation. *J Vasc Surg Venous Lymphat Disord*. 2017;5: 525−532.
80. Sinabulya H, Ostmyren R, Blomgren L. Editor's choice - mid-term outcomes of endovenous laser ablation in patients with active and healed venous ulcers: a follow-up study. *Eur J Vasc Endovasc Surg*. 2017;53:710−716.
81. Starodubtsev V, Lukyanenko M, Karpenko A, Ignatenko P. Endovenous laser ablation in patients with severe primary chronic venous insufficiency. *Int Angiol*. 2017;36:368−374.

82. Kim SY, Safir SR, Png CYM, et al. Mechanochemical ablation as an alternative to venous ulcer healing compared with thermal ablation. *J Vasc Surg Venous Lymphat Disord.* 2019;7:699−705.
83. Liu X, Zheng G, Ye B, Chen W, Xie H, Zhang T. Comparison of combined compression and surgery with high ligation-endovenous laser ablation-foam sclerotherapy with compression alone for active venous leg ulcers. *Sci Rep.* 2019;9:14021.
84. Sermsathanasawadi N, Jieamprasertbun J, Pruekprasert K, et al. Factors that influence venous leg ulcer healing and recurrence rate after endovenous radiofrequency ablation of incompetent saphenous vein. *J Vasc Surg Venous Lymphat Disord.* 2020;8:452−457.
85. Kuserli Y, Kavala AA, Turkyilmaz S. Comparison of high saphenous ligation and stripping, radiofrequency ablation, and subfascial endoscopic perforator surgery for the treatment of active venous ulcers: retrospective cohort with five-year follow-up. *Vascular.* 2022;30:375−383, 17085381211011356.
86. O'Banion LA, Reynolds KB, Kochubey M, et al. A comparison of cyanoacrylate glue and radiofrequency ablation techniques in the treatment of superficial venous reflux in CEAP 6 patients. *J Vasc Surg Venous Lymphat Disord.* 2021;9:1215−1221.
87. Yang X, Wu X, Peng Z, Yin M, Lu X, Ye K. Outcomes of endovenous laser ablation with additional iliac vein stenting of nonthrombotic lesions in patients presenting with active venous ulcers. *J Vasc Surg Venous Lymphat Disord.* 2021;9:1517−1525.
88. Marston WA, Owens LV, Davies S, Mendes RR, Farber MA, Keagy BA. Endovenous saphenous ablation corrects the hemodynamic abnormality in patients with CEAP clinical class 3-6 CVI due to superficial reflux. *Vasc Endovasc Surg.* 2006;40:125−130.
89. Goldschmidt E, Schafer K, Lurie F. A systematic review on the treatment of nonhealing venous ulcers following successful elimination of superficial venous reflux. *J Vasc Surg Venous Lymphat Disord.* 2021;9:1071−1076 e1.
90. Lawrence PF, Alktaifi A, Rigberg D, DeRubertis B, Gelabert H, Jimenez JC. Endovenous ablation of incompetent perforating veins is effective treatment for recalcitrant venous ulcers. *J Vasc Surg.* 2011;54:737−742.
91. Grover G, Tanase A, Elstone A, Ashley S. Chronic venous leg ulcers: effects of foam sclerotherapy on healing and recurrence. *Phlebology.* 2016;31:34−41.
92. Lloret P, Redondo P, Cabrera J, Sierra A. Treatment of venous leg ulcers with ultrasound-guided foam sclerotherapy: healing, long-term recurrence and quality of life evaluation. *Wound Repair Regen.* 2015;23:369−378.
93. Pang KH, Bate GR, Darvall KA, Adam DJ, Bradbury AW. Healing and recurrence rates following ultrasound-guided foam sclerotherapy of superficial venous reflux in patients with chronic venous ulceration. *Eur J Vasc Endovasc Surg.* 2010;40:790−795.
94. Poschinger-Figueiredo D, Virgini-Magalhaes CE, Porto LC, et al. Radiofrequency ablation for axial reflux associated with foam sclerotherapy for varicosities in one-step approach: a prospective cohort study comprising large diameters saphenous veins. *Vasc Health Risk Manag.* 2021;17:379−387.
95. Chan SSJ, Yap CJQ, Tan SG, Choke ETC, Chong TT, Tang TY. The utility of endovenous cyanoacrylate glue ablation for incompetent saphenous veins in the setting of venous leg ulcers. *J Vasc Surg Venous Lymphat Disord.* 2020;8:1041−1048.
96. Paravastu SC, Horne M, Dodd PD. Endovenous ablation therapy (laser or radiofrequency) or foam sclerotherapy versus conventional surgical repair for short saphenous varicose veins. *Cochrane Database Syst Rev.* 2016;11:CD010878.
97. Lin JC, Iafrati MD, O'Donnell Jr TF, Estes JM, Mackey WC. Correlation of duplex ultrasound scanning-derived valve closure time and clinical classification in patients with small saphenous vein reflux: is lesser saphenous vein truly lesser? *J Vasc Surg.* 2004;39:1053−1058.
98. Kerver AL, van der Ham AC, Theeuwes HP, et al. The surgical anatomy of the small saphenous vein and adjacent nerves in relation to endovenous thermal ablation. *J Vasc Surg.* 2012;56:181−188.
99. Boersma D, Kornmann VN, van Eekeren RR, et al. Treatment modalities for small saphenous vein insufficiency: systematic review and meta-analysis. *J Endovasc Ther.* 2016;23:199−211.
100. Moore HM, Lane TR, Franklin IJ, Davies AH. Retrograde mechanochemical ablation of the small saphenous vein for the treatment of a venous ulcer. *Vascular.* 2014;22:375−377.

101. Masuda E, Ozsvath K, Vossler J, et al. The 2020 appropriate use criteria for chronic lower extremity venous disease of the American venous Forum, the society for vascular surgery, the American vein and lymphatic society, and the society of interventional radiology. *J Vasc Surg Venous Lymphat Disord.* 2020;8:505–525 e4.
102. Yang GK, Parapini M, Gagnon J, Chen JC. Comparison of cyanoacrylate embolization and radiofrequency ablation for the treatment of varicose veins. *Phlebology.* 2019;34:278–283.
103. Gibson K, Khilnani N, Schul M, Meissner M, American College of Phlebology Guidelines C. American College of Phlebology Guidelines - treatment of refluxing accessory saphenous veins. *Phlebology.* 2017;32:448–452.
104. Hartmann K. Endovenous (minimally invasive) procedures for treatment of varicose veins : the gentle and effective alternative to high ligation and stripping operations. *Hautarzt.* 2020;71:67–73.
105. O'Donnell Jr TF, Passman MA, Marston WA, et al. Management of venous leg ulcers: clinical practice guidelines of the Society for Vascular Surgery (R) and the American Venous Forum. *J Vasc Surg.* 2014; 60:3S–59S.
106. Mauck KF, Asi N, Undavalli C, et al. Systematic review and meta-analysis of surgical interventions versus conservative therapy for venous ulcers. *J Vasc Surg.* 2014;60:60S–70S.
107. van Gent WB, Hop WC, van Praag MC, Mackaay AJ, de Boer EM, Wittens CH. Conservative versus surgical treatment of venous leg ulcers: a prospective, randomized, multicenter trial. *J Vasc Surg.* 2006; 44:563–571.
108. Gohel MS, Heatley F, Liu X, et al. Early versus deferred endovenous ablation of superficial venous reflux in patients with venous ulceration: the EVRA RCT. *Health Technol Assess.* 2019;23:1–96.

CHAPTER 23

Treatment of chronic venous insufficiency with foam sclerotherapy

Julianne Stoughton[1] and Sujin Lee[2]

[1]Harvard Medical School, Venous Program at Massachusetts General Hospital, Boston, MA, United States; [2]Harvard Medical School, Massachusetts General Hospital, Boston, MA, United States

Introduction

Injection sclerotherapy is commonly used in the treatment of superficial venous insufficiency, cutaneous telangiectasias, and reticular veins.[1] It involves introducing a sclerosing agent in the vessel lumen, which results in fibrosis of the venous intima and with occlusion of the dilated or superficial refluxing veins.[1] Its efficacy in obliterating abnormal veins has also led to its wide applicability, including in the treatment of pelvic venous reflux and venous malformations.[2–4] Success of the treatment depends on the ability of the sclerosant to cause full-thickness destruction of the vessel wall and to minimize thrombus formation.[5,6] Partial destruction of the vessel increases the risk of recanalization while local thrombosis causes pain and discoloration associated with perivascular inflammation and hemosiderin staining of the skin, respectively.[1,7] Therefore, the properties of the sclerosing agents and the attention to proper technique play an important role in the success of sclerotherapy.

Types of sclerosants

Historically, there have been many agents used for sclerotherapy (Table 23.1), yet currently the most commonly used sclerosants are sodium tetradecyl sulfate (STS) and polidocanol (POL).[7] STS and POL are FDA-approved detergents used in sclerotherapy that interfere with cell surface lipids and cause protein theft denaturation, resulting in

Table 23.1 Sclerotherapy agents and their appropriate target veins.

Agent	<1 mm veins	1–3 mm veins	3–6+ mm veins
Sodium tetradecyl sulfate (STS)	0.025%	0.5% Liquid of foam	1%–3% Foam best for larger veins
Polidocanol (POL)	0.5%	1.0% Liquid or foam	1%–3% Foam best for larger veins
Hypertonic saline (HS)	11.7%	23.4%	

irreversible damage to the endothelium and lysis of the cell wall.[8] Of note, both detergent sclerosing agents can be degraded by silicone contained in syringes and are inactivated by the serum albumin and other proteins in the blood.[9] Detergent sclerosants have the ability to be converted to foam solution, which has several advantages over traditional liquid agents.[6] Sclerosing foam is a mixture of gas bubbles in a liquid solution that has a very high surface area to volume ratio, which allows for its use at lower concentrations and volumes compared to liquid agents to achieve the desired effect.[10,11] Foam displaces the blood and has prolonged contact with the endothelium as opposed to liquid sclerosants, which become diluted after easily mixing with the blood in the vessel.[11] Additionally, foam is echogenic with ultrasound due to the presence of gas and therefore is able to be injected in a controlled fashion. To date, numerous randomized controlled trials have demonstrated the clinical efficacy of foam versus liquid sclerotherapy.[12] Although the studies have demonstrated an increased incidence of pain related to foam, there is no statistically significant difference in the rate of local inflammation, thrombophlebitis, or hyperpigmentation between foam and liquid agents.[12] These results support the continued use of foam sclerotherapy in the treatment of abnormal veins.

Composition of foam sclerosants

The efficacy of foam sclerotherapy depends on balancing the sclerosant chemistry and foam properties with the sclerosant side effect profiles. For instance, different types of gas have variable effects on the stability of the foam. Carbon dioxide is less stable than room air when mixed with a sclerosant.[13] However, room air foam has been associated with an increased incidence of neurologic injuries due to the embolism of insoluble nitrogen gas.[11] A high liquid:gas volume ratio increases the stability of foam, but higher concentrations of the surfactant agent result in an increased risk of side effects. Furthermore, the ideal foam would have a small bubble size and narrow bubble size distribution based on LaPlace's law, which states that the distending pressure in a bubble is inversely proportional to its radius.[5] A small uniformly distributed bubble would have increased viscosity and cohesion due to the higher yield stress required for the sample to flow with liquid-like behavior.[5,11] Therefore, precision and careful consideration of the mixture components are crucial to producing effective sclerosants that have consistent performance (Table 23.2).

Table 23.2 Composition and stability of foam.

Characteristics	More stable foam	Less stable foam
Variation of bubbles	Uniform bubble size	Highly variable bubble sizes
Size of bubbles	Smaller bubbles	Larger bubbles
Gas used	Nitrogen (*longer lasting, but increased embolic potential*)	CO_2/O_2 (*shorter duration and less embolic potential*)
Air:liquid ratio	<4 air: 1 liquid drug (wet foam)	>4 air: 1 liquid drug (dry foam)

Foam sclerosants can be classified into physician-compounded foams (PCFs) and proprietary endovenous microfoam (PEM).[14] PCFs are often created using the Double Syringe system, where the user vigorously squirts the liquid sclerosant and air mixture back and forth between two syringes joined by a straight connector.[1] An alternative method is the Tessari technique (Fig. 23.1), where the mixture is passed between two syringes

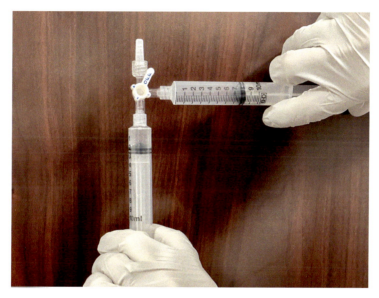

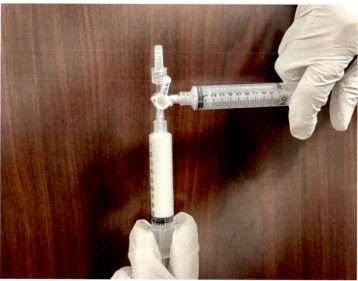

Figure 23.1 Tessari method for physician compounded foam (liquid to air ratio 1:4).

attached to the arms of a three-way stopcock where the aperture can be made smaller to create a smaller bubble size.[15] Characteristics of PCF may vary based on the type of syringe connector used and the number of passes made between the syringes in addition to the intrinsic properties of foam as discussed above.[16] Although PCFs are inexpensive and easy to produce, the variability in their composition and stability prevents their use from being standardized.[17] PEM was developed in an effort to market a consistent foam sclerosant. It is dispensed using a pressurized canister mechanism through which the gas and liquid mixture is passed through filters under pressure to produce a microstructurally consistent foam.[17,18] Due to the O_2/CO_2 mixture, the narrower distribution of bubble size, and overall smaller bubble size, PEM is generally more stable than PCF and is overall an effective treatment option for venous insufficiency with minimal complications.[17,19]

Complications and outcomes

Efficacy and side effects of foam versus liquid sclerotherapy

There have been several clinical trials comparing the effects of foam and liquid sclerotherapy. In a multicenter randomized controlled trial, Ouvry et al. demonstrated that sclerosant foam was more than twice as effective as a liquid from which the foam was prepared. 85% of patients who received foam therapy achieved complete elimination of great saphenous vein (GSV) reflux at three weeks compared to the 35% of patients who received liquid therapy ($P < .001$).[20] Respective rates at 2-year follow-up were 52% and 12%. However, those who received foam therapy had a higher incidence of venous spasms and length of sclerotic reaction. Similarly, Ukritmanoroat demonstrated that foam sclerotherapy had greater efficacy for treating varicose veins compared to conventional liquid sclerotherapy after 90 days of follow-up.[21] However, pain, inflammation, and hyperpigmentation appeared more often with foam POL therapy. Side effects such as local thrombosis, inflammation, and necrosis at the injection site can be uncommon side effects that are minimized by using the proper techniques and lowest effective concentrations for both foam and liquid sclerotherapy. Overall, many studies suggest that foam sclerotherapy is efficacious in treating all stages of superficial venous disease. However, there is insufficient evidence in the current literature that compares the efficacy of foam versus liquid sclerotherapy or surgery.

Microembolization

Foam sclerotherapy has also been associated with visual disturbances, but the overall rate is very low, as demonstrated in a large multicenter prospective study of 1025 patients that reported 2.6% of side effects.[22] A rare but significant side effect of foam sclerotherapy is foam microembolization with associated neurologic complications such as visual scotomata, headaches, paresthesia, TIA and rarely stroke has been reported following foam injection.[23] These events are likely due to the migration of bubbles associated with a right to left shunt as seen with patent foramen ovale (PFO), or intrapulmonary shunts.[24] In addition, there are potent vasoconstrictor substances released from the damaged endothelial cells which can lead to some of these neurologic events. Endothelin-1 has been shown to play a role and is released after both foam and liquid sclerotherapy.[25] Using transcranial doppler, high-intensity transient signals can be seen when small gas bubbles are entering the cerebral arterial circulation following foam sclerotherapy.[26] In a study by Raymond-Martimbeau, patients who had neurologic complications following foam sclerotherapy were screened and 70% of these patients had a PFO.[27] By contrast, the general population has approximately 26% incidence of PFO. In a study performed in the Netherlands, foam distribution was assessed by echocardiography in 33 patients undergoing foam sclerotherapy for chronic venous insufficiency (CVI).[28] The patients received a single 5cc injection of 1% polidocanol foam (air-to-liquid ratio of 4:1) into the great saphenous vein. This was performed with the leg elevated and while manual pressure was held at the saphenofemoral junction with a resultant decrease in blood flow velocity to zero. Despite these maneuvers, all patients demonstrated microemboli in the right atrium and ventricle between 45 s and 15 min after injection of the foam. There were five patients who were noted to have microemboli in the left atrium and ventricle. These patients who had left-sided emboli seen had no neurologic symptoms, but they were all noted to have a PFO. The authors also noted that before the initiation of this study, there were two patients who had developed neurologic symptoms, including transient scotomas and a migraine attack, who were each subsequently found to have PFOs. Current literature suggests that approximately 2% of the population who undergoes foam sclerotherapy develop serious neurologic symptoms, including scotomas, migraines, and strokes.[23,29] The vast majority of these neurologic events are transient and minor.

Data suggests that there is very little active drug present when the foam bubbles migrate. Tessari's investigations noted that the vast majority of sclerosant was trapped in the wall of the treated veins.[30] Additional data from the FDA application for PEM showed that the embolized bubbles contain minimal concentrations of the drug. Following direct intravenous injection of high doses of PEM with embolization of bubbles into the lung, they examined the acute and chronic cardiopulmonary effects in the dog model.[31] These data showed no clinically relevant elevations of pulmonary artery pressures, and no histopathologic changes in the lungs at 3 months follow-up. They also examined the concentration of the drug in the lungs and in the bubbles that had embolized to the lung vasculature and found no evidence to suggest that the systemically circulating gas bubbles contained more polidocanol than the surrounding plasma. Following PEM injection into the canine carotid artery at 8 times clinical doses there were no neurological effects. These pharmacologic studies provided reassurance of the safety of the embolizing microbubbles at lower doses of PEM.

In summary, microembolization of gas bubbles is commonly observed following foam injection, yet neurologic and cardiopulmonary complications are quite rare.

Despite the rarity of this complication, adequate risk prevention should be implemented to avoid a serious adverse neurologic effect. Risk and benefit analysis must be performed, and a known history of PFO should be considered as a relative contraindication to the routine use of foam sclerotherapy.[32]

Thrombosis following foam sclerotherapy

Another rare but important adverse event following sclerotherapy is thrombosis of the deep and superficial veins adjacent to the areas treated. Overall, the incidence of deep vein thrombosis (DVT) following foam sclerotherapy is low (0%—5.7%) in the literature. One single-center study carefully analyzed 1000 patients with a duplex scan at 2 weeks following ultrasound-guided foam sclerotherapy. They achieved complete occlusion in 84.5% of the target veins, and they had an overall DVT incidence of 1.5%. Most cases were extensions of thrombus into the deep system, which were labeled endovenous foam-induced thrombosis (EFIT). The extent of deep system involvement was broken down into the degree of occlusion in the deep veins (EFIT 1—4). Most of the cases of deep vein occlusion were found by ultrasound and were asymptomatic. Symptomatic DVT following foam sclerotherapy occurred only in 2 of 1166 treatments (0.2% of all cases).[33] In most studies of DVT following foam sclerotherapy, there is a direct correlation with the amount of foam injected, and in this large study, they found that doses >10 mL foam were at higher risk for ultrasound-detected DVT (Table 23.3).

Efficacy of foam sclerotherapy

Foam sclerotherapy is generally offered to patients with symptomatic venous insufficiency who fail compression therapy. Common measures of efficacy include complete occlusion of treated veins, elimination of reflux, and healing of venous ulcers. Venous

Table 23.3 Complications of foam sclerotherapy.

Common	Rare
Local pain, itching	Skin ulcerations
Inflammation	Dry cough
Trapped intraluminal thrombus	Chest pain
Hyperpigmentation	Migraine headache (true or ocular)
Telangiectatic matting	Visual disturbances
Bruising	Thrombosis (DVT, PE)
Edema	Stroke (with PFO)
	Arterial injection (ischemic necrosis)
	Nerve damage
	Allergic reaction

occlusion rates for foam sclerotherapy have been reported to be over 80%, with a higher occlusion rate compared to stab avulsion and liquid sclerotherapy.[34,35] Foam sclerotherapy has also been shown to be associated with an ulcer healing rate of 76%–100% at over 30 days.[24,36,37] In a prospective cohort study of 57 patients with active ulceration in the setting of recurrent superficial venous reflux, 80% of ulcers healed at a median of 5.3 months with an estimated 12-month recurrence rate of 9.2%.[38] Although the quality of evidence is low, there does appear to be a beneficial effect of foam sclerotherapy for patients with venous leg ulcers, both in healing time and ulcer recurrence.[39] Numerous studies confirm the role of foam sclerotherapy as a safe and effective option to treat superficial vein reflux that can help to control symptoms, and to heal and lower the recurrence of ulcerations.

Technique
Patient preparation

Preoperative planning with history, physical exam, and duplex evaluation is crucial for the best outcome following foam sclerotherapy. Full discussion must include the risks and benefits, the patient should be provided with pre- and postprocedural instructions, and all questions should be answered. As mentioned earlier, the presence of a right to left shunt or history of PFO may be a contraindication to foam treatments. Medication lists are reviewed, and in most cases, anticoagulants can be continued during the treatments as there is minimal risk of bleeding with percutaneous venopuncture. The risk of periprocedural thrombosis should be assessed using a Caprini score or other methods, and patients found to be at higher risk or those with thrombophilia should be given some appropriate prophylactic anticoagulation before and following the foam sclerotherapy procedures. Further study is needed to define exactly what prophylaxis is safest, but this author will give low-dose prophylaxis for patients at higher risk up to two weeks postprocedurally if needed. If patients require antibiotic prophylaxis, this author does recommend giving an appropriate dose 1–2 h before vein treatments. Most importantly, patients must plan for wearing postoperative compression following injections, they must be ambulatory, and they should avoid prolonged standing or sitting if possible following treatments.

Procedural details

Positioning of the patient for foam sclerotherapy is an important and somewhat controversial issue. The foam is directable, and the target vein(s) to be injected should always be toward the ceiling and above the deep vein system. For example, patients with target veins in the posterior leg should be positioned in the prone position. The knee should always be slightly flexed to avoid compression of the popliteal vein, which can occur with hyperextension or straight knee positioning.[40] Trendelenburg position offers the best clearance of blood from within the vein, which helps to increase effectiveness and decreases the dosage of the sclerosant needed to fill the lumen. However, obtaining intravenous access can be more difficult when the legs are elevated, and when the veins are emptied in this position. This author suggests a technique of first obtaining access in the flat or reverse Trendelenburg position, and then leaving a sheath or butterfly in place for later injection after the table has been tilted to elevate the legs. Studies have shown that compression at the saphenofemoral junction during foam injection is not effective in preventing microembolization to the right heart. In contrast, having the patients remain still and without conversation or Valsalva during or just after the injection does seem to reduce the rate of systemic microembolization.[32]

Another debated maneuver is whether to immediately compress the veins and have the patient perform calf muscle pumping following injections. Fegan reported no cases of DVT or pulmonary embolism in 13,352 patients when low doses of sclerosant were used with rapid compression of the veins immediately following injection.[41] Full dorsiflexion of the ankle can empty all the deep veins including the muscular veins in the calf, and Bergen et al. recommended leg elevation and calf muscle pumping following injections to clear the calf veins of sclerosant.[24] Others are proponents of keeping the patient still and not compressing the target veins until the sclerosant has had a chance to create vasospasm.[32] This author recommends the calf muscle pumping maneuver with leg elevation if there is rapid filling seen in the deep veins following injection. In addition, we often immediately compress the larger diameter superficial target veins following injection, and massage the foam sclerosant toward the peripheral vessels, particularly toward areas of ulceration.

Choosing target veins and careful planning before performing the injections is essential. Patients with venous ulcerations often have increased venous hypertension from multilevel reflux in saphenous vein trunks, tributaries, and incompetent perforators. In addition, the venous plexus underlying venous ulceration can be injected directly with ultrasound guidance which can help to obliterate the local venous hypertension to allow for better healing.[42] Combination treatments often work well with truncal reflux treated with endovenous thermal or adhesive methods, and tributaries or tortuous/recanalized veins that are not amenable to catheter ablation are treated with foam injection.

Some smaller superficial veins can be accessed with a transilluminating vein light, and most others are accessed and injected under ultrasound guidance. To prevent injection into the extravascular space, this author recommends the double syringe technique. The operator will use one hand on the ultrasound probe and the other guiding the needle either in the transverse or longitudinal plane. When injecting adjacent to perforator veins, the color flow should be analyzed to prevent inadvertent injection of an accompanying artery. Once the needle tip is thought to be intraluminal by ultrasound, the assistant will aspirate one syringe attached to a stopcock and extension tubing. When venous blood return is confirmed in the aspirating syringe, the stopcock is adjusted to allow injection of the foam with ultrasound guidance with the other syringes. This method can also be used to further aspirate and empty the majority of blood from the target vein just before injection, which leaves less blood to deactivate the sclerosant. The two syringes are used to create and to refresh the foam when using PCF, or the canister-derived PEM can be added directly to the injection site. The foam is highly echogenic and is easily seen by ultrasound, but the view of surrounding or deeper structures can be limited after the foam is introduced. The ultrasound can be used to externally massage or compress the veins until vasospasm has occurred. If reinjection of larger veins might be necessary, these veins could be accessed with a 4 French sheath, intracatheter, or butterfly needle to assist in remaining intravascular for subsequent injections.

Low injection pressure should be used to inject small volumes of sclerosant just enough to fill the lumen. Slower rates of injection are also recommended (0.5—1 cc per second), and maximal volume at a single site is recommended to be 5 cc. According to the instructions for use from PEM, the recommended maximum volume of PEM per session is 15 cc, and this author advises that PCF should be limited to 10cc per session in most cases. Low silicone syringes are recommended as the detergent sclerosants will degenerate the foam, which will become less stable in regular syringes. The foam should be prepared just before injection, and using the two-syringe technique (Tessari method), the foam can be refreshed just before additional injections.

When injecting larger veins, internal compression can be achieved by using ultrasound-guided infusion of tumescent local anesthesia into the perivenous space to compress the veins and to keep the walls coapted immediately following the intravascular injection of the foam sclerosant. This technique of tumescent anesthesia injection around the target veins has become more popular for larger veins, especially when used with catheter-directed sclerotherapy. The catheter-directed method is effective in assuring intraluminal injection despite the difficulty in visualization by ultrasound after tumescent injection. This Tumescent method has also been used for treatment with sclerotherapy of reticular veins as it was first described by Spitz.[43] This allows the veins to fibrose with a minimum of blood within the lumen and appears to result in less trapping of intraluminal thrombus and pigmentation postfoam sclerotherapy. Further study is needed to confirm these outcomes, but there does appear to be a beneficial effect when treating larger veins especially when postprocedure compression garments are not effective in compressing the target veins. This technique also may result in a superior closure rate for the foam, but further study is necessary (Fig. 23.2).

Postprocedure instructions

Compression of the veins following foam injection is an essential component of the treatment and can lead to increased success with fewer adverse events. There is a lack of consensus regarding the amount or duration of compression, but most studies show

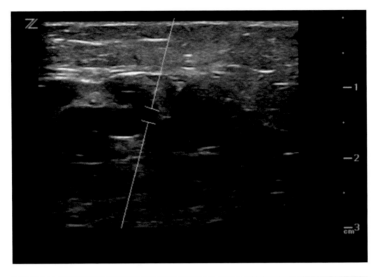

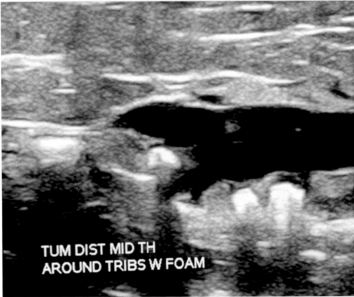

Figure 23.2 (A) Varicosities before injection. (B) Varicosities post foam injection with perivascular tumescent local anesthesia injected to create internal compression. (C) Before treatment leg varicosities photo. (D) Photo after one foam treatment with tumescent-assisted ultrasound-guided foam sclerotherapy.

improved efficacy and fewer side effects of thrombosis and intraluminal retained thrombus with some compression versus no compression following sclerotherapy.[44] This author recommends a minimum of 20–30 mm Hg compression stockings with an overlying ace wrap, which can be easily removed by the patient after discharge or

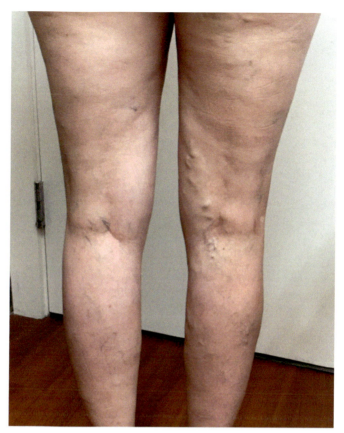

Figure 23.2 Cont'd

before bedtime. When treating a larger linear truncal vein, we will place eccentric compression pads over the injected vein. The primary effect of the sclerosant occurs within the initial 24 h, and we will recommend judicious use of the compression overnight for at least 1—2 days. Thereafter, we recommend 2 weeks of compression when upright. The layered use of compression pants, stocking liners, and external ace wraps is recommended for patients with larger veins and more advanced stages of CVI. When performing sclerotherapy associated with massive edema or venous ulceration, wound care is continued and is usually associated with multilayer wraps that are changed as frequently as needed.

Ambulation is an essential component of the postsclerotherapy instructions. Patients are ambulated right away and encouraged to continue to ambulate frequently following foam treatments. If PEM (O_2/CO_2 based) is used, the patients can be upright immediately after the compression is applied. If PCF (room air/Nitrogen based) is used, the patient can be left in the position in which they were injected for a time after injection to

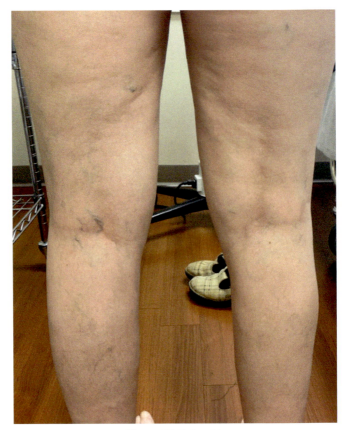
Figure 23.2 Cont'd

allow for the dissolution of the nitrogen bubbles. There is some debate on the length of time needed. Simka recommends at least 5 min,[32] and Coleridge Smith has recommended longer time periods of 10—30 min.[24] This author recommends 30 min of leg elevation without rolling and with occasional calf muscle pumps following PCF injection.

Follow-up appointments are scheduled in approximately 3—6 weeks for reevaluation and possible additional foam injections for residual veins as needed. Finally, the patient is made aware of the issue of postsclerotherapy inflammation from trapped intraluminal thrombus. If there are prominent superficial veins present, this author will schedule a follow-up visit to alleviate the inflammation and subsequent hyperpigmentation by performing a microthrombectomy to evacuate the thrombus. This can be done as early as 2—3 weeks, and up to months following injection if the thrombus remains liquefied. Villavincencio's group performed a prospective randomized trial of microthrombectomy for veins <1 mm, and a second group with veins <3 mm. They showed that there was a statistically significant improvement in hyperpigmentation with microthrombectomy for smaller veins. They also demonstrated a significant reduction in pain and inflammation in the patients with larger veins who had microthrombectomy at 1–3 weeks postsclerotherapy.[45] This author performs microthrombectomy using vein light or ultrasound guidance, starting with a sterile skin preparation, a small wheal of lidocaine is placed, and then an 18 gauge filter needle is used to puncture the occluded vein to allow evacuation of the partially liquefied intraluminal thrombus. This often results in immediate pain relief.

Summary

Foam sclerotherapy techniques are widely used in the treatment of CVI, and there are numerous preparations and technical issues that affect the safety and efficacy of these treatments. Sclerotherapy is an essential tool in treating patients, especially with multilevel advanced venous diseases with complex anatomy. Ultrasound guidance can be beneficial in gaining access and assuring intraluminal treatment. Foam can be directed toward the venous plexus adjacent to leg ulcerations which can be beneficial, and the sclerosant is uniquely able to treat both large and smaller veins using minimally invasive techniques. Many details can affect the outcomes including preoperative preparation, positioning, choice of sclerosant, method of delivery, and postoperative management. The clinician must be familiar with the properties of foam and techniques of administration to optimize success and to minimize complications.

References

1. Worthington-Kirsch RL. Injection sclerotherapy. *Semin Interv Radiol*. September 2005;22(3):209–217. https://doi.org/10.1055/s-2005-921954.
2. Lopez AJ. Female pelvic vein embolization: indications, techniques, and outcomes. *Cardiovasc Intervent Radiol*. August 2015;38(4):806–820. https://doi.org/10.1007/s00270-015-1074-7.
3. Daniels JP, Champaneria R, Shah L, Gupta JK, Birch J, Moss JG. Effectiveness of embolization or sclerotherapy of pelvic veins for reducing chronic pelvic pain: a systematic review. *J Vasc Interv Radiol*. October 2016;27(10):1478–1486. https://doi.org/10.1016/j.jvir.2016.04.016. e8.
4. Rabe E, Pannier F. For the guideline G. Indications, contraindications and performance: European guidelines for sclerotherapy in chronic venous disorders. *Phlebology*. May 2014;29(1 suppl):26–33. https://doi.org/10.1177/0268355514528127.
5. Roberts TG, Cox SJ, Lewis AL, Jones SA. Characterisation and optimisation of foams for varicose vein sclerotherapy. *Biorheology*. 2020;57(2–4):77–85. https://doi.org/10.3233/BIR-201004.
6. Bottaro E, Paterson JAJ, Quercia L, et al. In vitro and ex vivo evaluation of the biological performance of sclerosing foams. *Sci Rep*. July 8, 2019;9(1):9880. https://doi.org/10.1038/s41598-019-46262-0.
7. Albanese G, Kondo KL. Pharmacology of sclerotherapy. *Semin Interv Radiol*. December 2010;27(4):391–399. https://doi.org/10.1055/s-0030-1267848.
8. Rabe E, Schliephake D, Otto J, Breu FX, Pannier F. Sclerotherapy of telangiectases and reticular veins: a double-blind, randomized, comparative clinical trial of polidocanol, sodium tetradecyl sulphate and isotonic saline (EASI study). *Phlebology*. June 2010;25(3):124–131. https://doi.org/10.1258/phleb.2009.009043.
9. Lai SW, Goldman MP. Does the relative silicone content of different syringes affect the stability of foam in sclerotherapy? *J Drug Dermatol*. April 2008;7(4):399–400.

10. Bottaro E, Paterson J, Zhang X, et al. Physical vein models to quantify the flow performance of sclerosing foams. *Front Bioeng Biotechnol*. 2019;7:109. https://doi.org/10.3389/fbioe.2019.00109.
11. Meghdadi A, Jones SA, Patel VA, Lewis AL, Millar TM, Carugo D. Foam-in-vein: a review of rheological properties and characterization methods for optimization of sclerosing foams. *J Biomed Mater Res B Appl Biomater*. January 2021;109(1):69–91. https://doi.org/10.1002/jbm.b.34681.
12. Bi M, Li D, Chen Z, Wang Y, Ren J, Zhang W. Foam sclerotherapy compared with liquid sclerotherapy for the treatment of lower extremity varicose veins: a protocol for systematic review and meta analysis. *Medicine (Baltim)*. May 29, 2020;99(22):e20332. https://doi.org/10.1097/MD.0000000000020332.
13. Peterson JD, Goldman MP. An investigation into the influence of various gases and concentrations of sclerosants on foam stability. *Dermatol Surg*. January 2011;37(1):12–17. https://doi.org/10.1111/j.1524-4725.2010.01832.x.
14. Gibson KD, Ferris BL, Pepper D. Foam sclerotherapy for the treatment of superficial venous insufficiency. *Surg Clin North Am*. October 2007;87(5):1285–1295. https://doi.org/10.1016/j.suc.2007.07.001. xii–xiii.
15. Cavezzi A, Tessari L. Foam sclerotherapy techniques: different gases and methods of preparation, catheter versus direct injection. *Phlebology*. December 2009;24(6):247–251. https://doi.org/10.1258/phleb.2009.009061.
16. Rao J, Goldman MP. Stability of foam in sclerotherapy: differences between sodium tetradecyl sulfate and polidocanol and the type of connector used in the double-syringe system technique. *Dermatol Surg*. January 2005;31(1):19–22. https://doi.org/10.1111/j.1524-4725.2005.31008.
17. Carugo D, Ankrett DN, Zhao X, et al. Benefits of polidocanol endovenous microfoam (Varithena(R)) compared with physician-compounded foams. *Phlebology*. May 2016;31(4):283–295. https://doi.org/10.1177/0268355515589063.
18. Kim PS, Elias S, Gasparis A, Labropoulos N. Results of polidocanol endovenous microfoam in clinical practice. *J Vasc Surg Venous Lymphat Disord*. January 2021;9(1):122–127. https://doi.org/10.1016/j.jvsv.2020.04.015.
19. Eckmann DM. Polidocanol for endovenous microfoam sclerosant therapy. *Expert Opin Investig Drugs*. December 2009;18(12):1919–1927. https://doi.org/10.1517/13543780903376163.
20. Ouvry P, Allaert FA, Desnos P, Hamel-Desnos C. Efficacy of polidocanol foam versus liquid in sclerotherapy of the great saphenous vein: a multicentre randomised controlled trial with a 2-year follow-up. *Eur J Vasc Endovasc Surg*. September 2008;36(3):366–370. https://doi.org/10.1016/j.ejvs.2008.04.010.
21. Ukritmanoroat T. Comparison of efficacy and safety between foam sclerotherapy and conventional sclerotherapy: a controlled clinical trial. *J Med Assoc Thai*. March 2011;94(Suppl 2):S35–S40.
22. Gillet JL, Guedes JM, Guex JJ, et al. Side-effects and complications of foam sclerotherapy of the great and small saphenous veins: a controlled multicentre prospective study including 1,025 patients. *Phlebology*. June 2009;24(3):131–138. https://doi.org/10.1258/phleb.2008.008063.
23. Forlee MV, Grouden M, Moore DJ, Shanik G. Stroke after varicose vein foam injection sclerotherapy. *J Vasc Surg*. January 2006;43(1):162–164. https://doi.org/10.1016/j.jvs.2005.09.032.
24. Bergan J, Pascarella L, Mekenas L. Venous disorders: treatment with sclerosant foam. *J Cardiovasc Surg*. February 2006;47(1):9–18.
25. Frullini A, Felice F, Burchielli S, Di Stefano R. High production of endothelin after foam sclerotherapy: a new pathogenetic hypothesis for neurological and visual disturbances after sclerotherapy. *Phlebology*. August 2011;26(5):203–208. https://doi.org/10.1258/phleb.2010.010029.
26. Morrison N, Neuhardt DL. Foam sclerotherapy: cardiac and cerebral monitoring. *Phlebology*. December 2009;24(6):252–259. https://doi.org/10.1258/phleb.2009.009051.
27. Raymond-Martimbeau P. Transient adverse events positively associated with patent foramen ovale after ultrasound-guided foam sclerotherapy. *Phlebology*. June 2009;24(3):114–119. https://doi.org/10.1258/phleb.2008.008060.
28. Ceulen RP, Sommer A, Vernooy K. Microembolism during foam sclerotherapy of varicose veins. *N Engl J Med*. April 3, 2008;358(14):1525–1526. https://doi.org/10.1056/NEJMc0707265.
29. Guex JJ, Allaert FA, Gillet JL, Chleir F. Immediate and midterm complications of sclerotherapy: report of a prospective multicenter registry of 12,173 sclerotherapy sessions. *Dermatol Surg*. February 2005;31(2):123–128. https://doi.org/10.1111/j.1524-4725.2005.31030. discussion 128.

30. Tessari L. Special session on chronic venous disease: endotherapy for CVD. Is foam sclerotherapy safe? (9th annual meeting of the European Venous Forum 2008 Barcelona Spain).
31. *Department of Health and Human Services, Public Health Service, FDA. Center for Drug Evaluation and Research. Application Number: 205098Orig1s000 Pharmacology/toxicology NDA/BLA Review and Evaluation. NDA # 205098. Reference ID 3367720. William T. Link, PhD.* 2013.
32. Simka M. Principles and technique of foam sclerotherapy and its specific use in the treatment of venous leg ulcers. *Int J Low Extrem Wounds.* September 2011;10(3):138−145. https://doi.org/10.1177/1534734611418154.
33. Kulkarni SR, Messenger DE, Slim FJ, et al. The incidence and characterization of deep vein thrombosis following ultrasound-guided foam sclerotherapy in 1000 legs with superficial venous reflux. *J Vasc Surg Venous Lymphat Disord.* July 2013;1(3):231−238. https://doi.org/10.1016/j.jvsv.2012.10.060.
34. Belcaro G, Cesarone MR, Di Renzo A, et al. Foam-sclerotherapy, surgery, sclerotherapy, and combined treatment for varicose veins: a 10-year, prospective, randomized, controlled, trial (VEDICO trial). *Angiology.* May-Jun 2003;54(3):307−315. https://doi.org/10.1177/000331970305400306.
35. Hamel-Desnos C, Desnos P, Wollmann JC, Ouvry P, Mako S, Allaert FA. Evaluation of the efficacy of polidocanol in the form of foam compared with liquid form in sclerotherapy of the greater saphenous vein: initial results. *Dermatol Surg.* December 2003;29(12):1170−1175. https://doi.org/10.1111/j.1524-4725.2003.29398.x. discussion 1175.
36. Cabrera J, Cabrera Jr J, Garcia-Olmedo MA. Sclerosants in microfoam. A new approach in angiology. *Int Angiol.* December 2001;20(4):322−329.
37. Cabrera J, Redondo P, Becerra A, et al. Ultrasound-guided injection of polidocanol microfoam in the management of venous leg ulcers. *Arch Dermatol.* June 2004;140(6):667−673. https://doi.org/10.1001/archderm.140.6.667.
38. Grover G, Tanase A, Elstone A, Ashley S. Chronic venous leg ulcers: effects of foam sclerotherapy on healing and recurrence. *Phlebology.* February 2016;31(1):34−41. https://doi.org/10.1177/0268355514557854.
39. O'Donnell Jr TF, Passman MA. Clinical practice guidelines of the society for vascular surgery (SVS) and the American venous forum (AVF)–Management of venous leg ulcers. Introduction. *J Vasc Surg.* August 2014;60(2 Suppl):1S−2S. https://doi.org/10.1016/j.jvs.2014.04.058.
40. Spacil J. Popliteal vein compression with the limb extended. *Vasa.* September 2013;42(5):357−362. https://doi.org/10.1024/0301-1526/a000301.
41. Shami SK, Cheatle TR, Fegan G. *Fegan's Compression Sclerotherapy for Varicose Veins.* xvi. Springer; 2003:xii, 100.
42. Gschwandtner ME, Ehringer H. Microcirculation in chronic venous insufficiency. *Vasc Med.* 2001;6(3):169−179. https://doi.org/10.1177/1358836x0100600308.
43. Spitz G. *An Overview of Tumescent Enhanced Sclerotherapy as a Treatment for Superficial Veins.* Dec 2019.
44. Weiss RA, Sadick NS, Goldman MP, Weiss MA. Post-sclerotherapy compression: controlled comparative study of duration of compression and its effects on clinical outcome. *Dermatol Surg.* February 1999;25(2):105−108. https://doi.org/10.1046/j.1524-4725.1999.08180.x.
45. Scultetus AH, Villavicencio JL, Kao TC, et al. Microthrombectomy reduces postsclerotherapy pigmentation: multicenter randomized trial. *J Vasc Surg.* November 2003;38(5):896−903. https://doi.org/10.1016/s0741-5214(03)00920-0.

CHAPTER 24

Ultrasound guidance for endovenous treatment

Lisa Amatangelo, Kimberly Scherer, Vibhor Wadhwa, Jimmy Xia ScB and Neil Khilnani

Division of Interventional Radiology, Weill Cornell Medicine, New York Presbyterian Hospital, New York, NY, United States

Introduction

Duplex ultrasound provides physicians with practical information about venous anatomy and physiology. It is the established method to assess superficial, deep, and perforator veins of the lower extremity for competence, acute thrombotic conditions, and post-thrombotic changes. In addition, the use of duplex ultrasound has become an integral part of endovascular treatment, as it provides guidance for venous access and administration of tumescent anesthesia. It allows peri- and intraprocedural monitoring of venous treatments as well as postprocedural assessment of technical success and of thrombotic events or other complications after the procedure. This chapter will provide valuable details for the use of duplex ultrasound for preprocedure planning, venous access, periprocedural monitoring, and postprocedural evaluation. In the venous ulcer patient specifically, ultrasound guidance is vital for saphenous and perforator vein ablation, as well as the treatment of varicose vein tributaries leading to and beneath the ulcer bed.

Equipment

Lower extremity venous imaging requires ultrasonography machines with high-quality gray-scale imaging, pulsed wave Doppler, and a linear 7.5–10 MHz transducer. Color and color power Doppler can also be helpful; however, the flow dynamics and directionality documented by pulsed wave Doppler are more reliable and reproducible than that documented by color Doppler. Most superficial venous structures can be reliably imaged using a higher frequency linear transducer or occasionally, even higher frequency small footprint or L-shaped—"hockey stick" transducer (available up to 18 MHz). Higher frequency transducers generate sound waves with shorter wavelengths, thus producing images with better spatial resolution. However, there is a trade-off between resolution and beam penetration as shorter wavelength sound waves cannot penetrate deeper into tissues, such that deeper veins cannot be reliably imaged with higher frequency transducers. Because of that, lower-frequency transducers, with longer wavelengths, must be used to visualize deeper structures. 2–5 MHz curvilinear or phased array transducers would be appropriate for deeper imaging. These lower frequency transducers may be required in edematous states, in obese patients, and for deep veins of the legs (deep femoral vein) or pelvis (iliac veins or inferior vena cava).

Preprocedure planning

Immediately before an endovenous treatment for superficial reflux, a preoperative ultrasound is optimally performed with the patient in the standing position (or steep reverse Trendelenburg) to allow the veins to distend and to enhance the depiction of abnormally dilated and refluxing veins. The goals of preprocedure planning are to identify: (1) the vein(s) to be treated, (2) any other nontarget, yet important anatomical structures that need to be considered during the procedure, (3) the point and path to access, and (4) the deep vein connections of the target vein(s). To begin, the target vein(s) are identified and the course of the vein or veins to be treated can be traced on the skin with a marker. Junctions with deep veins and perforators are identified and marked as appropriate for the intended procedure. This is especially important when planning foam sclerotherapy as the pressure applied at these points can limit the entry of foam into the deep system. Planned and potential alternative access sites are marked. The angles needed to access the target vein(s) and avoid traversing structures such as an ulcer bed or severely indurated tissue are noted as well. The approximate length of the vein to be treated can also be estimated. Areas of vein dilation or tortuosity, tributary take-offs, adjacent nerves, and deep veins and/or arteries near the treatment area and access site(s) may also be noted, as pertinent to the procedure planned. Venous reflux as well as its extent can be reconfirmed before treatment. Ultrasound mapping involves additional considerations when planning perforator treatment. These veins are often tortuous, making the selection of the angle of approach for access more challenging.[1] As an example, if there is an accessible, straight segment of the perforator superficial to the fascia, it could be targeted longitudinally for treatment during pullback of the activated device, if thermal ablation is planned. If the perforator vein is tortuous, several transverse access points could be selected for several separate levels of treatment.

Once the patient is in the supine position and draped for the procedure, the locations of the aforementioned items can be reviewed, as there are sometimes subtle shifts in a location with the position change and more significant changes in target vein diameters. Advance planning of the positions of the patient, provider, assistants, and equipment during the procedure is also important for maintaining comfort, technical accuracy, efficiency, and sterility. Careful and detailed preprocedure planning is vital to the accuracy, success, and safety of endovenous treatment and should always be performed (Fig. 24.1).

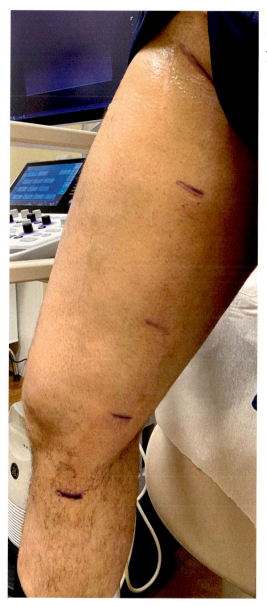

Figure 24.1 Preprocedure marking of the great saphenous vein noting the saphenofemoral junction, course of the vein, and potential access site, prior to ablation.

Venous access

In addition to preprocedure planning, ultrasound guidance is used to gain access during endovenous treatments. Before ablation, the target vein segment for establishing venous access is identified by ultrasound after the patient is placed in the reverse Trendelenburg position. There are two techniques commonly used to place a catheter for venous access, the transverse (short-axis) or the longitudinal (long-axis) approach. Using the transverse (short-axis) technique, the ultrasound probe is held perpendicular to the vein being accessed and the vein is placed in the center of the screen. The access needle is inserted into the skin adjacent to the middle of the probe and guided into the vein as the probe sweeps to try to identify the tip of the needle. Visualizing the tip of the needle with the short-axis technique requires experience to avoid advancing the needle tip beyond the beam of the probe. In this instance, the needle can be slowly withdrawn and/or the beam adjusted with sweeps to better visualize the needle tip. Occasional high frequency in and out needle movement makes it easier to find the tip. The long-axis approach is favored by some practitioners as more of the needle is visible with this technique. For this approach, the ultrasound probe is placed parallel to the target vein and the vein is displayed in a longitudinal plane. The access needle is inserted and can be seen approaching the vein until access is achieved.[2] Making the ultrasound beam as orthogonal to the needle length as possible takes advantage of increased sound reflection to improve needle conspicuousness. Accessing tortuous and small-diameter varicose veins can be challenging using a long-axis approach, therefore developing comfort with both the short and long-axis approach is imperative for the practitioner (Fig. 24.2A). After venous access is established, the patient is placed in the Trendelenburg position. If a catheter-based technique is being used, the catheter is advanced into the target vein through a sheath under continuous ultrasound visualization (Fig. 24.2B).

When accessing deep veins for iliofemoral or femoropopliteal treatments, ultrasound-guided access is also helpful. The access site is determined based on the segment of the vein to be treated. It is ideal to access a patent vein segment below an occlusion to provide good inflow and the use of ultrasound can help identify the best locations and pathways for safe access (Fig. 24.2C).

For common femoral, mid-femoral, and popliteal access, the transverse (short axis) approach is usually taken. For posterior tibial vein access, the longitudinal approach can be advantageous given the small caliber and access location. Ultrasound can help identify adjacent arteries and nerves that need to be avoided to reduce the risk of complications (Fig. 24.2D).

As discussed, perforator veins may be accessed using a transverse or longitudinal approach, depending upon the orientation of the perforator vein and less so upon adjacent structures or skin pathology. Ultrasound is used for needle access and if an ablation catheter is used, to appropriately position it within the perforator before ablation. With foam sclerotherpy, the ultrasound can help identify an optional perforating vein tributary to access and can help titrate the volume of sclerosing agent injected and determine when external compression of the perforator itself or adjacent veins is appropriate and when additional access points are needed. Using a "hockey stick" probe can be helpful for perforator vein access because of its higher frequency and smaller footprint and because it reduces the distance from the skin puncture to the perforator vein entry.[3]

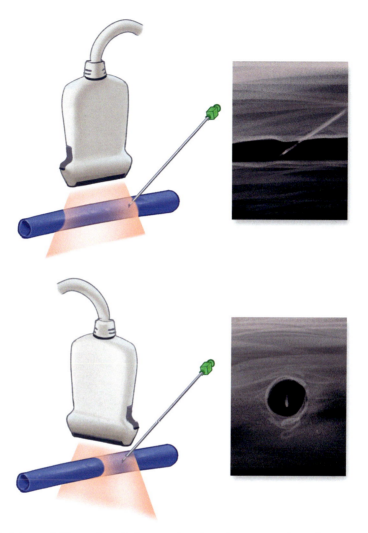

Figure 24.2A Schematic illustration of ultrasound probe orientation and needle position for longitudinal and transverse venous access.

Periprocedural monitoring

Truncal and tributary varicose veins can be treated using a variety of techniques, including endovenous ablation (thermal and nonthermal) and sclerotherapy. Real-time ultrasound monitoring is used to ensure the safety and effectiveness of the various treatments. When using catheter-based venous ablation techniques, ultrasound is used to confirm the proper placement of the catheter tip (Fig. 24.3). Ultrasound is also used to help navigate guidewires and/or catheters in areas of tortuosity, multiple branches, valve cusps, or

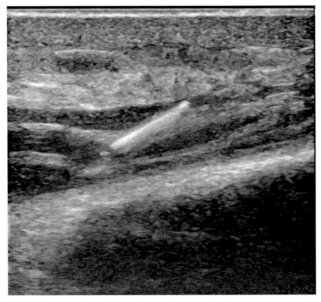

Figure 24.2B Ultrasound image of needle tip during longitudinal venous access. The tip of the needle is seen tenting the anterior wall of the vessel just before access.

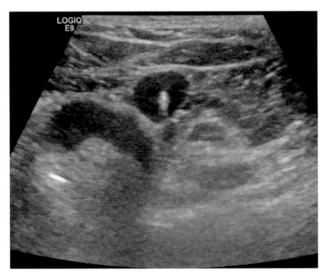

Figure 24.2C Ultrasound image of needle tip after vein puncture during transverse venous access.

thickened vein walls. Before delivering thermal energy or nonthermal product, ultrasound should be used to reconfirm tip placement and to locate the nearest deep junction so that the ultrasound probe can be used to compress just below the junction to prevent heat or

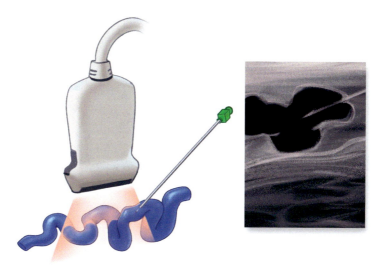

Figure 24.2D Schematic illustration of transverse imaging of tortuous varicosities while using a longitudinal access approach. This approach is not useful for catheter access, but is very useful for accessing tortuous veins during ultrasound-guided sclerotherapy, particularly those underlying ulcer beds.

product extension. Using continuous ultrasound surveillance, the catheter is advanced to approximately 0.5–1 cm (distal to the ostium of the superficial epigastric vein) and 1.5–2 cm distal to the saphenofemoral junction (SFJ) when radiofrequency and laser ablation are used, respectively.[4–8] Ultrasound can be used to confirm that thermal energy is being delivered and can be used to follow the catheter tip to allow the operator to use

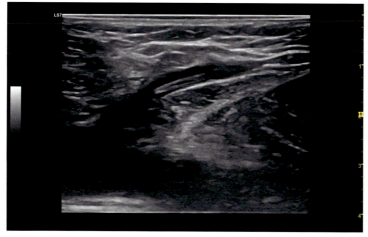

Figure 24.3 Ultrasound image of the saphenofemoral junction during positioning of a sheath in the great saphenous vein before ablation.

more energy or product in areas of dilation. In the case of cyanoacrylate, the probe is also used to compress along treated segments as the glue is dispensed and polymerized.

Sclerotherapy of varicose veins can be performed with physician-prepared or commercially produced foam (Varithena). Ultrasound-guided foam sclerotherapy requires real-time ultrasound guidance to ensure the foamed sclerosant is adequately distributed in all target veins as well as to minimize the flow of foam into the deep veins. Ultrasound can be used to provide real-time visualization of sclerosant administration, observe the filling of target veins, and avoid nontarget vein sclerosis. The echogenicity of the microfoam bubbles makes the foam visible on ultrasound surveillance, allowing the operator to ensure that the injection is intraluminal and to monitor where the foam is flowing. The ultrasound probe can be used to compress junctions with deep veins to prevent the advancement of the sclerosing foam into the deep venous system.[9] Ultrasound can also be used to confirm the spasm of veins, which suggests that the vein is adequately treated.

Tumescent anesthesia administration

Ultrasound guidance facilitates tumescent anesthesia delivery during thermal endovenous treatments. After catheter placement, perivenous tumescent anesthesia is percutaneously delivered under real-time ultrasound guidance along the entire length of the vein targeted for treatment. This creates a fluid layer around the target vein, which serves as an anesthetic, compresses the vein to the catheter to maximize energy delivered to the vein wall, and protects surrounding structures such as skin, nerves, nontarget veins, and arteries from thermal damage by physically widening the distance between the source of heat and surrounding anatomic structures and acting as an absorbent of thermal energy.[10,11] Adequate tumescent anesthesia appears as a 1-cm thick hypoechoic perivenous halo that compresses the vein lumen along the entire length of the vein to be treated[12] (Fig. 24.4). Tumescent anesthesia is created by using 50 cc of 1% lidocaine (with or without epinephrine) with 450 cc of normal saline to create a 0.1% lidocaine mix. Small aliquots of sodium bicarbonate (5–10 cc, 8.4%) are then added to neutralize the acidic effects of lidocaine. When treating saphenous veins, the ideal needle positioning is within the saphenous compartment (between the superficial and deep fascial layers).

Postprocedure monitoring

Immediately after the endovenous ablation is completed, ultrasonography may be used to assess the technical success of venous closure although the benefits of this may vary by the ablation technique used. To this end, the patient is returned to a reverse Trendelenburg (or horizontal) position, and ultrasonography is used for a careful and detailed evaluation of the treated venous segment for persistent venous flow. If ultrasonography detects flow, the procedure can be repeated.[13–15]

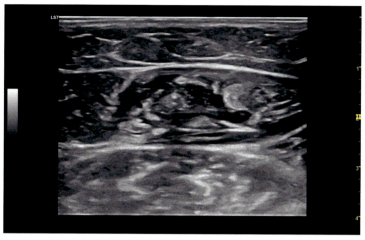

Figure 24.4 Ultrasound image of tumescent anesthetic circumferentially surrounding and compressing the target vein and catheter.

Procedural success is commonly assessed by ultrasound although the value of such investigations immediately after a procedure or at follow-up visits in asymptomatic patients is not supported by evidence. Immediately after an ablation procedure, ultrasound is used by some providers to confirm patency of deep veins—including the common femoral (if the great saphenous vein is treated) and the popliteal (if the small saphenous vein is treated). With thermal ablation, the vein should be smaller in diameter with wall thickening, sometimes referred to as having a "Cheerio" appearance (Fig. 24.5). A few days to

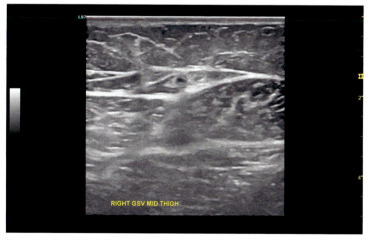

Figure 24.5 Ultrasound image of the great saphenous vein 1 week postablation—vein walls are thickened and vein diameter is decreased.

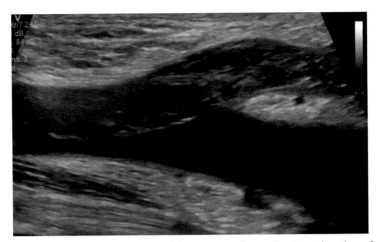

Figure 24.6 Longitudinal ultrasound image of the common femoral vein and saphenofemoral junction demonstrating thrombus extending from the great saphenous vein into the common femoral vein several days postablation. *(Image courtesy of Julie A. Cardoso, RVT, RPhS, RDCS.)*

1 month later, ultrasound is often used to confirm the absence of deep venous thrombosis, to evaluate for the presence of endovenous heat-induced thrombosis (EHIT), and demonstrate appropriate closure of the treated vein (absence of flow should be observed as well as thickened vein walls) (Fig. 24.6). EHIT is classified and indications for treatment can be determined using published criteria.[16] Complications from the procedure, such as hematoma, seroma, arterial-venous fistula, or pseudoaneurysm can also be identified, if present, at this point. After several months to a year, thermally or chemically treated vein segments should be difficult to identify. Cyanoacrylate will be visible in treated veins indefinitely. Ultrasound is also used following deep venous interventions to assess for stent and nonstented treated vein patency although accepted techniques and criteria for this assessment are not widely adopted. Color and pulsed wave Doppler are coupled with grayscale and compression (of nonstented accessible veins) for these assessments.[17,18]

References

1. Ozsvath K, Hager E, Harlander-Locke M, Masuda E, Elias S, Dillavou ED. Current techniques to treat pathologic perforator veins. *J Vasc Surg Venous Lymphat Disord*. 2017;5:293−296.
2. Zygmunt Jr JA. In: Zygmunt JA, ed. *Venous Ultrasound*. 2nd ed. CRC Press; 2020.
3. Mordhorst A, Yang GK, Chen JC, Lee S, Gagnon J. Ultrasound-guided cyanoacrylate injection for the treatment of incompetent perforator veins. *Phlebology*. 2021;36, 2683555211015564.
4. Puggioni A, Kalra M, Carmo M, Mozes G, Gloviczki P. Endovenous laser therapy and radiofrequency ablation of the great saphenous vein: analysis of early efficacy and complications. *J Vasc Surg*. 2005;42: 488−493.
5. Tzilinis A, Salles-Cunha SX, Dosick SM, Gale SS, Seiwert AJ, Comerota AJ. Chronic venous insufficiency due to great saphenous vein incompetence treated with radiofrequency ablation: an effective and safe procedure in the elderly. *Vasc Endovasc Surg*. 2005;39:341−345.

6. Proebstle TM, Vago B, Alm J, Gockeritz O, Lebard C, Pichot O. Treatment of the incompetent great saphenous vein by endovenous radiofrequency powered segmental thermal ablation: first clinical experience. *J Vasc Surg*. 2008;47:151–156.
7. Almeida JI, Kaufman J, Gockeritz O, et al. Radiofrequency endovenous ClosureFAST versus laser ablation for the treatment of great saphenous reflux: a multicenter, single-blinded, randomized study (RECOVERY study). *J Vasc Interv Radiol*. 2009;20:752–759.
8. Gale SS, Lee JN, Walsh ME, Wojnarowski DL, Comerota AJ. A randomized, controlled trial of endovenous thermal ablation using the 810-nm wavelength laser and the ClosurePLUS radiofrequency ablation methods for superficial venous insufficiency of the great saphenous vein. *J Vasc Surg*. 2010;52:645–650.
9. Hill D, Hamilton R, Fung T. Assessment of techniques to reduce sclerosant foam migration during ultrasound-guided sclerotherapy of the great saphenous vein. *J Vasc Surg*. 2008;48:934–939.
10. Markovic J, Shortell C. Varicose vein surgery. *Decker Med Surg*. 2014;24.
11. Conroy PH, O'Rourke J. Tumescent anaesthesia. *Surgeon*. 2013;11:210–221.
12. Khilnani NM, Min RJ. Imaging of venous insufficiency. *Semin Intervent Radiol*. 2005;22:178–184.
13. Salles-Cunha SX, Comerota AJ, Tzilinis A, et al. Ultrasound findings after radiofrequency ablation of the great saphenous vein: descriptive analysis. *J Vasc Surg*. 2004;40:1166–1173.
14. Sufian S, Arnez A, Labropoulos N, et al. Radiofrequency ablation of the great saphenous vein, comparing one versus two treatment cycles for the proximal vein segment. *Phlebology*. 2015;30:724–728.
15. Darvall KA, Bate GR, Adam DJ, Silverman SH, Bradbury AW. Duplex ultrasound outcomes following ultrasound-guided foam sclerotherapy of symptomatic primary great saphenous varicose veins. *Eur J Vasc Endovasc Surg*. 2010;40:534–539.
16. Kabnick LS, Sadek M, Bjarnason H, et al. Classification and treatment of endothermal heat-induced thrombosis: recommendations from the American venous forum and the society for vascular surgery. *J Vasc Surg Venous Lymphat Disord*. 2021;9:6–22.
17. Sebastian T, Barco S, Engelberger RP, et al. Duplex ultrasound investigation for the detection of obstructed iliocaval venous stents. *Eur J Vasc Endovasc Surg*. 2020;60:443–450.
18. Avgerinos ED, Labropoulos N. Duplex criteria for iliocaval stent obstruction: sounds of a cry for validated data. *Eur J Vasc Endovasc Surg*. 2020;60:451.

CHAPTER 25

Iliac vein stenting in chronic venous leg ulcers

Taimur Saleem and Seshadri Raju
The Rane Center for Venous and Lymphatic Diseases, Jackson, MS, United States

Introduction

Chronic venous leg ulcers (VLUs) have been known since antiquity. Some form of "bandaging" as a treatment option seems to have existed for these VLUs since ancient times. However, a pathological connection between VLUs and varicose veins was not discovered until the late 18th century. Saphenous vein stripping became the standard practice for VLUs and was influenced by the teachings of Trendelenberg. The Trendelenburg operation is the juxta-femoral flush ligation of the great saphenous vein (GSV) at its confluence with the common femoral vein.[1]

Involvement of deep venous disease, particularly deep reflux in ulcer causation, became known in the early 19th century. Dr. Gunnar Bauer, a Swedish surgeon, investigated the role of valve reflux in VLUs using venography.[2] Dr. Robert Linton Kistner described the technique of descending venography and introduced a grading system to assess reflux severity.[3,4] Kistner's "axial" reflux grading has since been extended to duplex technology. Duplex was noted to be more accurate in quantifying the degree and distribution of venous reflux when compared to descending phlebography.[5,6] Kistner also performed the first recorded valvuloplasty to correct femoral valve reflux.[7] These pioneering efforts had a significant influence in reviving interest in venous disease over the past several decades.[8]

A classification of chronic venous disease, Clinical, Etiological, Anatomical, and Pathophysiological (CEAP), was developed in 1993, adding to an adjunctive venous clinical severity scoring (VCSS) system for standardized clinical assessment.[9] CEAP was most recently revised and updated in 2020 (Table 25.1).[9] Stent technology that was being used in the arterial system was extended to the treatment of iliac vein stenosis (IVS). Surprisingly, high long-term patency in the thrombus-prone low-velocity venous system with good clinical improvement in a large cohort of patients was described by Neglen et al. in 2007.[10] Complete healing of chronic VLU in >50% of limbs after iliac vein stent placement was noted as well. This work also highlighted the importance and high prevalence of IVS and obstructive pathology in venous ulceration; previous research had exclusively focused on the role of reflux.

Table 25.1 Revised CEAP classification 2020.

CEAP[a] class	Description
C0	No visible or palpable signs of venous disease
C1	Telangiectasias or reticular veins
C2	Varicose veins
C2r	Recurrent varicose veins
C3	Edema
C4	Changes in skin and subcutaneous tissues secondary to chronic venous disease
C4a	Pigmentation or eczema
C4b	Lipodermatosclerosis or *atrophie blanche*
C4c	Corona phlebectatica
C5	Healed venous ulcer
C6	Active venous ulcer
C6r	Recurrent active venous ulcer

[a]CEAP—C (clinical manifestations), E (etiology), A (anatomic distribution), P (pathophysiology).
Adapted from Lurie F, Passman M, Mesiner M, et al. The 2020 update of the CEAP classification and reporting standards. *J Vasc Surg Venous Lymphat Disord.* 2020;8:342–352.

Iliac vein stent treatment has now largely replaced open surgery for the correction of deep vein obstruction with or without reflux. The treatment is minimally invasive with low morbidity and mortality.[11] Surprisingly, correction of obstruction alone in patients with combined obstruction and reflux appears to yield good results despite the presence of residual uncorrected deep or superficial reflux.[12,13]

Pathophysiology of venous ulcers

VLUs occur in the setting of ambulatory venous hypertension (AVH) secondary to IVS or valvular incompetence.[14] Chronic iliac venous obstruction (CIVO) can be due to postthrombotic syndrome, nonthrombotic iliac vein lesions (NIVL), or a combination of both (mixed type of lesions). Injury to the endothelium leading to alterations in the glycocalyx matrix, shear stress, and a complex interaction of various adhesion molecules is believed to be the first event in the formation of a VLU.[15] Over time, the accumulation of various inflammatory changes leads to changes in endothelial cell permeability, leukocyte proliferation with resultant increases in levels of cytokines, matrix metalloproteinases, various reactive oxygen species, and migration of many other different cell types. Deposition of hemosiderin in the subcutaneous tissues leads to hyperpigmentation around the ulcerated area.[15]

Clinical assessment

A complete history with the system-based review is necessary even though the presenting complaint is focused on an ulcer in the leg. A history of prior deep venous thrombosis (DVT) or diseases, trauma, surgeries, and hospitalizations that are known to be associated with DVT is important. A renal history with serum creatinine is relevant if contrast imaging is to be utilized in the work-up of a leg ulcer. Generally, such imaging is omitted if the serum creatinine is >1.5 mg/dL. Iliac vein stenting can be performed under local anesthesia but general anesthesia is preferred if the cardiopulmonary risk is not prohibitive. Balloon dilatation especially in recanalization procedures can be extremely uncomfortable even under conscious sedation. Also, some recanalization procedures can take a much longer time than routine stenting cases.

Chronic VLUs have a such characteristic appearance that a tentative diagnosis can often be made on the basis of physical examination alone (Fig. 25.1).[16] They are superficial, that is, seldom penetrate the deep fascia, and typically occur around the medial

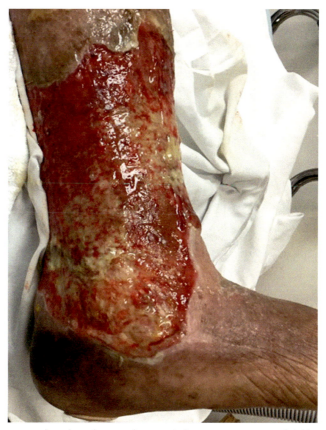

Fig. 25.1 Chronic venous leg ulcers have a characteristic appearance: superficial location in the lower third of the leg with granulation tissue and involving the medial malleolus. These ulcers are associated with frequent drainage. Hyperpigmentation may be seen around the ulcer edges.

malleolus. Lateral malleolar ulcers may occur if reflux involves the small saphenous vein. With rare exceptions, almost all VLUs are confined to the lower third of the leg—where the venous pressure in the erect position is the highest. Other signs of chronic venous insufficiency such as hyperpigmentation (Fig. 25.2), lipodermatoclerosis or *atrophie blanchie* (poorly vascularized scar) are frequently present. These signs are part of the CEAP classification detailed in Table 25.1. Chronic granulation tissue may be present in the ulcer bed (Fig. 25.1). Indolent infection is common in VLUs with often copious drainage. Cellulitis can occur episodically with the spread of erythema, local swelling, and pain beyond the ulcer. Oral antibiotics are usually sufficient to treat localized cellulitis but more severe cases warrant admission to the hospital with intravenous antibiotics. Associated septicemia is generally uncommon except in the elderly.[17,18]

CEAP classification, VCSS, and disability scoring are essential elements of evaluating VLUs. Pain and swelling are graded per subjective reporting by the patient in the VCSS system. We use a visual analog scale for pain assessment and an expanded grading of swelling based on physical examination to supplement VCSS. Swelling is graded as follows: Grade 0: none, Grade 1: pitting, Grade 2: ankle edema, Grade 3: gross involving

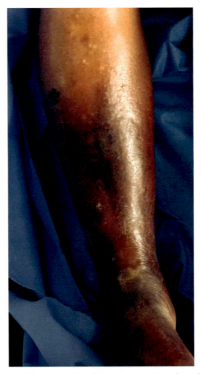

Fig. 25.2 Hyperpigmentation and lipodermatosclerosis associated with chronic venous insufficiency seen in the lower third of the leg. These skin changes represent CEAP Class C4.

the calf, Grade 4: involving the thigh. Letters A—D are appended if the swelling first appears late in the day, at noon, morning, or remains constant, respectively. The ChronIc Venous Insufficiency quality-of-life Questionnaire is an easily useable quality-of-life (QoL) instrument for outcome evaluation[19] (see Chapter 11).

Chronic VLUs need to be differentiated from arterial ulcers. Arterial ulcers tend to be more painful and typically penetrate the deep fascia exposing tendons and deep tissue; granulation is sparse or nonexistent. Strips of black gangrenous tissue may be visible in the ulcer. Other signs of ischemia such as absent pulses and cool extremities are also present. Diabetic ulcers, rheumatoid ulcers, and ulcers associated with vasculitis of immunological diseases such as scleroderma should be ruled out as well when appropriate. Marjolin ulcers may develop from long-standing venous ulcers.[20] They have a characteristic appearance with overgrowth of tissue around the edges. Kaposi's sarcoma has to be considered in HIV infections. A punch biopsy may be required to identify these and a number of dermatological ulcers that may mimic VLUs. An ankle-brachial index or toe pressure should be a routine part of the examination in patients in whom pulses are not readily palpable, and there are reasons in the history to suspect an arterial component, for example, arterial disease in other vascular beds such as coronary, carotid, or mesenteric vasculature. If a significant arterial component is present, its correction should take precedence over any venous intervention. Venous ulcers with an arterial component most often heal when the latter is corrected.[21,22]

Etiology of venous ulcers

Saphenous vein reflux and IVS are among the more common correctible pathologies found to cause VLUs. Both are currently amenable to the endovenous correction that is minimally invasive. Great saphenous vein reflux is well known to cause AVH. It has been recently shown that venous pressures monitored thorough the dorsal foot vein may not accurately reflect AVH in the GSV or the deep veins.[23] Measurement of column interruption duration (CID) in the GSV and tibial veins is more accurate.

Iliac vein stenosis is a surprisingly common lesion in the general population.[24,25] It remains silent in most and is an incidental finding in about 30% of autopsies.[26] Recent imaging studies suggest that as many as 75% of the population may have some element of the obstructive lesion ranging from mild to severe.[25] The lesion does produce symptoms in some patients, often precipitated by trauma (accidental or iatrogenic, commonly joint replacement and orthopedic interventions), infection/cellulitis, DVT, or onset of valve reflux with age.[27] This is typical behavior of a "permissive" lesion common in human pathologies (e.g., carotid stenosis and transient ischemic attack; ureteral reflux and pyelonephritis; gastric reflux and Barrett's esophagus). A well-known example of this phenomenon is patent foramen ovale (PFO), which is widely prevalent in the general population (25%) but symptomatic in only a small percentage (<3%). However, a

PFO is detected in >95% of the symptomatic subset with embolic stroke. The high prevalence of IVS in limbs with venous ulceration is likely due to a similar explanation.[27]

The pathology of IVS is due to congenital (rare) or more commonly primary or postthrombotic causes. The primary variety is eponymously known as May—Thurner syndrome or Cockett syndrome and generically as iliac vein compression syndrome. Traumatic pulsations of the closely overlying artery are thought to produce wall fibrosis, spurs, webs, trabeculations, and other forms of luminal compromise in the iliac vein. Therefore, the lesion can be complex morphologically and not merely attributable to "compression" by the artery.[27] If thrombosis occurs, its resolution is poor at these stenotic points. Postthrombotic lesions typically extend to involve the entire iliac vein segment(s) and may also involve additional venous segments proximally or distally. Rokitansky lesions are unique to the iliofemoral veins whereby there is a diffuse narrowing of the venous segment without a focal stenosis. These lesions are best identified by intravascular ultrasound (IVUS) and can be missed easily by venography or by using the adjacent venous segment as a comparator.

Investigations

Diagnostic investigations are directed to answer two primary questions:
1. Is there significant saphenous or deep reflux?
2. Is there a significant CIVO?

The results of these investigations will determine if GSV ablation or iliac vein stenting or both are likely to benefit the patient in the appropriate clinical context.

Saphenous and deep vein reflux

Current methodology to detect significant saphenous or deep reflux is to monitor, using ultrasound, reflux duration of the gateway valves (saphenofemoral, femoral, and popliteal). Guidelines suggest that a reflux duration >500 and 1000 ms at the saphenofemoral valve and the deep valves, respectively, is significant.[28] A study of valve action and venous pressure in normal subjects suggests that these three gateway valves open up within a few seconds after calf pump action allowing centripetal flow.[23] However, the common assumption that these gateway valves contribute to column segmentation and that reflux at these sites shorten column segmentation duration is incorrect. It is clear that column segmentation after calf pump contraction occurs at the level of tibial and saphenous valves in the lower third of the leg, not at the level of the gateway valves. Both great saphenous and deep reflux is better assessed by duplex measurement of CID (Figs. 25.3 and 25.4) in the saphenous and tibial veins in the lower third of the leg. This is a noninvasive measurement with a duplex probe after calf ejection with an automated cuff.[23]

Another way to gauge the presence and extent of saphenous reflux is by measuring the caliber of the vein in the proximal thigh, 15 cm below the groin (or 10 cm above

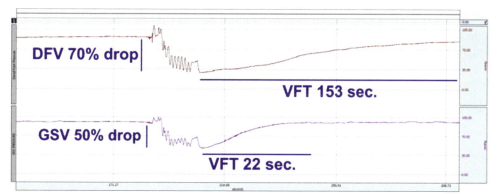

Fig. 25.3 Simultaneously recorded pressure curves in the dorsal foot vein (DFV) (top curve) and the great saphenous vein (bottom curve) in a healthy volunteer. There is a difference in percent decrease and venous refill time (VFT) between the two veins. *(By permission, JVS-VLD Raju S, Walker W, May C. Measurement of ambulatory venous pressure and column interruption duration in normal volunteers. J Vasc Surg Venous Lymphat Disord. 2020;8:127–136.)*

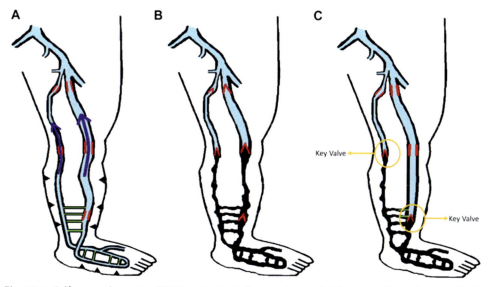

Fig. 25.4 Calf pump dynamics. (A) Blood in both deep and superficial systems flows due to calf ejection. Resistance to flow is higher in the great saphenous vein (GSV) compared to the deep axial veins, resulting is correspondingly lesser flow in the GSV. (B) At the completion of ejection process, valves in both outflow channels above the knee close transiently. (C) The saphenofemoral valve reopens soon after, allowing upward flow in the thigh portion of the GSV. The key valve in the GSV at or near the knee level remains closed with collapse of the saphenous segment below. In the deep system, the femoral valve reopens first, followed by popliteal valve a few seconds later, allowing flow in the upper femoral-popliteal axis. The key valve or valves in the lower tibial vein remain closed with collapse of the segments below. The dorsal foot vein (DFV) also remains collapsed *(By permission, JVS-VLD Raju S, Walker W, May C. Measurement of ambulatory venous pressure and column interruption duration in normal volunteers. J Vasc Surg Venous Lymphat Disord. 2020;8:127–136.)*

knee joint line).[29] It can be shown that a GSV <5.5 mm is anatomically incapable of refluxing >30 cc per calf pump action (Fig. 25.5).[30] The 30 cc threshold represents roughly half of ejected volume by calf pump action and is likely to shorten column segmentation duration. A more precise way to determine GSV reflux is to actually measure it from time-averaged reflux velocity, its duration, and the caliber of the GSV. About half of the GSV >5.5 mm in size reflux less than 30 cc per calf pump cycle, presumably because the size of the run-off perforators into the calf pump is restrictive. Navarro et al. also found the threshold GSV diameter of 5.5 mm was associated with the absence of reflux per air-plethysmography.[31]

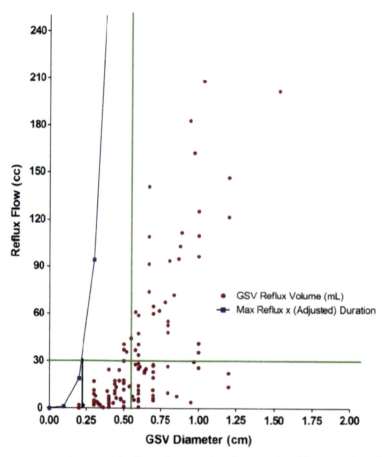

Fig. 25.5 Relationship of measured reflux volume and saphenous size. All except three limbs with a saphenous diameter <0.55 cm had reflux volumes ≤30 mL. Conversely, about 50% of limbs with saphenous diameter ≥5.5 mm had reflux volumes <30 cc. *(By permission, JVS-VLD Raju S, Ward M Jr, Jones TL. Quantifying saphenous reflux. J Vasc Surg Venous Lymphat Disord. 2015;3:8–17.)*

Diagnosis of iliac vein stenosis
Transfemoral venography and IVUS

Transfemoral venography has traditionally been the mainstay of iliac vein assessment (Fig. 25.6). It is particularly useful in postthrombotic disease whereby a panoramic picture of the lesion(s), collateral pathways, and inflow and outflow adequacy can all be obtained. Even decades ago, Negus who used the technique extensively noted that the diagnostic sensitivity was only about 50%.[24] This was mostly related to obscuration of stenotic lesions due to inadequate contrast volume or concentration. This fundamental problem has not been resolved despite improvements in instrumentation, technology, and technique.[32–34] In a blinded comparison of venography and IVUS in 155 IVS, venography failed to identify lesion existence in 19% of limbs. The median maximal area stenosis was significantly higher with IVUS than with venography (69% vs. 52%; $P < 0.0001$) (Fig. 25.7).[34] Furthermore, venography missed the location of maximal

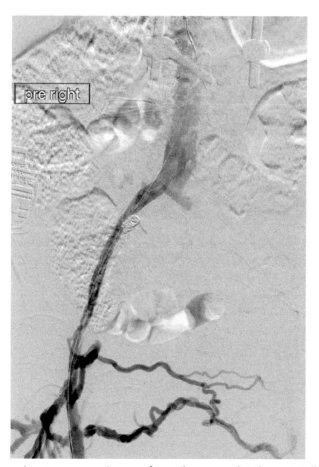

Fig. 25.6 Preintervention anteroposterior transfemoral venography showing collaterals and postthrombotic lesion in the iliofemoral segment in a patient with a prior history of deep venous thrombosis (DVTs) and pulmonary embolism (PE). Also noted in the image is spinal hardware and an inferior vena cava (IVC) filter.

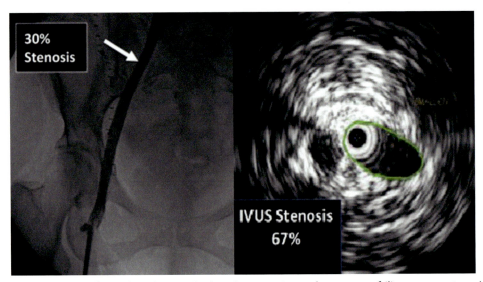

Fig. 25.7 Venography underestimates the location, severity, and presence of iliac venous stenosis compared to intravascular ultrasound (IVUS). In the example shown, common iliac vein stenosis was estimated at 30% by venography and measured at 67% by intravascular ultrasound. *(By permission, JVS-VLD Montminy ML, Thomasson JD, Tanaka GJ, Lamanilao LM, Crim W, Raju S. A comparison between intravascular ultrasound and venography in identifying key parameters essential for iliac vein stenting.* J Vasc Surg Venous Lymphat Disord. *2019;7:801–807.)*

stenosis in more than two-thirds of limbs. The iliac-caval confluence location on venography correlated with IVUS findings in only 15% of patients. In 74%, it was located higher by as much as one vertebral body with IVUS than with venography. The distal landing zone defined with IVUS was lower than with venography in 64% of limbs. Intravascular ultrasound is a better procedural guide than venography for stent placement and a better predictor of clinical outcomes.[32–34] In addition, there are other advantages with IVUS such as reduced radiation exposure, avoidance of contrast use, and related complications. Intravascular ultrasound planimetry also provides a more accurate stenosis measurement than multiplanar venography (Table 25.2).[35] In one study, IVUS had a 100% diagnostic yield in the identification of stenosis in at least one of three venous segments (common femoral vein, external iliac vein, and common iliac vein) in patients with renal failure and/or contrast allergy (Table 25.3).[36]

Two-segment CT venography

Some of the deficiencies of venography can be avoided by using CT venography. Routine contrast administration through the arm vein is adequate. Digital measurement of stenotic lesions appears to identify lesions with low false positives and negatives when compared to IVUS. This high accuracy is partly because the measurement of common

Table 25.2 Comparison of intravascular ultrasound and multiplanar venography.

IVUS[a]		MPV[c]	
Advantages	Disadvantages	Advantages	Disadvantages
No contrast needed	Invasive nature	Provides a roadmap	Less sensitive than IVUS
May be done in pregnancy with minimal radiation exposure	May not be able to clearly differentiate spurs from venous wall in severe compression	Helpful in chronic total venous occlusions	Uses contrast
Better intraluminal characterization of lesions (frozen valves, trabeculations, spurs)	Incomplete measurements due to missing borders—lack of centering mechanism	May provide hemodynamic information with presence of collaterals	Invasive nature
More sensitive than venography	Can miss lesions near confluences	More readily available and accessible than IVUS	Radiation exposure especially with multiple projections
Real-time continuous image	Less widely available than venography	Helpful in delineating pelvic anatomy and anatomic variants	May miss highly eccentric lesions
Can assess immediate technical outcome of intervention	Limitations with assessment of ipsilateral infrainguinal segments	Can assess immediate technical outcome of intervention	Lacks an internal scale
Safe in patients with renal failure or contrast allergy	Must pass through the obstruction to provide visualization	Can give some quantification of ISR[b]	Relative contraindication in patients with renal insufficiency or contrast allergy
Can assess complete or incomplete stent apposition to vessel wall, ISR[b]	Acoustic shadowing may be caused by calcification, stent struts, IVC filters	Can be done from a smaller sheath than IVUS	Lesions can be masked if contrast concentration or volume inadequate

[a]IVUS—intravascular ultrasound.
[b]ISR—instent restenosis.
[c]MPV—multiplanar venography.
Adapted from Saleem T, Raju S. Comparison of intravascular ultrasound and multidimensional contrast imaging modalities for characterization of chronic occlusive iliofemoral venous disease: a systematic review. *J Vasc Surg Venous Lymphat Disord.* 2021:S2213—S2333X(21)00213-4.

Table 25.3 Preintervention and postintervention areas noted by intravascular ultrasound (IVUS).

Vein segment	Preintervention area, mm^2	Postintervention area, mm^2	P-value
Common femoral vein	113 (40–197)	161 (81–234)	**<0.001**
External iliac vein	109 (25–206)	176 (91–259)	**<0.001**
Common iliac vein	140 (22–214)	217 (116–261)	**<0.001**

Significant P-value in bold text.
Adapted from Saleem T, Knight A, Raju S. Diagnostic yield of intravascular ultrasound in patients with clinical signs and symptoms of lower extremity venous disease. J Vasc Surg Venous Lymphat Disord. 2020;8:634–639.

iliac and external iliac veins yields two separate data points for the determination of stenosis. 84% of common iliac and 78% of external iliac veins were stenotic but one of the two was stenotic in 90% of the limbs, thus increasing the accuracy of the two-segment method (Table 25.4).[37]

Treatment of chronic venous ulceration

Trial of compression

Compression remains the first line of treatment for VLU before any type of intervention. It is a noninvasive treatment modality. Some insurance plans may cover the cost of compression for patients with venous ulcers. Latest Cochrane database review[38] shows the median time to ulcer healing following the application of compression with a four-layer bandage and short stretch bandage is 90 and 99 days, respectively (hazard ratio 1.31; 95% confidence interval 1.09–1.58). Multicomponent compression and high-compression stockings are more effective than single-component compression or short-stretch bandages in ulcer healing.[38] After the ulcer heals, patients with healed ulcers should be prescribed class II medical grade stockings with ≥30 mm Hg pressure.

If the ulcer fails to heal after an appropriate trial of compression over 3 months, procedural intervention should be considered. If saphenous reflux is present, it should be

Table 25.4 Superior diagnostic accuracy of two-segment method of computed tomography venography (CTV) assessment of iliac vein stenosis.

Stenosis threshold	No.	Sensitivity (%)	Specificity (%)	PPV (%)	NPV (%)	Accuracy (%)
CIV area <200 mm^2	83	83	62	98	40	80
EIV area <150 mm^2	91	79	70	90	48	77
CIV area <200 mm^2 OR EIV area <150 mm^2	79	97	38	93	60	91

PPV—positive predictive value, NPV—negative predictive value, CIV—common iliac vein, EIV—external iliac vein.
Adapted from Raju S, Walker W, Noel C, Kuykendall R, Jayaraj A. The two segment caliber method of diagnosing iliac vein stenosis on routine computed tomography with contrast enhancement. J Vasc Surg Venous Lymphat Disord. 2020;8:970–977.

carried out without undue delay. An avoidable confusion on this topic was generated by a design flaw in the famous ESCHAR trial.[39,40] As one of the earliest randomized trials to address chronic venous disease, the ESCHAR trial recommendations carried great influence. The authors compared the healing of VLU after saphenous interruption: both the test and control group had base line compression (which is in fact, a form of medical ablation of venous reflux). Therefore, it was not at all surprising that there was no difference in the ulcer healing rate in the two groups. A second control group without compression and treated with saphenous ablation alone would have likely resulted in rapid healing similar to the test group. The findings were misinterpreted to mean that saphenous vein ablation did not improve acute ulcer healing but rather reduced long-term recurrence only which is incongruous. Some practitioners have misinterpreted the recommendations to delay saphenous vein ablation while persisting with compression for more than 6–12 months or even years. Thus, long-term use of compression alone without additional intervention in the face of ulcer incidence or recurrences does not seem justified after a trial of conservative therapy beyond 3 months.[41]

Poor stocking compliance

Detailed history of stocking use was recorded in 3144 new patients seen in our clinic over an 8-year period.[42] As a tertiary referral venous practice, patients had been in the care of one or more physicians before referral. Only 21% reported daily usage, 16% intermittent use and 63% of patients did not use stockings at all (Fig. 25.8).[42] 25% had not been

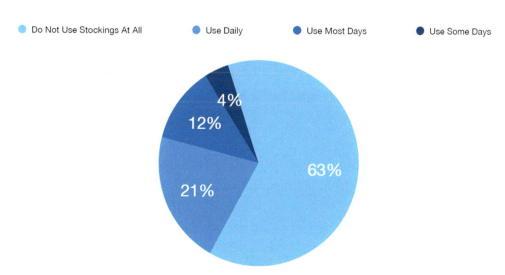

Fig. 25.8 Compliance with stockings among 3144 patients with chronic venous disease. *(By permission—Raju S, Hollis K, Neglen P. Use of compression stockings in chronic venous disease: patient compliance and efficacy. Ann Vasc Surg. 2007;21:790–795.)*

prescribed stockings, probably justified by local condition of the limb, comorbidities such as advanced congestive heart failure or arthritis, coexisting severe arterial disease or overall frailty. Stocking inefficacy and persistent symptoms were the cause of nonuse in about half the nonusers (Table 25.5).[42] A large fraction (30%) could not state a definitive reason or offered vague reasons for noncompliance. These noncompliance statistics are strikingly similar across the globe in cold as well as warm climates.[43,44] Compliance with stocking use was similar across all age groups (Fig. 25.9).[42]

"Patient education" and intense monitoring by health personnel have not improved compliance suggesting that the reasons maybe systemic and poorly understood.[45,46] It is important not to penalize such patients by withholding one-time intervention that may cure the ulcer.

Judicious wound care

Wound care should be incorporated judiciously into the management of VLUs. It is a common practice to frequently debride the wound while waiting for it to heal with compression. However, this should be avoided. Reepithelialization of venous ulcers occurs from the depth of hair follicles. Misguided frequent debridement only delays healing by destroying regenerative follicles forcing healing to occur more slowly from the periphery (Fig. 25.10).

Saphenous vein ablation versus iliac vein stenting in patients with chronic venous leg ulcers: determining the ideal procedure sequence

This is a very important practical question and the following algorithm is proposed in this regard:

Table 25.5 Reasons for nonuse of compression stockings.

Reason	%
Unable to state a specific reason	30
Not recommended by doctor	25
Ineffective, did not help	15
Too restrictive, poor fit	13
Too hot	7
Soreness	2
Needs application assistance	2
Cosmetic, poor appearance	2
Aggravating, itching, dermatitis	2
Other reasons	2

Adapted from Raju S, Hollis K, Neglen P. Use of compression stockings in chronic venous disease: patient compliance and efficacy. *Ann Vasc Surg.* 2007;21:790–795.

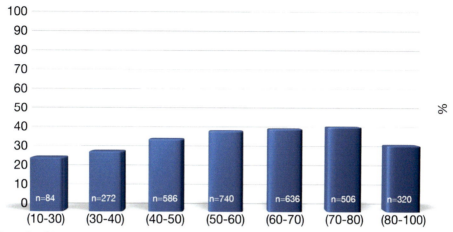

Fig. 25.9 Stocking use among various decile age groups *(By permission—Raju S, Hollis K, Neglen P. Use of compression stockings in chronic venous disease: patient compliance and efficacy. Ann Vasc Surg. 2007;21:790–795.)*

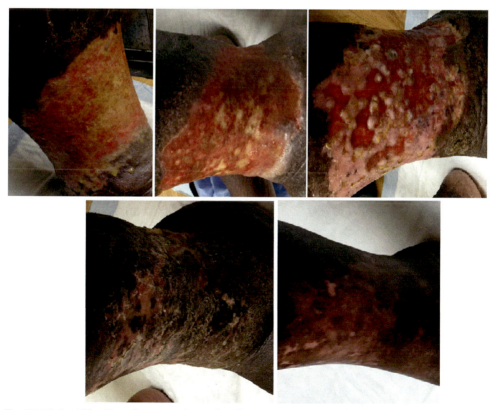

Fig. 25.10 Rapid healing of venous ulcers after iliac venous stenting. Islands of epithelial cells appear in the center of the wound from follicular remnants; this process may be significantly interrupted by repeated debridement.

1. When significant GSV reflux is present along with significant iliofemoral venous stenosis, it is our preference to perform the ablation first as it is more of a minimalist procedure than stenting which potentially carries greater morbidity over a longer duration.
2. If GSV ablation fails to heal the ulcer, an iliac vein stenting procedure is performed next. Patients should be informed beforehand of the possibility of sequential iliac venous stenting in case of failure of GSV ablation to heal the ulcer. Other authors have advocated a similar approach with good results.[47,48]
3. If the GSV is relatively small in caliber or the reflux duration is relatively trivial, ablation is easily combined with stent placement in a single setting.[49] Yang et al. recently reported outcomes of endovenous laser ablation with additional iliac vein stenting in patients with nonthrombotic iliac disease and venous ulceration. Addition of stenting to endovenous laser ablation improved ulcer healing from 66% to 87% ($P = 0.001$).[50] The DEVELOP trial is a prospective, multicenter, randomized controlled, feasibility trial, which is investigating whether stenting with superficial venous ablation in patients with active VLU will produce better results than standard therapy alone (graduated compression dressings).[51]
4. Some authors have recently suggested that iliac stenting should take precedence over saphenous ablation based on Guyton's theory of venous return control.[52,53] We believe that in cases of very large venous ulcers above the ankle/medial malleolar area, this approach may be justified. However, data to answer this important question is forthcoming.
5. Recurrence of venous stasis ulcers after iliac stenting was rare even in the presence of uncorrected reflux. Residual deep reflux rarely requires treatment after stenting.[12,54] After stenting, deterioration of axial reflux has not been observed.[55]

Other procedures

The need for leg amputation for VLU is extremely rare; it is used only if the ulcer fails to heal with the use of several backup procedures and the patient has recurrent bouts of deep tissue/bone infection or recalcitrant pain. These backup procedures include vascularized musculocutaneous flaps; thermal or sclerotherapy ablation of perforators located in the ulcer bed and the use of biogenic grafts.[56–59] If the ulcer does not heal despite venous stenting with or without superficial ablation, specialized techniques such as valve reconstruction or neovalve may be considered in carefully selected patients.[13]

Iliac vein stenting and chronic venous ulcers

Iliac vein stenting improves the rate of healing for VLUs in appropriately selected patients. Iliac vein stenting reduces venous hypertension and decompresses the venous system in the leg by providing blood a path of least resistance. Mostly collaterals will disappear once endovenous revascularization has been performed. In one study with 80 patients, deep venous intervention in the form of hyperdilation or endovenous stent placement was associated with symptomatic improvement and reduction of compartment pressure of the extremity undergoing intervention.[60]

In patients with iliac vein compression, stenting significantly shortened ulcer healing compared with those without stent treatment ($P < 0.001$).[61] Iliac vein stenting in chronic VLU can be performed via the popliteal, femoral, or internal jugular venous routes. Ye et al. described percutaneous transpopliteal iliac vein stent placement in 110 patients with postthrombotic chronic total venous occlusion (CTVO) with a 95% technical success rate. Ulcer healing was seen in 78% of patients (36/46).[62] In various other series, after transfemoral stenting, ulcer healing has been reported in 50%–100% of patients.[55,63–68]

In a series with 44 limbs that underwent venous stenting in the setting of venous ulceration, sustained ulcer healing was achieved in 60% of limbs. Additionally, 20% of ulcers among the remaining limbs had a reduction in the size of the ulcer. Ulcer recurrence was noted in 13% of limbs.[69]

In a series of 192 consecutive limbs with VLU, 158 (82.3%) were treated either with iliac stenting alone or with stenting and endovenous laser ablation. 38% of patients in this sample were not using stockings. Ulcer healing was defined as 100% epithelialization of the ulcer by clinical examination and visual inspection. After the intervention, 81% of small ulcers (<500 mm^2 or ≤ 1 inch in diameter) had completely healed at 14 weeks while only 15% of the large ulcers (≥ 500 mm^2) had healed by this time ($P < 0.001$). Stocking use and presence of uncorrected deep reflux were not found to have any effect on healing time. Long-term cumulative healing at 5 years was 75%; suggesting an overall low risk of recurrence.[13] While iliac vein stenting appears to help in the healing of all ulcers, this effect is more independent and pronounced for smaller ulcers (area <500 mm^2). For larger ulcers, adjuncts such as skin grafting or skin substitute biologic grafts may be considered in the healing process after endovenous stenting is performed to ensure complete healing of these larger ulcers.[13] In another study of 545 limbs, ulcers were present in 67 limbs (12.3%). 73% of the ulcers had healed over a median follow-up period of 26 months. Composite Z-stent and Wallstents were used for iliocaval stenting in this study.[70]

Sizing of stents

We have previously reported the optimal stent sizes in the common iliac, external iliac, and common femoral vein segments (Table 25.6).[36] These diameters and areas correspond to the minimal sizes needed for optimal flow calculated from flow equations such as Poiseuille's equation and Young's scaling ratios and IVUS observations of normal

Table 25.6 Optimal poststent diameter and areas in venous stenting by intravascular ultrasound.

Venous segment	Diameter (mm)	Area (mm^2)
Common femoral vein	12	125
External iliac vein	14	150
Common iliac vein	16	200
Inferior vena cava	18–24	300–400

Adapted from Saleem T, Knight A, Raju S. Diagnostic yield of intravascular ultrasound in patients with clinical signs and symptoms of lower extremity venous disease. *J Vasc Surg Venous Lymphat Disord.* 2020;8:634–639.

iliac vein segments. It is recommended to not use the adjacent "normal" segment as a comparator. Rather the IVUS measurements are indexed to a predetermined caliber for the segment based on measurements in healthy subjects without signs and symptoms of venous disease. The presence of long diffuse stenosis (Rokitansky stenosis) in the ilio-femoral venous segments poses a unique challenge during iliofemoral stenting and can lead to undertreatment of lesions if simply the adjacent venous segment is used as a comparator. A particular vein is considered to be stenosed if the size is smaller than the optimal sizes mentioned in Table 25.6.[36] The goal is to restore the caliber of the diseased venous segments to a level where relatively normal flow and pressure can be expected based on hemodynamic equations and observations made in healthy subjects. The grading of venous stenosis is different from arterial stenosis; venous pressure being a more important component than flow.[35,36,60,71–73]

Stenting technique for routine cases

Femoral vein access is typically and consistently obtained at the mid-thigh or lower thigh level followed by the placement of an 11 French sheath. Access at this level allows adequate visualization of the expected stent inflow (profunda-femoral vein confluence). The 11 French sheath is selected because it allows easy introduction and maneuverability of the large IVUS catheter (8.2 French, Visions EP 0.035; Volcano Corp, San Diego, CA) as well as easy delivery of balloons, catheters, and stents used in venous revascularization.[34,36]

The usual sequence is to perform an IVUS examination before balloon dilation and stent placement followed by a completion IVUS examination. Postdilation is also recommended. The smallest area (highest stenosis) in the common femoral, external iliac and common iliac veins is captured and stored with the IVUS planimetry software. For the performance of stenting, the ilio-caval confluence must be identified correctly for two reasons: (a) to prevent the undertreatment of proximal lesions and (b) avoid jailing of contralateral iliac vein (Fig. 25.11). Also, the distal landing zone should be carefully selected to avoid missing lesions and to avoid jailing of the profunda vein whenever possible.[34,36]

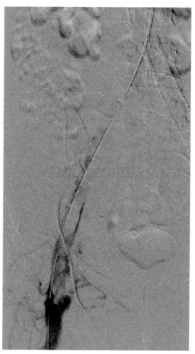

Fig. 25.11 Wallstent in left iliac vein leading to jailing of contralateral iliac venous orifice and resultant contralateral iliofemoral deep venous thrombosis (DVTs).

Intravascular ultrasound is used to confirm the position of the wire in the axial vessel. It also delineates the ilio-caval confluence and the ideal distal landing zone. The location of the ilio-caval confluence and the ideal distal landing zone identified by IVUS can also be correlated to the adjacent bony landmarks. These bony landmarks include the vertebral body, lower border of pubic ramus, bottom of femoral head, ischial crossing, and lesser trochanter. In the majority of the patients, the confluence has been found to be at the level of the L4 vertebral body.[34,36]

Intervention includes stent placement with or without other concomitant procedures such as percutaneous endovenous laser ablation of the saphenous vein. The most common type of stent used in our practice is the Wallstent (Boston Scientific, Marlborough, MA, USA) with the cephalad addition of Z-stent (Cook Medical, Bloomington, IN, USA) in most cases where additional radial strength is needed at the iliocaval confluence, which is a potential choke or stenosis point (Figs. 25.12 and 25.13).

Technical success of procedure

Procedural technical success is defined as successful treatment of the lesion (nonthrombotic or postthrombotic or mixed) that meets three important criteria:

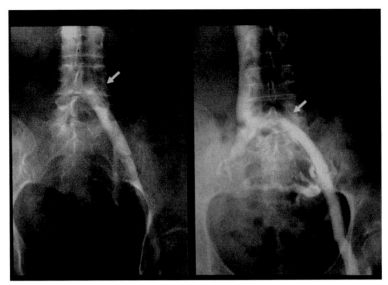

Fig. 25.12 Distal migration or "slip" of the Wallstent. The Wallstent was initially placed at the iliocaval junction which is a potential "choke" point.

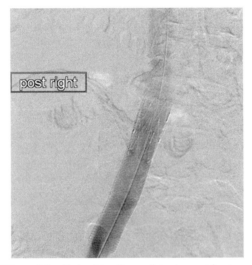

Fig. 25.13 Proximal addition of the Z-stent to the wallstent provides increased radial stent to this composite stent configuration and buttresses the wallstent in place as well.

1. Without intraoperative device complications
2. Establishment or restoration of in-line central venous flow
3. Less than 20% residual stenosis on completion of IVUS examination after postdilation.[36] However, the goal should be to restore the luminal caliber of the stented veins to cross-sectional areas listed in Table 25.6.[36]

Recanalization of chronic total venous occlusion

Percutaneous access is obtained in the ipsilateral femoral or profunda femoral vein in the middle to lower thigh. Popliteal or internal jugular venous access may also be required in certain patients. Common femoral vein access is generally not adequate in these patients as the common femoral vein itself is often diseased or completely occluded in continuity with occlusion of iliac and inferior vena cava segments. An initial venogram is performed to provide a visual road map. Chronic total venous occlusion crossing is performed with simple tools in the majority of cases using a 0.035-inch Glidewire (Terumo, Somerset, NJ, USA) and support catheter. Straight Glidewire without a loop is often preferred. Supporting sheaths may be needed in select cases. Perforations can occur but they are usually self-limiting unlike arterial perforations (Fig. 25.14).[64] Generally, the wire can be simply withdrawn and redirected away from the perforation to continue the CTVO recanalization without aborting the procedure.

Care should be taken to ensure that the wire is not in the spinal canal. There are case reports of the devastating complication where stents were placed in the spinal canal with resultant neurologic deficits.[74–76] A wire centered on the vertebral column requires further investigation. The correct location of the inferior vena cava (IVC) is to the right of the vertebral column. A lateral projection view may be utilized to see if the wire is in front of (correct position) or behind the vertebral column (incorrect position). Intravascular ultrasound and venography can be used to confirm the proper location of the wire in the true IVC lumen. In addition, the movement of the wire due to respiration is an adjunct confirmatory sign that the wire is in the true venous lumen and not in the exterior retroperitoneal space. Landing zones are decided based on IVUS measurements and imaging. Sometimes stenting may need to be extended across the renal or hepatic veins in

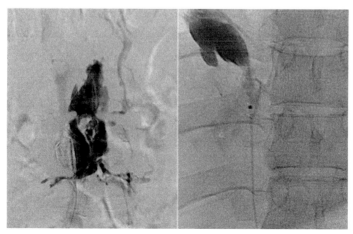

Fig. 25.14 Perforations during recanalizations of chronic total venous occlusions are usually innocuous and self-limiting; unlike their arterial counterparts. *(By permission, JVS-VLD Murphy EH, Johns B, Varney E, Raju S. Endovascular management of chronic total occlusions of the inferior vena cava and iliac veins. J Vasc Surg Venous Lymphat Disord. 2017;5:47–59.)*

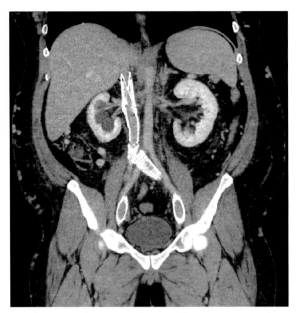

Fig. 25.15 Coronal view of computed tomography scan showing Z-stent placement across the renal veins during iliofemoral caval stenting in a patient with extensive postthrombotic syndrome.

extensive CTVOs. It is our practice to place Z stents at the orifices of the hepatic or renal veins if such extensive stenting is required (Fig. 25.15). Jailing the ostia of the renal or hepatic veins with Wallstents may potentially lead to occlusion of these veins due to gradual endothelialization of the struts of the stent. Current small series have not reported renal insufficiency due to such jailing, but the potential risk exists.[77] Multiple Wallstents are usually required for recanalizing extensive CTVOs. Minimum stent overlap of about 20–30 mm is recommended between the contiguous stents.[64]

Management of iliac-caval confluence

Iliac-caval reconstruction can be challenging due to two main reasons, (1) difficulty in the accurate delineation of the confluence, and (2) potential problems due to coning, collapse, or migration of stents landed in the area. The following two methods of iliac-caval reconstruction are commonly used to avoid these issues.

Composite Z-stent—Wallstent configuration

Wallstents are placed bilaterally in the iliac veins to within 1 cm of the confluence. Iliac Wallstents should not actually cross the confluence (Fig. 25.16).[78] This is important because landing the Wallstents further into the IVC can jail the contralateral side with ≈10% risk of DVT on the contralateral side.[78] Kissing Z-stents are then deployed from the right and left iliac veins to bridge the confluence. The Z-stents are oversized by

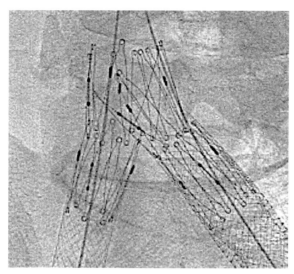

Fig. 25.16 Z-stent allows iliocaval confluence reconstruction and prevents contralateral deep venous thrombosis due to much wider interstices. Wallstents are landed to about 1 cm of the confluence. *(By permission—JVS-VLD Murphy ER, Johns B, Varney E, Buck W, Jayaraj A, Raju S. Deep venous thrombosis associated with caval extension of iliac stents. J Vasc Surg Venous Lymphat Disord. 2017;5:8—17.)*

10%—20% relative to the iliac vein Wallstents to prevent stent embolization.[64,70,78] By using this approach, the risk of contralateral DVT is reduced to ≤1%.[70,78] The IVC above the confluence can be stented with Wallstents or Wallstent covered Z stents. Typically, 24 mm Wallstents are used in the IVC to approximate relatively normal caliber IVC.

Double barrel Wallstent configuration

The confluence can also be reconstructed using the "double barrel" technique. This technique refers to the extension of the Wallstent from both iliac veins into the IVC for a distance of about 3 cm. We recently reported overall favorable outcomes with this technique in a series of 66 patients.[79] However, one unique complication that was noted in about 12% of the patient was that one stent barrel was compressing the other stent barrel, creating a type of iatrogenic outflow stenosis on one side (Fig. 25.17).[79] This may lead to recurrent or residual symptoms. Intravascular ultrasound can diagnose the problem accurately. The solution requires the creation of a large fenestration in between the two stent barrels that may be variably endothelialized (reducing cell porosity) (Fig. 25.18).[79] Z-stent can be deployed across the fenestration to support it long term.

Reinterventions in stented limbs

Reinterventions are usually required to correct residual or recurrent symptoms in about 10%—20% of limbs if Wallstents had been used.[80] About 80% require only a single reintervention; approximately 20% of patients require ≥ two reinterventions over time. The

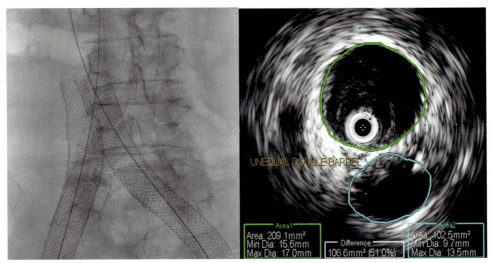

Fig. 25.17 Double barrel stent configuration with compression of the left barrel seen on intravascular ultrasound. *(By permission, Raju S, Powell T, Kuykendall R, Jayaraj A. A unique complication of double barrel Wallstent technique in iliac-caval stenting. J Vasc Surg Cases Innov Tech. 2021;7:211–214.)*

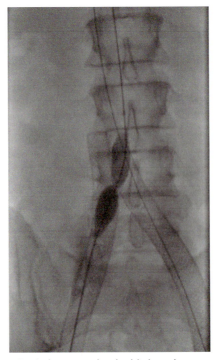

Fig. 25.18 Creation of fenestration in between the double barrel stents using angioplasty balloon. *(By permission, Raju S, Powell T, Kuykendall R, Jayaraj A. A unique complication of double barrel Wallstent technique in iliac-caval stenting. J Vasc Surg Cases Innov Tech. 2021;7:211–214.)*

reinterventional rate of venous-specific newer stents has not been well established yet. Overall, reinterventions can be broadly classified into the following five types:
1. Cephalad stent extension to correct stent outflow problems;
2. Caudad stent extension to correct inflow problems;
3. Balloon dilatation of in-stent restenosis (ISR);
4. Balloon dilatation of stent compression (SC);
5. Various combinations.

Two types of ISR have been encountered:
1. A "soft" lesion; likely due to reduced flow channel lined by thrombus within the stent from inflow/outflow problems.
2. A "hard" lesion that occurred independently, was resistant to dilatation and tended to recur unlike the "soft" lesion.

In one study, cumulative improvement in pain and swelling at 18 months following intervention was 67% and 72%, respectively. Complete cumulative healing of venous dermatitis and ulcer was 90% at 12-month postreintervention.[81]

Surveillance ultrasound protocol has been described elsewhere in this chapter and is important to detect stent malfunction in a timely manner. On duplex ultrasound, flow channel reduction by at least 50% due to either ISR or SC is usually considered significant (Fig. 25.19). However, the key point is that the decision to reintervene should be based

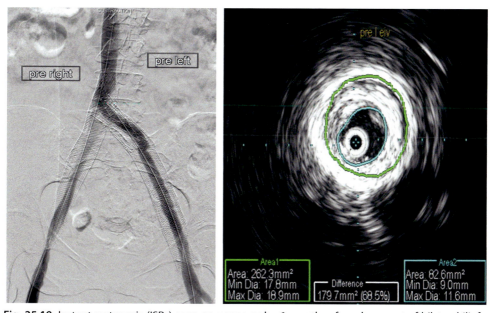

Fig. 25.19 Instent restenosis (ISRs) seen on venography 6 months after placement of bilateral iliofemoral caval stents in a patient with symptom recurrence. Intravascular ultrasound shows that the effective vein area had been reduced by over 60% since stent placement due to the ISR.

on the composite of duplex findings and the patient's symptoms. This generally includes recurrent or significant residual symptoms after initial stent placement.

Isodilation refers to the dilation of a stent up to its rated diameter. Hyperdilation, on the other hand, refers to balloon dilation of a venous stent to about 2–4 mm beyond its rated diameter. Restoration of the rated diameter of the stent is generally the goal but often times this is not possible to achieve because of complex factors such as SC. That is why the technique of hyperdilation with a larger caliber, high-pressure balloon may be required to restore the nominal diameter of the stent. In patients requiring hyperdilation, IVUS provides valuable information about SC or degree of ISR. We have recently shown that hyperdilation, rather than isodilation, provides more durable caliber improvement, better clinical outcomes, and a decrease in venous hypertension.[80]

Hyperdilation is performed with target caliber Atlas balloons (Bard Inc., Murray Hill, NJ, USA). In our experience, the Wallstent tolerates hyperdilation well up to 4 mm beyond the rated stent diameter without losing its structural integrity.[60,80,81] We have also recently described the use of laser atherectomy devices for the management of recalcitrant ISR ablation in stents in whom angioplasty alone does not produce a good result at the time of reintervention. The laser catheter is supported by an angled sheath. Although the patient subset was small ($n = 18$), laser atherectomy for recalcitrant ISR was noted to be safe and overall effective.[79]

Factors affecting ISR and stent compression

We recently analyzed our own experience with ISR and SC in 578 limbs.[82] In-stent restenosis was estimated from stent and flow channel diameters measured using duplex ultrasound. Stent compression was estimated from the rated stent diameter and actual stent diameter on duplex ultrasound. In-stent restenosis was noted to peak at 3 months and then plateaued thereafter. In contrast, SC peaked on day one postprocedure and then steadied thereafter. Stent inflow channel luminal area and shear rate were risk factors for the development of ISR, while asymmetric stent sizing was a risk factor for SC. Reintervention was performed for symptoms that significantly affected the quality of patient's life despite conservative therapy for at least 3–6 months. Median time to reintervention was 11 months.[82] Stent patencies after intervention are presented in Table 25.7. An

Table 25.7 Patency after reinterventions at 60 months.

Patency	After reintervention for ISR (%)	After reintervention for stent compression (%)
Primary	70	70
Primary-assisted	98	99
Secondary	84	84

Adapted from Jayaraj A, Fuller R, Raju S, Stafford J. In-stent restenosis and stent compression following stenting for chronic iliofemoral venous obstruction. *J Vasc Surg Venous Lymphat Disord*. 2021:S2213–S2333X(21)00304-8.

interesting observation in this study was that high-grade ISR (>50%) usually remained indolent with stable symptoms without progressing to total occlusion (rare). This suggests that there is no role for prophylactic balloon dilatation to prevent imminent occlusion especially when the patient is asymptomatic or only minimally symptomatic.

Anticoagulation protocol

All patients undergoing venous stenting receive preoperative subcutaneous enoxaparin (40 mg) and 75 mg of i.v. bivalirudin. For patients with a history of heparin-induced thrombocytopenia, fondaparinux or argatroban is used instead of lovenox. Postoperatively, low-dose daily aspirin (81 mg) is generally sufficient for nonthrombotic stented lesions. For postthrombotic stented lesions, we take into account patient's prior DVT history, extent of postthrombotic lesion, status of stent inflow on completion IVUS, flow of contrast through the stent column (sluggish vs. brisk), and results of thrombophilia panel to decide about long-term anticoagulation. Generally, long-term anticoagulation involves the use of one of the following agents: apixaban, rivaroxaban, warfarin, dabigatran, edoxaban, or enoxaparin. After recanalization and successful stenting of extensive CTVOs, our general protocol is to keep patients on therapeutic enoxaparin for the first 6 weeks before transitioning to an oral anticoagulation agent such as apixaban.

Thrombophilia and venous ulcers

Thrombophilia panel is obtained in carefully selected patients in whom the management is expected to be affected by the results of such a panel. We have found a high incidence of factor VIII elevation (>150 IU per deciliter) in patients with a history of DVT, especially recurrent DVT.[83] In a study of 306 patients, venous ulceration was noted more commonly in patients with elevated factor VIII levels.[83] It is hypothesized that patients with thrombophilia conditions such as factor VIII may have a higher incidence of valvular reflux (possibly related to damage from DVT episodes) that may contribute to microvascular injury and postthrombotic inflammation, leading to venous ulceration. In addition, they likely have postthrombotic IVS. Some of the thrombosis associated with thrombophilia conditions such as factor VIII elevation may be subclinical.[83,84] In a series of 44 limbs that underwent stenting for venous ulcers, the lesion was noted to be postthrombotic in 31 limbs (70%).[69] In another series with 192 consecutive limbs with venous ulcers, 60% of limbs were postthrombotic.[13]

Pentoxyfylline and venous ulcers

In the latest update on the subject from Cochrane Database (2012), 12 trials involving 864 patients were analyzed.[85] Pentoxifylline appeared to be more effective than placebo for complete ulcer healing or significant improvement (RR 1.70, 95% CI 1.30 to 2.24).[85] Similar results were reported by another systematic review on the subject.[86] The specific and adjunctive role of pentoxyfylline in the perioperative management of venous ulcers with iliac stenting has not been investigated. It is our practice to selectively use pentoxyfylline in patients with large ulcers perioperatively. Anecdotally, pentoxyfylline appears to be helpful in ulcer healing of large ulcers after the performance of stenting. The user has to be balanced against the risks of common adverse effects, which are predominantly gastrointestinal.

Stent surveillance in standard and recanalization cases

Stent surveillance is performed at postoperative day one and at 3 weeks initially. Patients are seen in the clinic at 3 weeks, at 3 months, and at six monthly or annual intervals thereafter. Typically, stent surveillance is performed at each of the above clinical visits in addition to the first postoperative month.

Stent surveillance is individualized and performed more frequently after recanalizations or when there is a rapid development of ISR during the first 6 months than in routine cases.[36]

Stent occlusions after recanalization

In one series, early in-stent thrombosis (22.5% vs. 6.1%, $P = 0.007$) was noted more frequently in limbs with visible remaining collateral circulation compared to limbs without visible remaining collaterals.[62] In our own experience, stent occlusion occurs rarely after stenting of non-occlusive stenosis. In an analysis of 3468 stents, 102 (3%) stent occlusions were noted.[87] Stent occlusions occurred at a median of 5.8 months after placement. 31% occlusions were acute (<30 days); hence, the importance of duplex surveillance beginning within the first month of stent placement. Not surprisingly, occlusions were more common (77%) in limbs with postthrombotic pathology. Technical success to recanalize occluded stents was >80% but suffered shortened secondary patency (25 ± 8 months) compared to nonoccluded stents (>5 years). A 40% ulcer healing rate was noted after recanalization during a median follow-up period of 17 months.[87] Patency after iliofemoral caval stenting at 60 months is shown in Table 25.8.[87]

Stent occlusion from ISR

In a large series on stent occlusions ($n = 102$), multivariable logistic regression did not reveal ISR as a predictor for stent occlusion.[87] In our experience, most patients with ISR >50% remain patent, and ISR shows slow progression during years of follow-up. On longer-term follow-up, <10% of patients with ISR go on to develop stent occlusions.[87]

Table 25.8 Patency after iliofemoral caval stenting at 60 months.

Patency	Initial stent % (for stenotic lesion)	Recanalized native vein, %	Recanalized stent, %
Primary	72	33	35
Primary-assisted	92	55	25
Secondary	95	65	18

Adapted from Jayaraj A, Crim W, Knight A, Raju S. Characteristics and outcomes of stent occlusion after iliocaval stenting. *J Vasc Surg Venous Lymphat Disord*. 2019;7:56–64.

Threshold stenosis for intervention

The VIDIO trial[32] showed a 50% IVUS stenosis threshold compared to the caliber of the adjacent "normal" segment' as being clinically relevant. Further analysis of the VIDIO trial showed that, per IVUS, a >54% stenosis was estimated to be the optimal threshold for interventional treatment; this threshold was higher (>61%) for postthrombotic patients.[32,33] However, it should be noted that a morphologic stenosis of this degree may not always be pathologic. Therefore, while the presence of a venous stenosis of 50% or greater is felt to be hemodynamically important by many authors, it is not strictly sufficient by itself to result in symptoms or warrant intervention. The criterion of 50% stenosis on IVUS for venous obstructive disease was likely extrapolated from arterial literature. However, it requires further clinical correlation.[35] We have recently demonstrated in a large series of 480 limbs that:

1. Degree of IVUS identified IVS does not appear to have a bearing on initial clinical presentation, CEAP clinical class, or supine foot venous pressures in patients presenting with QoL impairing venous outflow obstruction.
2. Poststenting, clinical improvement, QoL improvement, stent patency, and requirement of reintervention are all independent of the initial degree of venous stenosis.[88]

Therefore, the identification of iliofemoral venous stenosis in the appropriate clinical context (e.g., venous ulceration) is important. Patients presenting with QoL impairing symptoms such as venous ulcers who have failed conservative treatment merit consideration of correction of their venous outflow obstruction even if the degree of stenosis is less than 50%.[88,89]

Dedicated self-expanding nonbraided nitinol venous stents

Trials for four stents were carried out and they became available for use in the deep venous space in the last two and a half years in the United States. The trials for these stents included VIRTUS, VIVO, VERNACULAR, and ABRE.[86] Only the VICI stent venous stent (Boston Scientific Corporation, Marlborough, MA, USA) has a closed cell design matrix while the other three stents have an open cell design. Also, there is minimal foreshortening noted with these stents except the VICI stent (20% foreshortening reported).[90]

In a series of 75 patients, iliac stenting with the VICI venous stent resulted in ulcer healing in 75% of patients. The primary patency at 12 months was noted to be 100% for NIVL and 87% for postthrombotic obstructions.[91] In another series with 62 patients, endovascular iliocaval reconstruction was performed with self-expanding nitinol stents.[92] Ulcer healing was noted in 100% of patients ($n = 8$). However, reintervention rate to maintain stent patency was high (one-third of patients), particularly in those who had postthrombotic changes noted in the femoral veins.[92] Overall, it appears that the success of ulcer healing with dedicated venous stents is comparable to that of the Wallstent.[93] A third series reported 79 patients in whom the Venovo stent was used. All ulcers in this series healed (8/8) and none recurred at 2 years' follow-up.[94] Additional long-term data are necessary for proper evaluation of the new generation of venous stents.

The Sinus-Venous stent, available in Europe but not the United States, is another dedicated venous nitinol stent. In a series of 75 patients, low ulcer healing was reported (2/7; 28.6%) and none of the ulcers remained healed at the 12-month follow-up.[95]

Our largest experience among the newer stents has been with the Venovo stent ($n > 200$). The end of the Venovo stent has a flare, somewhat similar to a Z-stent (Fig. 25.20). We have not had to use a Z-stent when Venovo stents were placed at the iliac-caval confluence. It should be noted that VICI venous stent and Venovo (BD Interventional, Wokingham, UK) were recalled earlier this year due to issues with stent embolization (<1%, VICI) and problems with deployment delivery systems (>250 reports, Venovo).

Special considerations and techniques
Stent extension below the inguinal ligament

Arterial stenting across the inguinal ligament is not recommended because of the risk of focal stenosis or fracture of the stent. 177 iliofemoral venous Wallstents placed across the inguinal ligament were compared to 316 stents terminating in the iliac vein above the inguinal ligament.[96] Cumulative secondary patency rate was significantly worse for stents crossing the inguinal ligament ($P = 0.0001$). However, this appeared to be related to a larger number of chronic total occlusions (CTO) in the group crossing the ligament. The secondary patency in CTO recanalization was ≈20% worse than stenting for nonocclusive pathology. ISR at the inguinal ligament occurred in only 7% of limbs. None of the stents were compressed or fractured. Wallstents can be safely extended across the inguinal ligament if required to cover disease in the common femoral vein without adverse consequences. In another series, patients with stents extending below the inguinal ligament had a higher rate of ISR (hazard ratio = 1.77–6.5; $P = 0.0146$).[62]

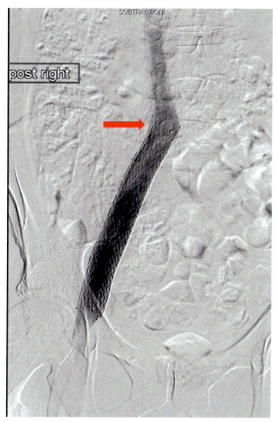

Fig. 25.20 Flare at the proximal end of the venovo stent functionally mimics the Z-stent at the iliocaval confluence.

Bilateral stenting

Patients with bilateral obstructive iliofemoral venous lesions often experience improvement of the contralateral limb symptoms (95%) after stenting of the worse ipsilateral limb.[97] This is probably due to off-loading of collaterals from the opposite limb. The opposite iliac vein is a major collateral pathway in unilateral IVS/occlusions. Only 15 of 304 (5%) symptomatic contralateral limbs required stenting during the follow-up period because of a worsening clinical picture. Therefore, a staged sequential approach to iliofemoral stenting in patients with bilateral symptoms focusing initially on the more symptomatic limb is strongly recommended rather than simultaneous bilateral stenting.[97] Simultaneous bilateral stenting poses additional risks inherent in the reconstruction of the iliac–caval confluence and may be unnecessary in many patients.

Femoral vein occlusion with iliac vein obstruction

39 patients with femoral vein occlusions underwent iliac venous stenting during a 13-year period.[98] Saphenous vein ablation was concurrently performed in 18% of these

patients to abolish reflux. Ulcer healing was noted in 54% of patients, and saphenous vein ablation was noted to be safe in these patients. This is likely due to the minimal collateral role of the saphenous vein in IVS in these patients. Predominantly, iliac obstructive pathology rather than saphenous reflux is responsible for the symptoms in these patients.[98]

Stenting across and below IVC filters

Generally, every effort should be made to remove IVC filters when their presence is no longer indicated and when their retrieval can be performed safely. Advanced techniques may be needed to remove IVC filters particularly if the filters that have been in place for several years.[99] In patients with iliac vein stents, IVC filters have been shown to increase the rate of DVT on the stented side. Additionally, an increased rate of reintervention due to ISR was also noted in stented patients with IVC filters.[100]

Alternatively, filters can be excluded by stenting across them (Fig. 25.21).[64,101,102] Placement of stents in the IVC across the IVC filters addresses and optimizes the outflow for the stent column. Excellent clinical and technical success as well as primary assisted and secondary patency have been reported with this technique.[101] When occluded filters are successfully crossed, they are crushed and displaced laterally during angioplasty. In an attempt to minimize complications, filters should be displaced to the right of the

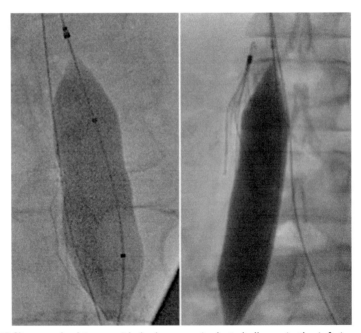

Fig. 25.21 IVC filters crushed to one side by large angioplasty balloons in the inferior vena cava. *(By permission, JVS-VLD Murphy EH, Johns B, Varney F, Raju S. Endovascular management of chronic total occlusions of the inferior vena cava and iliac veins. J Vasc Surg Venous Lymphat Disord. 2017;5:47–59.)*

abdomen, away from the aorta and duodenum when feasible.[64] In another study, clinical resolution of symptoms was higher after stenting if the IVC filter was removed.[103]

Obesity

In a series of 101 limbs in obese individuals, it was noted that bilateral chronic venous insufficiency signs and symptoms were twice as common as nonobese individuals.[104] The incidence of dermatitis and venous stasis ulceration was high (47%). After stenting, 58% of limbs were free of ulcer or dermatitis at 4 years. About half of the active ulcers healed and remained healed at 6 years. Weight loss management has the primary role in the management of obese patients with venous ulceration but iliac vein stenting should also be considered as an option on a case-to-case basis. Our analysis has shown that the mechanism of venous obstruction in obese individuals is often similar to those in nonobese. Primary or postthrombotic lesions were noted in 89% of patients on IVUS interrogation.[104] Central venous compressions from increased abdominal pressure may be the main mechanism in a minority of patients without intrinsic IVS.

Octogenarians

Self-application of compression devices is often difficult in elderly individuals because of advanced medical comorbidities such as severe arthritis or frailty. Chronic venous ulceration in these individuals poses a significant drain on financial, care-giver and emotional resources and can adversely affect the QoL of patients and their care-givers. These factors often force institutionalization of a patient; undesired though it may be. We found that iliac vein stenting is a safe and effective option in individuals with advanced age. In a series of 107 limbs in patients ≥80 years of age, active venous ulceration was present in 27% of patients.[18] Despite the use of stockings, ulcers had recurred in 26% and 31% of patients at 12 and 18 months, respectively. However, 61% of active ulcers healed after iliac venous stenting and about 40% of patients were able to stop the use of compression stockings altogether after intervention.[18]

Hybrid approaches

Rarely, a hybrid approach may be required in patients with VLUs. This can involve lower extremity venous bypasses and iliofemoral endovenectomy with or without the creation of an arteriovenous fistula (AVF) with concomitant iliac vein stenting. In particular, endovenectomy and AVF is done to improve stent inflow via the profunda or the femoral veins.[105–107] However, the risk of complications, such as surgical site infections, DVT, hematomas, and seromas, should be borne in mind with this procedure. Also, the long-term patency of venous bypasses is modest at best. Ulcer healing has been reported after patent open reconstructions but with a 50% recurrence rate.[108] Therefore, appropriate patient selection is important for these hybrid procedures. With the advent of venous stenting with its minimally invasive nature, low morbidity, and excellent long-term results, open surgical bypass reconstruction has taken a more secondary role.[109,110]

Conclusions

In summary, treatment of deep venous obstruction via iliac vein stenting facilitates the healing of chronic VLUs and has a role in preventing their recurrence. The healing effect may be complemented by the treatment of superficial venous reflux in select cases. Anticoagulation is used selectively in patients after iliac venous stenting. Iliac venous stenting should be considered in patients with VLUs who fail a trial of conservative therapy. Particular attention should be paid to a thorough clinical evaluation, careful patient selection, and the technical aspects of iliac venous stenting as outlined in detail earlier.

Financial disclosures

TS: none; SR: Royalty, Veniti Inc., US Patents, IVUS diagnostics and Iliac vein stent design.

References

1. Cumston CGVI. Varicose veins and their treatment by Trendelenburg's operation. *Ann Surg*. 1898;27: 626–639.
2. Eklof B. The dynamic approach to venous disease—following in the footsteps of Gunnar Bauer and Robert Kistner. *J Vasc Surg*. 2005;42:369–376.
3. Kistner RL. Diagnosis of chronic venous insufficiency. *J Vasc Surg*. 1986;3:185–188.
4. Kistner RL, Ferris EB, Randhawa G, Kamida C. A method of performing descending venography. *J Vasc Surg*. 1986;4:464–468.
5. Neglen P, Raju S. A comparison between descending phlebography and duplex Doppler investigation in the evaluation of reflux in chronic venous insufficiency: a challenge to phlebography as the "gold standard". *J Vasc Surg*. 1992;16:687–693.
6. Baker SR, Burnand KG, Sommerville KM, Thomas ML, Wilson NM, Browse NL. Comparison of venous reflux assessed by duplex scanning and descending phlebography in chronic venous disease. *Lancet*. 1993;341:400–403.
7. Kistner RL. Surgical repair of the incompetent femoral vein valve. *Arch Surg*. 1975;110:1336–1342.
8. Raju S. Robert Linton Kistner. *J Vasc Surg Venous Lymphat Disord*. 2020;8:297–298.
9. Lurie F, Passman M, Mesiner M, et al. The 2020 update of the CEAP classification and reporting standards. *J Vasc Surg Venous Lymphat Disord*. 2020;8:342–352.
10. Neglen P, Hollis KC, Olivier J, Raju S. Stenting of the venous outflow in chronic venous disease: long-term stent-related outcome, clinical, and hemodynamic result. *J Vasc Surg*. 2007;46:979–990.
11. Neglen P. Chronic venous obstruction: diagnostic considerations and therapeutic role of percutaneous iliac stenting. *Vascular*. 2007;15:273–280.
12. Raju S, Darcey R, Neglen P. Unexpected major role for venous stenting in deep reflux disease. *J Vasc Surg*. 2010;51:401–408.
13. Raju S, Kirk OK, Jones TL. Endovenous management of venous leg ulcers. *J Vasc Surg Venous Lymphat Disord*. 2013;1:165–172.
14. Youn YJ, Lee J. Chronic venous insufficiency and varicose veins of the lower extremities. *Korean J Intern Med*. March 2019;34(2):269–283.
15. Raffetto JD, Ligi D, Maniscalco R, Khalil RA, Mannello F. Why venous leg ulcers have difficulty healing: overview on pathophysiology, clinical consequences and treatment. *J Clin Med*. 2020;24:29.

16. Raju S, Neglen P. Clinical practice. Chronic venous insufficiency and varicose veins. *N Engl J Med.* 2009;360:2319−2327.
17. Raju S, Tackett Jr P, Neglen P. Spontaneous onset of bacterial cellulitis in lower limbs with chronic obstructive venous disease. *J Vasc Endovasc Surg.* 2008;36:606−610.
18. Raju S, Ward M. Utility of iliac vein stenting in elderly population older than 80 years. *J Vasc Surg Venous Lymphat Disord.* 2015;3:58−63.
19. Launois R, Reboul-Marty J, Henry B. Construction and validation of a quality of life questionnaire in chronic limb venous insufficiency (CIVIQ). *Qual Life Res.* 1996;5:539−554.
20. Eliassen A, Vandy F, McHugh J, Henke PK. Marjolin's ulcer in a patient with chronic venous stasis. *Ann Vasc Surg.* 2013;27:1182. e5−8.
21. Nelzen O, Bergqvist D, Lindhagen A. Leg ulcer etiology—a cross sectional population study. *J Vasc Surg.* 1991;14:557−564.
22. Georgopoulos S, Kouvelos GN, Koutsoumpelis A, et al. The effect of revascularization procedures on healing of mixed arterial and venous leg ulcers. *Int Angiol.* 2013;32:368−374.
23. Raju S, Walker W, May C. Measurement of ambulatory venous pressure and column interruption duration in normal volunteers. *J Vasc Surg Venous Lymphat Disord.* 2020;8:127−136.
24. Negus D, Fletcher EW, Cockett FB, Thomas ML. Compression and band formation at the mouth of the left common iliac vein. *Br J Surg.* 1968;55:369−374.
25. Kibbe MR, Ujiki M, Goodwin AL, Eskandari M, Yao J, Matsumura J. Iliac vein compression in an asymptomatic patient population. *J Vasc Surg.* 2004;39:937−943.
26. Ehrich WE, Krumbhaar EB. A frequent obstructive anomaly of the mouth of the left common iliac vein. *Am Heart J.* 1943;26:737−750.
27. Raju S, Neglen P. High prevalence of nonthrombotic iliac vein lesions in chronic venous disease: a permissive role in pathogenicity. *J Vasc Surg.* 2006;44:136−144.
28. Labropoulos N, Tiongson J, Pryor L, et al. Definition of venous reflux in lower-extremity veins. *J Vasc Surg.* 2003;38:793−798.
29. Mendoza E, Blattler W, Amsler F. Great saphenous vein diameter at the saphenofemoral junction and proximal thigh as parameters of venous disease class. *Eur J Vasc Endovasc Surg.* 2013;45:76−83.
30. Raju S, Ward Jr M, Jones TL. Quantifying saphenous reflux. *J Vasc Surg Venous Lymphat Disord.* 2015;3:8−17.
31. Navarro TP, Delis KT, Ribeiro AP. Clinical and hemodynamic significance of the greater saphenous vein diameter in chronic venous insufficiency. *Arch Surg.* 2002;137:1233−1237.
32. Gagne PJ, Tahara RW, Fastabend CP, et al. Venography versus intravascular ultrasound for diagnosing and treating iliofemoral vein obstruction. *J Vasc Surg Venous Lymphat Disord.* 2017;5:678−687.
33. Gagne PJ, Gasparis A, Black S, et al. Analysis of threshold stenosis by multiplanar venogram and intravascular ultrasound examination for predicting clinical improvement after iliofemoral vein stenting in the VIDIO trial. *J Vasc Surg Venous Lymphat Disord.* 2018;6:48−56. e1.
34. Montminy ML, Thomasson JD, Tanaka GJ, Lamanilao LM, Crim W, Raju S. A comparison between intravascular ultrasound and venography in identifying key parameters essential for iliac vein stenting. *J Vasc Surg Venous Lymphat Disord.* 2019;7:801−807.
35. Saleem T, Raju S. Comparison of intravascular ultrasound and multidimensional contrast imaging modalities for characterization of chronic occlusive iliofemoral venous disease: a systematic review. *J Vasc Surg Venous Lymphat Disord.* 2021;9(6):1545−1556.e2.
36. Saleem T, Knight A, Raju S. Diagnostic yield of intravascular ultrasound in patients with clinical signs and symptoms of lower extremity venous disease. *J Vasc Surg Venous Lymphat Disord.* 2020;8:634−639.
37. Raju S, Walker W, Noel C, Kuykendall R, Jayaraj A. The two segment caliber method of diagnosing iliac vein stenosis on routine computed tomography with contrast enhancement. *J Vasc Surg Venous Lymphat Disord.* 2020;8:970−977.
38. O'Meara S, Cullum N, Nelson EA, Dumville JC. Compression for venous leg ulcers. *Cochrane Database Syst Rev.* 2012;11:CD000265.
39. Barwell JR, Davies CE, Deacon J, et al. Comparison of surgery and compression with compression alone in chronic venous ulceration (ESCHAR study): randomised controlled trial. *Lancet.* 2004;363:1854−1859.

40. Wright DD. The ESCHAR trial: should it change practice? *Perspect Vasc Surg Endovasc Ther.* 2009;21:69–72.
41. Raju S, Lurie F, O'Donnell Jr TF. Compression use in the era of endovenous interventions and wound care centers. *J Vasc Surg Venous Lymphat Disord.* 2016;4:346–354.
42. Raju S, Hollis K, Neglen P. Use of compression stockings in chronic venous disease: patient compliance and efficacy. *Ann Vasc Surg.* 2007;21:790–795.
43. Jull AB, Mitchell N, Arroll J, Jones M, Waters J, Latta A. Factors influencing concordance with compression stockings after venous leg ulcer healing. *J Wound Care.* 2004;13:90–92.
44. Kerstein MD, Gahtan V. Outcomes of venous ulcer care: results of a longitudinal study. *Ostomy/Wound Manag.* 2000;46:22–26, 28–9.
45. Kahn SR, Shapiro S, Wells PS, et al. Compression stockings to prevent post-thrombotic syndrome: a randomised placebo-controlled trial. *Lancet.* 2014;383:880–888.
46. Moffatt CJ, Franks PJ. Implementation of a leg ulcer strategy. *Br J Dermatol.* 2004;151:857–867.
47. Alhalbouni S, Hingorani A, Shiferson A, et al. Iliac-femoral venous stenting for lower extremity venous stasis symptoms. *Ann Vasc Surg.* 2012;26:185–189.
48. Lawrence PF, Hager ES, Harlander-Locke MP, et al. Treatment of superficial and perforator reflux and deep venous stenosis improves healing of chronic venous leg ulcers. *J Vasc Surg Venous Lymphat Disord.* 2020;8:601–609.
49. Neglen P, Hollis KC, Raju S. Combined saphenous ablation and iliac stent placement for complex severe chronic venous disease. *J Vasc Surg.* 2006;44:828–833.
50. Yang X, Wu X, Peng Z, Yin M, Lu X, Ye K. Outcomes of endovenous laser ablation with additional iliac vein stenting of non-thrombotic lesions in patients presenting with active venous ulcers. *J Vasc Surg Venous Lymphat Disord.* 2021;9(6):1517–1525.
51. Aherne TM, Keohane C, Mullins M, et al. DEep VEin Lesion OPtimisation (DEVELOP) trial: protocol for a randomised, assessor-blinded feasibility trial of iliac vein intervention for venous leg ulcers. *Pilot Feasibility Stud.* 2021;7:42.
52. Normahani P, Shalhoub J, Narayanan S. Repurposing the systemic venous return model for conceptualization of chronic venous insufficiency and its management. *Phlebology.* 2020;35:749–751.
53. Normahani P, Shalhoub J, Narayanan SA. Guytonian explanation for hemodynamic responses to interventions in superficial venous disease. *Phlebology.* 2021;36:245–250.
54. Raju S, McAllister S, Neglen P. Recanalization of totally occluded iliac and adjacent venous segments. *J Vasc Surg.* 2002;36:903–911.
55. Neglen P, Thrasher TL, Raju S. Venous outflow obstruction: an underestimated contributor to chronic venous disease. *J Vasc Surg.* 2003;38:879–885.
56. Harlander-Locke M, Lawrence PF, Alktaifi A, Jimenez JC, Rigberg D, Derubertis B. The impact of ablation of incompetent superficial and perforator veins on ulcer healing rates. *J Vasc Surg.* 2012;55:458–464.
57. Harlander-Locke M, Lawrence P, Jimenez JC, Rigberg D, Derubertis B, Gelabert H. Combined treatment with compression therapy and ablation of incompetent superficial and perforating veins reduces ulcer recurrence in patients with CEAP 5 venous disease. *J Vasc Surg.* 2012;55:446–450.
58. Kumins NH, Weinzweig N, Schuler JJ. Free tissue transfer provides durable treatment for large non-healing venous ulcers. *J Vasc Surg.* 2000;32:848–854.
59. Masuda EM, Kessler DM, Lurie F, Puggioni A, Kistner RL, Eklof B. The effect of ultrasound-guided sclerotherapy of incompetent perforator veins on venous clinical severity and disability scores. *J Vasc Surg.* 2006;43:551–557.
60. Saleem T, Knight A, Raju S. Effect of iliofemoral-caval venous intervention on lower extremity compartment pressure in patients with chronic venous insufficiency. *J Vasc Surg Venous Lymphat Disord.* 2020;8:769–774.
61. Liu P, Peng J, Zheng L, et al. Application of computed tomography venography in the diagnosis and severity assessment of iliac vein compression syndrome: a retrospective study. *Medicine (Baltim).* 2018;97:e12002.

62. Ye K, Lu X, Jiang M, et al. Technical details and clinical outcomes of transpopliteal venous stent placement for postthrombotic chronic total occlusion of iliofemoral vein. *J Vasc Intervent Radiol*. 2014;25: 925–932.
63. Neglen P, Berry MA, Raju S. Endovascular surgery in the treatment of chronic primary and post-thrombotic iliac vein obstruction. *Eur J Vasc Endovasc Surg*. 2000;20:560–571.
64. Murphy EH, Johns B, Varney E, Raju S. Endovascular management of chronic total occlusions of the inferior vena cava and iliac veins. *J Vasc Surg Venous Lymphat Disord*. 2017;5:47–59.
65. Ye K, Lu X, Li W, et al. Long-term outcomes of stent placement for symptomatic nonthrombotic iliac vein compression lesions in chronic venous disease. *J Vasc Intervent Radiol*. 2012;23:497–502.
66. Raju S. Best management options for chronic iliac vein stenosis and occlusion. *J Vasc Surg*. 2013;57: 1163–1169.
67. Raju S. Treatment of iliac-caval outflow obstruction. *Semin Vasc Surg*. 2015;28:47–53.
68. Mousa AY, Broce M, Yacoub M, AbuRahma AF. Iliac vein interrogation augments venous ulcer healing in patients who have failed standard compression therapy along with pathological venous closure. *Ann Vasc Surg*. July 2016;34:144–151.
69. George R, Verma H, Ram B, Tripathi R. The effect of deep venous stenting on healing of lower limb venous ulcers. *Eur J Vasc Endovasc Surg*. 2014;48:330–336.
70. Jayaraj A, Noel C, Kuykendall R, Raju S. Long-term outcomes following use of a composite Wallstent-Z stent approach to iliofemoral venous stenting. *J Vasc Surg Venous Lymphat Disord*. 2021;9:393–400. e2.
71. Raju S, Buck WJ, Crim W, Jayaraj A. Optimal sizing of iliac vein stents. *Phlebology*. August 2018;33(7): 451–457.
72. Kassab G, Raju S. Grading venous stenosis is different from arterial lesions. *J Vasc Surg Venous Lymphat Disord*. 2019;7:151–152.
73. Raju S, Kirk O, Davis M, Olivier J. Hemodynamics of "critical" venous stenosis and stent treatment. *J Vasc Surg Venous Lymphat Disord*. 2014;2:52–59.
74. Di Santo M, Belhaj A, Rondelet B, Gustin T. Intraspinal iliac venous stent migration with lumbar nerve root compression. *World Neurosurg*. 2020;137:372–375.
75. Schwartz C, Hafez A, Lönnrot K, et al. Microsurgical removal of a misplaced intraspinal venous stent in a patient with inferior vena cava atresia. *J Neurosurg Spine*. 2020:1–5.
76. Zaldivar Jolissaint JF, de Schlichting E, Haller C, Morard M. Foot drop after iliocaval vein stenting: radicular syndrome from stent misplacement in spinal canal. *World Neurosurg*. 2020;137:43–45.
77. O'Sullivan GJ, Lohan DA, Cronin CG, Delappe E, Gough NA. Stent implantation across the ostia of the renal veins does not necessarily cause renal impairment when treating inferior vena cava occlusion. *J Vasc Intervent Radiol*. 2007;18:905–908.
78. Murphy ER, Johns B, Varney E, Buck W, Jayaraj A, Raju S. Deep venous thrombosis associated with caval extension of iliac stents. *J Vasc Surg Venous Lymphat Disord*. 2017;5:8–17.
79. Raju S, Powell T, Kuykendall R, Jayaraj A. A unique complication of double barrel Wallstent technique in iliac-caval stenting. *J Vasc Surg Cases Innov Tech*. 2021;7:211–214.
80. Raju S, Knight A, Buck W, Mary C, Jayaraj A. Caliber-targeted reinterventional overdilation of iliac vein Wallstents. *J Vasc Surg Venous Lymphat Disord*. 2019;7(2):184–194.
81. Raju S, Tackett Jr P, Neglen P. Reinterventions for nonocclusive iliofemoral venous stent malfunctions. *J Vasc Surg*. 2009;49:511–518.
82. Jayaraj A, Fuller R, Raju S, Stafford J. In-stent restenosis and stent compression following stenting for chronic iliofemoral venous obstruction. *J Vasc Surg Venous Lymphat Disord*. 2022;10(1):42–51.
83. Saleem T, Burr B, Robinson J, et al. Elevated plasma factor VIII levels in a mixed patient population on anticoagulation and past venous thrombosis. *J Vasc Surg Venous Lymphat Disord*. 2021;9(5): 1119–1127.
84. Darvall KA, Sam RC, Adam DJ, Silverman SH, Fegan CD, Bradbury AW. Higher prevalence of thrombophilia in patients with varicose veins and venous ulcers than controls. *J Vasc Surg*. 2009;49: 1235–1241.
85. Jull AB, Arroll B, Parag V, Waters J. Pentoxyfylline for treating venous leg ulcers. *Cochrane Database Syst Rev*. 2012;12:CD001733.

86. Jull A, Waters J, Arroll B. Pentoxyfylline for treatment of venous leg ulcers: a systematic review. *Lancet*. 2002;359:1550−1554.
87. Jayaraj A, Crim W, Knight A, Raju S. Characteristics and outcomes of stent occlusion after iliocaval stenting. *J Vasc Surg Venous Lymphat Disord*. 2019;7:56−64.
88. Jayaraj A, Powell T, Raju S. Utility of the 50% stenosis criterion in patients undergoing stenting for chronic iliofemoral venous obstruction. *J Vasc Surg Venous Lymphat Disord*. 2021;9(6):1408−1415.
89. Joh M, Desai KR. Treatment of nonthrombotic iliac vein lesions. *Semin Intervent Radiol*. 2021;38:155−159.
90. Murphy EM. Surveying the 2019 venous stent landscape. Endovascular Today. URL: https://evtoday.com/articles/2019-july/surveying-the-2019-venous-stent-landscape [accessed on 6/13/2021].
91. Lichtenberg M, Breuckmann F, Stahlhoff WF, Neglen P, Rick G. Placement of closed-cell designed venous stents in a mixed cohort of patients with chronic venous outflow obstructions − short-term safety, patency, and clinical outcomes. *Vasa*. 2018;47:475−481.
92. Sebastian T, Dopheide JF, Enelberger RP, Spirk D, Kucher N. Outcomes of endovascular reconstruction of the inferior vena cava with self-expanding nitinol stents. *J Vasc Surg Venous Lymphat Disord*. 2018;6:312−320.
93. Badesha AS, Singh Bains PR, Singh Bains BR, Khan T. A systematic review and meta-analysis of the treatment of obstructive chronic deep venous disease using dedicated venous stents. *J Vasc Surg Venous Lymphat Disord*. 2021;S2213−S2333X(21), 00210-9.
94. Lichtenberg MKW, Stahlhoff WF, Stahlhoff S, Özkapi A, Breuckmann F, de Graaf R. Venovo venous stent for treatment of non-thrombotic or post-thrombotic iliac vein lesions—long term efficacy and safety results from the Arnsberg venous registry. *Vasa*. 2021;50:528.
95. de Wolf MA, de Graaf R, Kurstjens RL, Penninx S, Jalaie H, Wittens CH. Short-term clinical experience with a dedicated venous nitinol stent: initial results with the sinus-venous stent. *Eur J Vasc Endovasc Surg*. 2015;50:518−526.
96. Neglen P, Tackett Jr TP, Raju S. Venous stenting across the inguinal ligament. *J Vasc Surg*. 2008;48:1255−1261.
97. Jayaraj A, Noel C, Raju S. Contralateral limb improvement after unilateral iliac vein stenting argues against simultaneous bilateral stenting. *J Vasc Surg Venous Lymphat Disord*. July 2020;8(4):565−571.
98. Raju S, Ward Jr M, Davis M. Relative importance of iliac vein obstruction in patients with post-thrombotic femoral vein occlusion. *J Vasc Surg Venous Lymphat Disord*. 2015;3:161−167.
99. Daye D, Walker TG. Novel and advanced techniques for complex IVC filter retrieval. *Curr Treat Options Cardiovasc Med*. 2017;19:28.
100. Jayaraj A, Noel C, Raju S. Impact of presence of inferior vena cava filter on iliocaval stent outcomes. *Ann Vasc Surg*. 2020;68:166−171.
101. Chick JFB, Jo A, Meadows JM, et al. Endovascular iliocaval stent reconstruction for inferior vena cava filter-associated iliocaval thrombosis: approach, technical success, safety and two year outcomes in 120 patients. *J Vasc Intervent Radiol*. 2017;28:933−939.
102. Ko H, Ahn S, Min S, Hur S, Jae HJ, Min SK. Recanalization of an occluded vena cava filter and iliac veins with kissing stents to treat postthrombotic syndrome with a venous stasis ulcer. *Vasc Specialist Int*. 2020;36:116−121.
103. Rollo JC, Farley SM, Jimenez JC, Woo K, Lawrence PF, Derubertis BG. *J Vasc Surg Venous Lymphat Disord*. 2017;5:789−799.
104. Raju S, Darcey R, Neglen P. Iliac-caval stenting in the obese. *J Vasc Surg*. 2009;50:114−120.
105. Verma H, Tripathi RK. Common femoral endovenectomy in conjunction with iliac vein stenting to improve venous inflow in severe post-thrombotic obstruction. *J Vasc Surg Venous Lymphat Disord*. 2017;5:138−142.
106. Pokrovsky A, Ignatyev I, Gradusov E. First experience performing hybrid operations in chronic venous obstructions of iliofemoral segments in patients with postthrombotic syndrome. *Vasc Endovasc Surg*. 2017;51:447−452.

107. Comerota AJ, Grewal NK, Thakur S, Assi Z. Endovenectomy of the common femoral vein and intraoperative iliac vein recanalization for chronic iliofemoral venous occlusion. *J Vasc Surg*. 2010;52:243−247.
108. Garg N, Gloviczki P, Karimi KM, et al. Factors affecting outcome of open and hybrid reconstructions for nonmalignant obstruction of iliofemoral veins and inferior vena cava. *J Vasc Surg*. 2011;53:383−393.
109. Meissner MH, Elkof B, Smith PC, et al. Secondary chronic venous disorders. *J Vasc Surg*. 2007;46(Suppl S):68S−83S.
110. Gloviczki P, Gloviczki ML. Evidence on efficacy of treatments of venous ulcers and on prevention of ulcer recurrence. *Perspect Vasc Surg Endovasc Ther*. December 2009;21(4):259−268.

CHAPTER 26

Venous valve reconstructions in patients with severe chronic venous insufficiency

Oscar Maleti[1], Marzia Lugli[1] and Michel Perrin[2]
[1]Department of Cardiovascular Surgery, International Centre of Deep Venous Surgery, Hesperia Hospital, Modena, Italy;
[2]Unité de Pathologie Vasculaire Jean Kunlin, Chassieu, France

Introduction

Chronivenous insufficiency (CVI) is a term used to define severe clinical findings related to lower limb venous system dysfunction. The diagnosis of CVI, according to the Clinical, Etiological, Anatomical, and Pathophysiological (CEAP) classification,[1,2] is based on the detection of clinical signs such as edema (C3), skin pigmentation (C4a), trophic skin changes and subcutaneous tissue changes (C4b), and venous ulcers either healed (C5) or active (C6), combined with ultrasonographic or radiologic evidence of venous incompetence. Venous ulceration is distressing and disabling CVI sign, and its management is costly for both patients and society.[3,4] As terms describing deep vein disorders hemodynamics have often been used inconsistently, we begin with a set of definitions according to The Vein Glossary (Table 26.1).[5]

Deep venous reflux can be caused by different etiologies originating from various anatomical lesions or anomalies. Valve repair is rarely used for treating superficial reflux. Valve repair for deep vein reflux is indicated after failure of conservative therapies (medical compression, lifestyle modification, venoactive drugs) or in patients nonresponding to technically successful treatment of deep venous obstruction or superficial venous reflux. It is essential to stent veins with proximal obstruction before treating deep vein reflux, as stenting results in long-term ulcer healing in 68%−80% of cases.[6,7] Superficial and perforator vein(s) insufficiency is generally managed using percutaneous ablation and can provide good results without prior deep vein reflux treatment; however, ulcer recurrence is observed in 30%−50% of cases.[8]

Prolonged marked superficial venous insufficiency can lead to functional deep venous insufficiency based on "overload theory,"[9] which explains how the deep venous system can potentially regain its competence following superficial ablation. However, this reversibility does not always occur, probably due to the concomitant presence of congenital valve anomalies such as cusp asymmetry, in which case direct valve repair is required.[9]

Table 26.1 Definition of terms.

Term	Definition
Venous obstruction	Hemodynamic concept caused by venous obliteration or occlusion.
Venous obliteration	Anatomical concept: narrowing of the venous lumen related to postthrombotic lesions, extrinsic compression, venous parietal lesion.
Venous occlusion	Anatomical concept: total obliteration of the venous lumen related to the lesions described in the term venous obliteration and in the presence of acute venous thrombosis.

Deep venous reflux etiology

Valvular dysfunction has three main etiologies. Primary (CEAP—Ep),[10] secondary (CEAP—Es) valvular dysfunctions (wherein the valves are destroyed or damaged, are related to deep venous thrombosis (DVT)) and congenital (CEAP—Ec) valvular dysfunction that involves atrophic or completely absent (Ec) valves. Valve injury or destruction (CEAP—Es) is by far the most frequent etiology of valvular dysfunction, followed by valve malfunction (CEAP—Ep); the congenital absence (CEAP—Ec) of valves (agenesia) is rare.[11] The main cause of valve injury is DVT, which can result in postthrombotic syndrome (PTS). During the recanalization phase of DVT, inflammation occurs followed by fibrosis, which leads to a more or less complete destruction of the cusp.

Hemodynamic and diagnostic evaluation

The combination of reflux and obstruction renders hemodynamic evaluation difficult, approximately in 40%—70% of patients.[12–14] Reflux directly results from the lack of valve physiologic function[14]; nonetheless, the main mechanism of reflux is the lack of volume and pressure reductions that normally occur in the calf during walking. This increases the residual volume, thereby creating microcirculatory dysfunction and resulting in microcirculatory venous hypertension. All the venous systems involved contribute to this residual volume. Hence, the correction of superficial and/or perforator vein reflux can improve hemodynamic conditions by restoring the efficiency of the venous pump. In contrast, the axial competence reconstruction that neglects associated superficial or parallel vein refluxes through the femoral profunda vein does not correct the hemodynamic damage resulting from reflux.[15–18]

Patient history and physical examination are crucial to eliminate the presence of other diseases. A complete venous duplex evaluation is required when a clinical diagnosis of CVI is established.[1,19,20] The duplex evaluation provides morphological information that clarifies the etiology: postthrombotic parietal lesions, lumen obliteration or occlusion, vein compression, double channel, valve thickening, and valve damage; the duplex evaluation also provides hemodynamic information on axial and segmental reflux. Moreover, the evaluation provides information pertinent to proximal obstruction as well as obstruction in the femoral and popliteal veins.

However, if venous reflux or pathological findings are not detected by duplex evaluation in patients with signs of CVI, the diagnostic protocol described below should be followed as these patients can be candidates for deep vein anomalies correction. The same investigation protocol is recommended when superficial vein reflux is correctly treated without clinical improvement (Fig. 26.1). The diagnostic protocol for CVI assessment is shown in Fig. 26.2.

Air plethysmography (dynamic and occlusion plethysmography) can be used to quantify the impact of deep venous insufficiency before and after deep venous valve reconstruction. Air plethysmography provides a quantitative assessment of venous insufficiency by means of the venous filling index (VFI) (abnormal >2 mL/s) and residual volume fraction (RVF) (abnormal >35%); the latter has a linear correlation with ambulatory venous pressure measurements. Air plethysmography is crucial for evaluating the muscle pump function.[21] The ejection fraction (EF) (abnormal <60%) is correlated with calf muscle pump efficiency, which is a critical factor in obtaining good results after deep venous reconstruction.

Ascending and descending transfemoral venography is used to define the deep venous anatomy, the reflux extent,[22] profunda femoral vein competence, and the extent of iliac vein obliteration or occlusion.[23] Complementary investigations include intravascular

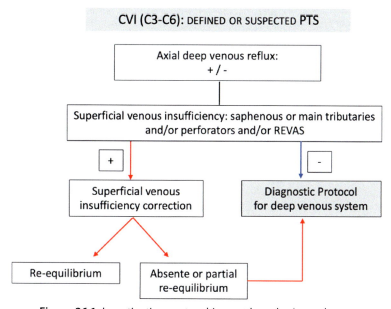

Figure 26.1 Investigation protocol in postthrombotic syndrome.

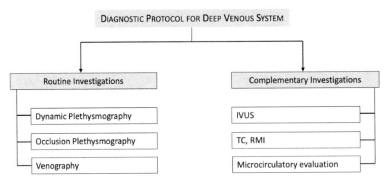

Figure 26.2 Investigation protocol for deep venous system.

ultrasound (IVUS), computed tomography (CT), magnetic resonance imaging (MRI), and microcirculatory evaluation. IVUS represents the gold standard[24] for the identification and correct interpretation of suprainguinal obliterative lesions. MRI and CT are most frequently used to evaluate proximal venous obliteration, particularly in detecting extrinsic compression.[19] Microcirculatory evaluations, performed by means of laser Doppler and/or capillaroscopy, provide further information about the severity of tissue damage.

Indication for deep venous reflux correction

When CVI is not improved by conservative treatment or procedures that treat superficial reflux the full investigation protocol is recommended, particularly in young patients. The surgical indication for deep venous system reflux correction distinguishes between two groups of patients: those with and without an identified proximal obstruction (external and common iliac veins, vena cava). Fig. 26.3 shows the treatment protocol for patients with an identified proximal obliteration responsible for obstruction. Deep venous reflux correction is considered when obstruction correction yields poor results, with a failure of muscular pump improvement or an increase of RVF and VFI. In patients without proximal obstruction, who are eligible for deep venous surgery, the algorithm reported in Fig. 26.4 can be applied. The technique used for correcting the deep venous reflux is based on etiology and pathology of each individual case.

Surgical techniques
Internal valvuloplasty

Internal valvuloplasty, proposed by Kistner in 1968,[25] represents the first-choice intervention for the correction of reflux related to valve malfunction incompetence. The malfunction may be due to multiple variations in the cusps, principally cusp elongation and

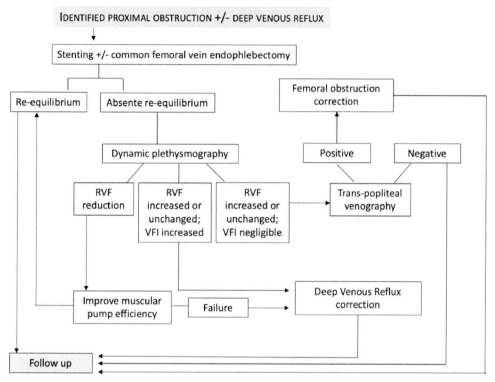

Figure 26.3 Treatment algorithm in postthrombotic syndrome with proximal obstruction.

asymmetry (Fig. 26.5), which do not adhere adequately to each other during the valve closure phase, thus allowing reflux. The technique involved the performance of longitudinal venotomy to adequately visualize the cusps, which were reshaped by applying interrupted 6/0 or 7/0 prolene sutures. Subsequently, several technical variants were made including transverse supracommissural venotomy by Raju,[26] hybrid T-shaped venotomy by Sottiurai,[27] and internal "trap door" venotomy by Tripathi.[28]

Thereafter, another technique was proposed wherein valvuloplasty was performed with the application of transparietal and transcommissural sutures, without performing venotomy.[29] This technique presumes a perfect identification of the parietal line of insertion of the cusps to reduce the angle between them using interrupted or continuous sutures.[30] Moreover, the technique can be performed with or without the use of angioscopic monitoring.[31,32] Furthermore, the technique is applied only in selected cases owing to the risk of creating a stenosis and valve malfunction due to the reduction of the valve sinus, a crucial anatomical structure involved in the correct closure of the cusps.

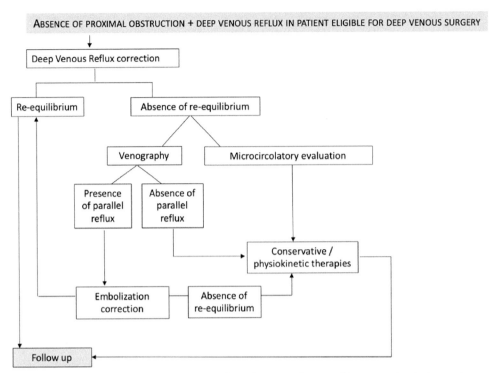

Figure 26.4 Treatment algorithm in postthrombotic syndrome without proximal obstruction.

Transposition

When a valvulated parallel axis is adjacent to the main devalvulated axis, transposition can be performed by dissecting the devalvulated axis and anastomosing it below the competent valve of the parallel system.[33] This procedure can be performed when the deep femoral vein has a competent valve at its termination. Alternatively, the great saphenous vein can be used when its proximal valves are competent. When using a branch of the deep femoral vein to perform side-to-end anastomosis, it is necessary to carefully evaluate the calibers of the donor and recipient axes to avoid subsequent dilation of the recipient axis, which predisposes to premature valvular incompetence. When using the great saphenous vein, it is more advantageous, from the hemodynamic point of view, to dissect the proximal tributaries of the extended saphenous vein and insert the saphenous vein in the subfascial area, where it will be anastomosed end-to-end to the femoral vein sectioned below the confluence with the profunda femoral vein. Transposition can be used only in approximately 20% of patients with PTS.

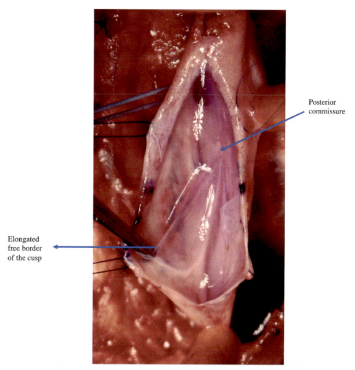

Figure 26.5 Primary incompetence.

Valve transplantation

Valve transplantation consists of inserting a segment of an autologous vein with a competent valve in the incompetent popliteal femoral axis.[34] The donor vein is generally the axillary vein, and rarely the contralateral femoral vein or a segment of the large or small saphenous vein. The segment with valves is inserted at the femoral or popliteal level, and two interrupted sutures are applied to respect the congruence of the caliber of the vein. In cases where the axillary vein is incompetent, it may be repaired using valvuloplasty.[35]

Neovalve

Despite numerous attempts, the use of xenografts and allografts has not been successful, for treating deep vein reflux and the reasons for this remain unclear. Therefore, we focused on the use of venous autologous tissue.[30,36,37] In patients with PTS, it is possible to apply techniques such as valve transposition or transplantation; however, in most cases, it is technically impossible to correct the reflux. The creation of a neovalve, although a nonstandardized and difficult technique, can solve this problem. A neovalve is an antireflux mechanism obtained by creating a parietal flap,[38] using the fibrotic thickening of a postthrombotic venous wall (Fig. 26.6). The neovalve constitutes one or two opposing flaps used to imitate a valve. Fibrosis is usually asymmetrical in patients with PTS, consequently thickening is found only on one part of the vein; hence, it is easier to construct a monocuspid valve.[39,40]

Subsequent modifications in the creation of a neovalve have been proposed[41,42]; nonetheless, it is fundamental to clarify the mechanism of action of a neovalve. The neovalve does not reduce the walking or hydrostatic pressure. These parameters can only be reduced by performing Kistner's valvuloplasty in patients with primary CVI, with monoaxial reflux, and without parallel refluxes. However, the reduction of hydrostatic pressure is not possible in PTS with parallel refluxes, and when the neovalve is not 100% competent.

Since the use of the neovalve yields good clinical results[39] that are not related to venous pressure, the valve probably has a different mechanism of action. Hemodynamic studies have shown that the fundamental parameter in a postthrombotic limb is the volume of blood within the venous system; when this volume does not decrease as a result of muscle contractions (residual volume), symptoms and signs of the CVI ensue.

Deep axial venous reflux prevents the emptying caused by extrinsic muscular compression, which determines the premise for creating microcirculatory hypertension. Reflux volume reduction allows the venous system to partially empty itself and renders other treatment modalities, such as compression therapy, more effective. In fact, before neovalve creation, when a patient stands up and begins to walk, the venous system immediately fills up with blood. Consequently, after neovalve creation, the venous system

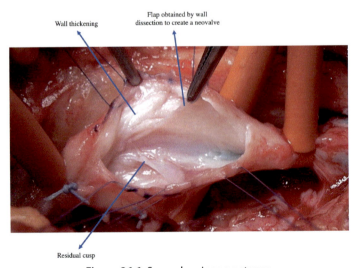

Figure 26.6 Secondary incompetence.

volume takes time to reach a plateau, which is still lower than the volume when it is fully filled. Through this mechanism, the neovalve improves the symptoms and signs of PTS.

Banding

When an external sleeve is placed around the vein at the valve location,[43] the sleeve reduces the vein circumference just enough to restore valve competence. The cuff can be made of synthetic (Dacron, PTFE) or biological (bovine pericardium) material and is fixed to the wall of the vein using interrupted sutures. The same goal can be achieved by applying a spiral around the valve site.[44] Polyester banding can be applied to the popliteal side to reduce the vein circumference by one-third.[45] Generally, distal venous hypertension compensates for this reduction in circumference; therefore, it is necessary to evaluate the entire hemodynamic variation introduced by banding, and not just the abolition of reflux. The application of banding on a segment without competent valves (PTS or valve agenesia) would create a stenosis point that reduces reflux due to the reduction in the fluid static pressure when a fluid flows through a constricted part of a catheter/tube. An increased velocity through a constriction is balanced by a drop in pressure that can be used as a suction mechanism (Venturi effect). In reality, this presumed action is only theoretical, as the creation of a stenosis point with a consequent increase in resistance can negatively affect the cumulative hemodynamics of the limb.

Results

The outcomes of deep venous reflux surgery must be distinguished on the basis of etiology.

In patients with primary deep venous reflux, the first-choice procedure is valvuloplasty. The results of some series[9,18,46–53] are shown in Table 26.2.

Generally, the success rate of valvuloplasty exceeds 70% based on the absence of ulcer recurrence and improvement in hemodynamic parameters for a period of more than 5 years. External valvuloplasty (transparietal and transcommissural) did not show the same results as those of internal valvuloplasty with phlebotomy. Series using banding reported satisfactory results only after a short follow-up period,[54] with the exception of the Lane et al.[43] series which had satisfactory long-term results.

In the secondary etiology responsible for deep vein reflux (PTS), where valvuloplasty is rarely applicable, the efficacy rates are low (Tables 26.3–26.5). This is because the entire venous system is devoid of competent valves and has concomitant parallel refluxes that are more or less associated with obliteration processes at various levels in the limb. In patients with PTS, we achieved good results in approximately 50% of cases over a period of more than 5 years, with a poor correlation between clinical and hemodynamic results.

Table 26.2 Internal valvuloplasty results.

Author, Year	Number of limbs (number of valves repaired)	Etiology PVI/all other etiologies	Number of months of follow-up (mean)	Ulcer recurrence or nonhealed ulcer (%)	Hemodynamic results	
					Competent valve (%)	☐ AVP ■ VRT
Ferris, Kistner 1982[46]	32		12–156 (72)	6/32	16/22	7/8
Eriksson 1988[47]	27	27/27	(49)	/	19/27 (70)	☐ ↗ 81% (av) ■ ↗ 50% (av)
Masuda 1994[48]	32	27/32	48–252 (127)	9/32 (28)	24/31 (77)	☐ ↗ 81% (av) ■ ↗ 56% (av)
Raju 1996[18]	68 (71)		12–144	16/68 (26)	30/71 (42)	/
Sottiurai 1996[49]	143		9–168 (81)	9/42 (21)	107/143 (75)	/
Lurie 1997[50]	49		36–108 (74)	18/49		
Perrin 2000[51]	85 (94)	65/85	12–96 (58)	10/35 (29)	72/94 (77)	■ normalized 63% (av)
Tripathi 2004[52]	90 (144)	102/102	(24)	29/90 (23.3)	115/144 (79.8)	/
Lehtola 2008[53]	12	5/12	24–78	/	/	/
Maleti 2017[9]	13	13/13	(40)	1/13	13/13	12/13

Table 26.3 Transposition results.

Author, Year	Number of extremities treated	Etiology PTS/all other etiologies	Number of months of follow-up	Ulcer recurrence or nonhealed ulcer (%)	Hemodynamic results	
					Competent valve (%)	☐ AVP ■ VRT
Johnson 1981[55]	12	12/12	12	4/12 (33)	/	☐ unchanged ■ unchanged
Masuda - Kistner 1994[48]	14	/	48–252	7/14 (50)	10/13 (77)	☐ ↗ 70% (av) ■ ↗ 70% (av)
Sottiurai 1996[49]	20	/	9–149	9/16 (56)	8/20 (40)	/
Cardon 1999[56]	16	16/16	24–120	4/9 (44)	12/16 (75)	/
Perrin 2000[51]	17	16/17	12–168	2/8 (25)	9/17 (53)	/
Lehtola 2008[53]	14	12/14	24–78	/	(43)	/

Table 26.4 Transplantation results.

Author Year	Number of extremities treated	Site	Etiology PTS/all other etiologies	Number of months of follow-up (mean)	Ulcer recurrence or nonhealed ulcer (%)	Hemodynamic results	
						Competent valve (%)	☐ AVP ■ VRT
Taheri 1982[57]	71	F, P	/	/	1/18 (6)	28/31 (90)	☐ ↗ 15% (av)
Eriksson 1988[47]	35	F, P	35/35	6–60	/	11/35 (31)	■ unchanged
Nash 1988[58]	25	P	25/25	/	3/17 (18)	18/23 (77)	☐ ↗ 18% (av)
Bry 1995[59]	15	P	/	15/132	3/14 (21)	7/8 (87)	☐ unchanged ■ unchanged
Mackiewicz 1995[60]	18	F	/	43/69	5/14 (36)	/	■ improved
Sottiurai 1996[49]	18	F, P	/	7–144	6/9 (67)	6/18 (33)	/
Raju 1996[18]	54	F	/	12–180	/	16/44 (36)	/
Raju 1999[61]	83	F, P, T	83/83	12–180	(40) 6 yrs	(38) 4 yrs	☐ unchanged
Perrin 2000[51]	32	F	31/32	12–124 (66)	9/22 (41)	8/32 (25)	■ ↗ 19% (av)
Tripathi 2004[52]	35	F, P	35/35	(24)	(45)	(41)	/
Lehtola 2008[53]	29	F, P	25/29	24–78 (54)	/	(16)	/
Rosales 2008[62]	22 including 2 double Tr. Tr.+ other procedure	F, P	22/22	6–108	/	GSV Tr. 14/26 AV Tr. 3/6	/
Kabbani 2011[63]	19	FC, P, GSV	12/18	(37)	6/8 (80)	8/19 (42)	

Table 26.5 Neovalve results.

Author Year	Technique	Number of limbs	Etiology PTS/all other etiologies	Number of months of follow-up (mean)	Ulcer recurrence or nonhealed ulcer (%)	Hemodynamic results	
						Competent valve (%)	☐ AVP ■ VRT
Plagnol 1999[36]	Bicuspid neovalve	44	44/44	6–47 (17)	3/32 (17)	38/44 (86)	
Opie 2008[42]	Monocuspid neovalve	14	/	(48)	0/6	13/14 (92)	
Maleti 2009[39]	Monocuspid or bicuspid neovalve	19 + 21 = 40	36/40	2–78 (28.5)	7/40 (17)	13/19 (68) 21/21 (100)	75 VRT improved

Abbreviations (Tables 26.1–26.5).
PVI = Primary Venous Insufficiency.
AVP = Ambulatory Venous Pressure.
VRT = Venous Refill Time.
av = average.
↗ = Improved.
PTS = Post-Thrombotic Syndrome.
Tr. = transposition.
GSV = Great Saphenous vein.
F = Femoral vein.
FC = Common femoral vein.
P = Popliteal vein.
T = Tibial (Posterior) vein.

Clinical results in patients with PTS deteriorate over time due to the development of new superficial refluxes resulting from recurrent varicose veins or perforator incompetence and the deterioration of the deep valve procedure outcome. The hemodynamic assessment of these elements must be performed through careful follow-up. Moreover, in patients with PTS, reflux repair surgery rarely reduces the hydrostatic pressure while walking; hence, it is based on the improvement of other parameters, such as residual volume reduction. Therefore, reflux surgery in patients with PTS should be followed by adequate muscle pump strengthening and the use of an elastic compression, even if the effect of compression is lower than that of pretreatment.

Conclusion

Creating a valve competence mechanism still constitutes a surgical challenge due to the technical difficulties involved and the hemodynamic complexity of lower limbs. Surgery can improve the hemodynamics of the PTS-affected limb in most cases, although the results are maintained only through close follow-up with the implementation of additional therapeutic actions. Nonetheless, deep venous system surgery is safe when performed in centers with high competence in the field, and therefore should be performed when necessary. Reflux surgery should be performed while considering the treatment of obstructive components and pathologies affecting the superficial venous system. Based on the incidence of ulcers in the general population, and despite the success rates of superficial and perforator venous and stenting interventions, there remains a considerable number of patients who could potentially benefit from deep valve reconstruction, and therefore would require establishing a considerable number of centers that are dedicated for this purpose.

References

1. Eklöf B, Bergan JJ, Carpentier PH, et al. For the American venous forum's international Ad Hoc committee for revision of the CEAP classification. Revision of the CEAP classification for chronic venous disorders. A consensus statement. *J Vasc Surg*. 2004;40:1248–1252.
2. Lurie F, Passman M, Meisner M, et al. The 2020 update of the CEAP classification system and reporting standards. *J Vasc Surg: Venous and Lym Dis*. 2020;8:342–352.
3. Criqui MH, Jamosmos M, Fronek A, et al. Chronic venous disease in an ethnically diverse population: the San Diego Population Study. *Am J Epidemiol*. 2003;158:448–456.
4. Rabe E, Guex JJ, Puskas A, Scuderi A, Quesada F, Coordinators VCP. Epidemiology of chronic venous disorders in geographically diverse populations: results from the Vein Consult Program. *Int Angiol*. 2012;31:105–115.
5. Perrin M, Eklof B, Maleti O. *The Vein Glossary*. Goussainville: JPA Imprimeurs; 2018.
6. Raju S, Owen S, Neglen P. The clinical impact of iliac venous stents in the management of chronic venous insufficiency. *J Vasc Surg*. 2002;35:8–15.
7. Williams ZF, Dillavou ED. A systematic review of venous stents for iliac and venacaval occlusive disease. *J Vasc Surg: Venous and Lym Dis*. 2020;8:145–153.
8. Montminy ML, Jayaraj A, Raju S. A systematic review of the efficacy and limitations of venous intervention in stasis ulceration. *J Vasc Surg: Venous and Lym Dis*. 2018;6:376–389.
9. Maleti O, Lugli M, Perrin M. After superficial ablation for superficial reflux associated with primary deep axial reflux, can variable outcomes be caused by deep venous valve anomalies? *Eur J Vasc Endovasc Surg*. 2017;53:229–236.
10. Kistner RL, Eklof B. Classification and etiology of chronic venous disease. In: Gloviczki P, ed. *Handbook of Venous and Lymphatic Disorders*. 4th ed. Boca Raton, FL: CRC Press, Taylor & Francis Group; 2017:39–49.
11. Plate G, Brudin L, Eklof B, Jensen R, Ohlin P. Physiologic and therapeutic aspects in congenital vein valve aplasia of the lower limb. *Ann Surg*. 1983;198(2):229–233.
12. Raju S, Fredericks RK, Hudson CA, et al. Venous valve station changes in "primary" and post-thrombotic reflux: an analysis of 149 cases. *Ann Vasc Surg*. 2000;14:193–199.
13. O'Donnell TF. Chronic venous insufficiency: an overview of epidemiology, classification and anatomic considerations. *Semin Vasc Surg*. 1988;1:60–65.
14. Danielsson G, Arfvidsson B, Eklöf B, Kistner RL, Masuda EM, Sato DT. Reflux from thigh to calf, the major pathology in chronic venous ulcer disease: surgery indicated in the majority of patients. *Vasc Endovasc Surg*. 2004;38:209–219.
15. Eriksson I, Almgren B. Influence of the profunda femoris vein on venous hemodynamics of the limb. Experience from thirty-one deep vein valve reconstructions. *J Vasc Surg*. 1986;4:390–395.
16. Raju S, Fountain T, Neglen P, Devidas M. Axial transformation of the profunda femoris. *J Vasc Surg*. 1998;27:651–659.

17. O'Donnell Jr TF, Mackey WC, Shepard AD, Callow AD. Clinical, hemodynamic and anatomic follow-up of direct venous reconstruction. *Arch Surg*. 1987;122:474–482.
18. Raju S, Fredericks RK, Neglen P, Bass D. Durability of venous valve reconstrcution techniques for primary and postthrombotic reflux. *J Vasc Surg*. 1996;23:357–367.
19. O'Donnell Jr TF, Passman MA, Marston WA, et al. Management of venous leg ulcers: clinical practice guidelines of the Society for Vascular Surgery and the American Venous Forum. *J Vasc Surg*. 2014;60: 3S–59S.
20. Gloviczki P, Comerota AJ, Dalsing MC, et al. The care of patients with varicose veins and associated chronic venous disease: clinical practise guidelines of the Society for Vascular Surgery and the American Venous Forum. *J Vasc Surg*. 2011;53:2S–48S.
21. Araki CT, Back TL, Pagberg FT, et al. The significance of calf muscle function in venous ulceration. *J Vasc Surg*. 1994;20:872–879.
22. Kistner RL, Ferris EB, Raudhawa G, Kamida C. A method of performing descending venography. *J Vasc Surg*. 1986;4:464–468.
23. Neglen P, Raju S. Proximal lower extremity chronic venous outflow obstruction: recognition and treatment. *Semin Vasc Surg*. 2002;15:57–64.
24. Lugli M. IVUS. In: *Guex JJ. Ultrasons et Phlébologie. Les Éditions Phlébologiques Françaises Ed*. 2016. Cap. 20.
25. Kistner RL. Surgical repair of a venous valve. *Straub Clin Proc*. 1968;24:41–43.
26. Raju S. Venous insufficiency of the lower limb and stasis ulceration. Changing concepts and management. *Ann Surg*. 1983;197:688–697.
27. Sottiurai VS. Technique in direct venous valvuloplasty. *J Vasc Surg*. 1988;8:646–648.
28. Tripathi R, Ktenedis KD. Trapdoor internal valvuloplasty – a new technique for primary deep vein valvular incompetence. *Eur J Vasc Endovasc Surg*. 2001;22:86–89.
29. Kistner RL. Surgical technique of external venous valve repair. *Straub Found Proc*. 1990;55:15–16.
30. Raju S, Hardy JD. Technical options in venous valve reconstruction. *Am J Surg*. 1997;173(4):301–307.
31. Nishibe T, Kudo F, Miyazaki K, et al. Intermediate-term results of angioscopy-assisted anterior valve sinus plication for primary deep venous insufficiency. *J Cardiovasc Surg*. 2007;48:21–25.
32. Gloviczki P, Merrell SW, Bower TC. Femoral vein valve repair under direct vision without venotomy: a modified technique with use of angioscopy. *J Vasc Surg*. 1991;14:645–648.
33. Kistner RL. Transposition techniques. In: Bergan JJ, Kistner RL, eds. *Atlas of Venous Surgery*. Philadelphia: W.B. Saunders; 1992:153–156.
34. Taheri SA, Lazar L, Elias SM, Marchand P. Vein valve transplant. *Surgery*. 1982;91:28–33.
35. Raju S, Fredericks R. Valve reconstruction procedures for nonobstructive venous insufficiency: rationale, techniques, and results in 107 procedures with two- to eight- year follow-up. *J Vasc Surg*. 1988;7: 301–310.
36. Plagnol P, Ciostek P, Grimaud JP, Prokopowicz SC. Autogenous valve reconstruction technique for post-thrombotic reflux. *Ann Vasc Surg*. 1999;13:339–342.
37. Maleti O. Venous valvular reconstruction in post-thrombotic syndrome. A new technique. *J Mal Vasc*. 2002;27:218–221.
38. Maleti O, Lugli M. Neovalve construction in postthrombotic syndrome. *J Vasc Surg*. 2006;43(4): 794–799.
39. Lugli M, Guerzoni S, Garofalo M, Smedile G, Maleti O. Neovalve construction in deep venous incompetence. *J Vasc Surg*. 2009;49(1):156–162.
40. Maleti O, Perrin M. Reconstructive surgery for deep vein reflux in the lower limbs: techniques, results and indications. *Eur J Vasc Endovasc Surg*. 2011;41:837–848.
41. Corcos L, Peruzzi G, Procacci T, Spina T, Cavina C, De Anna D. A new autologous venous valve by intimal flap. One case report. *Minerva Cardioangiol*. 2003;51:395–404.
42. Opie JC. Monocusp-novel common femoral vein monocusp surgery uncorrectable chronic venous insufficiency with aplastic/dysplastic valves. *Phlebology*. 2008;23:158–171.
43. Lane RJ, Cuzzilla ML, McMahon CG. Intermediate to long-term results of repairing incompetent multiple deep venous valves using external stenting. *Aust N Z J Surg*. 2003;73:267–274.

44. Makhatilov G, Askerkhanov G, Kazakmurzaev MA, Ismailov I. Endoscopically directed external support of femoral vein valves. *J Vasc Surg*. 2009;49:676−680.
45. Ma T, Fu W, Ma J. Popliteal vein external banding at the valve-free segment to treat severe chronic venous insufficiency. *J Vasc Surg*. 2016;64:438−445.
46. Ferris EB, Kistner RL. Femoral vein reconstruction in the management of chronic venous insufficiency. A 14-year experience. *Arch Surg*. 1982;117:1571−1579.
47. Eriksson I, Almgren B. Surgical reconstruction of incompetent deep vein valves. *J Med Sci*. 1988;93:139−143.
48. Masuda EM, Kistner RL. Long-term results of venous valve reconstruction: a four-to-twenty-one-year follow-up. *J Vasc Surg*. 1994;19:391−403.
49. Sottiurai VS. Current surgical approaches to venous hypertension and valvular reflux. *J Int Angiol*. 1996;5:49−54.
50. Lurie F, Makarova NP, Hmelnicher SM, et al. Results of deep-vein reconstruction. *Vasc Surg*. 1997;31:275−276.
51. Perrin M. Reconstructive surgery for deep venous reflux: a report on 144 cases. *Cardiovasc Surg*. 2000;8:246−255.
52. Tripathi R, Sieunarine K, Abbas M, Durrani N. Deep venous valve reconstruction for non-healing leg ulcers: techniques and results. *Aust N Z J Surg*. 2004;74:34−39.
53. Lehtola A, Oinonen A, Sugano N, Alback A, Lepantalo M. Deep venous reconstructions:long- term outcome in patients with primary or post- thrombotic deep venous incompetence. *Eur J Vasc Endovasc Surg*. 2008;35:487−493.
54. Wang SM, Hu ZJ, Li SQ, Huang XL, Ye CS. Effect of external valvuloplasty of the deep vein in the treatment of chronic venous insufficiency of the lower extremity. *J Vasc Surg*. 2006;44:1296−1300.
55. Johnson ND, Queral LA, Flinn WR, Yao JS, Bergan JJ. Late objective assessment of venous value surgery. *Arch Surg*. 1981;116:1461−1466.
56. Cardon JM, Cardon A, Joyeux A, et al. Use of ipsilateral greater saphenous vein as a valved transplant in management of post-thrombotic deep venous insufficiency: long-term results. *Ann Vasc Surg*. 1999;13:284−289.
57. Taheri SA, Lazar L, Elias S. Status of vein transplant after 12 months. *Arch Surg*. 1982;117:1313−1317.
58. Nash T. Long-term results of vein valve transplants placed in the popliteal vein for intractable post-phlebitic venous ulcers and pre-ulcer skin changes. *J Cardiovasc Surg*. 1988;29:712−716.
59. Bry JD, Muto PA, O'Donnell TF, Isaacson LA. The clinical and hemodynamic results after axillary-to-popliteal vein valve transplantation. *J Vasc Surg*. 1995;21:110−119.
60. Mackiewicz Z, Molski S, Jundzill W, Stankiewicz W. Treatment of postphlebitic syndrome with valve transplantation: five year experience. *Eurosurgery 1995. Bologna Monduzzi*. 1999:305−310.
61. Raju S, Neglen P, Doolittle J, Meydrech EF. Axillary vein transfer in trabeculated postthrombotic veins. *J Vasc Surg*. 1999;29:1050−1064.
62. Rosales A, Jorgensen JJ, Slagsvold CE, Stranden E, Risum O, Kroese AJ. Venous valve reconstruction in patients with secondary chronic venous insufficiency. *Eur J Vasc Endovasc Surg*. 2008;36:466−472.
63. Kabbani L, Escobar GA, Mansour F, Wakefield TW, Henke PK. Longevity and outcomes of axillary valve transplantation for severe lower extremity chronic venous insufficiency. *Ann Vasc Surg*. 2011;25:496−501.

SECTION 5

Special considerations

CHAPTER 27

Treatment of recalcitrant venous ulcers with free tissue transfer for limb salvage

Grant R. Darner[1] and David A. Brown[2]

[1]Duke University School of Medicine, Durham, NC, United States; [2]Division of Plastic, Maxillofacial, and Oral Surgery, Duke University School of Medicine, Durham, NC, United States

Introduction

Venous leg ulcers (VLUs) are open lesions of the lower limb and are the most severe sequelae of chronic venous insufficiency.[1] VLUs represent between 60% and 80% of all leg ulcerations that occur in the presence of venous disease.[2] Healing is slow and unpredictable with only 60% of ulcers healing by 12 weeks and once healed, a recurrence rate of 75% within three weeks has been documented.[2] VLUs typically occur in the gaiter region of the leg and consist of a full-thickness defect surrounded by diseased lipodermatosclerotic (LDS) tissue (Fig. 27.1A and B).

VLUs are typically treated via a conservative multimodal approach including compression therapy, wound care, and possibly pharmacologic intervention.[4] While most patients can be treated conservatively, a subset of patients will not respond to

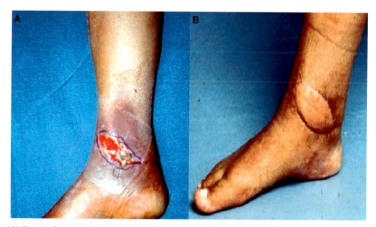

Figure 27.1 (A) Typical-appearing venous ulcer on medial aspect of lower third of extremity. Note the surrounding area of LDS that must be considered as part of the pathologic process of venous ulceration. (B). Appearance of the venous ulcer following excision of ulcer surrounding area of LDS and coverage with a fasciocutaneous RFF. *(Images modified from Kovach SJ, Levin LS. Treatment of recalcitrant venous ulcers with free tissue transfer for limb salvage. In:* Venous Ulcers; *2007:261–274.)*

medical therapy, and treatment modalities (both invasive and minimally invasive) aimed at treating underlying venous hypertension become necessary. Examples of current techniques aimed at surgically correcting venous hypertension include venous angioplasty, venous stenting, venous ablation, venous bypass, venous ligation/stripping, and venous excision.[5] For patients who cannot be successfully treated with conventional medical and surgical management, recurrent and recalcitrant venous ulceration may become a limb-threatening condition. It is at this point that reconstructive techniques should be considered. It' worth emphasizing that the above-mentioned treatment approach sequence is partly due to the current mandate by many United States insurers that compression must be attempted before intervention is considered.[5]

Reconstructive methods of wound closure can be arranged in a hierarchical form increasing from simple to complex. This hierarchical paradigm is commonly employed by reconstructive surgeons and is known as the reconstructive ladder. The reconstructive ladder is utilized as an incremental stepwise approach to select the most appropriate method of achieving wound closure. The lowest rungs on the reconstructive ladder consist of allowing the wound to heal by secondary intention, primary closure, and closure with assisted negative pressure wound therapy (NPWT). As previously mentioned, rates of healing by secondary are poor, with less than 40% of new ulcers healing spontaneously and a recurrence rate of 75% for ulcers that do heal.[2] By the time patients present to a reconstructive surgeon, they will have already undergone lengthy attempts at healing by secondary intention while undergoing conventional therapy. Primary closure is of little use in this population as venous stasis ulcers are often too large to be closed primarily, primary closure does not address the underlying pathology, and the defects that result from the excision of the ulcer and diseased tissue substantially increase the area to be reconstructed. NPWT refers to wound dressing systems that continuously or intermittently apply subatmospheric pressure to the system. NPWT is thought to facilitate healing by accelerating angiogenesis, increasing beneficial growth factors, decreasing wound exudate, lowering bacterial concentration, and generating prohealing mechanical forces.[6] Current data suggest that NPWT may be more effective than conventional wound care techniques at reducing time to complete healing in people with venous ulcers of at least 6 months duration. However, it has also been shown that NPWT may be no more effective at reducing median time to or rate of ulcer recurrence compared with conventional wound care techniques for patients with VLUs (see Chapter 14).[7]

Higher on the reconstructive ladder is skin grafting. For patients with VLUs, skin grafting has been met with limited success because chronic venous ulcers have poor wound beds for the imbibition and vascular ingrowth needed to support grafting. The dermal microcirculation has been shown to have pathologic changes that are not conducive to the healing of skin grafts.[8] According to a 2013 Cochrane review, there is insufficient evidence for the effectiveness of autografts or allografts for the treatment of VLUs. However, there is some evidence that bilayer tissue-engineered skin replacement used with compression increases the rate of healing of VLUs compared with simple dressing used with compression.[9]

Local flaps represent the next rung of the reconstructive ladder. Local flaps are notoriously limited in the distal third of the leg with the reverse sural and propeller flap being used most frequently. More importantly, the local tissue available for reconstruction is typically involved in the disease process and is sclerotic, which limits its mobility and utility. Data regarding the use of local flaps for coverage of VLUs is exceedingly sparse and limited to three patients who underwent coverage via pedicled soleus muscle flaps as part of a larger series.[10] In this series, the recurrence rate of local flaps was comparable to that of free flaps. The authors of this series concluded that the pedicled soleus flap is an easy and safe procedure that can be used when the defects are not very large.

Free tissue transfer is the most complex reconstructive strategy that can be employed and has emerged as a valuable means of soft tissue reconstruction of the lower extremity.[11] Among the reasons for the failure of all free flaps, venous thrombosis is the leading etiology.[12] Therefore, the underlying venous pathology in the extremity is of obvious concern. The literature that comprises free tissue transfer for recalcitrant VLUs is limited; however, from peer-reviewed studies (Table 27.1), success is reported to range from 90%

Table 27.1 Published series of free tissue transfer for recalcitrant venous ulcers.

Series	Pts/Flaps	Flap type	Flap failure[a]	Recurrence[b]	Mean F/U (y)
Dunn et al.[16]	6/8	7 Parascapular	0/7	0/7	1.3
		1 Anterolateral thigh	0/1	0/1	7.6
Weinzweig[17]	18/20	13 Rectus abdominis	1/13	0/12	2.7
		5 Latissimus dorsi	0/5	0/5	2.6
		1 Serratus	1/1	—	—
		1 Gracilis	0/1	0/1	2.9
Steffe et al.[18]	11/14	8 Latissimus dorsi	1/8	7/7	5.6
		3 Serratus muscle	0/3	3/3	3.5
		1 Serratus fascia	0/1	1/1	1.5
		1 Parascapular	0/1	1/1	—
		1 Deltoid skin flap	1/1	—	—
Kumins et al.[19]	22/25	16 Rectus Abdominis			
		5 Latissimus dorsi			
		2 Omentum	1/25	0/24	4.8
		1 Gracilis			
		1 Serratus muscle			
Isenberg et al.[20]	9/9	9 Radial forearm	0/9	0/9	3.2
Kawamura et al.[10]	8/8	5 Parascapular	0/5	2/5	11.8
		3 Latissimus dorsi	0/3	1/3	8.0

[a]Flap failure is defined as total flap necrosis.
[b]Recurrence is defined as recurrent ulceration within the flap or at the flap/wound border.

to 100%. The surgical technique of free tissue transfer for coverage of venous ulcers began as limited case reports[13–15] and now has emerged as a clinically effective means of reconstruction in properly selected patients.[10,16–20]

Indications and preoperative planning

When evaluating a patient with recalcitrant VLU as a candidate for free tissue transfer, the first step is to ensure that the etiology of chronic ulceration is in fact venous insufficiency.[3] As outlined in other chapters of this text, the diagnosis of venous insufficiency can be made by physical exam and venous duplex ultrasound demonstrating the presence of venous reflux.[21] Patients who are referred to the reconstructive surgeon as candidates for free tissue transfer generally will have already failed lengthy attempts at conservative therapy. In addition, such patients have also usually undergone surgical interventions to correct their venous hypertension and may have undergone prior attempts at closure through means such as excision and skin grafting. Ideally, patients should be free of risk factors and underlying comorbidities that may complicate a significant surgical undertaking. Because of the duration of general anesthesia required for most free flaps— generally 4–8 h——a careful cardiovascular risk assessment is required for all patients. Patients at significantly elevated risk for major adverse cardiovascular events in the perioperative period should be excluded from consideration for these procedures. Once the wound has progressed to a state where free tissue transfer is considered, a thorough risk/benefit analysis of the limb salvage attempt versus amputation should be discussed with the patient. In the event that the patient has modifiable risk factors such as smoking, uncontrolled diabetes, and malnutrition, potential surgery should be delayed until these risk factors are addressed and corrected. Additionally, the patient should be counseled on the likelihood of flap compromise requiring a return to the operating room, which may be as high as 15% of flaps in this patient population. Flap loss can also occur, which would be followed by another attempt at free tissue transfer, a suboptimal secondary attempt at reconstruction, or amputation.

Free tissue transfer requires suitable inflow and outflow vessels for microvascular anastomosis. In many cases, the presence or quality of these vessels determines whether the procedure may be safely performed. Because of the coincidence of venous stasis, peripheral arterial disease, and/or diabetes in some patients, arteries of the lower extremity may be calcified and/or occluded. A preoperative CT angiogram or peripheral angiogram should be strongly considered in patients with suspected abnormal anatomy or vasculopathy. Recently, a retrospective review was performed consisting of 57 patients with chronic limb-threatening lower extremity wounds who underwent lower extremity arteriography and 59 free flap operations for limb salvage.[22] Angiographic abnormalities were observed in 67.8% of patients and 27.5% of patients required endovascular intervention. Stenosis/occlusion was detected in 15.3% of patients with no previously known arterial disease, leading to a newly diagnosed peripheral vascular disease. The authors used arteriographic findings to guide flap recipient vessel selection and reported a flap survival rate of 98.3% with 10.2% of patients progressing to amputation. The results of this study suggest that preoperative lower extremity arteriography is useful in optimizing flap recipient vessel selection, preventing flap loss, and in detecting unknown arterial abnormality and allowing for direct endovascular intervention.

Additionally, free tissue transfer to the lower extremity benefits from optimization of venous outflow, which is usually produced by prolonged leg elevation and a strict protocol for progressive gravity dependence, or dangling, of the extremity. The prescribed duration of strict leg elevation varies between surgeons but is typically on the order of 7–14 days, followed by a 4–6 week period of mostly bedrest with progressive dangling of the leg. Even one episode of early prolonged dangling can result in thrombosis within the venous outflow of the flap and ultimately flap failure. Patients at significant risk of pressure injury, thromboembolism, or those without resources to complete the leg elevation and dangling protocol should not be considered for lower extremity-free flaps.

Flap selection

There are several factors to be considered when choosing the appropriate free flap for VLUs, which are almost exclusively located in the lower 1/3 of the leg. The number of free flaps available to reconstructive surgeons has continued to grow throughout the years, with over 100 potential donor sites reported.[23] Today, the most commonly used free flaps for reconstruction in the distal third of the leg are the anterolateral thigh flap (ALT), the radial forearm flap (RFF), the lateral arm flap, the rectus abdominis flap (RAF), the gracilis flap (GF), and the latissimus dorsi flap (LDF).[24]

It is important to note that the role of free tissue transfer is not only soft tissue coverage, but also improvement in venous hemodynamics resulting from functional venous valves within the venous system of the transferred flaps. Watterson et al. conducted an anatomic study of the venous system in 10 different muscles and revealed the presence of microvenous valves in all of the muscles.[25] Other studies have explored the venous system of several common free flaps, such as the RFF and the parascapular flap, and demonstrated that the veins contain numerous valves with anatomic and histologic characteristics of veins in the lower extremities.[16,26,27] These anatomic findings and reported clinical outcomes suggest that any type of flap may be used for improving venous hemodynamics and that the most important point when selecting the flap is that it has the potential to cover the defect after complete removal of the ulcer and surrounding LDS tissue.

If complete removal of the LDS tissue and coverage of the resulting defect with healthy tissue are not accomplished, the remaining LDS tissue may cause new ulceration in the future (Fig. 27.2). It is crucial to consider that there is typically a large area of LDS tissue underneath the visible ulcer that will require wide debridement, thus the flap will

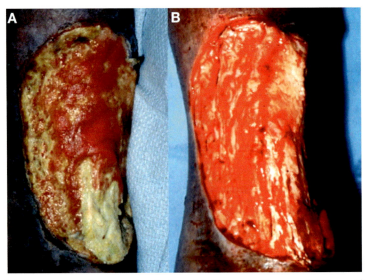

Figure 27.2 (A). Appearance of venous ulcer before debridement. All areas of LDS must be debrided if at all possible. Debridement should include the underlying fascia, and all ectatic veins that are encountered should be ligated. (B). Appearance of venous ulcer in (A) after radical debridement. *(Images modified from Kovach SJ, Levin LS. Treatment of recalcitrant venous ulcers with free tissue transfer for limb salvage. In: Venous Ulcers; 2007:261–274.)*

need to be larger than expected from the size of the ulcer alone. In fact, it appears that almost all cases of ulcer recurrence occur at the flap/wound margin, which has been attributed to underestimating the amount of diseased tissue that must be removed.[10,18]

The patient must have a suitable source of donor tissue to cover the defect. As described in the proceeding section, each free flap has unique considerations for donor site morbidity. The flap of choice should avoid areas of prior surgery and attempt to minimize functional and aesthetic morbidity. For example, patients who have undergone upper extremity hemodialysis access procedures should not be considered for RFF in that arm, while patients with prior ventral hernia repairs should not be offered RAFs. Flaps harvested from the lower extremities such as the ALT and the GF will be complicated in the presence of significant venous stasis, obesity, or lymphedema of the thigh and should be reconsidered in these patients. One must also consider if the patient will rely on crutches following reconstruction, if so, the LDF should be avoided. Another common scenario is reconstruction in the obese patient. The RAF causes weakening of the abdominal wall and, if possible, is best avoided in obese patients.[28]

Surgical technique

If possible, the patient is positioned on the table to allow simultaneous excision of the ulcer and surrounding area of LDS and harvesting of the flap. Before tissue transfer, the ulcer and LDS are excised under tourniquet control down to and including the free fascia. It is paramount to remove all LDS otherwise the patient will likely require additional surgical intervention in the future (Fig. 27.2). Typical donor vessels are the posterior tibial and anterior tibial arteries for posterior/medial and anteriorly based defects, respectively. Regardless of the flap chosen, general microsurgical principles should be applied. Free tissue transfer in the chronic wound population is often complicated with limited and compromised vascularity and some high-volume groups have adopted institutional use of a longitudinal slit arteriotomy end-to-side arterial anastomoses to minimize intimal insult.[29] In this technique, the flap artery is prepared with a 70 degree bevel in the direction of the flap to obtain a final 20 degree resting angle. The final incision length should be 1 × 3 the flap artery diameter and made in the area of least calcification (if present). Anastomosis should be made using an inside-to-outside approach utilizing a continuous stitch. This approach will tack plaques against the vessel wall, prevent intimal shearing, and allow for maximal visualization during the anastomosis.[29] The venous anastomosis is performed in an end-to-end fashion to the venous comitantes. If the venous comitantes are deemed inadequate, a suitable vein graft is harvested from the contralateral saphenous vein or cephalic vein for an interposition graft to the popliteal vein. The flap is inset with absorbable sutures to the surrounding skin under a moderate amount of stretch. Attempts at aggressive flap inset easily can impinge on the low-pressure venous system and therefore an implantable Cook-Swartz Doppler Probe (Cook Medical, Bloomington, New Jersey) should be used for real-time feedback during inset. If a muscle flap is employed, it is skin grafted with a split-thickness graft that is meshed at a ratio of 1.5−1.

Postoperative care

Postoperative care should be tailored to the individual. Generally, flap monitoring should include a combination of clinical examination (color, temperature, capillary refill) and Doppler assessment of the arterial and venous system. Flap checks should occur at regular intervals with decreasing frequency over several days. Starting around postoperative day 5−7, patients may begin a graduated dangling protocol. Once the flap shows stability after 45 min of continuous gravitational dependence, patients may be safely discharged. Prolonged standing is not allowed for 4−6 weeks at which point physical rehabilitation should begin. Routine skin graft care is employed if a muscle flap and skin graft were used.

Discussion

Dunn et al.[16] were the first to publish a series of successful free tissue transfers in patients with intractable venous ulcers in 1994. Their series consists of seven parascapular flaps and one ALT in six patients over a 7-year period. There were no recurrent ulcers and no flap loss at a follow-up of 24 months. Only minor wound healing issues were reported in two patients.

Weinzweig et al.[30] reported a series of 20 free muscle flaps in 18 patients. In contrast to Dunn et al., all patients underwent muscle flaps with the RAF being the most common, followed by the latissimus dorsi flap, GF, and serratus muscle flap. There were two instances of total flap loss, both of which were attributed to intractable vasospasm. Five patients had partial skin graft loss but there were no ulcer recurrences at an average follow-up of 32.7 months.

Steffe et al.[18] reported a series of 14 free flaps to the lower extremities of 11 patients with over a 10 year period. The authors performed 11 muscle flaps (8 LDF and 3 serratus muscle), and three fascial flaps (1 serratus fascia, 1 parascapular, and 1 deltoid skin flap). Two patients had partial flap loss and two flaps were lost secondary to venous thrombosis. Notably, 100% of patients developed new ulcers at the flap margins 1–72 months from the time of reconstruction. All but two recurrent ulcers required operative treatment. All patients required skin grafting and three patients underwent amputations. Although the reason for the poor outcomes in this study is not clear, it is most likely secondary to incomplete excision of involved LDS areas.

Kumins et al.[19] reported a series of 25 free flaps in 22 patients. This consisted of 23 muscle flaps (16 RAF, 5 LDF, 1 GF, and 1 serratus muscle flap) and two omental flaps in the same patient with bilateral ulcers. All patients had previously failed grafting and had large areas of ulceration (average 237 cm^2). 46% of patients had exposed tendon, bone, or joint. They reported only one flap loss secondary to vasospasm. Seven patients experienced a partial loss of flap or graft and three of these patients required additional procedures. In the successful free flap patients, no ulceration was noted in areas covered by the flap at a mean follow-up of 58 months. However, three patients had new ulcers arise in the same extremity but outside the reconstructed area after 6–77 months.

Isenberg et al.[20] published a series of nine free flaps in nine patients. All patients had multiple prior failed attempts at the closure of their wounds. All nine patients underwent reconstruction with a fasciocutaneous RFF. There were no instances of flap loss, either partial or complete. However, two patients had delayed healing at the flap-wound margin resulting in an overall complete wound healing rate of 78%.

The most recent case series was published in 2007 by Kawamura et al.[10] This series consists of eight flaps in eight patients with five fasciocutaneous parascapular flaps and three myocutaneous LDFs. All transferred flaps survived completely and there were four instances of ulcer recurrence. None of the recurrent ulcers involved the territory of the transferred flap. Three of four ulcers recurred around the flap perimeter, and the other one recurred at the popliteal region apart from the transferred flap. Three of four recurrent ulcers were healed with resection and primary closing, conventional skin graft, and transposition of the transferred flap, but the other one developed largely around the flap and did not heal with continued dressing changes.

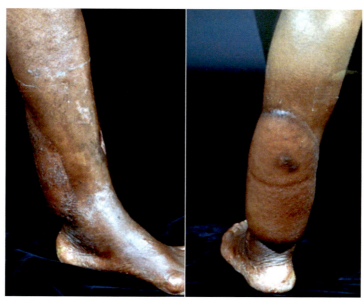

Figure 27.3 Long-term follow-up of patient in Fig. 27.2 after undergoing free myocutaneous LDF. *(Images modified from Kovach SJ, Levin LS. Treatment of recalcitrant venous ulcers with free tissue transfer for limb salvage. In: Venous Ulcers; 2007:261–274.)*

The data presented here represent all series of free tissue transfer for recurrent venous ulceration that has been published. It is of interest why there has been a paucity of studies published in recent years. It is unclear whether this reflects a decrease in demand for free flap reconstructions for VLU management or simply due to a lack of publishing activity. Given the scarcity of data, there are several outstanding questions surrounding the use of free tissue transfer for the treatment of venous stasis ulcers: (1) What type of flap provides the best outcomes? (2) How much tissue surrounding the ulcer needs to be removed to faithfully prevent recurrence? (3) What is the longevity of these free flaps and is long-term healing accomplished in these cases? What can be stated with confidence is that free tissue transfer has been shown to be a valuable tool for achieving soft tissue coverage of intractable venous ulcer and is potentially a limb-saving procedure for patients who have exhausted all other options (Fig. 27.3).

Conclusion

Patients who have failed to heal venous ulcers after long trials of conservative therapy, surgical or endovenous correction of venous hemodynamics, and lesser attempts at wound closure become candidates for free tissue transfer. There have been six published case series comprising a total of 74 patients and 84 free flaps for recalcitrant venous ulceration with an overall flap survival rate of 94%. This number is comparable to accepted rates of success in other anatomic sites and for lower extremity reconstruction or limb salvage for other etiologies. However, we advise the surgeon to proceed with caution in implementing free flaps for vascular disease of the lower extremity, as these procedures require a lengthy duration of general anesthesia, are relatively fragile in the immediate postoperative period, and require at least one week in the hospital followed by several postoperative visits. While there are no larger studies comparing limb salvage versus amputation in the venous stasis ulcer population, we can infer from the trauma population that functional outcomes may be similar in the long term.[31] Nonetheless, in cases of recurrent venous ulceration and exposed vital structures, free tissue transfer may be the most suitable option for the individual patient.

References

1. Lurie F. *Advanced Stages of Chronic Venous Disease: Evolution of Surgical Techniques and Advantages of Associated Medical Treatment*; 2020. https://link.springer.com/content/pdf/10.1007/s12325-019-01216-w.pdf.
2. Probst S, Weller CD, Bobbink P, et al. Prevalence and incidence of venous leg ulcers—a protocol for a systematic review. *Syst Rev*. 2021;10:148.
3. Kovach SJ, Levin LS. Treatment of recalcitrant venous ulcers with free tissue transfer for limb salvage. *Venous Ulcer*. 2007;36:261—274.
4. Raffetto JD, Ligi D, Maniscalco R, Khalil RA, Mannello F. Why venous leg ulcers have difficulty healing: overview on pathophysiology, clinical consequences, and treatment. *J Clin Med*. 2020;10:29.
5. Gloviczki P, Dalsing MC, Henke P, et al. Report of the society for vascular surgery and the American venous forum on the july 20, 2016 meeting of the medicare evidence development and coverage advisory committee panel on lower extremity chronic venous disease. *J Vasc Surg Venous Lymphatic Disord*. 2017;5:378—398.
6. Kunze KN, Hamid KS, Lee S, Halvorson JJ, Earhart JS, Bohl DD. Negative-pressure wound therapy in foot and ankle surgery. *Foot Ankle Int*. 2020;41:364—372.
7. Vuerstaek JDD, Vainas T, Wuite J, Nelemans P, Heumann M, Veraart J. State-of-the-art treatment of chronic leg ulcers: a randomized controlled trial comparing vacuum-assisted closure (V.A.C.) with modern wound dressings. *J Vasc Surg*. 2006;44:1029—1037.
8. Pappas PJ, Lal BK, Padberg FT, Zickler RW, Duran WN. *The Vein Book*. 2007:89—101. https://doi.org/10.1016/b978-012369515-4/50012-0.
9. Jones JE, Nelson EA, Al-Hity A. Skin grafting for venous leg ulcers. *Cochrane Db Syst Rev*. 2013;1:CD001737.
10. Kawamura K, Yajima H, Kobata Y, Shigematsu K, Takakura Y. Long-term outcomes of flap transfer for treatment of intractable venous stasis ulcers in the lower extremity. *J Reconstr Microsurg*. 2007;23:175—179.
11. Pederson W, Grome L. Microsurgical reconstruction of the lower extremity. *Semin Plast Surg*. 2019;33:054—058.
12. Ahmadi I, Herle P, Miller G, Hunter-Smith DJ, Leong J, Rozen WM. End-to-End versus end-to-side microvascular anastomosis: a meta-analysis of free flap outcomes. *J Reconstr Microsurg*. 2017;33:402—411.
13. RJ A, R C, C D, TF C. Management of chronic venous insufficiency ulcers with free flaps. *Wounds*. 1989;17:193—197.
14. OM R. The effectiveness of the free muscle flap in the treatment of the recalcitrant venous stasis ulceration. *Plast Surg Forum*. 1992:77—78.
15. WM S. Free tissue transfers for intractable chronic venous ulcerations: a long term evaluation. In: *Proceedings of the Annual Meeting of the American Association of Plastic Surgeons*. 1989.

16. Dunn R, GM F, Walton R, Malhorta R. Free flap valvular transplantation for refractory venous ulceration. *J Vasc Surg*. 1994;19:525–531.
17. Weinzweig N, Schuler J. Free tissue transfer in treatment of the recalcitrant chronic venous ulcer. *Ann Plast Surg*. 1997;38:611–619.
18. Steffe TJ, Caffee HH. Long-term results following free tissue transfer for venous stasis ulcers. *Ann Plast Surg*. 1998;41:131–139.
19. Kumins NH, Weinzweig N, Schuler JJ. Free tissue transfer provides durable treatment for large non-healing venous ulcers. *J Vasc Surg*. 2000;32:848–854.
20. Isenberg JS. Additional follow-up with microvascular transfer in the treatment of chronic venous stasis ulcers. *J Reconstr Microsurg*. 2001;17:603–606.
21. Coleridge-Smith P, Labropoulos N, Partsch H, Myers K, Nicolaides A, Cavezzi A. Duplex ultrasound investigation of the veins in chronic venous disease of the lower limbs—UIP consensus document. Part I. Basic principles. *Eur J Vasc Endovasc*. 2006;31:83–92.
22. Janhofer DE, Lakhiani C, Kim PJ, et al. The utility of preoperative arteriography for free flap planning in patients with chronic lower extremity wounds. *Plast Reconstr Surg*. 2019;143:604–613.
23. Shaw WW. Microvascular free flaps the first decade. *Clin Plast Surg*. 1983;10:3–20.
24. Kozusko SD, Liu X, Riccio CA, et al. Selecting a free flap for soft tissue coverage in lower extremity reconstruction. *Injury*. 2019;50:S32–S39.
25. Watterson PA, Taylor GI, Crock JG. The venous territories of muscles: anatomical study and clinical implications. *Br J Plast Surg*. 1988;41:569–585.
26. Butz PC, Smahel J. [Morphology and topography of valves of the superficial venous system of the forearm and dorsum of the foot (a microsurgical dissection study using scanning electron microscopy]. *Handchirurgie Mikrochirurgie Plastische Chir Organ Der Deutschsprachigen Arbeitsgemeinschaft Für Handchirurgie Organ Der Deutschsprachigen Arbeitsgemeinschaft Für Mikrochirurgie Der Peripher Nerven Und Gefässe Organ Der Vereinigung Der Deutschen Plastischen Chir*. 1985;17(Suppl):3–7.
27. Aharinejad S, Dunn RM, Nourani F, Vernadakis AJ, Marks SC. Morphological and clinical aspects of scapular fasciocutaneous free flap transfer for treatment of venous insufficiency in the lower extremity. *Clin Anat*. 1998;11:38–46.
28. Lee K-T, Mun G-H. Effects of obesity on postoperative complications after breast reconstruction using free muscle-sparing transverse rectus abdominis myocutaneous, deep inferior epigastric perforator, and superficial inferior epigastric artery flap. *Ann Plast Surg*. 2016;76:576–584.
29. Bekeny JC, Zolper EG, Steinberg JS, Attinger CE, Fan KL, Evans KK. Free tissue transfer for patients with chronic lower extremity wounds. *Clin Plast Surg*. 2021;48:321–329.
30. Weinzweig N, Schlechter B, Baraniewski H, Schuler J. Lower-limb salvage in a patient with recalcitrant venous ulcerations. *J Reconstr Microsurg*. 1997;13:431–437.
31. Higgins TF, Klatt JB, Beals TC. Lower extremity assessment project (LEAP) – the best available evidence on limb-threatening lower extremity trauma. *Orthop Clin N Am*. 2010;41:233–239.

CHAPTER 28

Management of venous ulcers in patients with congenital vascular malformations

Jovan N. Markovic[1] and Byung-Boong Lee[2]

[1]Department of Surgery, Division of Vascular Surgery, Duke University School of Medicine, Durham, NC, United States;
[2]Department of Surgery, Division of Vascular Surgery, George Washington University, Washington, DC, United States

Congenital vascular malformations (CVMs) result from dysmorphogenesis during various stages of the embryologic development of the vascular system.[1-3] It is estimated that CVMs affect approximately 1.2%–1.4% of the general population.[4] Predominantly venous malformations (VMs) (Fig. 28.1) are the most common subtype of CVMs representing nearly 2/3 of all CVMs, with an estimated incidence of 0.8%–1%.[5] Most VMs occur as sporadic lesions, but in rare cases (1%–2%), they are familial, with an autosomal dominant inheritance pattern.[6,7] In recognition that most CVMs do not have familial distribution, it has been conjectured that they result from somatic mutations that would be lethal if they affected the germline. A mounting amount of data is becoming available to support this hypothesis and provide a fertile environment for developing novel treatment modalities. Relatively recently, it has become evident that mutations in the PIK3CA gene are present in a significant number of CVM patients.[8-10] In addition, Klippel–Trenaunay syndrome (KTS) is characterized by phenotypic features that overlap with phenotypes of patients with PI3K-AKT pathway mutations. Hotspot mutations in the PIK3CA gene include E542K and E545K as well as H1047R.[11] In addition, RAS/MAPK/MEK signaling pathway mutations have been observed in patients with complex vascular anomalies.[12,13]

CVM appearance or progression can be precipitated by trauma, infection, the effects of hormones (during puberty or pregnancy), and treatment; however, they may develop without any identified triggering factors. These lesions are present at birth but may not become clinically apparent until later in life. When apparent at birth, their clinical course is characterized by the absence of spontaneous regression. This is an important clinical characteristic used to differentiate between CVMs and vascular tumors, most notably hemangiomas.[1,2,14-16]

CVMs represent a complex group of vascular lesions histologically characterized by nonproliferative endothelial cellular growth pattern that leads to structural and hemodynamic abnormalities of the affected vascular segments, which subsequently can become clinically characterized by unpredictable course and a broad range of symptoms, including bleeding, pain, hemorrhage, ulcerations as well as negative affect on patient's appearance, psychological well-being, and quality of life.[17]

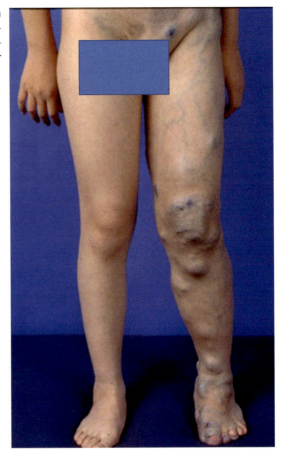

Figure 28.1 Extensive venous malformation involving pelvis and left lower extremity. Significant venous hypertension represents hemodynamic environment responsible for development of leg ulcers.

As mentioned earlier, CVM can lead to the development of leg ulcers that can appear similar to venous leg ulcers caused by venous hypertension secondary to valvular incompetence and/or venous outflow obstruction. However, in patients with CVM, additional pathophysiologic mechanisms for altered hemodynamics have to be considered and appropriately evaluated. Most notably, the presence of arterio-venous (AV) shunts in patients with arterio-venous malformations (AVM) can cause an increased influx of blood into the venous system with subsequent elevation of venous pressure and a well-known cascade of events associated with resulting venous hypertension that leads to signs of chronic venous insufficiency (CVI), including leg ulceration. Secondly, the prevalence of aplasia or hypoplasia of deep venous structure is considerable in CVM patients. This leads to the obstruction of venous outflow and resulting venous hypertension. Furthermore, the lack of involution of primordial veins, including lateral marginal vein (LMV), provides a hemodynamic environment for the development of severe venous hypertension as there is the absence of valves and often there are numerous A-V. Lastly, CVM can negatively affect patients' functional capacity and mobility, which in turn represents a risk factor for chronic venous insufficiency. A multifactorial underlying etiology can be involved in the development of venous hypertension and subsequent CVI that can result in the development of venous ulcers. These findings underscore the importance of detailed, comprehensive, and systematic evaluation of CVM patients by well-trained practitioners in well-structured healthcare system settings. It is worth emphasizing that compared to CVI patients who frequently develop venous ulcerations in advanced age, ulcerations caused by CVMs can involve pediatric patients who can represent a treatment challenge to even very experienced physicians.

Multidisciplinary approach

Often, the inexorable development and progression of CVMs can result in substantial morbidity and, in some cases, premature mortality. Despite this most clinicians, from primary care doctors to subspecialists (including vascular surgeons) lack the expertise and training required for appropriate diagnosis and management of congenital vascular anomalies in general. Management of these patients is reserved for referral centers with specialized expertise in this area. Konez and Burrows reported more than half of patients with congenital vascular anomalies referred to the vascular anomalies clinic had been previously misdiagnosed.[18] Consequently, misdiagnosis can lead to inappropriate treatment, pertinent not only to deferral of necessary treatment in a timely fashion but also to inappropriate use of different treatment modalities including pharmacotherapy, radiation, surgery, and/or embolization.[18] The most frequent diagnostic error in patients with congenital vascular anomalies is the misclassification of CVM as hemangiomas. Another common diagnostic error is erroneous differentiation between different subtypes of CVM including VMs, AVMs, and lymphatic malformation (LM). An additional challenge, in correctly diagnosed patients with confirmed CVM, represents the selection of the most appropriate treatment option based on anatomic characteristics, embryology, and morphology of particular, individual lesions.[19]

During the last 2 decades, a significant effort has been made to elucidate the etiology and pathophysiology of CVM, to mitigate the tenacity of utilizing an obsolete eponym-based nomenclature, and to provide treatment guidelines in which absence has traditionally obstructed accurate diagnostic workup and selection of lesion-appropriate treatment modalities.[19–22] These efforts combined with the growing recognition of the need for a multidisciplinary team approach, sophisticated imaging strategies, as well as therapeutic innovations have significantly improved the management of CVM patients.

In recognition of the fact that the management of CVMs falls within the purview of numerous medical specialties, it was realized that the establishment of a multidisciplinary approach by combining the expertise of each specialty, structured as Multidisciplinary

Congenital Vascular Anomalies Clinics, is obligatory.[23–27] These include vascular, reconstructive, orthopedic and pediatric surgery, adult and pediatric dermatology, adult and pediatric ophthalmology; adult and pediatric hematology; and diagnostic and interventional radiology.[26] A dedicated team coordinator and advance practice providers are essential to assist with care logistics and interface with patients and their families. A relatively recent study showed that approximately 23% with KTS had a diagnosed psychiatric disorder, most commonly depression and anxiety, which occurred in 15.1% and 5.1% of patients, respectively.[28] Furthermore, psychosocial stressors pertinent to finances, work difficulty, and relationships were also identified. Awareness of the psychosocial impact of CVMs and proper screening is important.[29,30] Therefore, a close collaboration and efficient referral pathways between multidisciplinary teams and psychiatrists, psychologists, support groups, and social workers are becoming increasingly important. A multidisciplinary approach affords the opportunity not only to streamline the clinical evaluation and diagnostics but also to coordinate the care of these patients and to treat them comprehensively by mitigating the need for multiple visits to different clinics. The approach mentioned earlier is essential in managing CVM patients who present with ulcerations, since these patients differ from patients typically seen with venous leg ulcers, in that they have comorbidities and unique aspects of management pertinent to their underlying congenital pathology.

Classification

Before any treatment is planned and considered, it is critically important to classify the lesions correctly. Not correctly classified lesion can result in a treatment selection that is contraindicated based on a hemodynamic and morphologic characteristic of an actual lesion and the outcome can be catastrophic. Historically, the classification and nomenclature used to describe CVMs have not been primarily based on lesion physiology, leading to confusion and inconsistency about the true nature of these lesions. In addition, the use of eponymous designations of certain lesions, which characterized much of the early literature, does not provide information about lesion type, leading to further confusion. Numerous attempts have been made to correctly classify CVMs. The Hamburg classification system and the International Society for the Study of Vascular Anomalies (ISSVA) system are the most widely applied classifications.[31–33] The ISSVA system divides all vascular anomalies into two major categories: CVMs and vascular tumors. This separation permitted more efficient communication between different medical specialists and emphasized the critical point that in contrast to vascular tumors, histologically and physiologically CVMs are not proliferative, neoplastic lesions. Based on hemodynamic characteristics and vessels involved, CMVs can be classified as capillary malformations, VM, AVMs, and LMs.[34] They can appear as isolated lesions or combined, and they also can be part of a syndrome, most notably KTS. However, the ISSVA classification has been modified numerous times to address concerns about clinical applicability, especially with regard to the differentiation between truncular and extratruncular subtypes of CVMs. Based on the embryonic stage of developmental arrest, this differentiation is clinically important because these two groups of CVMs are significantly different with regard to morphology, clinical severity, response to treatment, and recurrence rates. In 2013, the International Union of Phlebology (IUP) commissioned a group of experts to create the newest and most comprehensive document outlining both a new classification scheme and detailed guidelines for management.[22] The authors of the Consensus Document of the IUP for the diagnosis and treatment of venous malformations used both the ISSVA and modified Hamburg classification systems as the basis for the development of a new Integrated Classification System for the management of CVMs.

Diagnostic modalities

Although clinical evaluation is an important initial step in the management of CVMs, it frequently underestimates the involvement of deep anatomic structures and is not sufficient to differentiate between different types of lesions. Therefore, evaluation by duplex ultrasonography, magnetic resonance imaging (MRI), and angiography is essential. Malformations with the arterial flow are characterized on ultrasonography by multidirectional blood flow and high-amplitude arterial waveform with spectral broadening. In contrast to this, venous and lymphatic malformations reveal on ultrasonography mixed venous waveform and complete absence of flow signal, respectively.[35–38] In patients presenting with ulcers, it is critical to evaluate the entire superficial and deep venous system for the presence of reflux, obstruction, or deep vein thrombosis (please refer to Chapters 6 and 7). Due to the relatively high prevalence of deep vein abnormalities, deep veins of lower extremities, as well as pelvic veins, need to be carefully examined (Fig. 28.2). However, in patients with CVM, ultrasonography is not adequate to delineate the extension of the lesions and their infiltration of surrounding anatomic structures. Therefore, initial ultrasonography is followed by other imaging modalities. Bright MRI hypersignal on T2-weighted spin echo sequences delineates the extent of the lesions throughout the involved tissues, and it shows the relation between the lesion and normal vascular and nonvascular structures and provides good soft tissue definition. Venous malformations are characterized by the high signal intensity in T2-weighted MRI.[39–42] In complex cases, differentiation between subtypes of malformations can remain difficult by these radiologic characteristics because other factors may affect these findings. For example, a blood vessel that courses within the imaging plane may give falsely negative results by suggesting a low-flow lesion despite fairly fast flow.[26,43] Diagnostic angiograph especially in patients requiring invasive treatment is considered to be a gold standard for accurate diagnosis of CVMs. However, catheter-based procedures in general are not entirely benign and entail some risk to the patient. To this end, dynamic contrast-

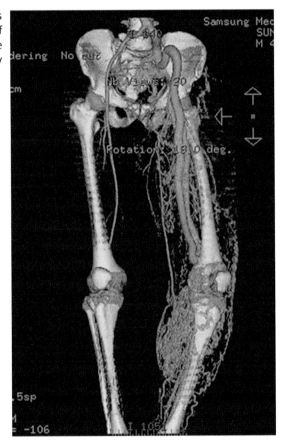

Figure 28.2 Evaluation of the entire venous system is critical in the management of CVM patients. Treatment plan should be determined based on underlying pathology for each individual lesion.

enhanced magnetic resonance imaging (dceMRI) can be used as it does not require catheter placement and can definitively distinguish AVMs from LMs and VMs with an accuracy of approximately 84%.[44] However, in inconclusive cases, confirmatory angiography is still required.[41]

Treatment

The management of CVM is often characterized by multimodal treatment requiring multiple office visits. It has to be emphasized that the treatment of extensive lesions is palliative and goal-oriented, rather than curative. One of the landmark studies regarding the treatment of venous ulcers in patients with VMs was published by Bergan and Cheng.[45] The authors evaluated 1427 patients for venous disorders during a 30-month period and found 14 (9 women and 5 men; 1%) to have VMs. A prevalence of 1% was expected based on the previous studies, and this study also emphasized that CVM patients in contrast to typical CVI patients can present at an early age. The youngest patient was 15 years old. Mean age (±SD) was 30.8 ± 18.6 years. All the patients were treated with ultrasound-guided foam sclerotherapy. Sclerosant used was a 1% or 2% concentration of polidocanol and the foam was produced using the Tessari technique.[46] The mean number of treatments (±SD) was 3.6 ± 2.8 ranging from 1 to 10 treatment sessions. A treatment goal of pain-free healing of ulcers or cosmetic improvement was set for each patient before starting the treatment. The present goals were achieved in all the patients. There were no major side effects. Although polidocanol has been effective, there are other sclerosing agents that can be used with comparable efficacy and acceptable complication rates including sodium tetradecyl sulfate (STS), bleomycin, and doxycycline. Ethanol has been extensively used in the past as it is associated with high efficacy; however, high complication rates and morbidity associated with ethanol made its use less frequent.[47,48] It has to be emphasized that ethanol with or without coiling remains a commonly used endovascular ablative technique to AVMs by some practitioners. To reduce the risk of complications and to reduce the amount of ethanol needed, percutaneous or retrograde transvenous coiling preceding sclerotherapy is recommended for the treatment of AVMs where multiple arteries connect to a single enlarged or aneurysmal draining vein.[49] Correct delivery of ethanol to the AVM nidus, as well as multiple treatment sessions, is recommended to reduce the risk of complications. The choice of embolic agent and sclerosants remains at the discretion of the treating surgeon/physician.

Cabrera et al. used ultrasound-guided foam sclerotherapy with polidocanol (0.25%–4%) to treat venous ulcers in patients with VMs. 46 of the 50 patients (92%) with VMs responded to therapy. In the subgroup of five patients with chronic nonhealing ulcers, the efficacy rate was 100%.[50]

As mentioned earlier, the treatment of ulcers in patients with CVM can involve the pediatric population and patients who were previously misdiagnosed. In 2022, Atamulu et al. reported a case of a 4-year-old male with Parkes Weber syndrome (PWS) who was initially misdiagnosed and treated as KTS for 2 years.[51] The patient presented with significant right lower extremity hypertrophy and two ulcerations, one affecting the anterior and the other medial peri-malleolar region with significant varicose veins visible adjacent to the ulcer. Doppler ultrasonography demonstrated arterial flow in the right common femoral vein. The presence of arterial flow is used to differentiate between PWS and KTS. Further diagnostic work-up with CT-guided angiography demonstrated an A-V fistula between right common femoral artery and vein. The patient was treated with open surgical repair of an AV fistula. The treatment was successful and at the 3-month follow-up, the ulcers were completely healed. Given the difference in arterial and venous pressures and hemodynamics, the existence of AV fistula resulted in an influx of arterial blood flow into the venous system and consequently increased venous pressure, which is a well-established etiology for the development of venous ulcers. Phair et al. reported another pediatric case that involved a 16-year-old male with bilateral lower extremity ulcerations in the settings of agenesis of infrarenal inferior vena cava and iliofemoral veins (truncal CVMs).[52] The patient has been managed successfully with compression therapy and wound care and will likely require more definitive treatment in the future as this case has all the hemodynamic characteristics required for the development of venous hypertension in the lower extremities. Notably, the IVC and iliofemoral veins agenesis significantly impact venous return by an obstruction, which is also suggestive with the presence of extensive lumbar and pelvic collaterals visualized on magnetic resonance venography.

It is important to fully understand underlying anatomic and hemodynamic pathology when treating CVM patients with leg ulcers as they significantly differ from the underlying etiology of superficial venous disease. Although both, CVI patients and CVM patients, have venous hypertension in common, they are two entirely different entities in regards to underlying causes that lead to venous hypertension. Etiology in CVI patients is discussed elsewhere in the book and it is mainly related to valvular incompetence and/or obstruction of venous outflow. However, in patients with AVM influx of blood from high-flow, high-pressure arteries is the main cause for the development of venous hypertension. Therefore, the treatment of superficial venous disease in patients with AVM is characterized by a high rate of recurrences as the main underlying etiology for venous hypertension remains untreated, which is the existence of shunts between arterial and venous blood flow. This principle was emphasized in a retrospective analysis from 2021 by Vuillemin et al.[53] evaluating 15 patients with microfistulous variants of AVMs. Studies showed that even a microfistolus shunting of arterial blood into the venous system leads to venous hypertension and the development of leg ulcers. Not surprisingly patients who were treated following treatment guidelines for superficial venous reflux disease had persisting symptoms and nonhealing ulcers, due to uncorrected underlying vascular pathology. Therefore, in patients with atypical presentation including tissue hypertrophy, atypical pain, increased overlying skin temperature, and presence of vascular "birthmarks" should be approached with caution and their underlying pathology should be identified before starting a treatment for CVI.

Conway et al. evaluated 14 patients with PWS with a mean age of 19.9 years (range, 4.7–68.8). The lower extremity was affected in 12 (86%) patients. All patients presented with pain and swelling in the affected limb. Six patients presented with lower extremity ulcers. There were also five patients with echocardiographic evidence of high-output cardiac failure.[54] 11 (79%) patients underwent transcatheter embolization of the AVM's arterial inflow. Embolization with n-butyl-2-cyanoacrylate (nBCA) adhesive, microspheres, and combination of coils and nBCA adhesive was used in 22 (69%), 8 (25%), and 2 (6%) cases, respectively. Six (55%) patients also had interventions to treat the VM with either radiofrequency ablation (17%), coil embolization (17%), STS sclerotherapy (33%), and a combination of STS, coil embolization, and vein stripping (33%). The authors reported no complications and technical success in all the cases. As expected, there was no complete clinical response. Two patients with complex wounds required major amputations at 128 and 66 months after the index procedure.

Bernhard et al. analyzed 31 AVM patients ranging in age from 1 month to 72 years (mean age 18 years). The most common signs and symptoms were pain, edema, and soft tissue hypertrophy with respective rates of 81%, 68%, and 42%.[55] Notably, ulcerations were present in 89% (eight of nine) patients where AVM was syndrome associated or combined with other types of CVM. Lymphedema and venous hypertension were found to be the underlying etiology in six patients, and the ischemic steal phenomenon was the underlying etiology in two patients who presented with ulceration. Mortality due to major bleeding from ulcerations, wound sepsis, and complications of disseminated intravascular coagulation was documented in three patients, two of whom had above-knee amputations. The authors emphasized the necessity to accurately differentiate between simple nonsyndromic AVM and combined or syndromic AVM, because their clinical courses can significantly differ despite hemodynamic and presenting similarities.

Shahbahrami et al. published their experience with treating leg ulceration located over the medial malleolus, which was unresponsive to compression and wound care therapy in a patient with KTS with aplasia of the right iliac vein (truncular VM).[56] The absence of the main venous outflow caused significant outflow obstruction with resulting venous hypertension and ulcer formation. The patient underwent surgical reconstruction of the affected vasculature to decompress the affected venous system. The right common femoral vein to inferior vena cava bypass was created using polytetrafluoroethylene grafts. To reduce the risk of graft thrombosis by increasing flow rates and preventing stasis, the authors created AV fistula (superficial femoral artery to common femoral vein). At the 7-month follow-up, the patient had complete resolution of the pain, swelling, and ulcerations. Nassiri et al. successfully treated chronic refractory left lateral malleolar and distal leg stasis ulceration initially measuring 4.3 × 4 × 0.2 cm with angiographically confirmed congenital pelvic AVM. A superselective microcatheterization with Onyx 34 of the AVM nidus was performed through various arterial feeders arising from the main bifurcation of the left internal iliac artery.[57] The patient remained ulcer free at a 1-year follow-up. This approach suggests that the elimination of arterial influx into venous circulation via AVM reduces (or eliminates) venous hypertension and provides an adequate hemodynamic environment for wound healing.

Komai et al. emphasized the importance of appropriate diagnostic workup and classification of AVM in the setting of a massive life-threatening hemorrhage from a leg venous ulcer. The patient was successfully treated with coil embolization of a major feeding artery arising from the peroneal artery followed by stripping of the great saphenous vein. In patients with massive and/or recurrent bleeding from ulcers or varicose, AVM should be considered as the underlying etiology of AV communication until proven otherwise. Equally important is to determine if the communication between artery and vein is direct which would be indicative of "truncular" lesion or indirect, through the nidus that would be indicative of "extratruncular" lesion. This differentiation is very important, as "extra-truncluar" lesions have a very high recurrence rate and the treatment itself can be a trigger to stimulate the further growth of this lesion.[58]

In 2020, Anderson et al. published a retrospective study from the Mayo Clinic that evaluated 410 KTS.[59] 83 (20%) of patients had ulcers. The study showed ulcerations most commonly affected the lower extremities, feet, and buttocks/perineum/genitalia in 66%, 34%, and 17%, respectively. Previous case series and retrospective reviews showed that the prevalence of ulcers ranged from 3% to 6%.[60,61] In a larger cohort of 252 patients, ankle ulcers occurred in 6% of patients; however, it remains unclear if all the patients were correctly diagnosed as approximately 63% of patients fulfilled the criteria for KTS diagnosis which possibly skewed the results. In a cohort of 40 patients where strict criteria were used to define KTS, 23% had cellulitis, 15% had bleeding, and 3% had ulcers.[60,61]

Special consideration

From a diagnostic standpoint and treatment planning evaluation of the deep venous system deserves special consideration. Eifert et al. documented (in a study of CVM 392 patients) that 8% of CVM patients have either aplastic or hypoplastic deep venous trunks.[62] It is important to recognize that in these cases venous blood flow return from the affected leg depends on superficial, abnormal vessels and collaterals. Treatment of these vessels can compromise the entire venous circulation of the affected limb. Therefore, patency assessment and assessment of anatomic variations of both, deep (including pelvis) and superficial veins, is essential in these patients. In KTS patients (Fig. 28.3), the prevalence of the anomalies of deep veins is even higher with an estimated rate of 18%.[63] This is immensely important in patients with LMV as the presence of LMV can be erroneously diagnosed as a superficial varicose vein. LMV is a truncular CVM and is characterized by the absence of valves and by existing A-V fistulas that combined provide a hemodynamic and anatomic environment for the development of significant venous hypertension of the affected extremity. Primordial venous blood return during embryonic takes place through the sciatic and LMV, which are developed in early gestation.[64,65] Physiologically these two veins involute and differentiate into the mature venous systems of the lower extremities.[66] LMV is important as it may be the only drainage for the lower limb.[67–70]

Although the LMV primarily exists in patients with KTS, it can also be found in patients with other phosphatidylinositol 4,5-bisphosphate 3-kinase catalytic subunit alpha (PIK3CA)-related overgrowth syndromes.[68] Ultrasonography is used as an initial

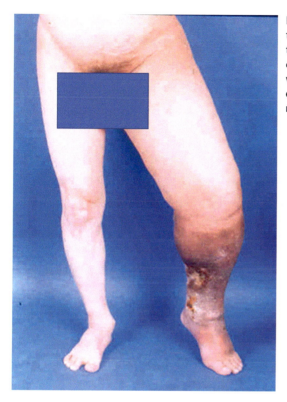

Figure 28.3 KTS patient with left lower extremity medial perimalleolar ulcer. KTS patients have relatively high prevalence of deep venous system hypoplasia or aplasia, which obstructs blood return from affected extremity. Venous abnormalities are a hallmark of KTS.

diagnostic evaluation for clinically suspected LMV. However, this is followed by venography as the evaluation of the entire vein system including the visualization of LMV, its inflow and outflow, and precise assessment for the presence and patency of the deep vein system.[71] It is very important to emphasize that visualization of the deep venous system can frequently be difficult in cases where blood return is predominantly conducted via malformed veins. Following the LMV removal, the venous return is redirected to the deep venous system, which becomes more easily detected in imaging.

The relationship of the LMV to the deep venous system is important for the clinically applicable assessment of these patients. In the settings of a fully developed and patent deep venous system, this vein is named LMV. In the settings of aplastic deep venous system, this vein is named the embryonic vein. Therefore, it becomes obvious that treatment of the embryonic vein is contraindicated as it can jeopardize the venous return since the embryonic vein is the only draining vein of the affected extremity. Treatment of the embryonic vein is palliative rather than curative, and it is directed toward a gradual decrease in venous hypertension. This is achieved by staged elimination of arterial blood influx by surgical ligation of the A-V fistulas, ideally in early childhood. In contrast, LMV can be treated (if the deep venous system is patent), preferably in early childhood to reduce the risk for the development of venous hypertension and the consequent development of venous ulcers. When selecting the treatment of LMV, numerous A-V fistulas should be taken into consideration as they carry a significant risk for bleeding and thromboembolic events during stripping and sclerotherapy, respectively. If treatment is indicated, endovenous ablation with EVLA or RFA is an appropriate treatment option.[65,72,73] In cases with hypoplastic deep venous system, resection of the LMV is recommended in a staged approach. This approach leads to gradual hemodynamic changes in the veins of the affected extremity with resulting gradual adaptation of the hypoplastic venous system to the redirection of blood flow from the LMV and newly increased venous blood return through the deep venous system.

In summary, CVMs represent potentially life- and limb-threatening congenital lesions. The treatment of leg ulcers depends on the underlying pathology which may be different than in patients with CVI. Multidisciplinary approach and diagnostic algorithm utilized to distinguish arterial from lymphatic and venous lesions have been validated as clinically applicable for making an accurate anatomical and hemodynamic diagnosis of VMs, and they serve as a basis for proper treatment selection and significantly facilitate communication among different medical specialists.[74] Implementation of the previously summarized diagnostic protocols and therapeutic algorithms in a multidisciplinary setting results in favorable outcomes with acceptable complication rates in this challenging patient population.

References

1. Enjolras O. Classification and management of the various superficial vascular anomalies: hemangiomas and vascular malformations. *J Dermatol*. 1997;24(11):701–710.
2. Enjolras O, Mulliken JB. Vascular cutaneous anomalies in children: malformations and hemangiomas. *Pediatr Surg Int*. 1996;11(5–6):290–295.
3. Cohen Jr MM. Vasculogenesis, angiogenesis, hemangiomas, and vascular malformations. *Am J Med Genet*. 2002;108(4):265–274.
4. Tasnádi G. Epidemiology and etiology of congenital vascular malformations. *Semin Vasc Surg*. 1993;6(4):200–203.
5. Garzon MC, Huang JT, Enjolras O, Frieden IJ. Vascular malformations: Part I. *J Am Acad Dermatol*. 2007;56(3):353–370. quiz 71-4.
6. Blei F, Walter J, Orlow SJ, Marchuk DA. Familial segregation of hemangiomas and vascular malformations as an autosomal dominant trait. *Arch Dermatol*. 1998;134(6):718–722.
7. Nguyen HL, Boon LM, Vikkula M. Genetics of vascular malformations. *Semin Pediatr Surg*. 2014;23(4):221–226.
8. Castillo SD, Baselga E, Graupera M. PIK3CA mutations in vascular malformations. *Curr Opin Hematol*. 2019;26(3):170–178.
9. di Blasio L, Puliafito A, Gagliardi PA, et al. PI3K/mTOR inhibition promotes the regression of experimental vascular malformations driven by PIK3CA-activating mutations. *Cell Death Dis*. 2018;9(2):45.
10. Venot Q, Blanc T, Rabia SH, et al. Targeted therapy in patients with PIK3CA-related overgrowth syndrome. *Nature*. 2018;558(7711):540–546.
11. Keppler-Noreuil KM, Rios JJ, Parker VE, et al. PIK3CA-related overgrowth spectrum (PROS): diagnostic and testing eligibility criteria, differential diagnosis, and evaluation. *Am J Med Genet*. 2015;167a(2):287–295.

12. Shimano KA, Eng W, Adams DM. How we approach the use of sirolimus and new agents: medical therapy to treat vascular anomalies. *Pediatr Blood Cancer*. 2022;69(Suppl 3):e29603.
13. Al-Samkari H, Eng W. A precision medicine approach to hereditary hemorrhagic telangiectasia and complex vascular anomalies. *J Thromb Haemostasis*. 2022;20(5):1077−1088.
14. Burrows PE. Hemangiomas and vascular malformations. *Can Assoc Radiol J*. 1995;46(2):143.
15. Arbiser JL, Bonner MY, Berrios RL. Hemangiomas, angiosarcomas, and vascular malformations represent the signaling abnormalities of pathogenic angiogenesis. *Curr Mol Med*. 2009;9(8):929−934.
16. Enjolras O, Herbreteau D, Lemarchand F, et al. [Hemangiomas and superficial vascular malformations: classification]. *J Mal Vasc*. 1992;17(1):2−19.
17. Lokhorst MM, Horbach SER, Waner M, et al. Responsiveness of quality of life measures in children with peripheral vascular malformations: the OVAMA project. *JPRAS Open*. 2021;27:70−79.
18. Konez O, Burrows PE. Magnetic resonance of vascular anomalies. *Magn Reson Imag Clin N Am*. 2002;10(2):363−388 (vii).
19. Lee BB, Antignani PL, Baraldini V, et al. ISVI-IUA consensus document diagnostic guidelines of vascular anomalies: vascular malformations and hemangiomas. *Int Angiol*. 2015;34(4):333−374.
20. Lee BB, Bergan J, Gloviczki P, et al. Diagnosis and treatment of vascular malformations. Consensus document of the International Union of Phlebology (IUP)-2009. *Int Angiol*. 2009;28(6):434−451.
21. Paolacci S, Zulian A, Bruson A, et al. Vascular anomalies: molecular bases, genetic testing and therapeutic approaches. *Int Angiol*. 2019;38(2):157−170.
22. Lee BB, Baumgartner I, Berlien P, et al. Diagnosis and treatment of venous malformations. Consensus document of the international union of phlebology (IUP): updated 2013. *Int Angiol*. 2015;34(2):97−149.
23. Chandrasekhar SS. Multidisciplinary approach to vascular anomalies maximizes outcomes. *Otolaryngol Clin*. 2018;51(1):xv−xvi.
24. Donnelly LF, Adams DM, Bisset 3rd GS. Vascular malformations and hemangiomas: a practical approach in a multidisciplinary clinic. *AJR Am J Roentgenol*. 2000;174(3):597−608.
25. Gloviczki P, Duncan A, Kalra M, et al. Vascular malformations: an update. *Perspect Vasc Surg Endovasc Ther*. 2009;21(2):133−148.
26. Markovic JN, Shortell CE. Multidisciplinary treatment of extremity arteriovenous malformations. *J Vasc Surg Venous Lymphat Disord*. 2015;3(2):209−218.
27. Timbang MR, Richter GT. Update on extracranial arteriovenous malformations: a staged multidisciplinary approach. *Semin Pediatr Surg*. 2020;29(5):150965.
28. Harvey JA, Nguyen J, Anderson KR, et al. Pain, psychiatric comorbidities, and psychosocial stressors associated with Klippel-Trenaunay syndrome. *J Am Acad Dermatol*. 2018;79(5):899−903.
29. van der Ploeg HM, van der Ploeg MN, van der Ploeg-Stapert JD. Psychological aspects of the Klippel-Trenaunay syndrome. *J Psychosom Res*. 1995;39(2):183−191.
30. Oduber CE, Khemlani K, Sillevis Smitt JH, Hennekam RC, van der Horst CM. Baseline quality of life in patients with klippel-trenaunay syndrome. *J Plast Reconstr Aesthetic Surg*. 2010;63(4):603−609.
31. Dasgupta R, Fishman SJ. ISSVA classification. *Semin Pediatr Surg*. 2014;23(4):158−161.
32. Wassef M, Borsik M, Cerceau P, et al. [Classification of vascular tumours and vascular malformations. Contribution of the ISSVA 2014/2018 classification]. *Ann Pathol*. 2021;41(1):58−70.
33. Lee BB, Laredo J, Lee TS, Huh S, Neville R. Terminology and classification of congenital vascular malformations. *Phlebology*. 2007;22(6):249−252.
34. Lowe LH, Marchant TC, Rivard DC, Scherbel AJ. Vascular malformations: classification and terminology the radiologist needs to know. *Semin Roentgenol*. 2012;47(2):106−117.
35. Danahey J, Seip R, Lee B, et al. Imaging of vascular malformations with a high-intensity focused ultrasound probe for treatment planning. *J Vasc Surg Venous Lymphat Disord*. 2021;9(6):1467−1472. e2.
36. Flors L, Park AW, Norton PT, Hagspiel KD, Leiva-Salinas C. Soft-tissue vascular malformations and tumors. Part 1: classification, role of imaging and high-flow lesions. *Radiologia (Engl Ed)*. 2019;61(1):4−15.
37. Reis 3rd J, Koo KSH, Monroe EJ, et al. Ultrasound evaluation of pediatric slow-flow vascular malformations: practical diagnostic reporting to guide interventional management. *AJR Am J Roentgenol*. 2021;216(2):494−506.

38. Sadick M, Müller-Wille R, Wildgruber M, Wohlgemuth WA. Vascular anomalies (Part I): classification and diagnostics of vascular anomalies. *Röfo*. 2018;190(9):825—835.
39. Carqueja IM, Sousa J, Mansilha A. Vascular malformations: classification, diagnosis and treatment. *Int Angiol*. 2018;37(2):127—142.
40. Flors L, Leiva-Salinas C, Maged IM, et al. MR imaging of soft-tissue vascular malformations: diagnosis, classification, and therapy follow-up. *Radiographics*. 2011;31(5):1321—1340. discussion 40-1.
41. Hussein A, Malguria N. Imaging of vascular malformations. *Radiol Clin*. 2020;58(4):815—830.
42. McCafferty IJ, Jones RG. Imaging and management of vascular malformations. *Clin Radiol*. 2011;66(12):1208—1218.
43. Turley RS, Lidsky ME, Markovic JN, Shortell CK. Emerging role of contrast-enhanced MRI in diagnosing vascular malformations. *Future Cardiol*. 2014;10(4):479—486.
44. Lidsky ME, Spritzer CE, Shortell CK. The role of dynamic contrast-enhanced magnetic resonance imaging in the diagnosis and management of patients with vascular malformations. *J Vasc Surg*. 2012;56(3):757—764.e1.
45. Bergan J, Cheng V. Foam sclerotherapy of venous malformations. *Phlebology*. 2007;22(6):299—302.
46. Bunke N, Brown K, Bergan J. Foam sclerotherapy: techniques and uses. *Perspect Vasc Surg Endovasc Ther*. 2009;21(2):91—93.
47. Lee BB, Do YS, Byun HS, Choo IW, Kim DI, Huh SH. Advanced management of venous malformation with ethanol sclerotherapy: mid-term results. *J Vasc Surg*. 2003;37(3):533—538.
48. Lee BB, Kim DI, Huh S, et al. New experiences with absolute ethanol sclerotherapy in the management of a complex form of congenital venous malformation. *J Vasc Surg*. 2001;33(4):764—772.
49. Lee BB, Baumgartner I, Berlien HP, et al. Consensus document of the international union of angiology (IUA)-2013. Current concept on the management of arterio-venous management. *Int Angiol*. 2013;32(1):9—36.
50. Cabrera J, Cabrera Jr J, García-Olmedo MA, Redondo P. Treatment of venous malformations with sclerosant in microfoam form. *Arch Dermatol*. 2003;139(11):1409—1416.
51. Yüce Atamulu K, Yaşar Durmuş S, Uylar Seber T. A rare cause of chronic leg ulcer in childhood: parkes-weber syndrome. *Int J Low Extrem Wounds*. 2022:1—5. Online ahead of print.
52. Phair J, Trestman E, Stableford J. Venous status ulcers due to congenital agenesis of the inferior vena cava in a 16-year-old male. *Vascular*. 2016;24(1):106—108.
53. Vuillemin N, Bernhard S, Haine A, et al. Capillary-venule malformation is a microfistulous variant of arteriovenous malformation. *J Vasc Surg Venous Lymphat Disord*. 2021;9(1):220—225.
54. Conway AM, Qato K, Nguyen Tran NT, et al. Embolization techniques for arteriovenous malformations in parkes-weber syndrome. *Ann Vasc Surg*. 2020;69:224—231.
55. Bernhard SM, Tuleja A, Laine JE, et al. Clinical presentation of simple and combined or syndromic arteriovenous malformations. *J Vasc Surg Venous Lymphat Disord*. 2022;10(3):705—712.
56. Shahbahrami K, Resnikoff M, Shah AY, Lydon RP, Lazar A, Cavallo G. Chronic lower extremity wounds in a patient with Klippel Trenaunay syndrome. *J Vasc Surg Cases Innov Tech*. 2019;5(1):45—48.
57. Nassiri N, Crystal DT, Hoyt C, Shafritz R. Chronic refractory venous ulcer exacerbated by a congenital pelvic arteriovenous malformation successfully treated by transarterial Onyx embolization. *J Vasc Surg Venous Lymphat Disord*. 2017;5(3):417—420.
58. Lee BB. New approaches to the treatment of congenital vascular malformations (CVMs)–a single centre experience. *Eur J Vasc Endovasc Surg*. 2005;30(2):184—197.
59. Anderson KR, Nguyen H, Schoch JJ, Lohse CM, Driscoll DJ, Tollefson MM. Skin-Related complications of Klippel-Trenaunay Syndrome: a retrospective review of 410 patients. *J Eur Acad Dermatol Venereol*. 2021;35(2):517—522.
60. Maari C, Frieden IJ. Klippel-Trenaunay syndrome: the importance of "geographic stains" in identifying lymphatic disease and risk of complications. *J Am Acad Dermatol*. 2004;51(3):391—398.
61. Jacob AG, Driscoll DJ, Shaughnessy WJ, Stanson AW, Clay RP, Gloviczki P. Klippel-Trenaunay syndrome: spectrum and management. *Mayo Clin Proc*. 1998;73(1):28—36.
62. Eifert S, Villavicencio JL, Kao TC, Taute BM, Rich NM. Prevalence of deep venous anomalies in congenital vascular malformations of venous predominance. *J Vasc Surg*. 2000;31(3):462—471.

63. Browse NLBK, Lea Thomas M. The Klippel trenaunay syndrome. In: Browse NL, Burnand KG, Thomas ML, eds. *Diseases of the Veins: Pathology, Diagnosis and Treatment*. London, UK: Edward Arnold; 1988:609−625.
64. Oduber CE, Young-Afat DA, van der Wal AC, van Steensel MA, Hennekam RC, van der Horst CM. The persistent embryonic vein in Klippel-Trenaunay syndrome. *Vasc Med*. 2013;18(4):185−191.
65. Garg L, Mittal UK, Puri SK, Rissam HK. Klippel-Trenaunay syndrome, an unusual association with persistent lateral marginal vein of Servelle: colour Doppler and 256 dual-source MDCT evaluation. *BMJ Case Rep*. 2015;2015.
66. Rojas Martinez R, Puech-Leão P, Guimarães PM, Netto BM. Persistence of the embryonic lateral marginal vein: report of two cases. *Rev Hosp Clin Fac Med Sao Paulo*. 2001;56(5):159−162.
67. Dahal S, Karmacharya RM, Vaidya S, Gautam K, Bhatt S, Bhandari N. A rare case of persistent lateral marginal vein of Servelle in Klippel Trenaunay Syndrome: a successful surgical management. *Int J Surg Case Rep*. 2022;94:107052.
68. Fereydooni A, Nassiri N. Evaluation and management of the lateral marginal vein in Klippel-Trénaunay and other PIK3CA-related overgrowth syndromes. *J Vasc Surg Venous Lymphat Disord*. 2020;8(3):482−493.
69. Kota AA, Agarwal S. Significance of lateral marginal vein in Klippel-Trenaunay syndrome. *ANZ J Surg*. 2021;91(1−2). E61-e2.
70. Mattassi R, Vaghi M. Management of the marginal vein: current issues. *Phlebology*. 2007;22(6):283−286.
71. Ochoco G, Enriquez CAG, Urgel RJL, Catibog JS. Multimodality imaging approach in a patient with Klippel-Trenaunay syndrome. *BMJ Case Rep*. 2019;12(8).
72. Kim YW, Lee BB, Cho JH, Do YS, Kim DI, Kim ES. Haemodynamic and clinical assessment of lateral marginal vein excision in patients with a predominantly venous malformation of the lower extremity. *Eur J Vasc Endovasc Surg*. 2007;33(1):122−127.
73. Uller W, Hammer S, Wildgruber M, Müller-Wille R, Goessmann H, Wohlgemuth WA. Radiofrequency ablation of the marginal venous system in patients with venous malformations. *Cardiovasc Intervent Radiol*. 2019;42(2):213−219.
74. Markovic JN, Kim CY, Lidsky ME, Shortell CK. A 6-year experience treating vascular malformations with foam sclerotherapy. *Perspect Vasc Surg Endovasc Ther*. 2012;24(2):70−79.

CHAPTER 29

Lymphatic disorders in the pathogenesis of chronic venous insufficiency

Stanley G. Rockson
Allan and Tina Neill Professor of Lymphatic Research and Medicine, Stanford University School of Medicine, Stanford, CA, United States

In humans and other mammals, the lymphatic conduits bear the responsibility for the return transport of interstitial fluid, macromolecules, and immune cells into the central venous circulation. Accordingly, failure of the lymphatic system to prevent the stigmata of chronic edema plays a vital role in the pathogenesis and natural history of chronic venous insufficiency.

Lymphedema is the collective term that describes a variety of pathologies in which the uninterrupted accumulation of protein-enriched interstitial fluid leads to functional and structural dysfunction. Lymphedema, therefore, represents the pathological end result, over time, of an efflux of lymph that is insufficient to offset the rate of interstitial fluid production.

Lymphedema is a common, complex, and inexplicably underappreciated human disease.[1] While it is most commonly recognized and acknowledged by practitioners in the context of cancer therapeutics (where cancer containment leads to an obstructive anatomic defect in lymphatic structure and function), it is less commonly acknowledged that an indistinguishable presentation of lymphedema can arise through high input failure, as occurs in the setting chronic venous insufficiency. When lymphatic transport reserve is diminished, a state of subclinical lymphedema can prevail.[2] In such circumstances, venous hypertension can precipitate the progression to overt edema.[3,4] Undiagnosed venous disease represents an important substrate in the presentation of chronic lymphedema, both in the context of cancer therapeutics and elsewhere.[5]

To fully appreciate the role of the lymphatic system in the prevention of chronic edema, it is essential to appreciate recent insights into the function of the blood vascular glycocalyx.[6] While microcirculatory fluid flux is traditionally ascribed to the balance of hydrostatic and oncotic forces that govern the gradients that prevail across the blood capillary membrane, it is, in fact, the impact of the glycocalyx to determine the central role of the lymphatic circulation in the prevention of tissue edema. Accumulation of capillary filtrate in the tissue spaces is avoided mainly through lymphatic clearance and not, as was previously thought, through reabsorption.[6] The transcapillary pressure gradients dictate net filtration, except in states of dynamic disequilibrium. Under steady-state conditions, at the blood capillary glycocalyx interface, the ultrafiltration of absorbed interstitial fluid increases the plasma protein concentration immediately beneath the glycocalyx; this, in combination with declining interstitial hydrostatic pressure as interstitial fluid is removed, dictates a steady state of sustained slight filtration along the entire length of the blood capillary.

Chronic venous insufficiency, with its attendant increase in postcapillary hydrostatic pressure, leads to ever-increasing ultrafiltration of the blood by the blood vascular capillary. This, in turn, produces a filtration edema that, with persistence, can lead to the same pathological outcome as the chronic edema that is created by the obstructive defects in the lymphatics that produce edema through a diminution in lymphatic circulatory function. With venous edema, increased hydrostatic pressure at the venous end of the capillary results in excess production of interstitial fluid through increased capillary filtration; when the production of lymph exceeds the maximal transport capacity of the lymphatic conduits, lymphedema supervenes, even if these structures are anatomically and functionally normal. In summary, both high- and low-output lymphedema occur; their clinical presentations are indistinguishable.

Inadequate lymphatic transport engenders the accumulation of glycoproteins within the extracellular space. Chronic inflammation leads to a hypercellular population of fibroblasts, keratinocytes, and adipocytes, along with macrophages and other mononuclear cells.[7] Deposition of collagen accompanies hypertrophy of subcutaneous adipose tissues.[8]

Clinical staging of lymphedema

Historically, guidelines for the staging of lymphedema have been limited in scope.[9] The most widely acknowledged staging schema is that proposed by the International Society of Lymphedema. A latent, subclinical stage and three clinical grades have been identified and accepted[10]; each grade is subclassified as mild, moderate, or severe:

Latent phase: Excess fluid accumulates and fibrosis occurs around the lymphatics, but no edema is apparent clinically.

Grade I: Edema pits on pressure and is reduced largely or completely by elevation; there is no clinical evidence of fibrosis.

Grade II: Edema does not pit on pressure and is not reduced by elevation; moderate-to-severe fibrosis is evident on clinical examination.

Grade III: Edema is irreversible and develops from repeated inflammatory attacks, fibrosis, and sclerosis of the skin and subcutaneous tissue. This is the stage of lymphostatic elephantiasis.

When utilized, this classification permits the evaluation of treatment effectiveness and different treatment modalities can theoretically be compared.

Clinical presentation
History

In the context of chronic venous insufficiency, with or without the presence of ulceration, the clinical scenario helps to establish the diagnosis of putative phlebolymphedema. Although the pathogenesis of obstructive and filtration lymphedemas is different, the clinical presentation and characteristic physical findings are frequently similar. When phlebolymphedema is under consideration, the clinician will look for concomitant stigmata of both lymphedema and venous hypertension (Fig. 29.1).

Signs and symptoms
Pain

The presence and quality of pain will reflect the contribution and severity of the venous and lymphatic components, respectively. While it is a misconception to assume that

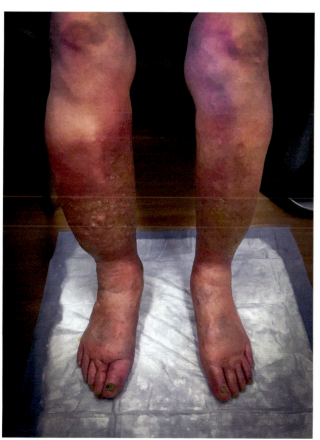

Figure. 29.1 Phlebolymphedema. The legs display the combined features of chronic venous insufficiency and lymphedema. The venous insufficiency produces hyperpigmentation and dependent rubor, while the lymphedema is responsible for cutaneous thickening and hyperkeratosis, with edema of the digits. The persistent edema reflects the presence of both conditions.

lymphedema causes "painless swelling" of the limb, intense pain (in the absence of venous ulceration and superimposed infection) is rare. With either venous varicosities or lymphatic hypertension, some aching or heaviness of the limb is a frequent complaint.

The clinical signs and symptoms of phlebolymphedema substantially depend upon the duration and severity of chronic venous insufficiency. Beyond the veno-specific findings, the remainder of the physical examination will reflect the degree to which the presence of lymphatic hypertension has altered the architecture of the tissues.

Edema

All edema, with or without stigmata of lymphedema, initially entails the physical find of pitting that signals the presence of excess, relatively protein-rich interstitial fluid. The edema is typically soft, easily displaced with manual pressure, and, in its earliest stages, resolves with limb elevation. *Stemmer's* sign is considered to be pathognomonic for the lymphatic contribution to the edema, as it reflects the dermal thickening that is characteristic of lymphedema: the skin at the base of the second toe becomes inelastic to the point that the examiner's fingers cannot elicit tenting of the skin.[11] The dorsum of the forefoot and/or the digits can be involved, but, in phlebolymphedema, the foot is often spared. Over longer durations, the involved portions of the leg may acquire a woody texture as the surrounding tissue becomes indurated and fibrotic.

Skin changes

In early lymphedema, the skin usually has a pinkish-red color and a mildly elevated temperature owing to the increased capillary blood flow in the skin, but this typical finding of phlebolymphedema may be altered by the presence of the dependent rubor that accompanies venous hypertension. In more well-established lymphedema, the skin thickens and manifests areas of *peau d'orange*, hyperkeratosis, and lichenification.[8] Excoriation or chronic eczematous dermatitis may occur. Disruption of the circulation to the skin and subcutaneous tissue will reflect the degree to which the venous component contributes to the net clinical presentation. In the late stages, the chronic skin changes of lymph stasis may include the presence of verrucae or of small vesicles. The latter frequently drain clear lymph (lymphorrhea).

Infection

Recurrent soft tissue infection is a hallmark of lymphedema.[12] Contributing factors to infection include the physically altered and impaired cutaneous barrier to microbial invasion, the impaired local immune responses that reflect the loss of normal lymphatic immune traffic, and the presence of excess, proteinaceous interstitial fluid in the limb that provides a fertile substrate for bacterial growth. In the context of chronic lower extremity edema, the lifetime prevalence of cellulitis exceeds 35%.[13]

In phlebolymphedema, the clinical presentation of soft tissue infection can pose diagnostic challenges. The severity can be quite variable. At one extreme, there can be an acute presentation of rapidly progressive infection, but it is also common to encounter only a modest increase in pain or edema, associated with a subtle increase in cutaneous erythema without any accompanying fever or other systemic manifestations. Recurrent attacks of cellulitis can further damage the cutaneous lymphatics, exacerbate skin pathology, and initiate worsening edema.

Malignant tumors

The appearance of malignant tumors within the limb represents a rare, recognized complication of long-standing lymphedema. Lymphangiosarcoma after long-standing secondary lymphedema was originally described by Stewart et al.[14] This rare multicentric neoplasm can feature bluish nodules, sclerotic plaques, or bullous changes. Additional malignant tumors that appear with some frequency in late lymphedema include squamous cell carcinoma, melanoma, Kaposi's sarcoma, and lymphoma.

Diagnosis

In established phlebolymphedema, the characteristic clinical presentation within the context of recognized chronic venous insufficiency will establish the diagnosis, based on the history, and associated physical findings.[15] In more subtle presentations, it may be difficult to distinguish the lymphedematous contribution to the edema manifestations. The diagnosis is more difficult to establish in the early stages of the disease, particularly when the lymphedema component is within its earliest stages.

Physical examination

The physical examination for suspected lymphedema must include inspection for cutaneous and subcutaneous fibrosis and *peau d'orange*. An attempt should be made to establish a positive Stemmer sign, if present.

Imaging

Objective documentation of lymphatic dysfunction is sometimes useful. A variety of imaging modalities can be considered; of these, radionuclide lymphoscintigraphy is considered to be the diagnostic gold standard.[1] Additional modalities, such as magnetic resonance imaging (MRI), axial tomography, and ultrasonography, enjoy increasing utilization. Direct contrast lymphography was the first available imaging modality for lymphedema, but it is rarely used today. In addition to the accepted modalities to image structure and/or function of the lymphatics, it is often helpful to perform venous duplex ultrasonography whenever phlebolymphedema is under diagnostic consideration.

Lymphoscintigraphy

Radionuclide lymphoscintigraphy can reliably confirm the diagnosis of lymphedema.[2,16] A radiolabeled macromolecular tracer is injected intradermally or subdermally within the interdigital spaces of the affected limb. Technetium Tc 99 m-labeled human serum albumin, the most frequently utilized imaging agent in the United States,[16–19] will appear within the pelvic nodes after an interval as short as 10 min.[20] Activity should distinctly be evident in the inguinal nodes by 1 h (range, 15–60 min), along with uptake in the bladder. Faint hepatic and para-abdominal nodal uptake can be visible by 1 h after injection. At 3 h, there will be intense uptake in the liver, with symmetrical staining of the inguinal, pelvic and abdominal lymph nodes, and, less commonly, tracer activity might be faintly detected in the region of lymphaticovenous valve of the thoracic duct.

Qualitative interpretation of lymphoscintigraphic images provides moderate sensitivity and excellent specificity for the diagnosis of lymphedema.[18] Characteristic lymphoscintigraphic findings of lymphedema include dermal backflow, absent or delayed transport of tracer, absent or delayed visualization of lymph nodes, and crossover filling with retrograde backflow. The asymmetry or delayed appearance of radiocontrast material within the draining lymph node bed serves as a semiquantifiable index of lymphatic vascular dysfunction. Spatial resolution by lymphoscintigraphy is not ideal,[21] yet this imaging modality has the requisite high specificity and sensitivity to reliably detect the presence of lymphedema.[22]

Computed tomography and magnetic resonance imaging

Computed tomography and MRI can be used to discriminate lymphedema as distinct from other causes of edema. In lymphedema, there is a characteristic honeycomb pattern of edema that is confined to the epifascial tissue planes, accompanied by cutaneous thickening. In venous edema (and, therefore, in phlebolymphedema), both the epifascial and subfascial compartments are affected. In addition, MRI can facilitate the identification of lymph nodes and enlarged lymphatic trunks. Anatomic delineation by MRI can complement the functional data provided by lymphoscintigraphy.[23] Magnetic resonance imaging without contrast enhancement[24–27] utilizes T2-weighted sequences to image the fluid content of the lymphatic vasculature, with simultaneous suppression of tissue signaling.[26,28]

Contrast-enhanced MR lymphography is also used for the evaluation of lymphedema. In a technique analogous to the one employed for radionuclide lymphoscintigraphy, a T1-weighted gadolinium contrast agent is injected into the interstitial space, thus permitting the detection both of dilated lymphatics and of dermal backflow.[29] In comparison to lymphoscintigraphy, MRI provides more favorable imaging depth limits and permits three-dimensional volumetric imaging. However, venous signal enhancement is a somewhat more undesirable attribute, as is the higher cost of this imaging technique.[21]

Near-infrared lymphography

Near-infrared lymphography (NIR) is the technique most recently added to the diagnostic armamentarium for imaging evaluation of lymphedema. This technique has had its most robust application in the preoperative evaluation of lymphedema patients for microsurgical intervention,[30] an approach that will generally not be applicable to the phlebolymphedema patient. However, in addition to its surgical applications, NIR can be used to identify compensatory drainage pathways in the lymphedema patient[31] and thereby can be used to guide conservative management techniques for the lymphedema patient, including those with phlebolymphedema.

Management of lymphedema
Medical management

At present, there is no useful pharmacology for the medical management of lymphedema, although active investigation of pharmacological interventions for lymphedema is underway,[32,33] at present, these are not clinically available. In modern practice, the stabilization and amelioration of chronic lymphedema rely upon proven physical techniques that can reduce excess edema volume and stimulate available lymphatic contractility and function.

Complex decongestive physiotherapy (CDP) is a multimodal physical intervention to minimize the stigmata of lymphedema. CDP typically entails manual lymphatic massage, multilayer bandaging, exercise, and skincare. Available guidelines suggest that this approach is central to the management of lymphedema,[34] having been accorded a 1B level of evidence.[35] With or without the component of venous incompetence, the dependence upon the use of graduated compression for the maintenance phase is widely acknowledged.[36] The International Society of Lymphology has endorsed the use of the highest tolerated compression in the range of 20–60 mm Hg.[37] In cases of phlebolymphedema with venous reflux, the conservative management of the secondary lymphedema is undertaken once definitive management of venous reflux has been established.

Intermittent sequential compression

Intermittent sequential compression is a frequently employed, elective component of decongestive physiotherapy. The incorporation of this approach into a multimodal physical approach to the lymphedema patient has long been advocated by some proponents.[38] Sequential compression is efficacious, well tolerated, and, when properly undertaken, is demonstrably free of complications.[39,40] The use of sequential compression devices provides a reduction both in the utilization of medical resources and of healthcare costs.[41–43] This is particularly true for patients with phlebolymphedema.[44]

Various multichamber devices are commercially available, but those that employ sequential distal-to-proximal graduated compression are the most efficacious. It is important to ensure that the patient who will undergo compression therapy has adequate arterial inflow into the affected limb inasmuch as sustained compression can further compromise arterial blood flow (see Chapter 10).

Intermittent sequential compression is effective but should not be considered to be a stand-alone therapeutic intervention. The published literature to support this component of therapy is supportive but heterogeneous.[34]

Surgical management

Suction-assisted protein lipectomy

In well-established, chronic lymphedema, without regard to pathogenesis, the volume excess in the limb predominantly reflects progressive hypertrophy of the subcutaneous adipose tissues. In many cases, the effective application of external compression and palliation of symptoms can be facilitated through surgical debulking of the edematous limb(s).

Suction-assisted protein lipectomy (SAPL) affords stable, significant reduction volume reduction for lower limb lymphedema. In one recent series, SAPL produced a median volume reduction of 101% in secondary disease of the lower limb.[45] The salutary surgical outcome is not influenced by BMI or other patient characteristics, but for SAPL, sustained volume reduction will not be successful without the maintenance of chronic compression after surgical intervention.[46] Liposuction combined with long-term decongestive compression therapy can more effectively reduce limb volume than compression therapy alone.

Microsurgical tenchiques

Microsurgical procedures for lymphedema fall into two broad categories: lymphaticovenous anastomosis (LVA) and autologous, vascularized lymph node transfer (VLNT). Various individual strategies appear to be efficacious,[47] but published analyses entail small patient numbers, and there is a general absence of long-term follow-up.[48]

Lymphaticovenous anastomosis is typically considered in patients with early or mid-stage lymphedema.[49] Anatomic considerations also govern the decision to entertain LVA within the treatment approach,[50] inasmuch as lymphatic imaging must demonstrate the presence of patent, functional lymphatic channels for LVA to be feasible, either alone or in combination with VLNT. VLNT, more appropriate for more advanced stages of lymphedema, entails a substantially higher risk profile.[34] It should be noted that, in phlebolymphedema, the presence of chronic venous incompetence will often render LVA undesirable and may also hamper the ability to successfully undertake VLNT.

In the surgical management of lymphedema, the American Venous Forum has proposed that all surgical interventions for chronic lymphedema require at least 6 months of antecedent nonoperative conservative management (IC recommendation). Suction-assisted protein lipectomy should be reserved for late-stage nonpitting lymphedema (2C recommendation).

References

1. Rockson SG. Advances in lymphedema. *Circ Res*. 2021;128:2003—2016.
2. Szuba A, Shin WS, Strauss HW, Rockson S. The third circulation: radionuclide lymphoscintigraphy in the evaluation of lymphedema. *J Nucl Med*. 2003;44:43—57.
3. Bollinger A, Leu AJ, Hoffmann U, Franzeck UK. Microvascular changes in venous disease: an update. *Angiolo*. 1997;48:27—32.
4. Brautigam P, Vanscheidt W, Foldi E, Krause T, Moser E. [Involvement of the lymphatic system in primary non-lymphogenic edema of the leg. Studies with 2-compartment lymphoscintigraphy]. *Hautarzt*. 1997;48:556—567.
5. Szuba A, Razavi M, Rockson SG. Diagnosis and treatment of concomitant venous obstruction in patients with secondary lymphedema. *J Vasc Intervent Radiol*. 2002;13:799—803.
6. Mortimer PS, Rockson SG. New developments in clinical aspects of lymphatic disease. *J Clin Invest*. 2014;124:915—921.
7. Szuba A, Rockson S. Lymphedema: anatomy, physiology and pathogenesis. *Vasc Med*. 1997;2:321—326.
8. Schirger A. Lymphedema. *Cardiovasc Clin*. 1983;13:293—305.
9. O'Donnell Jr TF, Allison GM, Iafrati MD. A systematic review of guidelines for lymphedema and the need for contemporary intersocietal guidelines for the management of lymphedema. *J Vasc Surg Venous Lymphat Disord*. 2020;8:676—684.
10. Casley-Smith JR. International society for lymphology. *J R Soc Med*. 1985;78:271.
11. Stemmer R. [A clinical symptom for the early and differential diagnosis of lymphedema]. *Vasa*. 1976;5:261—262.
12. Quirke M, Ayoub F, McCabe A, et al. Risk factors for nonpurulent leg cellulitis: a systematic review and meta-analysis. *Br J Dermatol*. 2017;177:382—394.
13. Burian EA, Karlsmark T, Franks PJ, et al. Cellulitis in chronic oedema of the lower leg: an international cross-sectional study. *Br J Dermatol*. 2021;185(1):110—118.
14. Stewart NJ, Pritchard DJ, Nascimento AG, Kang YK. Lymphangiosarcoma following mastectomy. *Clin Orthop*. 1995;135—141.
15. Rockson SG, Miller LT, Senie R, et al. American cancer society lymphedema workshop. Workgroup III: diagnosis and management of lymphedema. *Cancer*. 1998;83:2882—2885.
16. Weissleder H, Weissleder R. Lymphedema: evaluation of qualitative and quantitative lymphoscintigraphy in 238 patients. *Radiology*. 1988;167:729—735.
17. Vaqueiro M, Gloviczki P, Fisher J, Hollier LH, Schirger A, Wahner HW. Lymphoscintigraphy in lymphedema: an aid to microsurgery. *J Nucl Med*. 1986;27:1125—1130.
18. Gloviczki P, Calcagno D, Schirger A, et al. Noninvasive evaluation of the swollen extremity: experiences with 190 lymphoscintigraphic examinations. *J Vasc Surg*. 1989;9:683—689. discussion 690.
19. Cambria RA, Gloviczki P, Naessens JM, Wahner HW. Noninvasive evaluation of the lymphatic system with lymphoscintigraphy: a prospective, semiquantitative analysis in 386 extremities. *J Vasc Surg*. 1993;18:773—782.
20. Devoogdt N, Pans S, De Groef A, et al. Postoperative evolution of thickness and echogenicity of cutis and subcutis of patients with and without breast cancer-related lymphedema. *Lymphatic Res Biol*. 2014;12:23—31.
21. Polomska AK, Proulx ST. Imaging technology of the lymphatic system. *Adv Drug Deliv Rev*. 2021;170:294—311.
22. Hassanein AH, Maclellan RA, Grant FD, Greene AK. Diagnostic accuracy of lymphoscintigraphy for lymphedema and analysis of false-negative tests. *Plast Reconstr Surg Glob Open*. 2017;5:e1396.
23. Mitsumori LM, McDonald ES, Wilson GJ, Neligan PC, Minoshima S, Maki JH. MR lymphangiography: how i do it. *J Magn Reson Imag*. 2015;42:1465—1477.
24. Yu DX, Ma XX, Wang Q, Zhang Y, Li CF. Morphological changes of the thoracic duct and accessory lymphatic channels in patients with chylothorax: detection with unenhanced magnetic resonance imaging. *Eur Radiol*. 2013;23:702—711.

25. Kim EY, Hwang HS, Lee HY, et al. Anatomic and functional evaluation of central lymphatics with noninvasive magnetic resonance lymphangiography. *Med (Baltim)*. 2016;95:e3109.
26. Arrive L, Derhy S, El Mouhadi S, Monnier-Cholley L, Menu Y, Becker C. Noncontrast magnetic resonance lymphography. *J Reconstr Microsurg*. 2016;32:80–86.
27. Cellina M, Oliva G, Menozzi A, Soresina M, Martinenghi C, Gibelli D. Non-contrast Magnetic Resonance Lymphangiography: an emerging technique for the study of lymphedema. *Clin Imag*. 2019;53:126–133.
28. Arrive L, Derhy S, Dlimi C, El Mouhadi S, Monnier-Cholley L, Becker C. Noncontrast magnetic resonance lymphography for evaluation of lymph node transfer for secondary upper limb lymphedema. *Plast Reconstr Surg*. 2017;140:806e–811e.
29. Mitsumori LM, McDonald ES, Neligan PC, Maki JH. Peripheral magnetic resonance lymphangiography: techniques and applications. *Tech Vasc Intervent Radiol*. 2016;19:262–272.
30. Rockson SG. A role for near infrared fluorescent imaging in the evaluation of lymphatic function. *Lymphatic Res Biol*. 2017;15:203.
31. Koelmeyer LA, Thompson BM, Mackie H, et al. Personalizing conservative lymphedema management using indocyanine green-guided manual lymphatic drainage. *Lymphatic Res Biol*. 2021;19(1):56–65.
32. Tian W, Rockson SG, Jiang X, et al. Leukotriene B4 antagonism ameliorates experimental lymphedema. *Sci Transl Med*. 2017;9.
33. Rockson SG, Tian W, Jiang X, et al. Pilot studies demonstrate the potential benefits of antiinflammatory therapy in human lymphedema. *JCI Insight*. 2018;3.
34. Gianesini S, Obi A, Onida S, et al. Global guidelines trends and controversies in lower limb venous and lymphatic disease. *Phlebol: J Ven Dis*. 2019;34:4–66.
35. Lee BB, Antignani PL, Baroncelli TA, et al. IUA-ISVI consensus for diagnosis guideline of chronic lymphedema of the limbs. *Int Angiol*. 2015;34:311–332.
36. Rabe E, Partsch H, Hafner J, et al. Indications for medical compression stockings in venous and lymphatic disorders: an evidence-based consensus statement. *Phlebology*. 2018;33:163–184.
37. Executive C. The diagnosis and treatment of peripheral lymphedema: 2016 consensus document of the international society of lymphology. *Lymphol*. 2016;49:170–184.
38. Leduc O, Leduc A, Bourgeois P, Belgrado JP. The physical treatment of upper limb edema. *Cancer*. 1998;83:2835–2839.
39. Szuba A, Achalu R, Rockson SG. Decongestive lymphatic therapy for patients with breast carcinoma-associated lymphedema. A randomized, prospective study of a role for adjunctive intermittent pneumatic compression. *Cancer*. 2002;95:2260–2267.
40. Mayrovitz HN. The standard of care for lymphedema: current concepts and physiological considerations. *Lymphatic Res Biol*. 2009;7:101–108.
41. Brayton KM, Hirsch AT, PJ OB, Cheville A, Karaca-Mandic P, Rockson SG. Lymphedema prevalence and treatment benefits in cancer: impact of a therapeutic intervention on health outcomes and costs. *PLoS One*. 2014;9:e114597.
42. Karaca-Mandic P, Hirsch AT, Rockson SG, Ridner SH. The cutaneous, net clinical, and health economic benefits of advanced pneumatic compression devices in patients with lymphedema. *JAMA Dermatol*. 2015;151:1187–1193.
43. Karaca-Mandic P, Hirsch AT, Rockson SG, et al. A comparison of programmable and nonprogrammable compression devices for treatment of lymphedema using an administrative health outcomes dataset. *Br J Dermatol*. 2017;177(6):1699–1707.
44. Lerman M, Gaebler JA, Hoy S, et al. Health and economic benefits of advanced pneumatic compression devices in patients with phlebolymphedema. *J Vasc Surg*. 2019;69:571–580.
45. Lamprou DA, Voesten HG, Damstra RJ, Wikkeling OR. Circumferential suction-assisted lipectomy in the treatment of primary and secondary end-stage lymphoedema of the leg. *Br J Surg*. 2017;104:84–89.
46. Brorson H, Ohlin K, Olsson G, Karlsson M. Breast cancer-related chronic arm lymphedema is associated with excess adipose and muscle tissue. *Lymphatic Res Biol*. 2009;7.
47. Cormier JN, Rourke L, Crosby M, Chang D, Armer J. The surgical treatment of lymphedema: a systematic review of the contemporary literature (2004-2010). *Ann Surg Oncol*. 2012;19:642–651.

48. Hadamitzky C, Pabst R, Gordon K, Vogt PM. Surgical procedures in lymphedema management. *J Vasc Surg Venous Lymphat Disord.* 2014;2:461–468.
49. Carl HM, Walia G, Bello R, et al. Systematic review of the surgical treatment of extremity lymphedema. *J Reconstr Microsurg.* 2017;33:412–425.
50. Neligan PC, Kung TA, Maki JH. MR lymphangiography in the treatment of lymphedema. *J Surg Oncol.* 2017;115:18–22.

Index

Note: 'Page numbers followed by "*f*" indicate figures and "*t*" indicate tables.'

A

Aberdeen Varicose Vein Questionnaire (AVVQ), 186
ABI, 193—196
Abnormal plantar load, 318
Abnormal superficial, perforator, and deep veins, 366
Acellular human tissue, 338—339
Acellular xenograft tissue, 339—341
Acetylsalicylic acid, 274
Activated protein C resistance, 145
Active (biologic) dressings, 232—233
 growth factor therapy, 233
Acute venous thrombosis, 114
Adherence to therapy, 240—241
Adjusted hazard ratio (AHR), 55
Age, 149
Alginates, 231—232
American Venous Forum study, 52—53
Anatomic classification, 182t
Anatomic considerations, 298—299
Ankle motility, 319
Anterior accessory saphenous vein treatment, 367
Anterior accessory vein, endovenous ablation for, 396
Antibiotics, 274—275
Anticoagulation, 301—302
Anticoagulation protocol, 459
Antiphospholipid syndrome, 146—147
Antithrombin, 142—144, 143t—144t
Apoptosis, 10—11
AROM, 320
Arterial circulation assessment, 201
Arterial claudication, 88—89
Arterial hypertension, 316—317
Arterial testing, 129—130
Arteriography, 170
Arterio-venous (AV) shunts, 504—505
Atrophie blanche, 93f

B

Banding, 481
Bilateral stenting, 463

Bilayered skin equivalent (BSE), 328
Biologic debridement, 226
Biologic wound care products, 275—276
Biomarkers cytokines/proteolytic, 12—13
Biosynthetic, 341—342
Bonn Vein study, 51—52
Bottom line on compression, 265
Branched-chain amino acids (BCAA), 10—11

C

Calcium dobesilate, 272—273
Calf
 muscle deficiency, 320
 perforating veins of, 26—27, 27f—28f
Cancer, 150
Capillary exchange, 301
Capillary hydrostatic pressure, regulation of, 35—36
Cellular and tissue-based products (CTPs), 327, 335
Cellular changes, 7—9
Cerebral autosomal dominant arteriopathy with subcortical infarcts and leukoencephalopathy (CADASIL), 4—6
Charing Cross Venous Ulcerations Questionnaire (CCVUQ), 186
Chronically swollen leg with ulcers
 diagnostic studies, 163—165, 165f
  goals, 164—165
  imaging, 164
  laboratory test, 163—164
 edema assessment, 161—163
 epidemiology, 159—161
 metalloproteinases, 163
 physical examination, 163
Chronic infectious diseases, 86—87
Chronic total venous occlusion, recanalization of, 453—454, 453f—454f
Chronic venous disease (CVD), 139
 apoptosis, 10—11
 biomarkers cytokines/proteolytic, 12—13
 connexins, 10—11
 endothelium, 6—7

Chronic venous disease (CVD) (*Continued*)
   genetics, 12—13
   glycocalyx, 6—7
   hypoxia, 10—11
   inflammatory cells, 6—7
   iron, 12—13
   machine learning GWAS, 12—13
   matrix metalloproteinases (MMP), 4
   metabolic abnormalities, 10—11
   microvenous valves, 10
   molecules, 6—7
   MT-MMP/ADAMTS, 12—13
   nitrogen species, 12—13
   oxidative stress, 12—13
   pathophysiology, 5f
   reactive oxygen, 12—13
   varicose veins (VV), 4
   venous leg ulcers (VLU), 4
      cellular changes, 7—9
      genetic predisposition, 4—6
      inflammation and advanced skin changes, 8f
      MMPs regulation, 7—9
      structural proteins and alterations, 9—10
      venous structure, 6—7
Chronic venous insufficiency (CVI), 65—66
   banding, 481
   blood flow and inflammation, 36
   calf, perforating veins of, 26—27, 27f—28f
   capillary hydrostatic pressure, regulation of, 35—36
   deep venous dysfunction, 23—26, 24f—25f
   deep venous reflux correction, 476, 477f
   deep venous reflux etiology, 474
   deep venous system, 476f
   diagnostic modalities, 129—136
      arterial testing, 129—130
      CT venography (CTV), 131—132
      diagnostic algorithm, 134—136, 135f
      duplex ultrasonography, 130—131
      IVUS, 134
      MRV, 132—133
      venography, 133—134
   epidemiological data on, 49—53
   hemodynamics, 127—128
      anatomy, 20—22
      diagnostic evaluation, 474—476
   history and physical, 128—129
   iliac and pelvic veins anatomy, 127—128
   internal valvuloplasty results, 482t
   interstitial edema, 34—35
   lower extremity, ultrasound evaluation of
      acute venous thrombosis, 114
      chronic venous obstruction, 114—116, 115f
      examination protocol, 109—114, 110f
      imaging during/after intervention, 117—120, 118f—119f
      obstruction diagnosis, 113—120
      principles, 103—104
      recurrent venous thrombosis, 116—117
      reflux, diagnosis of, 104—106, 105f
      superficial, deep, and perforator veins, 106—109, 108f
   microcirculation, 34—35
   neovalve, 479—481, 480f, 484t
   pathophysiology, 22—23
   postthrombotic syndrome, 475f
   signs of, 91—92, 91f—94f
   superficial venous incompetence, 22—23
   surgical techniques, 476—477
      internal valvuloplasty, 476—477, 478f
   symptoms of, 88—89, 88t
   transposition, 478, 479f
   valve transplantation, 479, 483t
   venous hemodynamics, foot and calf pump function, 27—34, 28f—29f, 31f—34f
   venous hypertension, 128—129
   venous valves repair
      venous obliteration, 474t
      venous obstruction, 474t
      venous occlusion, 474t
Chronic Venous Insufficiency Questionnaire (CIVIQ), 186
Chronic venous leg ulcers, 448—449
Chronic venous obstruction, 114—116, 115f
Chronic venous ulceration, 292t, 444—445
   judicious wound care, 446
   poor stocking compliance, 445—446, 445f
   trial of compression, 444—445
Chronic venous ulcers, 217—218
Cigarette smoking, 320
Cleansers, irrigation, 226—227
Cleansing, 223—228
Clinical-Etiology-Anatomy-Pathophysiology (CEAP), 85—86, 179—189, 180t, 433, 434t
   anatomic classification, 182t
   etiologic classification, 181t
   pathophysiologic classification, 182t

QoL
  Aberdeen Varicose Vein Questionnaire (AVVQ), 186
  Charing Cross Venous Ulcerations Questionnaire (CCVUQ), 186
  Chronic Venous Insufficiency Questionnaire (CIVIQ), 186
  measures, 185
  SF-36, 185
  VEINES-QoL/SYM, 186
 revised venous clinical severity score (rVCSS), 181–184, 183t
 venous disability score, 184–185, 184t–185t
 venous segmental disease score, 184
 venous severity score, 181
 wound documentation, 186–189, 188f–190f
Coexisting comorbidities
 abnormal plantar load, 318
 ankle motility, 319
 aROM, 320
 arterial hypertension, 316–317
 calf muscle deficiency, 320
 cigarette smoking, 320
 congestive heart failure, 317
 coronary artery disease, 317
 diabetes mellitus, 316
 gait disorders, 318
 leg muscle strength, 319
 malnutrition, 321
 obesity, 321–322
 physical activity, 318–319
 psychosocial condition, 322
 reduced muscle pumping function, 317–320
 venous leg ulcers (VLUs), 315
Combination therapy, 334
Combined oral contraceptive pill (COCP), 149
Composite Z-stent, 454–455, 455f
Compression, 285
 bandages, 204–205, 204t
 devices, 202, 202t
 socks, 263
 techniques, 211–213
 therapy, 263
  arterial circulation assessment, 201
  compression bandages, 204–205, 204t
  compression devices, 202, 202t
  compression techniques, 211–213
  elastic and inelastic bandages, 205–207, 205t, 206f
  intermittent pneumatic compression, 209–210
  leg, interface pressure and stiffness measured on, 205
  local therapy, 201
  medical compression stockings, 202–203, 204t
  pads, 210, 210f–211f
  pelottes, 210, 210f–211f
  practical guidelines, 211–213
  single-layer and multilayer bandages, 207–208, 208f–209f
  ulcer recurrence, prevention of, 210
  venous reflux, correction of, 201
Computed tomography (CT), 192, 524
Congenital vascular malformations (CVMs), 504f
 arterio-venous (AV) shunts, 504–505
 classification, 506–507
 diagnostic modalities, 507–508
 lateral marginal vein (LMV), 504–505
 multidisciplinary approach, 505–506
 special consideration, 512–514, 513f
 treatment, 508–512
Congestive heart failure, 317
Connexins, 10–11
Coronary artery disease, 317
Cryopreserved and other viable PDT, 343
CVD, CEAP etiologies of, 41–49
Cytokines, 7–9

D

Debridement, 223–226, 286–288, 287f, 332–333
Dedicated self-expanding nonbraided nitinol venous stents, 461–462
Deep venous dysfunction, 23–26, 24f–25f
Deep venous reflux
 correction, 476, 477f
 etiology, 474
Deep venous system, 476f
Deep venous thrombosis (DVT), 23, 77, 298
 collateralization and flow patterns, 300–301
 phlegmasia cerulea dolens, 300–301
 treatment, randomized trials of, 47
Dehydrated PDT, 342–343
Diabetes mellitus, 316
Diagnostic modalities, 129–136, 507–508
 arterial testing, 129–130
 CT venography (CTV), 131–132
 diagnostic algorithm, 134–136, 135f
 duplex ultrasonography, 130–131

Diagnostic modalities (*Continued*)
    IVUS, 134
    MRV, 132–133
    venography, 133–134
Dietary supplements, 266
Differential diagnosis, 171–174
Direct oral anticoagulant (DOAC), 150
Double barrel Wallstent configuration, 455
Dressings, 288–289
Duplex ultrasound (DUS), 43, 421
Dysfibrinogenemia, 146

E
Early Venous Reflux Ablation (EVRA) study, 355–359, 378
    background, 355
    challenges to implementation, 359–360
    design and population, 356
    driving change, potential levers for, 360–361
    healthcare professionals education, 359
    health economic outcomes, 358
    limitations of, 358–359
    patient population, 356–357
    patient-reported outcomes, 357
    resource implications, 360
    ulcer healing, 357
    ulcer recurrence, 357
    updated policies and guidelines, 360
Edema assessment, 161–163
Edinburgh Vein Study, 49–51
Effect of Surgery and Compression on Healing And Recurrence (ESCHAR) trial, 354–355, 375
    challenges to implementation, 359–360
    driving change, potential levers for, 360–361
    healthcare professionals education, 359
    resource implications, 360
    updated policies and guidelines, 360
Elastic and inelastic bandages, 205–207, 205t, 206f
Elastic compression therapy, 302–303
Electrical stimulation, 238
Endothelial cells, 6–7
Endothelium, 6–7
Endovenous heat-induced thrombosis (EHIT), 375, 429–430
Endovenous laser ablation, 386
    anatomic success, 389–390
    early results of, 389
    general principles, 386

long-term results of, 389–390
    operative technique, 386–387
    postoperative care, 387–388
Endovenous thermal ablation techniques, 383–388
    ClosureFast, operative technique with, 384–385
    general principles, 384
    radiofrequency ablation, 383–385
Endovenous thermal *vs.* nonthermal ablations, 390
Enzymatic debridement, 225–226
Epidemiologic studies, 46–47
Etiologic classification, 181t
Etiologic spectrum, 67–68, 67t
European Society of Vascular Surgery (ESVS), 150
External stimulants, 289–290
Extracellular matrix therapy (ECM), 337–338
Extravenous etiology, 48–49

F
Factor V Leiden, 144–145
Femoral vein occlusion with iliac vein obstruction, 463–464
Flap selection, 495–496, 496f
Foam sclerosants, 406–408, 406t, 407f
Foam sclerotherapy, chronic venous insufficiency with
    efficacy, 410–411
    foam sclerosants, 406–408, 406t, 407f
    liquid sclerotherapy *vs.*, 408
    microembolization, 408–410
    patient preparation, 411
    postprocedure instructions, 413–417, 414f
    procedural details, 411–413, 414f
    sclerosants, 405–406, 405t
    thrombosis following, 410, 410t
Food and Drug Administration (FDA), 328
FOXC2 gene, 22
Free tissue transfer, 493–494, 493t
    flap selection, 495–496, 496f
    free tissue transfer, 493–494, 493t
    indications and preoperative planning, 494–495
    local flaps, 493
    negative pressure wound therapy (NPWT), 492
    postoperative care, 497
    surgical technique, 496–497
    wound closure, reconstructive methods of, 492
French study, 52

G

Gait disorders, 318
Gaiter area, hyperpigmentation, 92f
Genetic disorders, 4–6
Genetic predisposition, 4–6
Genetic risk factors, 56–57
Genetics, 12–13
Genome-wide association study (GWAS), 4–6
Glycocalyx, 6–7
Great saphenous vein (GSV), 367, 381, 433

H

Healing, 77–78
Hemochromatosis C282Y (HFE) gene mutation, 4–6
Hemodynamics, 127–128, 299–300
    anatomy, 20–22
    diagnostic evaluation, 474–476
Hormone replacement therapy, 149
Horse chestnut, 274
Hybrid approaches, 465–466
Hydrocellular foam dressings, 232
Hydrocolloid dressings, 231
Hydrogel dressings, 230
Hydrostatic capillary pressure, 35
Hydrosurgical debridement, 224–225
Hypercoagulable state
    activated protein C resistance, 145
    age, 149
    antiphospholipid syndrome, 146–147
    antithrombin, 142–144, 143t–144t
    cancer, 150
    chronic venous disease (CVD), 139
    combined oral contraceptive pill, 149
    dysfibrinogenemia, 146
    European Society of Vascular Surgery (ESVS), 150
    factor V Leiden, 144–145
    hormone replacement therapy, 149
    myeloproliferative neoplasms, 148
    paroxysmal nocturnal hemoglobinuria (PNH), 147–148
    postthrombotic syndrome (PTS), 139
    potential subsequent venous ulceration, 149–154
    protein C, 141–142, 143t–144t
    protein S, 141–142, 143t–144t
    prothrombin G20210A gene mutations, 145–146
    puerperium, 149
    thrombophilia, 152
        post-thrombotic syndrome, 152
        venous ulceration, 152–154
    VTE, 149–154
Hypoxia, 10–11

I

Iliac and pelvic veins anatomy, 127–128
Iliac-caval confluence, 454–455
Iliac vein stenting
    anticoagulation protocol, 459
    bilateral stenting, 463
    chronic total venous occlusion, recanalization of, 453–454, 453f–454f
    chronic venous leg ulcers, 448–449
    chronic venous ulceration, 444–445
        judicious wound care, 446
        poor stocking compliance, 445–446, 445f
        trial of compression, 444–445
    clinical assessment, 435–437, 435f
    Clinical, Etiological, Anatomical, and Pathophysiological (CEAP), 433, 434t
    composite Z-stent, 454–455, 455f
    dedicated self-expanding nonbraided nitinol venous stents, 461–462
    diagnosis of, 441–444
        transfemoral venography, 441–442, 441f–442f
        two-segment CT venography, 442–444, 443t–444t
    double barrel Wallstent configuration, 455
    factors affecting ISR, 458–459
    femoral vein occlusion with iliac vein obstruction, 463–464
    great saphenous vein (GSV), 433
    hybrid approaches, 465–466
    iliac-caval confluence, 454–455
    inguinal ligament, stent extension below, 462
    intervention, threshold stenosis for, 461
    investigations, 438–440
    ISR, stent occlusion from, 460
    obesity, 465
    octogenarians, 465
    other procedures, 448
    pentoxyfylline, 459–460
    procedure, technical success of, 451–452, 452f
    recanalization, stent occlusions after, 460, 461t

Iliac vein stenting (*Continued*)
  reinterventions in stented limbs, 455–458, 456f–457f
  routine cases, stenting technique for, 450–451, 451f
  saphenous and deep vein reflux, 438–440, 439f–440f
  saphenous vein ablation, 446–448, 446t, 447f
  sizing of stents, 449–450, 450t
  special considerations and techniques, 462–465
  standard and recanalization cases, stent surveillance in, 460
  stent compression, 458–459
  stenting across and below IVC filters, 464–465, 464f
  thrombophilia, 459
  venous ulcers
    etiology of, 437–438
    pathophysiology of, 434
  Wallstent configuration, 454–455
Incompetent calf perforating veins, 26–27
Incompetent perforator veins, treatment options for, 366–378
Incompetent superficial truncal veins, treatment options for, 366
Inelastic compression garments, 263–264
Infectious ulcers, 172
Inflammatory cells, 6–7
Inflammatory ulcers, 171–172
Inguinal ligament, stent extension below, 462
Interactive dressings, 229–232
Intermittent pneumatic compression, 209–210
Intermittent sequential compression, 525–526
Internal valvuloplasty, 476–477, 478f, 482t
Interstitial edema, 34–35
Intraluminal hydrostatic pressure, 22–23
Iron, 12–13
ISR, stent occlusion from, 460
Italian study, 52

L

Lateral marginal vein (LMV), 504–505
Leg, interface pressure and stiffness measured on, 205
Leg muscle strength, 319
Lifestyle considerations, 262–263
Lifestyle modification, 283–284
Lipodermatosclerosis, 93f
Living skin cellular products, 335–337

Local flaps, 493
Local therapy, 201
Lower extremity wounds, 86–88, 87t
  acute venous thrombosis, 114
  arteriography, 170
  chronic venous obstruction, 114–116, 115f
  differential diagnosis, 171–174
  examination protocol, 109–114, 110f
  history, 167–169
  imaging during/after intervention, 117–120, 118f–119f
  infectious ulcers, 172
  inflammatory ulcers, 171–172
  malignant ulcers, 172–173, 173f, 175f
  microvascular occlusive ulcers, 173, 175f
  obstruction diagnosis, 113–120
  physical examination, 167–169
  principles, 103–104
  pyoderma gangrenosum (PG), 174, 176f
  recurrent venous thrombosis, 116–117
  reflux, diagnosis of, 104–106, 105f
  superficial, deep, and perforator veins, 106–109, 108f
  treatment, 170–171
  vascular laboratory testing, 169
Lymphatic disorders
  clinical presentation, 521–523
    computed tomography, 524
    diagnosis, 523–525
    edema, 522
    history, 521
    imaging, 523
    infection, 522–523
    lymphoscintigraphy, 524
    magnetic resonance imaging, 524
    malignant tumors, 523
    near-infrared lymphography (NIR), 525
    pain, 521–522
    physical examination, 523
    signs and symptoms, 521–522
    skin changes, 522
  lymphedema management, 525–526
    clinical staging of, 520
    intermittent sequential compression, 525–526
    medical management, 525
    microsurgical tenchiques, 526
    suction-assisted protein lipectomy, 526
Lymphatic involvement, 57–58
Lymphoscintigraphy, 524

M

Machine learning GWAS, 12–13
Macrocirculatory venous hypertension, 32–34
Magnetic resonance imaging, 524
Malignant tumors, 523
Malignant ulcers, 172–173, 173f, 175f
Malnutrition, 321
Matrix metalloprotease modulation, 331–332
Matrix metalloproteinases (MMP), 4
Mechanical-chemical ablation (MOCA), 381
Medical compression stockings, 202–203, 204t
Medical management, 525
Medical therapies
  acetylsalicylic acid, 274
  animal models, 267–269
  antibiotics, 274–275
  biologic wound care products, 275–276
  bottom line on compression, 265
  calcium dobesilate, 272–273
  compression socks, 263
  compression therapy, 263
  dietary supplements, 266
  guidelines, 276
  horse chestnut, 274
  inelastic compression garments, 263–264
  lifestyle considerations, 262–263
  medications, 267
  micronized purified flavonoid fraction, 269–272
  multilayer compression bandages, 264–265
  paste boots, 264
  pathophysiology, 261–262
  pentoxifylline, 273–274
  physical activity, 266–267
  pneumatic compression, 265–266
  ruscus extract, 271–272
  sulodexide, 273–274
Medical therapy, 284
Metabolic abnormalities, 10–11
Metabolomics, 10–11
Metalloproteinases, 163
Microcirculation, 34–35, 301
Microembolization, 408–410
Micronized purified flavonoid fraction, 269–272
Microsurgical tenchiques, 526
Microvascular occlusive ulcers, 173, 175f
Microvenous valves, 10
MMPs regulation, 7–9
Molecules, 6–7

Monocyte chemoattractant protein-1 (MCP-1), 36
MT-MMP/ADAMTS, 12–13
Multidisciplinary approach, 505–506
Multilayer compression bandages, 264–265
Myeloproliferative neoplasms, 148

N

Natural history, 75–76, 75f
Near-infrared lymphography (NIR), 525
Negative pressure wound therapy (NPWT), 330–331, 492
  case studies, 253–255, 253f–255f
  description, 250–251
  mechanism of action, 250–251
  VLUs, 251–253
Neovalve, 479–481, 480f, 484t
Nitrogen species, 12–13
Nonhealing wounds, treatment modalities for
  chronic venous ulcer, 292t
  compression, 285
  debridement, 286–288, 287f
  dressings, 288–289
  external stimulants, 289–290
  identification and treatment of infection, 285–286
  lifestyle modification, 283–284
  local agents, 288–289
  medical therapy, 284
  skin grafting, 291
  skin substitutes, 291
  soft tissue substitutes, 291
  untreated arterial/deep venous disease, 284–285
Nonthermal endovenous ablation techniques, 388
  endovenous cyanoacrylate embolization, 388
  mechanical-chemical ablation procedure, 388
  ultrasound guided foam sclerotherapy, 388

O

Obesity, 76, 321–322, 465
Octogenarians, 465
Open surgical reconstructions, 306–307
Operating room (OR), 253–255
Optimizing care delivery, 238–241
  wound care clinics, 238–239
Organizational fibrotic process, 25–26
Oxidative stress, 12–13

P

Pads, 210, 210f—211f
Pain management, 227—228
Paroxysmal nocturnal hemoglobinuria (PNH), 147—148
Paste boots, 264
Pathophysiologic classification, 182t
Patient and wound assessment, 218—222, 219t—222t
Pelottes, 210, 210f—211f
Perforator vein closure technique, 369—371, 370f—371f
Perforator vein treatment, 369
Periprocedural monitoring, 425—428, 427f
Peri-wound management, 234
Perturbed microcirculation, 6—7
Physical activity, 266—267, 318—319
Physical examination, 523
Placenta-derived tissue, 342
Pneumatic compression, 265—266
Point-of-care assessment of wound perfusion, 193—196
Point-of-care wound measurements, apps and cameras used for, 192, 193t
Polish study, 52
Postoperative care, 497
Postprocedure instructions, 413—417, 414f
Postprocedure monitoring, 428—430, 429f—430f
Postthrombotic syndrome (PTS), 139, 475f
   anatomic considerations, 298—299
   anticoagulation, 301—302
   capillary exchange, 301
   case studies, 305—306, 306f
   deep venous thrombosis (DVT), 298
      collateralization and flow patterns, 300—301
      phlegmasia cerulea dolens, 300—301
   definition, 297
   elastic compression therapy, 302—303
   experience, 304—305
   hemodynamics, 299—300
   microcirculation, 301
   open surgical reconstructions, 306—307
   surgical thrombectomy, 304
   thrombolysis/endovascular therapies, 303
   ultrasound-accelerated thrombolysis, 307
   venoactive drugs, 302
   venous ulcer care, 307—308
Potential subsequent venous ulceration, 149—154

Preoperative planning, 381—383, 382f
Preprocedure planning, 422, 423f
Protein C, 141—142, 143t—144t
Protein S, 141—142, 143t—144t
Prothrombin G20210A gene mutations, 145—146
Psychosocial condition, 322
Puerperium, 149
Pyoderma gangrenosum (PG), 174, 176f

Q

Quality of life (QOL), 251
   Aberdeen Varicose Vein Questionnaire (AVVQ), 186
   Charing Cross Venous Ulcerations Questionnaire (CCVUQ), 186
   Chronic Venous Insufficiency Questionnaire (CIVIQ), 186
   measures, 185
   SF-36, 185
   VEINES-QoL/SYM, 186

R

Radiofrequency ablation, early results of, 388—397
Randomized controlled trials, 45t
Reactive oxygen, 12—13
Recanalization, stent occlusions after, 460, 461t
Recurrent venous thrombosis, 116—117
Reduced muscle pumping function, 317—320
Reflux, diagnosis of, 104—106, 105f
Reinterventions in stented limbs, 455—458, 456f—457f
Revised venous clinical severity score (rVCSS), 181—184, 183t
Routine cases, stenting technique for, 450—451, 451f
Ruscus extract, 271—272

S

San Diego population study, 51
Saphenous and deep vein reflux, 438—440, 439f—440f
Saphenous vein ablation, 446—448, 446t, 447f
Sclerosants, 405—406, 405t
Secondary valvular dysfunction, 22
Semipermeable film dressings, 232
SF-36, 185
Sharp debridement, 224
SIMPLE strategies, 240t

Single-layer and multilayer bandages, 207–208, 208f–209f
Sizing of stents, 449–450, 450t
Skin
　changes, 522
　grafting, 236–237, 291
　protection products, 235t
　substitutes, 291
Small saphenous vein
　closure, 367
　endovenous ablation of, 396
Socioeconomic aspects, 73–75, 74f
Sodium tetradecyl sulfate (STS), 381
Soft tissue substitutes, 291
SPY angiography, 193–196
Standard and recanalization cases, stent surveillance in, 460
Stasis dermatitis, 91f
Stent compression, 458–459
Stenting across and below IVC filters, 464–465, 464f
Suction-assisted protein lipectomy, 526
Sulodexide, 273–274
Superficial, deep, and perforator veins, 106–109, 108f
Superficial saphenous vein treatments, 367
Superficial surgery/perforator interruption
　abnormal superficial, perforator, and deep veins, 366
　complications, 375
　diagnosis, 363–365, 364f–365f
　facility requirements for treatment, 366
　follow-up recommendations, 371–374, 372f–374f
　incompetent perforator veins, treatment options for, 366–378
　incompetent superficial truncal veins, treatment options for, 366
　literature review for venous ulcers, 375–378, 376f–377f
　perforator vein closure technique, 369–371, 370f–371f
　perforator vein treatment, 369
　superficial saphenous vein treatments, 367
　technique, 367
Superficial vein ablation, endovenous techniques for
　anterior accessory vein, endovenous ablation for, 396
　endovenous laser ablation, 386
　　anatomic success, 389–390
　　early results of, 389
　　general principles, 386
　　long-term results of, 389–390
　　operative technique, 386–387
　　postoperative care, 387–388
　endovenous thermal ablation techniques, 383–388
　　ClosureFast, operative technique with, 384–385
　　general principles, 384
　　radiofrequency ablation, 383–385
　endovenous thermal vs. nonthermal ablations, 390
　great saphenous vein (GSV), 381
　mechanical-chemical ablation (MOCA), 381
　nonthermal endovenous ablation techniques, 388
　　endovenous cyanoacrylate embolization, 388
　　mechanical-chemical ablation procedure, 388
　　ultrasound guided foam sclerotherapy, 388
　preoperative planning, 381–383, 382f
　radiofrequency ablation, early results of, 388–397
　small saphenous vein, endovenous ablation of, 396
　sodium tetradecyl sulfate (STS), 381
　SVS/AVF guidelines, recommendations of, 397
　venous ulcer patients, clinical success in, 390–391, 392t–395t
　venous ulcers treatment, other endovenous procedures for, 397
Superficial venous incompetence, 22–23
Superficial venous intervention
　Early Venous Reflux Ablation (EVRA) study, 355–359
　　background, 355
　　challenges to implementation, 359–360
　　design and population, 356
　　driving change, potential levers for, 360–361
　　healthcare professionals education, 359
　　health economic outcomes, 358
　　limitations of, 358–359
　　patient population, 356–357
　　patient-reported outcomes, 357
　　resource implications, 360
　　ulcer healing, 357
　　ulcer recurrence, 357
　　updated policies and guidelines, 360

Superficial venous intervention (*Continued*)
   Effect of Surgery and Compression on Healing And Recurrence (ESCHAR) trial, 354–355
      challenges to implementation, 359–360
      driving change, potential levers for, 360–361
      healthcare professionals education, 359
      resource implications, 360
      updated policies and guidelines, 360
   epidemiology, 353–354, 353f
   problem extent, 353–354, 353f
   rationale for treating, 354
Surgical techniques, 476–477
Surgical thrombectomy, 304
Swedish Skaraborg County studies, 53

T

TCPO2, 193–196
Thrombolysis/endovascular therapies, 303
Thrombophilia, 152, 459
   post-thrombotic syndrome, 152
   venous ulceration, 152–154
Tissue inhibitors of metalloproteases (TIMPs), 7–9
Topical antimicrobial therapy, 333–334
Transfemoral venography, 441–442, 441f–442f
Transposition, 478, 479f
Tumescent anesthesia administration, 428
Two-segment CT venography, 442–444, 443t–444t

U

Ulcer recurrence, prevention of, 210
Ultrasound-accelerated thrombolysis, 307
Ultrasound guidance, endovenous treatment
   equipment, 421–422
   periprocedural monitoring, 425–428, 427f
   postprocedure monitoring, 428–430, 429f–430f
   preprocedure planning, 422, 423f
   tumescent anesthesia administration, 428
   venous access, 423–424, 425f–427f
Ultrasound therapy, 237
Untreated arterial/deep venous disease, 284–285

V

Vacuum-assisted cover (VAC), 253–255
Valve transplantation, 479, 483t
Varicose veins (VV), 4
Vascular laboratory testing, 169
Vascular smooth muscle (VSM), 7–9

VEINES-QoL/SYM, 186
Venoactive drugs, 302
Venous access, 423–424, 425f–427f
Venous Clinical Severity Score (VCSS), 85–86
Venous disability score, 184–185, 184t–185t
Venous disease, progression of, 53–54
Venous hemodynamics, foot and calf pump function, 27–34, 28f–29f, 31f–34f
Venous hypertension, 128–129
Venous leg ulcers (VLU), 4
   age and sex distribution, 72–73, 72f–74f
   cellular changes, 7–9
   chronic venous insufficiency, 65–66
   defining, 66
   DVT, 77
   epidemiology, 63–64
   etiologic spectrum, 67–68, 67t
   future trends, 78–79, 78f
   genetic predisposition, 4–6
   healing, 77–78
   incidence of, 64–65, 65t, 71–72
   inflammation and advanced skin changes, 8f
   MMPs regulation, 7–9
   natural history, 75–76, 75f
   obesity, 76
   prevalence, 66–71, 69t–71t
   socioeconomic aspects, 73–75, 74f
   structural proteins and alterations, 9–10
   survival, 77–78
Venous leg ulcers (VLUs), 251–253, 315
   acellular human tissue, 338–339
   acellular xenograft tissue, 339–341
   adjusted hazard ratio (AHR), 55
   adjuvant therapy for, 234–238
   American Venous Forum study, 52–53
   assessment of, 94–95, 95f, 96t
   bilayered skin equivalent (BSE), 328
   biosynthetic, 341–342
   Bonn Vein study, 51–52
   care, 307–308
   cellular and tissue-based products (CTPs), 327, 335
   characteristics, 44t
   classification and venous severity scoring, 97–98, 97t–99t
   clinical risk factors for, 54–56, 55t
   combination therapy, 334
   cryopreserved and other viable PDT, 343
   CVD, CEAP etiologies of, 41–49

CVI
  epidemiological data on, 49–53
  signs of, 91–92, 91f–94f
  symptoms of, 88–89, 88t
debridement, 332–333
definitions, 42t
dehydrated PDT, 342–343
duplex ultrasound (DUS), 43
DVT treatment, randomized trials of, 47
Edinburgh Vein Study, 49–51
epidemiologic studies, 46–47
etiology of, 437–438
extracellular matrix therapy (ECM), 337–338
extravenous etiology, 48–49
Food and Drug Administration (FDA), 328
French study, 52
genetic risk factors, 56–57
history of prior, 89
Italian study, 52
living skin cellular products, 335–337
lower extremity ulcerations, differential diagnosis of, 86–88, 87t
lymphatic involvement, 57–58
matrix metalloprotease modulation, 331–332
measurements
  ABI, 193–196
  point-of-care assessment of wound perfusion, 193–196
  point-of-care wound measurements, apps and cameras used for, 192, 193t
  SPY angiography, 193–196
  TCPO2, 193–196
  wound assessment and measurement, imaging modalities for, 192
mechanism of action, 329–330
medical history, 86–89
negative pressure wound therapy (NPWT), 330–331
past treatment history, 89
pathophysiology of, 434
patients, clinical success in, 390–391, 392t–395t
physical exam, 89–95, 90f
placenta-derived tissue, 342
Polish study, 52
prospective natural history studies, 46
quality of life assessment, 96–97
randomized controlled trials, 45t
risk factors, 86
San Diego population study, 51
Swedish Skaraborg County studies, 53
topical antimicrobial therapy, 333–334
treatment algorithm, 343–345, 345f
venous disease, progression of, 53–54
wound bed preparation, 330
Venous reflux, correction of, 201
Venous segmental disease score, 184
Venous severity score, 181
Venous structure, 6–7
Venous ulcers treatment, other endovenous procedures for, 397
Venous valves repair
  venous obliteration, 474t
  venous obstruction, 474t
  venous occlusion, 474t
Venous wounds, 217
VTE, 149–154

W

Wallstent configuration, 454–455
Wet-to-dry dressings, 230
Wound assessment
  measurement, imaging modalities for, 192
  tools, 179–196
Wound bed management
  biologic debridement, 226
  cleansers and irrigation, 226–227
  cleansing, 223–228
  debridement, 223–226
  enzymatic debridement, 225–226
  general principles, 223
  hydrosurgical debridement, 224–225
  pain management during, 227–228
  sharp debridement, 224
  standard therapies for, 223–228
Wound bed preparation, 330
Wound closure, reconstructive methods of, 492
Wound documentation, 186–189, 188f–190f
Wound dressings, 228–234, 228f
  alginates, 231–232
  hydrocellular foam dressings, 232
  hydrocolloid dressings, 231
  hydrogel dressings, 230
  interactive dressings, 229–232
  occlusive and semi-occlusive, 229–232, 230t
  semipermeable film dressings, 232
  wet-to-dry dressings, 230
Wound healing
  active (biologic) dressings, 232–233

Wound healing (*Continued*)
  growth factor therapy, 233
  adherence to therapy, 240–241
  chronic venous ulcers, 217–218
  electrical stimulation, 238
  optimizing care delivery, 238–241
    wound care clinics, 238–239
  patient and wound assessment, 218–222, 219t–222t
  peri-wound management, 234
  SIMPLE strategies, 240t
  skin grafting, 236–237
  skin protection products, 235t
  treatment algorithm, 241, 241f
  ultrasound therapy, 237
  venous ulcer, adjuvant therapy for, 234–238
  venous wounds, 217
  wound bed management
    biologic debridement, 226
    cleansers and irrigation, 226–227
    cleansing, 223–228
    debridement, 223–226
    enzymatic debridement, 225–226
    general principles, 223
    hydrosurgical debridement, 224–225
    pain management during, 227–228
    sharp debridement, 224
    standard therapies for, 223–228
  wound dressings, 228–234, 228f
    alginates, 231–232
    hydrocellular foam dressings, 232
    hydrocolloid dressings, 231
    hydrogel dressings, 230
    interactive dressings, 229–232
    occlusive and semi-occlusive, 229–232, 230t
    semipermeable film dressings, 232
    wet-to-dry dressings, 230

Z

Z-stent, 454–455, 455f

Printed in the United States
by Baker & Taylor Publisher Services